THE NORTON HISTORY OF
ASTRONOMY
AND COSMOLOGY

JOHN NORTH moved in 1977 from Oxford to the University of Groningen, The Netherlands, where he is Professor of History of the Exact Sciences and Dean of the Philosophy Faculty. He is a member of the Royal Netherlands Academy of Arts and Sciences, the Royal Danish Academy of Sciences, the Academia Leopoldina, and a Fellow of the British Academy. His books include *The Measure of the Universe* (1965, 1990), *Richard of Wallingford* (3 vols, 1976) and *Chaucer's Universe* (1988, 1990).

Frontispiece to Kepler's *Rudolphine Tables* (1627), symbolizing the history
of astronomy in a temple of Urania. Note especially Hipparchus, Copernicus, Tycho Brahe and Ptolemy and their instruments, the map of
Tycho's island of Hven *(centre panel below)*, and the telescope *(upper left)*.
The Imperial Eagle *(top)* is disgorging gold pieces for Kepler's work. He is
dimly seen calculating by candlelight in the lower left panel. Tycho's
heirs objected to an earlier sketch for this, which placed Regiomontanus
to the fore with Tycho, and included Albategni but not Hipparchus.

OF SCIENCE
?orter)

THE NORTON HISTORY OF

ASTRONOMY
and
COSMOLOGY

W·W· NORTON & COMPANY
New York   London

Originally published in England under the title
The Fontana History of Astronomy and Cosmology

Printed in the United States of America

The text of this book is composed in Linotron Meridien.
Manufacturing by The Courier Companies.

ISBN 0-393-03656-1   (cl)
ISBN 0-393-31193-7   (pa)

W. W. Norton & Company, Inc.
500 Fifth Avenue, New York, N.Y. 10110
W. W. Norton & Company Ltd.
10 Coptic Street, London WC1A 1PU

1 2 3 4 5 6 7 8 9 0

PREFACE TO
THE NORTON HISTORY OF SCIENCE

Academic study of the history of science has advanced dramatically, in depth and sophistication, during the last generation. More people than ever are taking courses in the history of science at all levels, from the specialized degree to the introductory survey; and, with science playing an ever more crucial part in our lives, its history commands an influential place in the media and in the public eye.

Over the past two decades particularly, scholars have developed major new interpretations of science's history. The great bulk of such work, however, has been published in detailed research monographs and learned periodicals, and has remained hard of access, hard to interpret. Pressures of specialization have meant that fewer survey works have been written that have synthesized detailed research and brought out wider significance.

It is to rectify this situation that the Norton History of Science has been set up. Each of these wide-ranging volumes examines the history, from its roots to the present, of a particular field of science. Targeted at students and the general educated reader, their aim is to communicate, in simple and direct language intelligible to non-specialists, well-digested and vivid accounts of scientific theory and practice as viewed by the best modern scholarship. The most eminent scholars in the discipline, academics well-known for their skills as communicators, have been commissioned.

The volumes in this series survey the field and offer powerful overviews. They are intended to be interpretative, though not primarily polemical. They do not pretend to a timeless, definitive quality or suppress differences of viewpoint, but are meant to be books of and for their time; their authors offer their own interpretations of contested issues as part of a wider, unified story and a coherent outlook.

Carefully avoiding a dreary recitation of facts, each volume develops a sufficient framework of basic information to ensure that the beginner finds his or her feet and to enable student readers to use such books as their prime course-book. They rely upon chronology as an organizing framework, while stressing the importance of themes, and avoiding the narrowness of anachronistic 'tunnel history'. They incorporate the best up-to-the-minute research, but within a larger framework of analysis and without the need for a clutter of footnotes – though an attractive feature of the volumes is their substantial bibliographical essays. Authors have been given space to amplify their arguments and to make the personalities and problems come alive. Each volume is self-contained, though authors have collaborated with each other and a certain degree of cross-referencing is indicated. Each volume covers the whole chronological span of the science in question. The prime focus is upon Western science, but other scientific traditions are discussed where relevant.

This series, it is hoped, will become the key synthesis of the history of science for the next generation, interpreting the history of science for scientists, historians and the general public living in a uniquely science-oriented epoch.

ROY PORTER
Series Editor

CONTENTS

PREFACE

It is no exaggeration to say that astronomy has existed as an exact science for more than five millennia. Throughout all of that time it has touched on matters of deep human concern. Writing its history presents us with innumerable problems. We begin with a period known to us largely by inference; we continue into times from which much of the evidence is known to have been lost; and we end with the last decades of a century that has provided astronomers with unprecedented attention and economic resources. From a typical century in the Hellenistic period, a golden age of astronomy, we might be left with a mere handful of texts. By contrast, there are now more than twenty thousand astronomical articles being published every year, and over a five-year period the number of astronomers under whose names they appear is of the order of forty thousand.

If this history begins as a sketch, therefore, of necessity it ends as a silhouette, defining its material as much by what it leaves out as by what it includes. It gathers pace in such a way that the space devoted to a dozen recent books of the highest importance might be a small fraction of that given at the outset to what would now seem a quite trivial piece of doctrine. This is not by accident. I have tried to pace the book in part by reference to the intellectual challenge facing successive generations of astronomers, and in part by the rational and social repercussions of their work, so that by the time I reach the twentieth century a certain sleight of hand is required. The design of a rocket gyroscope that would have been a technological miracle in antiquity will pass entirely unnoticed, as will, alas, every one of the hundred people who made it possible. And similarly at a theoretical level. There is no royal road to an understanding of Einstein, Eddington and Hawking, for instance, short of reading their scientific writings.

This book is a history, and cannot even begin to serve as a substitute for astronomical treatises, although I should not feel cheated if it were to increase the five-year phalanx to forty thousand and one. It is a history, and histories are usually written, if not for historians, at least for readers in a historical frame of mind. Of course no professional historian ever admits to being pleased by universal history, but then, the sort of history that can only satisfy professional historians is hardly worth writing.

The book owes its existence to the intuition of my wife, Marion North. She persuaded me that it was needed, and hers was the crucial discovery of a way of creating the time needed to write it, that is, of prising open crevices in the rocks of faculty administration.

J. D. NORTH
University of Groningen
January 1993

ILLUSTRATIONS

All illustrations furnished by the author, unless otherwise stated

FIGURES

ON NUMBERS AND UNITS

Many very large numbers, and a few very small numbers, are here expressed only in words, the word 'billion' always denoting a *thousand million*. The metric system has usually been preferred for units of mass, length and time, except in a few cases where doing so might have obscured a historical point. (As an example of the problem: many telescopes, for instance the famous 200-inch at Palomar, are widely known by names quoting the diameters of their mirrors in inches.) Familiarity takes precedence over consistency, and no rigorous attempt has been made, for example, to introduce so-called 'SI units'.

Readers will no doubt be familiar with the division of degrees (angular measure) into sixtieths. The Babylonian invention of the sexagesimal system is discussed in chapter 3. Sixtieths of a degree are minutes of arc, sixtieths of these are seconds of arc, and so forth. The old notation for an angle of, say, 23 degrees 5 minutes and 14 seconds would have been 235'14". Such a notation may be extended in an obvious way to thirds (sixtieths of seconds), fourths (sixtieths of thirds) and so on, but then it becomes very cumbersome. It is now conventional to replace it (to take this same example) with 23;05,14. The same notation can be used with our familiar and rather inconsistent units of time (as in the mixed example 5d 21;56,07,16,34h, where here 'd' denotes days and 'h' hours).

The following astronomical constants might prove useful:

1 astronomical unit (mean geocentric distance of the Sun)
= 149 674 000 kilometres = 93 003 000 miles

1 parsec (distance corresponding to a parallax of 1")
= 206 265 astr. units

1 light year (the distance light travels in a year)
= 9460 billion kilometres = 5878 billion miles

INTRODUCTION

When calculation called for personal rather than electronic powers, and numbers had a greater capacity to mystify, there was a widespread feeling that a science that quoted its results to ten places of decimals certainly merited the description 'exact'. It was thought that astronomy must in some sense be superior to sciences that counted petals, that mixed *A* and *B* to get *C*, or that predicted the death of a patient 'within a week or two'. Judged in this limited way, for well over two thousand years astronomy has had no equal among the empirical sciences. In a far more important sense, however, astronomy has been an exact science for a much longer period of time, for it has been set out in a highly logical and systematic way, with its patterns of argument modelled on – and even helping to create – those of mathematics. So highly regarded were they, in the past, that astronomy and its sister geometry became accepted as models, prototypes for the empirical sciences generally, helping to provide them with form and structure. If astronomers are to take a special pride in the ancient origins of their subject's exactness, then it should be in this second sense of the word.

Astronomy was not altogether fitted for such a responsible role. After all, it differs from most other sciences in at least one important respect. It studies objects that cannot for the most part be manipulated for purposes of experiment. The astronomer observes, analyses what is seen, and devises principles to explain what has been seen, and what will be seen tomorrow. Even in these days of interplanetary rockets, the subject remains analytical rather than experimental. This quality no doubt explains in part why astronomy became the first highly formalized science.

How and where this came about we cannot say. The answer will depend to some extent on how generously we define our terms. It has been claimed that sequences of moon-shaped marks cut into bone artefacts found from cultures as widely separated in time as 36 000 and 10 000 BC represent the days of the month. The length of the month, from new moon to new moon, is approximately twenty-nine and a half days, but any primitive tally would naturally have introduced extra days when the Moon was invisible. Since in some cases counts might have been made from the new crescent to the last visible crescent, and in others up to the next new crescent, we should not be too disdainful of the fact that groupings found on these pieces of bone, ranging from twenty-seven to thirty-one, have been claimed as evidence for lunar counting. There is much variation in the numbers of marks in groupings that have been said to distinguish between the four quarters of the month. Such evidence is intrinsically difficult to handle, even statistically. The thesis is not implausible; the marks on these bones do often seem to have been gouged out to resemble a lunar crescent; and more than that we cannot say with confidence.

It is tempting to see in the counting of lunar days a first step towards a mathematics of the heavens. Sooner or later such a step had to be taken: quite apart from its possible religious meaning, and its connection with human fertility, keeping a tally of days of the month would have been useful to anyone who valued the Moon's light by night. But we must be wary of introducing our own preconceptions into prehistory. It is commonly assumed that a primitive calendar requires the counting of days, and that the Sun was first intensively studied for its usefulness in establishing the sort of calendar the earliest agricultural peoples needed. The seasons, however, obviously mattered and were known to the hunter long before the introduction of farming, and one may be fairly certain that solar calendars – in the broadest sense of devices for keeping a check on the seasons – did not originally have anything to do with a *counting* of days. They were based rather on the changing pattern of rising and setting of the Sun over the horizon throughout the year.

The epoch at which the change from a hunting to a farming culture took place varied considerably according to geographical region. There were settled agricultural communities in south-west Asia before 8000 BC, and such communities may be just as old in south-east Asia. Farming spread into the Mediterranean well over a thousand years before it reached Britain, at the outer edge of Europe, which it did around 4400 BC. No doubt some astronomical ideas were developed locally in all regions at one time or another. Some were undoubtedly transmitted from one centre to another, and not always with a corresponding migration of peoples. It is difficult to decide on the relative importance of these tendencies, but it does seem that in the fourth millennium BC some remarkable changes were taking place in the sheer intensity with which the heavens were being studied in northern Europe. The evidence for these changes comes from archaeological remains. Perhaps time will prove them to have been anticipated or paralleled elsewhere, but they are remarkable enough in themselves to be taken as a starting point for our history.

There is every reason for introducing this early European astronomical activity into our account. *It was truly scientific, in the sense that it reduced what was observed to a series of rules.* This is not to say that its motivation resembled our own. I have no doubt that the main reasons for rationalizing the appearances of the heavens were religious or mystical in character, and that they remained so, well into historical times. Most peoples have in the course of their development treated the Sun, Moon and stars as *alive*, and often as *human* in character. They have told stories of the heavenly bodies with a strong reliance on an *analogy* between them and human existence. In such analogies we might claim to be able to detect the beginnings of science, but let us not exaggerate this element. Astronomy has always been strongly allied to religion – in the first place, one suspects, because both were concerned with the same objects. The Sun, Moon and stars *were* divinities in many societies. To have done full justice to this alliance with religion I should need to have written a very different sort of book from this, but to have omitted it entirely in favour of the 'exact' parts of the science would have been to mistake the wood for the trees.

Throughout history, omens and signs have been carefully sought out as a means of foretelling, or even forestalling, those superhuman powers that seem to govern the Universe. It is hardly surprising that the heavenly bodies were among the objects to which most attention was paid. They were assigned divine natures in part because they were of such obvious importance in themselves – the Sun in particular. As material for omens they have the virtue of behaving with a certain degree of regularity. Here it is surely the magical conception of Nature that comes first, to be followed at a later stage by a realization that it is *easier* to be systematic and precise about the stars than about the livers of sheep, the weather and flocks of birds. Astrology flourished in the Near East, for instance, once this principle was recognized. This may not entirely explain the very first systematic astronomical theories, but it would be foolish to pretend that astronomy is somehow too noble a subject to have depended on anything of the sort.

The science of the stars was not at first pursued for its own sake, but so potentially strong is the human feeling for system and order that in time astronomy did acquire a measure of self-sufficiency. Astronomers managed eventually to tear up most, if not all, of its ancient roots. It was to be studied increasingly as an independent system of explanations, a system that did not have to be justified by its usefulness – whether in astrology, or navigation, or elsewhere, or by its relevance to God's creation of the world. Admittedly, in comparison with the science itself, its non-astronomical context was often far more important to common humanity. I hope that I have not ignored it unduly in favour of the more formal side of the science, but if I have done so, this is because astrology and cosmic religion are almost commonplace, viewed as symptoms of the human condition. The long record of achievement in astronomy, on the other hand, has very few intellectual parallels in the whole of human history.

Prehistory

PREHISTORIC ASTRONOMY

The patterns of rising and setting of the Sun, Moon, and stars over the horizon played a central part in astronomy from prehistoric times well into our own era. As seen from a particular place, the stars rise and set over the same points of the horizon every day of the year. (We may for the time being ignore long-term changes in their positions.) The risings and settings may not be visible to us, however, at certain seasons – that is, when the Sun is in such a position relative to them that the phenomena in question are during the hours of daylight. The places of the Sun's own risings and settings change from day to day. In the northern hemisphere the Sun rises in the east and sets in the west at the spring and autumn equinoxes. Its points of rising and setting move progressively further to the north as spring advances, reaching a maximum at the summer solstice. The points then move southwards, through the autumn equinox until a maximum southern azimuth (horizon direction) is reached at the winter solstice. The points finally move northwards again until the spring equinox is reached and the year is completed.

The very language with which we today describe these events carries with it certain information – such as that the day and night are of equal length at certain times of the year, the 'equinoxes' – that was not known to early peoples, but the broad truth of the cyclical movements of the points of rising and setting, and their correlation with the cycle of growth in Nature, was known from very early times. This we can see from the many prehistoric archaeological remains that are directed towards the points of midsummer and midwinter rising and setting. (The extremes were evidently what counted; the equinoxes were seemingly of lesser importance to early peoples.) In the earliest periods during which structures were oriented on the heavens, however, directions seem to have been towards the risings and settings of a handful of bright stars, rather than towards the Sun and Moon.

From say half a millennium before 4000 BC, long barrows – elongated burial mounds of earth – were arranged so as to be aligned on the risings and settings of bright stars. It is often extremely difficult to decide when and where a star crosses the distant natural horizon, if that horizon is covered by forest. Neolithic people surmounted this difficulty in numerous ways, all showing extraordinary intelligence and dedication. To begin with the long barrows: these wedge-shaped communal graves, many of them flanked by ditches, were often so fitted into the landscape that by looking along their slope one's line of sight was elevated very slightly above the natural horizon beyond. The smooth back of the barrow thus provided an artificial horizon. Such long barrows could also serve as artificial horizons when viewed transversely. There is good reason to think that viewing of risings and settings was often done by people standing in the ditches to the side of the barrow, perhaps viewing along the lengths of the ditches too.

Neolithic people built other ditched monuments mainly of Earth, often on a very grand scale, and they too followed similar principles of alignment. One example is the so-called 'Cursus' at Stonehenge, now barely visible, but there are many others. When we reach the third millennium, however, there is a clear sign that more attention is being paid to the Sun. A new vogue for circular monuments developed, embodying the Sun's solsticial positions. Stonehenge is the best-known example of these, but the familiar structures there are late, and from the end of a millennium that had seen comparable activity – although usually on a smaller scale – in numerous other centres in Britain and northern Europe. Such circles of stones were merely a more permanent version of circles of timber posts, which carried cross-members (lintels) in much the same way. In the middle of this third millennium we find the technique being transferred to circular grave architecture, the so-called round barrows, and in this form it lasted for well over a thousand years.

Why this particular phase in prehistoric astronomical activity is so important to us is that it testifies to the marriage of astronomy with geometry. The key to an understanding of these circular monuments is their intricate three-dimensional structural form. In plan they may show concentric circles of posts, but often rings that were of oval outline. Their elevations, however, are no less important than their plans, and only by considering the paths of lines of sight in three dimensions can they be properly understood. There are very many different varieties, but always the technique was to mark the line of sight to the Sun's extreme position (summer or winter solstice) by trapping the Sun's image in a 'window' created by at least two uprights, one far and one near, and at least two lintels, one far and one near. Strictly speaking, it was not just any part of the Sun that was observed but (usually) its upper limb, so that the first glint of the Sun was seen at sunrise and the last glint at sunset, midsummer or midwinter.

These lintelled structures were thus, like the long barrows and other earthen structures before them, artificial horizons. Viewing was from carefully prepared places, often in a circular ditch surrounding the monument, occasionally over the bank that surrounded it. A standard unit of length was often used in the design, not only for the plan but for the heights of the posts or stones. This unit was derived from measurements on relatively late stone monuments by Alexander Thom, who called it the 'megalithic yard', but the use of the unit antedates the period of the great megalithic (stone) monuments. There is even reason to think that *angles* of solar altitude were set by ratios of this unit. At Woodhenge, for instance, a long-vanished monument near Stonehenge comprising six ovals of timber posts, a favoured angle appropriate to the rising of the solsticial Sun in winter was set by a rise of one in sixteen.

At some monuments there is clear evidence that the Moon was also observed. The Moon has a somewhat similar pattern of behaviour to the Sun's, except that its extremes of rising and setting follow several cycles, the most important of course being the monthly cycle. Briefly, one may say that each month it reaches to certain extremes that may themselves be treated as objects, with their own movements along the horizon. They in turn reach to certain extremes. These absolute extremes seem to have attracted prehistoric interest. In all strictness they are not correctly described as 'absolute' extremes, since they also fluctuate, and can be treated as objects at a still higher level, with their own extremes. It has been held that even this third level of fluctuation was known to the people of the Bronze Age, say from the end of the third millennium onwards.

There is no doubt that the Moon's extremes of rising and setting were recorded in monumental form, presumably for ritual and religious reasons rather than for the creation of a calendar in any complex sense. The religious dimension is all-important. A grave of a three-year-old girl was found at Woodhenge, her skull cleft so as to suggest a ritual sacrifice. It turns out that there are three key rays set by the monument that cross at the centre of the child's grave. A line to midsummer sunrise is at right angles to it; a line to midwinter sunrise is very nearly at right angles to the first ray; and a line to the Moon's extreme northern setting is in exactly the reverse direction (in plan) of the midwinter ray. Those with a partial awareness of the underlying astronomy may suspect a mistake here, for are the lunar and solar lines of rising and setting not in markedly different directions, at the latitude of Woodhenge? The answer is that if suitable choices are made of the altitudes set by the artificial horizons of the monument, the directions (azimuths) of the rays can brought into exact agreement. The builders of Woodhenge did this, and they related what they had done to the grave of the sacrificial victim.

There is much more to be said about the intellectual achievements of these northern peoples of the Neolithic and Bronze Ages. We can say far more of their purely astronomical and geometrical achievements than of the way they fitted their view of the heavens into their social and religious life. Among the stars that seem to have attracted most attention were Deneb, which set in a fairly constant direction to the north, Rigel, the foot of the constellation we know as Orion, and Aldebaran, which for long has been the eye of Taurus, the Bull. It is probable that the sacrifice of bulls was for long related to the risings and settings of Aldebaran. Whether the entire constellation of Taurus was viewed as a bull it is impossible to say. Perhaps an early constellation figure was the precursor of a figure cut into the chalk downs at Uffington in southern England. The 'Uffington White Horse' is a creature of indeterminate species, but that the star Aldebaran rose over it, as seen from a curved gallery on a nearby track, is determinate enough. And this raises a question that must be introduced sooner or later.

The places over which the stars rise change very slowly, but in most cases must have become appreciable over a century or two. The cause is the so-called precession of the equinoxes, which would be discovered – although described differently – by Hipparchus, in the second century BC. At many Neolithic sites, religious activity lasted for well over a millennium, and it cannot be doubted that the drift in question was sooner or later vaguely appreciated.

This is of course not to say that the precession of the equinoxes was discovered, much less that it was quantified, in Neolithic times. There was a 'Pan-Babylonian' movement in the early years of the twentieth century, some of whose followers wished to ascribe the discovery to their own heroes, but again the same thing has to be said: Hipparchus was the first not only to recognize what was happening but to describe it in terms of a precise astronomical co-ordinate system.

Many attempts have been made to reconstruct the belief systems of these early northern peoples, from surviving artefacts and later written accounts – for instance from Roman accounts of the Druids, and early Scandinavian literature. There are numerous indications of cults of the Sun and Moon, one of the most interesting finds being that in Trundholm (Zealand, Denmark) of a Bronze Age horse-drawn solar disc. Many engraved symbols are no doubt solar or lunar, but most can be interpreted in numerous different ways, and the modern cult of Freud does not make matters any easier. The literary exercise in particular is fraught with difficulties. The Romans interpreted what they found in terms of their own experience, and medieval Scandinavian sources are often contaminated with Christian prejudice. And all these are, of course, very late testimony indeed.

Ancient Egypt

While in northern Europe we are forced to rely mainly on archaeological remains for our knowledge of prehistoric culture in general, and astronomy in particular, in Egypt there are written historical records taking us back three millennia before the Christian era – that is, to the period of perhaps even the first important phase of activity on the Stonehenge site. Our familiar image of Egypt belongs to the third and second millennia BC. It is an image of the Pharaohs, the pyramids and the sphinx, the treasures of Tutankhamen and the Egyptian gods – Osiris, Isis, Ptah, Horus and Anubis, for instance. There are no technical astronomical writings from this period to be compared with those produced in Babylonia in later centuries. Egypt had to wait for the Persian conquests of the first millennium BC for the stimulus that cosmological ideas from the Near East could provide.

More intensive activity belongs to a period after Alexander the Great won the Egyptian throne from Persia in 332 BC. This last period falls under the dynasty of the Ptolemies, who were Macedonian Greek in origin, and who gained control of the country in the wars that divided Alexander's empire after his death in 323. This is not to say that there was no native astronomy. Some concern with astronomy had already been shown in a *Cosmology* associated with the rulers Seti I (1318–1304 BC) and Ramses IV (1166–1160 BC). The Egyptians had by then long been adept at measuring time and designing calendars, using simple astronomical techniques. And text or no text, they too aligned their buildings on the heavens.

ORIENTATION AND THE PYRAMIDS

Some early Egyptian sources speak of a cult relating the sun-god Re and an earlier creator-god Atum. At first this was centred mainly on a temple to the north of the old Egyptian capital of Memphis. The place was known by the Greeks as Heliopolis, 'City of the Sun', but by the Egyptians as On. (In the book of Genesis, Potipherah is said to be a priest of On.) By historical times, the priests of Heliopolis had laid down a cosmogony that held Re-Atum to have generated himself out of Nun, the primordial ocean. His offspring were the gods of air and moisture, and only after them, and as their offspring, were Geb, the Earth god, and Nut, the sky goddess, created. The nine deities of Heliopolis (the Great Ennead) were made up with Osiris, Seth, Isis and Nephthys, the offspring of Geb and Nut.

The Sun was worshipped at Heliopolis in other forms than Re-Atum – for instance, as 'Horus of the Horizon' (Horakhti) and as Khepri, in the form of a scarab-beetle. These different personalities are interesting, and not only because of the fine analogy between the sun-god, pushing the Sun, and the beetle, rolling its ball of dung along the ground. Khepri was the Sun at its morning rising, and Re-Atum at its evening setting. These different solar characters mirrored precisely the crucial solar observations that were made throughout the prehistoric world.

The cult of Re-Atum was well established by the time of the first great pyramids, that is, about 2800 B C. For all their outward simplicity, there seems to have been a relationship of sorts between the pyramids and the Sun and stars. The architect of the first great stone pyramid, the 'Step Pyramid' built for King Zoser, was Imhotep, whose name passed into Egyptian history as an astronomer, as well as magician and physician. He was later deified, not without reason. When he built the Step Pyramid, it was without equal anywhere in the world: its complex subterranean structure distinguishes it even from its successors. Some of these, however, particularly the Great Pyramid in the Giza group, were much more accurately sun-orientated in the simple sense of being directed towards the four points of the compass. This is something they had in common with Egyptian temples. One of the most interesting series of plans relate to the Osiris temples at Abydos, where a series of reconstructions can be traced from the first to the twenty-sixth dynasty. At first the entrance faced south; in the next four structures it faced north; and in the last three east.

The pyramids are more astonishing as works of engineering than as examples of complex orientation. The sheer accuracy of their simple alignments is remarkable, even so: in some cases, average errors are of the order of only a few minutes of arc, and the levelling of the bed of rock on which the Great Pyramid sits is such that the difference in average levels between the north-west corner and the south-east is only a couple of centimetres or so.

The pyramids are in several cases pierced by upward-slanting shafts that are usually described as ventilation shafts, but that some writers regard as having been directed towards selected stars at their upper culminations. The Great Pyramid at Giza has an entrance passage on the northern side sloping down towards the centre (in this case an underground chamber) at an angle of 26;31,23°, and apart from a slight variation of a fraction of a degree in the angle, it shares this property with six of the nine remaining pyramids at Giza, and with the only two well preserved pyramids at Abu Sir. At the Chephren pyramid the angle is 25;55°. (We are here dealing with pyramids within a century or two on either side of 2500 BC.) Was there any important star culminating at these altitudes at the time? (Of course the latitudes all differ by great or small amounts from that of Giza.) If so, whatever star it was, it had to be a star circling the north pole and at *lower* culmination, since the altitude of the pole above the horizon is nearly 30°. The brightest star in the neighbourhood of the true pole was then alpha Draconis (Thuban), and this, at lower culmination, was very probably the star towards which the corridors were directed. Why was a star observed at its *lower* culmination? Perhaps because the constellation we know as the Great Bear, then portrayed as the foreleg of a bull, was in a reasonably upright position to the side of it. The Egyptians identified several constellations round the pole: the most significant seems to have been the foreleg of the bull (not connected with Taurus), but others include a hippopotamus, a crocodile and a mooring post.

SOLAR RITUAL

The ruins of the temple of Amun-Re at Karnak, across the Nile from Thebes, have a corridor running north-west and south-east through the middle of the buildings on the site for more than four hundred metres. The central courtyard and chambers of this date from the time of the Middle Kingdom (2052 to 1756 BC), but the most impressive parts were due to Thutmose III (1490–1436 BC), and additions were regularly made right into the Christian era. There has been much controversy about the significance of the precise direction of this axis, but here too it seems likely that use was made of an artificial horizon, marginally higher than the line of the distant hills, rather as had been northern practice so long before. For well over a millennium, the last glint of the setting Sun would have been visible from the Holy of Holies without the need to move far from a position on the axis of the temple. So again religion was a proven stimulus to astronomy, albeit of a simple sort.

The Egyptian pantheon grew steadily over the centuries, with the incorporation of different local traditions into the official religion. The behaviour ascribed to the gods became as a result filled with inconsistencies. There was always a strict hierarchy of importance, though, and by the fourteenth century BC the dynastic god Amun, 'the Hidden One', was supreme. When the heretical king Amenophis IV decided that the nation should be converted to the worship of something more visible than Amun, he settled on the Globe of the Sun, Aten, as deserving of elevation to the status of the one true god, and built a new capital, Amarna, filled with sun-inspired art. And when he built his own tomb at Tell el Amarna, all corridors and rooms leading up to the burial chamber were along a single axis (directed south-east and north-west). This 'pure' solar convention, reminiscent of those followed in the chambered tombs of northern Europe two millennia earlier, was in violation of a tradition of the burials of his forebears and successors, whereby either the burial chamber was entered by a corridor at right angles to the main corridor of approach, or the main corridor was staggered, so as to break the directness of the approach.

Amenophis renamed himself 'Akhenaten', meaning 'pleasing to Aten'. The extreme form of his solar religion was short-lived, but solar rituals lived on in many ways, for instance in the installation of rulers. We are very fortunate in the Egyptian remains from this historical period, for we have a number of actual illustrations of Sun worship, especially from the walls of the royal tombs at Tell el Amarna (the modern name for Akhenaten). Some idea of Egyptian solar symbolism may be had from the walls of the entrance passage to Tutankhamen's tomb and four of its chambers. All are strictly north-to-south and east-to-west. Entry is from east to west. Each of the four chambers has a known ritual purpose, depicted on the walls, and no doubt these have something in common with their more ancient counterparts. One (keyed to the south) was a chamber of eternal kingship, another (east) a chamber of rebirth, a third (west) a chamber of 'departure to sepulchral destinies', and the fourth (north) that of the reconstitution of the body. The dead Osiris would, it was thought, after his arduous quest for rebirth, reappear in the form of the rising Sun, Re. Those who were responsible for ordering solar rituals must have had a sound intuitive awareness of the Sun's behaviour, but there are no signs here of sophistication in a study of the Moon comparable with that evident in northern Europe at roughly the same time.

THE CALENDAR

Some time after nomadic tribes of North Africa first settled as farmers in the Nile valley, they realized that there was a correlation between the pattern of the river's behaviour and that of the star Sirius (called Sothis), the brightest in the sky. The rising of the Nile, important because its flood waters irrigated the valley, was seen to coincide with the first sighting of Sirius on the eastern horizon shortly before sunrise, after a long period of invisibility. This event, now known as its 'heliacal rising', took place in mid July, and so not at a notable point in the solar year. The three seasons for the Egyptians were related to the behaviour of the river, and so the names of the months that began them were 'Flood', 'Emergence', and 'Low Water' or 'Harvest'. The other month names came from lunar festivals. The Sun was at first important for them, it seems, mainly as an indicator of the yearly cycle. It is unlikely that the solstices could have been as important to the Egyptians as to northern peoples, whose views on a year with only three seasons it would be amusing to learn.

To reconcile the three interlocking systems – stellar, lunar and solar – was one of the main tasks of Egyptian astronomy, and it remained at the centre of astronomy until modern times. The festival of Sirius/Sothis more or less followed the solar year of about $365\frac{1}{4}$ days, but the length of twelve months, each of twenty-nine or thirty days, averaged only about 354 days. From at least the middle of the third millennium, therefore, the Egyptians devised one of the earliest calendar rules known to history: an extra ('intercalary') month named *Thoth*, a lunar deity, was added to the year only if Sirius/Sothis rose heliacally in month twelve.

As Egyptian society became steadily better organized, the calendar was developed further. The length of the year was determined as 365 days, and the 'months' were standardized at thirty days, each of them divided into three 'weeks' of ten days. The system, which dates from perhaps the twenty-ninth or thirtieth century B C, has many advantages. The week is a conventional matter of no great astronomical importance, but there is an obvious ambiguity in the word 'month' here. To those for whom the visible Moon is a matter of religious importance, the only problems that this calendar solves are essentially problems of book-keeping. It is hardly surprising that the Egyptian year of 365 days is one that astronomers have found attractive, for it simplifies the conversion of long periods of time to days. Even Copernicus, following Hellenistic practice, used it for his astronomical tables.

The new calendar inevitably ran into difficulties as the slight error in the length of the year accumulated (that is, the odd quarter of a day or so). The solution was to devise a new lunar year to run in harness with the civil year. New rules of intercalation were drawn up around 2500 B C, and for over two millennia Egypt had three calendars in use, side by side.

THE HOURS OF DAY AND NIGHT

The Egyptians' wish to divide the night into smaller parts combined in a curious way with their civil calendar to give us our division of the day into twenty-four hours. Any society that carries out ritual acts by night is likely to devise ways of judging the passing of the night. The Egyptians wrote copiously on the passage of the sun-god Re on his night-boat through the Other World, between sunset and sunrise, and the stages were marked by the movements of the stars. To find how they were marked we must turn back for a moment to the calendar.

Heliacal rising (first morning visibility after a period during which the star rose only in daylight) was an important Egyptian calendar notion. A bright star was chosen as a marker – we have already seen that Sirius/Sothis was the most important marker of all. Each day subsequent to its heliacal rising, a star rose a little more in advance of the Sun than before, until another suitable marker star rose heliacally. But who is to select those stars, and on what principle? The solution was reasonably straightforward. We have seen that the civil calendar divided the year into thirty-six 'weeks' of ten days each. Thirty-six stars or constellations were therefore sought that marked, by their heliacal risings, the beginnings of those thirty-six weeks. (It seems that they were chosen to be as much like Sirius/Sothis as possible, each invisible for about seventy days in the year.)

Forget, now, the reasons for which the stars were chosen, that is, reasons having to do with the division of the year, that is, with the *calendar*. We simply have thirty-six stars, or groups of stars, recognized as being of great importance, and during any night their risings will occur at moderately regular intervals. Turning as they do once in roughly a single solar day, one might imagine that eighteen of the thirty-six stars would rise successively during an average night, but in fact the problem is complicated in several respects. During most of twilight the stars are invisible: *total* darkness was what was deemed to matter. The stars chosen were to the south of the celestial equator, that is, did not cross the ideal horizon due east. (They were in fact in a belt roughly parallel to, and to the south of, the Sun's path through the stars, the ecliptic.) Not all nights are in any case of equal length – at Egyptian latitudes a midwinter's night is nearly half as long again as a midsummer's night. We need not go further into the theoretical reasons for it: even though during much of the year the night was not divided into exactly twelve parts by these stars, *it was finally regarded as being so divided.* Evidence for this way of dividing the night comes from diagrams on the inside of coffin-lids from the Eleventh Dynasty (twenty-second century B C).

Daylight was later divided into twelve hours, by analogy with the night. And so we were given the twenty-four hours of our day, from which even the 'rational' endeavours of Revolutionary France did not manage to disengage us.

It is assumed that the coffin-lids are concise versions of representations on the ceilings of tombs of contemporary rulers of the Middle Kingdom. The earliest of these is in an unfinished tomb of Senmut, vizier of Queen Hatshepsut, and it was followed by those in the underground cenotaphs of Seti I, Ramses IV and later rulers. Almost a thousand years separate the Seti ceiling from a notable papyrus manuscript, Carlsberg 1, that amounts to a commentary on it. This funerary text contains instructions for making a shadow-clock, with four divisions on its base, illustrated in figure 2.1 in two positions, that on the left for afternoon use, and that on the right for mornings. Another from the time of Thutmose III (1490–1436 B C) has five divisions.

How aware early peoples were of the pattern of variation in the lengths of day and night it is impossible to say, but

FIG: 2.1 *Egyptian shadow clock*

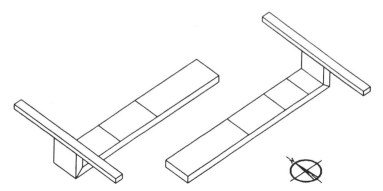

this awareness must clearly have been heightened with the invention of the water clock. The earliest example dates from the time of Amenhotep III (1397–1360 BC), but relates to calendar-reckoning during the reign of Amenhotep I (1545–1525 BC), and it is from an inscription in a tomb from that earlier period that we know of attempts to express the ratio of the lengths of longest and shortest nights. The ratio given (14 : 12) is not particularly accurate, but it is the principle that counts, and we are in the unusual position of knowing the name of the man responsible, Amenemhet.

After the Alexandrian conquests succeeded in Hellenizing Egypt, the Babylonian zodiac – that is, the set of constellations centred on the ecliptic – was introduced into Egyptian astronomy. When this happened, the thirty-six star divisions were simply transformed into thirty-six sections of the zodiac, each of ten degrees. These are the 'decans' or 'faces' of Greek (Hellenistic) and later astrology.

Around the fifteenth century BC, it was recognized that the risings of the decanal stars were a poor way of regulating civil and religious life, and another set of stars was chosen. These were observed, not at the horizon, but as they crossed the meridian. Again this was no exact matter: these meridian transits were observed with reference to the head, ears and shoulders of a sitting man. Three royal Ramesside tombs (*c.* 1300–1100 BC) were decorated each with twenty-four tables (two to a month) that made it possible to judge the hour in this way. It has been conjectured that these star clocks were drawn up with the help of water clocks, for several water clocks survive with astronomical material engraved on them.

Despite the great cultural wealth and length of time over which the heavens were scrutinized by the Egyptians, not to mention the respect in which they held many celestial objects, except in the case of the calendar it does not seem to have occurred to them to seek for any deeply systematic explanation of what they observed. For all that they were in possession of a script, they seem to have produced no regular records of planetary movements, eclipses, or other phenomena of a plainly irregular sort. The Egyptians read legends more easily than mathematics in the stars. Decorated monuments, of which more than eighty are known that could be somehow classified as astronomical, represent the cosmic deities of mythology, including the solar and lunar deities, the planets, the winds, the constellations, Earth, air, sky, the cardinal points, and so forth. They testify to great familiarity with the constellation patterns – theirs were not identical to ours, of course. The great reputation the Egyptians have enjoyed for most of the last two thousand years is based, however, on a confusion.

To the Romans, 'the Egyptians' were those who lived in Egypt, and so often included those of Greek culture. Almost invariably, when Egyptian astronomy or astrology is mentioned, *Hellenistic* Egyptian astronomy is intended. Zodiacs in temples and tombs are Hellenistic, and take over much from Mesopotamian cultures. The first known Egyptian zodiac was a ceiling from the temple of Esna, no earlier than 246 BC, copied down during the Napoleonic expedition to Egypt but since destroyed. The best known, now in the Louvre, is from the Dendera temple, and was once part of the ceiling of a chapel on the roof of the temple of Hathor (before 30 BC). The origins of these zodiacs are clear from the constellation images they contain, such as our Capricorn and Sagittarius, which are found on much more ancient Babylonian boundary stones.

If we are to narrow our focus and consider only mathematical, theoretical, progress, then we may take calendar schemes as the strongest native tradition, so that it comes as no surprise to find that in perhaps the fourth century BC the Egyptians recognized a twenty-five-year lunar cycle. This equated twenty-five Egyptian years (9125 days exactly) with 309 months ('synodic' months, new moon to new moon). This is an excellent approximation, which slips by a day in only about five centuries. It is a finding worthy to be placed side by side with the discoveries of the mathematical astronomers of Mesopotamia. From the sixth century BC, a time of Persian domination in Egypt, the Egyptians had been adapting the Babylonian lunar months to their own civil calendar, and this fact goes some way towards explaining a result that is conspicuous by its rarity.

Ptolemy, the greatest astronomer of antiquity, was an Alexandrian, and belongs to a later period. After the collapse of the Roman empire, of which Egypt was for long a part, the country became largely Christianized, and Coptic literature, using a form of Greek script for native dialects, carried a very superficial form of astronomical knowledge down to Ethiopia; but again this was at heart Hellenistic, and did not have more ancient Egyptian roots. For a more truly scientific study of the heavens we must look further east, to Mesopotamia.

Mesopotamia

MESOPOTAMIAN CIVILIZATION

The sort of astronomy developed in Mesopotamia during the long period from which we have written evidence differs so radically from that of the Egyptians before it, and of the Greeks and Hindus after, that it seems natural enough to separate them. To do so is to risk encouraging the mistaken belief that these cultures were themselves simple and monolithic. Matters of general cosmological concern enter into the mythology of all peoples, even though local variations are legion. Language forms a natural obstacle to the diffusion of ideas, and the Near East knew many languages, but such was their pride that even neighbouring cities sharing a common language would often worship different planetary deities. Under these circumstances it is impossible to do more than rough historical justice to the region in a brief space, but one point, at least, we must keep constantly in mind. All those sciences in the modern world that make use of mathematical methods are indebted to the astronomers of Babylon, from the last five or six centuries before the Christian era, which saw the culmination of the Mesopotamian achievement. This being so, we must look beneath the diversity for what was peculiar to Mesopotamian astronomical practice.

In the sixth century BC, Babylon enjoyed a brief period of independence after having long been dominated by the Assyrians. With the overthrow of the Assyrian empire, the Babylonians set to work to revive their own lost glories – that is, from a thousand years before, when the great Hammurabi had been ruler. The power of the Persians was also growing rapidly, however, and in 539 BC Cyrus the Great succeeded in defeating Nabonidus, the last Babylonian king. The repercussions of these events are familiar to many through the Biblical account of the release of the Israelites who had been brought to Babylon after the destruction of Jerusalem by Nebuchadnezzar.

At length the Persian empire was defeated by Alexander the Great, and links with Greek culture became greatly strengthened. Alexander entered Babylon in 331 BC. After his death in 323, and the division of his empire, Babylon came after various reverses to fall under the rule of Alexander's most politically able successor, Seleucus, the founder of a new dynasty. Seleucus ensured a steady stream of Greek immigration into this new Greek and Macedonian empire in Asia, and to speak generally of 'Greek learning' after this time it is necessary to take into account also this vast cultural area in Asia Minor – not to mention a Hellenized Egypt, including – as its name tells us – Alexandria.

With these events in mind, it is useful to distinguish between four historical periods: first that of the Hammurabi dynasty, and what went before it, secondly the Assyrian period (1000–612), thirdly the period of independence (612–539) followed by subservience to Persia (539–331), and last the Seleucid period (331–247).

THE HAMMURABI DYNASTY

Over the last century and a half, archaeologists have excavated several Mesopotamian temples with their massive associated towers, known as 'ziggurats'. At Uruk, for instance, one complex goes back to the fourth millennium, while at the old Sumerian city of Eridu some of the building might be as old as 5000 BC. The oldest written records are from the temples, and it seems likely that political power was largely centred on them. They owned most arable land, and played a central role in the control of irrigation.

Although it is not altogether clear how the religious power base was related to the power of the secular rulers, there is no doubt that an extraordinarily complex bureaucracy grew up within the priesthood. The Sumerians invented the cuneiform script, where marks were impressed by a wedge-shaped stylus in soft clay, which was then allowed to dry out in the Sun. The Babylonians – a semitic people – used this, and adapted it to their own language, using a mixture of phonetic spelling and Sumerian ideograms.

They also took over the Sumerian number-system, a fact of great scientific importance, since this used a place-value notation, just as does our own system (unlike that of Roman numerals). The difference is that where we work to a scale of 10, they took a scale of 60. This 'sexagesimal' system goes back to the third millennium. Numbers up to 60 were built up in a simple way, reminiscent of our Roman numerals, based on repetition of a wedge-mark for 10 and a vertical stroke for unity. (Thus $>>|||$ would denote 23.) Beyond 60, numbers were separated with spaces (we can represent these with commas) so that, for instance, 2,9,14 could signify $2 \times 60^2 + 9 \times 60^1 + 14$. Potential complications arose when the same arrangement was used for *fractions*, in this system. It is as though we, in our decimal notation, were to use a string of numbers like 3546 to mean, indifferently, 3546, 354.6, 35.46, and so on. We need some sort of punctuation to separate the fractional part from the rest, and the convention among modern writers using sexagesimal fractions is to use the semicolon. Thus the string of numbers 2,7,17;52,13 is now written to represent the number $2 \times 60^2 + 7 \times 60^1 + 17 + 52/60 + 13/60^2$.

We perhaps feel uneasy at grafting our decimal notation on such a powerful system as theirs, but the mixture ought to give rise to no great confusion, for we use the Babylonian legacy whenever we write a time in hours, minutes and seconds, or an angle in degrees, minutes, etc. As for describing their system as 'powerful', it is so simply because 60 has so many prime factors. (It was a great oversight on the part of our creator not to give us twelve fingers.)

The Sumerians, and the Babylonians after them, became expert calculators within this system of numeration, and to help them they devised one of the most useful of all scientific inventions, that of the *table of numbers*. They had multiplication tables, tables for reciprocals, for squares, and even for square roots. The Babylonians were adept at carrying out procedures that we can only describe as algebraic. They could solve linear and quadratic equations, and even restricted examples of equations of higher degrees. They used geometrical proofs of algebraic formulae, much as do we when we follow in the Greek tradition. These techniques flourished under the first Babylonian dynasty, which began with Hammurabi 'the Lawgiver', and lasted for three centuries. (It ended no later than 1531, and some would argue as early as 1651 BC.) All the evidence points to relatively many of the educated priesthood being highly expert in arithmetic, and this fact is of great importance for the astronomical style they gave to the world.

The earliest documented period of Mesopotamian history speaks for a great variety of local cults, spreading with changes in political power, even from the older Sumerian population to their Semitic successors. Many of the gods had nothing to do with the heavens, but the city of Ur favoured a local deity called Sin, a moon-god; the cities of Larsa and Sippar both followed Shamash, a sun-god; and several accepted Ishtar, later the planet Venus, but perhaps once a personification of fertility. Babylon itself was loyal to Marduk, a god given supremacy as 'Creator', when Hammurabi combined the gods of the city-states into a single pantheon.

Not all gods, by any means, can be identified with stars: the three highest – Anu, Enlil and Ea – corresponded to heaven, Earth and water. There is, however, a good case to be made out for claiming most of the old Babylonian gods as in some way *cosmic*. A deep concern with cosmic affairs is apparent in the oldest literature to have come down to us. The Babylonian Gilgamesh epic has in it hints of a ritual of observing the Sun, Moon and planets over the tops of distant peaks.

After the trauma of his companion's death, Gilgamesh arrived on his journey at a mountain, Mâshu, 'whose peaks reach to the banks of heaven, and whose breast reaches down to the underworld'. He there prayed to the Moon (Sin) to preserve him. There was on the mountain a gate through which the Sun passed on its daily journey, guarded by two scorpion-people, man and wife, who allowed him to pass. Gilgamesh travelled eleven hours in utter darkness, and a twelfth in what clearly represented twilight, before coming out on the other side in a beautiful garden with jewelled shrubs. He discovered a plant that can bestow immortality, but it was snatched away by a serpent that ate it, shed its skin and renewed its life – presumably an allegory of the Sun's motion. The Sun, like the serpent, was supposed to have the power of renewing its life, but Gilgamesh was refused the option. The gloomy message of the epic as a whole is that humankind must die.

Old Babylonian seal cylinders – engraved cylinders about the size of a pen-top that are rolled over soft clay to reproduce a picture – support the idea that the elements of this sort of story were well known. They often show the sun-god stepping through a mountain pass between two gate-posts, sometimes brandishing the key to the gate. There are lions on the gates, a common solar attribute, and solar rays issuing from the god's arms. There are no mountains in Babylonia visible from the valley of the Tigris, and any mountain sun-god worshipped there might have come from Elam in the east (and the region around the city of Susa) or from the ranges that run northwestwards from there through present-day Turkey. In the earliest period the sun-god is depicted holding a serrated weapon, and if not lifting himself by his hands, then he stands with one foot on a mountain. As represented later in the Middle Empire – say

2800 BC, but still earlier than the literary sources – a stool takes the place of the mountain, in keeping with the new terrain. This shows the importance of local circumstance. Just as the mythology is partly determined by it, so it must have been with the religious techniques for observing the heavens.

Under Hammurabi, not only the pantheon but also the calendar was unified, and Babylonian names were imposed on the months. Since the length of the month is near to, but not exactly, 29.5 days, and since the number of months in a solar year is not a whole number (it is about 12.4), rules of intercalation are needed in any calendar that is to take both into account. Such rules were found, decreeing whether a month was to be counted as twenty-nine or thirty days, or whether a year should have twelve or thirteen months. Although they were far from perfect and not uniformly applied, they remained in use until 528 BC.

Not always for the best of reasons, the Babylonians have always been associated with astrology. The subject usually understood by this word made its appearance at a relatively late date, and owes most to *Greek* influence, but it is true that Babylonian divination took on a cosmic character at a very early date. There is a large collection of omens, about seven thousand in all, containing originally many thousands of phenomena that were open to interpretation. The series is known from its opening words as 'Enuma Anu Enlil'. The omens survive to us in tablets from the time of the Cassite rule over the Babylonians (say 1500–1250 BC), but many of them were very probably taken from sources as early as the Akkadian dynasty (about 2300 BC). Many of these omens concern the astrological significance of the position and appearance of the planet Venus. The sixty-third tablet of the series of omens is one of the most important of all early astronomical documents, for it deals with methods for calculating the appearance and disappearance of Venus, and of interpreting astrologically those phenomena.

The Venus tablets (various copies survive) are associated with the reign of the king Ammizaduga, whose rule began 146 years after the beginning of Hammurabi's. We earlier spoke of heliacal risings and settings of fixed stars. In the Ammizaduga tablet a virtually complete list of heliacal risings and settings of the planet Venus is given for a period of twenty-one years, each event being offered an astrological interpretation concerning the fortunes of climate and war, famine and disease, kings and nations.

PLANETARY MOTIONS: A DIGRESSION

To discuss these remarkable tablets we must first make a digression, to explain from a modern point of view the motions that are involved. Although not true to an ancient view of these matters, to remember the pattern of the motions it is helpful to think of them in terms of a sun-centred planetary system. The general sense of rotation round the Sun of all the planets visible to the naked eye is the same: looking down on the solar system from the northern half of the sky, the rotations are anticlockwise. (These are the long-term motions, and have nothing to do with the apparent daily motions, risings and settings that are due to the spin of the Earth on its axis.) A planet that is observed from a place in the northern hemisphere of the Earth will of course rise and set daily, but seen against the background of fixed stars over a long period of time it will have a tendency to move slowly leftwards, against the sense of the daily motion. It will rise slightly later each day. This is true even of the Sun, for – as a moment's thought will show – just as the Earth goes anticlockwise round the Sun, so, with reference to the Earth, the Sun goes anticlockwise round the Earth. Figure 3.1 may help to illustrate these general points. This type of motion is called 'direct'. For future reference, it is the sense in which co-ordinates of the 'longitude' type are always taken, when the positions of the

FIG: 3.1 *A heliocentric or geocentric system?*

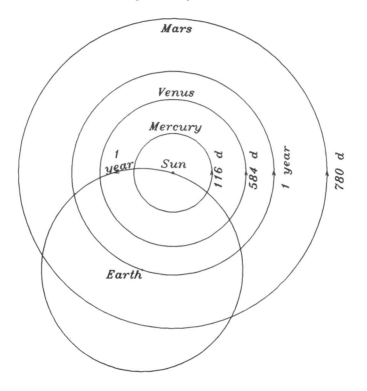

planets and stars are recorded in relation to the ecliptic or equator.

Planets move in general in relation to the stars with direct motion, but there are times for any planet when it moves in the opposite sense, with what is called 'retrograde' motion. Venus is nearer to the Sun than are we, and so may lie between us and the Sun, or beyond it. When Venus is beyond, it moves with direct motion, for it shares the Sun's direct motion and has its own in addition. When it is nearer, however, it can have a retrograde motion of its own (movement from left to right, for our northern observer) faster than the Sun's direct motion (right to left), which as viewed from the Earth it shares. There are certain points at which, from the Earth, it will seem to have no motion at all. They are marked as *MS* and *ES* on figure 3.2, these denoting *morning* and *evening stationary* points, respectively.

FIG: 3.2 *Risings and settings of Venus*

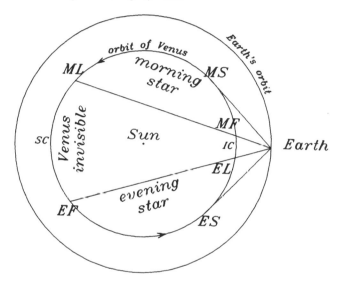

If we think of the planet as a satellite to the rising and setting Sun, a moment's consideration will show that when Venus is at *MS*, the planet might be seen in the morning, rising before the Sun, but that in the evening it will have moved below the horizon before the Sun, and so will not be visible then. When it is at its other stationary point, *ES*, it will be seen only in the evening.

When Venus is in the neighbourhood of *superior conjunction*, namely *SC* in the figure, or at *inferior conjunction*, *IC*, it will not be seen, for it will be lost in the Sun's rays. It has to be of the order of 10° distance from the Sun to be seen at all. (This angle depends on many factors, and we need not go into more detail here.) The points *MF* and *ML* in the figure are those at which Venus is seen for the *first* time and the *last* time in its cycle, as a *morning* star; and *EF* and *EL* are its points of *first* and *last* appearance as an *evening* star.

Figure 3.2 is drawn approximately to scale. Venus takes 584 days in its complete circuit of the Sun. How long it is lost to view depends on various factors, such as geographical latitude and the time of year. As a very rough guide, a dozen weeks of invisibility at superior conjunction, and a fortnight at inferior conjunction, are not unusual.

Of the planets known before the discovery of Uranus in the eighteenth century, the only other planet between us and the Sun, Mercury, has a pattern of behaviour similar to that of Venus. The superior planets (Mars, Jupiter and Saturn), which have orbits outside our own, behave differently. They can be seen overhead at midnight, for example, which is clearly out of the question for Mercury and Venus; but they too can have retrograde motions, and they too have periods of invisibility, that is, when their angular separation from the Sun at superior conjunction is small.

The phenomena of first and last sightings of the planets were of great interest to the Babylonians. Presumably attention was drawn to them only after a tradition had been established of observing the heliacal risings and settings of the fixed stars. They were to prove to be of great importance for the foundation of theories capable of predicting planetary positions, for they provided the first reference points with respect to which planetary positions could be specified at particular times.

THE VENUS TABLETS AND ASTRAL RELIGION

The Venus tablets of Ammizaduga give the years, months, and day-numbers on which the planet reached the positions labelled *EL*, *MF*, *ML* and *EF*, over a period of about twenty-one years. It might have been evident then, as it certainly was later, that the sequence of phenomena began to repeat itself almost perfectly every eight years (more precisely, every ninety-nine Babylonian months less four days). It did so after five complete cycles of the four phenomena. We ourselves can see this equivalence as a consequence of the fact that the synodic period of Venus (the time taken to orbit the Sun once, with reference to the earth–Sun line) is 583.92 days: five such periods is 2919.6 days, while eight years of 365.25 days is 2922.0 days. Astronomers made much use of this near-equivalence throughout later history, recognizing that it is relatively easy to draw up a tolerably good ephemeris for the planet Venus after the first eight years, since the planet's co-ordinates then repeat themselves on almost the same (solar) calendar dates. There are similar cyclical relationships for all the planets, but none so simple as this.

In one version of the tablets it is clear that we are no longer dealing with a set of unmodified observations, for there is a recognizable pattern in the data. Periods of invisibility are always put at either three months or seven days, and it can be seen that the dates are grouped in sets involving equal increments from set to set. (Thus it goes from one *MF* at month 1, day 2, to a later *EF* at month 2, day 3, to a later *MF* at month 3, day 4, and so on.) This is a clear proof that the *periodicity* of the phenomena was recognized, and so is a very important milestone in the history of scientific astronomy. It is unfortunate that the tablet cannot be dated with certainty. It antedates the destruction of Assurbanipal's library by the Medes in 612 BC, but might be as much as eight or nine centuries earlier.

The precise sequence of dates in the basic Ammizaduga tablets allowed B. L. van der Waerden to date them, and consequently to reject two out of the three Babylonian chronologies previously favoured by historians. Hammurabi's dynasty is now usually set at 1830 to 1531, his own reign at 1728 to 1685, and Ammizaduga's at 1582 to 1562 BC.

The Venus tablets were copied repeatedly over many centuries for religious and astrological reasons. The planet was identified with the goddess Ishtar, as mentioned earlier, and the calendar that charted her appearances was that of a Sun and Moon that were in turn related to the gods Sin and Shamash. Astrological and religious motives cannot be separated at this early date. The deities that were *worshipped* were, through their celestial behaviour, thought to be capable of *determining* what happened in matters of love, war, and so forth. ('When Venus is high in the sky, copulation will be pleasurable . . .', and so on.) The third aspect of the case, the *foretelling* of events by the planets, was one that followed naturally as soon as the prediction of their positions became scientifically possible. Of course one had to believe in their efficacy as gods. With the passage of time, this religious belief dwindled or was lost entirely, leaving behind, however, a belief in the possibility of *prediction on the basis of planetary positions alone*. Astrology without religion might have become more scientific, but it had become less rational. In classical antiquity, and then in the middle ages and after, astrologers worked hard to try to restore some of the subject's lost rationality, by inventing theories of celestial influence.

THE ASSYRIAN PERIOD

Babylon fell to the Hittites in 1530 BC, but was soon incorporated in the Cassite empire. This period came to an end around 1160 BC, but during that time astronomical traditions were gradually strengthened. The astrological series giving lists of omens, and known as *Enuma Anu Enlil*, was written. (This is the series of tablets that we mentioned earlier because it preserves for us the Venus material from the time of Ammizaduga.) Babylon already had a reputation for learning, and it is interesting to see how, even before the advent of Assyrian supremacy, the Assyrians were using the Babylonian dialect in their inscriptions. Star lists were developed that connected the passage of the months with the heliacal risings of stars. They are often misleadingly called 'astrolabes', but are better known by their Assyrian name, the 'Three stars each'. They contained typically thirty-six stars in all, twelve 'stars of Anu' that were near the celestial equator, twelve 'stars of Ea' that were to the south of it, and twelve 'stars of Enlil' to the north. (One peculiarity is that planets' names also occasionally occur, which is odd since their heliacal risings are not at fixed dates.) It now seems probable that the 'paths' of the three gods are not actually bands of the sky, but sections of the eastern horizon within which the stars rise. Known examples of the 'Three stars each' come from Babylon, Assur, Nineveh and Uruk. Some of these particular clay tablets were rectangular in form, but others were circular, resembling a dartboard with twelve divisions.

Certain numerals on two of them (one rectangular and one circular) have been interpreted as showing that the Babylonians divided daylight into twelve equal parts, and night likewise. (Compare what we said of the same Egyptian division.) This division, which if strictly interpreted makes the lengths of hours vary with the season of the year, became customary throughout the Near East, and eventually throughout Europe. Writing in the fifth century BC, the Greek historian Herodotus tells us that the Greeks had taken their twelve-part day from the Babylonians. It became the method of time-reckoning of the populace as a whole, despite its probable astronomical ancestry. By contrast, there is the method of 'equal hours', for instance as measured precisely by the daily rotation of the heavens or by a water clock. An ivory prism from Nineveh in a later period (eighth century BC or later), now in the British Museum, shows that conversion between the two systems of time-reckoning was then conceived of as an astronomical problem. The astronomers used a system of twelve double-hours (bēru), each divided into thirty parts (USH), one of which was thus four minutes of our time.

Out of the 'Three stars each' tradition grew another, a series of tablets known as mul-Apin. What is known of this has been pieced together from texts spanning several centuries, but the basic set seems to have been just one pair of tablets. From the scores of surviving fragments, almost all of the first and most of the second is known. They contain what seems to have amounted to a summary of most of Babylonian astronomical knowledge, exclusive of omens, from before the seventh century BC. There are improvements to the older lists of the 'stars of Ea, Anu and Enlil', there are lists of stars that rise while others set, and of periods of visibility of certain stars. The tablets allow a picture to be built up of the constellation names and patterns adopted by the Babylonians. They are not quite ours, although ours ultimately owe much to them, through Greek intermediaries. (Apin, for instance, meaning 'plough', is our constellation Triangulum together with the star gamma Andromedae.)

The stars do not rise and set over the desert in a clear-cut way, as they tend to do over the standard Hollywood rendering of Araby. Just as water-vapour, over and above other atmospheric absorption, affects visibility appreciably in cooler latitudes, so in the Near East the horizon is often obscured by atmospheric turbulence and dust. Since the observation of risings was so often a hit-and-miss affair, and might in any case have been obscured by buildings, the mul-Apin tablets gave lists of secondary stars (the *ziqpu* stars) that culminated (crossed the meridian) at the same time as the more fundamental stars were rising. By 'fundamental star' here we mean the star by whose horizon position time and the calendar were reckoned. This list of *ziqpu* stars is scientifically important, for it represents a step towards a more reliable measure of time. There are reports of lunar eclipses, for instance, no later than the seventh century BC, that measure the time with reference to the culmination of *ziqpu* stars, supplemented by water clocks for the slight difference. (The time unit used was the USH, 4 minutes on our reckoning.)

The mul-Apin series does not make use of the signs of the zodiac as a way of dividing up the path of the Sun through the stars into twelfths, but it does use a system that is somewhat similar, for it lists the constellations in the path of the *moon*. There seem to be eighteen constellations named, the gods in the path of the Moon, rather than the later twelve of the zodiac. Of course the Moon's path through the sky is more or less coincident with the Sun's — the paths are inclined at five degrees or so — so that the longer list contains the shorter. The text makes it plain that the Sun, Moon and five planets were regarded as moving on the same path. For all the existence of an eighteen-part division of the sky, the solar year was at this date certainly divided into twelve months.

Another example of the way the mul-Apin series testifies to a growing mathematical involvement in astronomy is its list of the times when the shadow of a vertical rod (gnomon) one cubit high is 1, 2, 3, 4, 5, 6, 7, 8, 9, or 10 cubits long on the ground, at various seasons. Again, this is not strictly a report on observations, but rather a rationalized list, embodying certain rough rules of proportionality between time from sunrise and shadow length. In short, although the list is primitive, it is genuinely scientific, and we find the same systematic reduction of observation to rational order elsewhere in the tablets. They give rules, for instance, for calculating the times of rising and setting of the Moon, in relation to its phase. (Thus at new moon, setting is just before sunset; thereafter setting is delayed by one fifteenth part of the night each successive day; and after the fifteenth night there are parallel rules for rising.) The same rules were still in use in Roman times. As well as being based on arithmetical procedures, this material is highly practical. One simple scheme for inserting extra days into the calendar is based on the date in the first month of the year on which the Moon passes the Pleiades. In an 'ideal year' – so goes the rule – this happens on day 1. If it turns out to be day 3, then intercalation must be performed.

By comparison, the planets are accorded much slighter treatment, and even the lunar theory contained in this work is elementary by comparison with what was available at the time of copying of many of the surviving tablets. They had a long history, but probably stemmed from somewhere near the latitude of Nineveh at the beginning of the first millennium. Whatever their precise date, they are witness to a deep concern with a theoretical framework for astronomy.

BABYLONIAN INDEPENDENCE AND PERSIAN RULE

Throughout the period that saw the fall of the Assyrian empire, the revival of Babylon's fortunes under Chaldaean rule, and then the Persian conquests, intellectual life and religious pursuits seem to have continued much as before. Texts continued to be written in cuneiform script in the Sumerian and Akkadian languages even after both had given way to the Aramaic language and alphabet in much of civil life. Omens gave way to a new style of divination based on the horoscope, that is, on the arrangement of the heavens at whatever was deemed the significant moment – the commencement of a journey, a battle, the birth of an individual, or whatever. There was systematic observation of planetary phenomena, and of the Moon, and eclipses, not sporadically, but without serious interruption until late Seleucid times. The records that were deemed valuable survive in texts from Seleucid archives. The two main categories are what A. Sachs called the 'astronomical diaries' and (much less common) collections of data for particular types of astronomical phenomena as observed over several years.

The diaries (known from 568 BC onwards) record many different sorts of 'meaningful' events, planetary positions in

relation to the fixed stars, the weather, solar haloes, earthquakes, epidemics, water levels, even market prices. Angular positions were specified in 'fingers' and 'cubits' (each of
twenty-four fingers), common measures of length. Here it
seems that a finger is 2 or 2.5 degrees.

The collections of observations were to prove of great
importance in the long term. In fact the Alexandrian astronomer Ptolemy chose the year of Nabonassar's accession
(748 BC) as the starting point of his calendar, expressly
because the old observations were preserved from that time
onwards. Some of the eclipse data in collections available
today go back to 731 BC. Lunar and solar eclipses, both
arranged in eighteen-year periods, observations of Jupiter
(twelve-year periods), Venus (eight-year periods, for
reasons explained earlier), and others for Mercury and
Saturn, are here included. As time progressed, more detail
was added – for instance in recording conjunctions with,
and distances from, fixed stars. The recording of planetary
periodicities was also a new occurrence, for although clearly
known much earlier, not even in the mul-Apin texts were
they explicitly recorded. They are now found increasingly
as an aid to prediction, short periods for rough use, and
longer periods where more accuracy was called for.

The eighteen-year period for the Sun and Moon is of
great importance in the later history of calendar-reckoning.
The Babylonians discovered a period over which the cycle
of eclipses begins to repeat itself, more or less, not only in
character but in the time of day. They discovered that this
happens after 18 years, or more exactly 6585⅓ days (18.03
years), which is a whole number (223) of *synodic months*
(new moon to new moon). They were fortunate in this
discovery, for the time of day of an eclipse depends heavily
on the motion of the Moon 'in anomaly', to use the modern
technical term. The *anomalistic month* is reckoned from perigee to perigee (perigee is the nearest point of the Moon's
orbit to the earth), and it so happens that 223 synodic
months are approximately 239 anomalistic months.

This is a suitable point to introduce the *draconitic month*, which is measured with reference to the Moon's nodes, the points in the sky where the Moon's path crosses the Sun's. Bearing in mind the nature of either sort of eclipse, it should be obvious that the Sun and Moon must be near to a node simultaneously for an eclipse to take place. (The Moon can wander five degrees or so away from the Sun's path, and the apparent angular sizes of the two bodies are only about half a degree.) Now it so happens that 223 synodic months are approximately 242 draconitic months, and the fact that again we have a whole number is what makes the period of eighteen years work so well in eclipse-reckoning.

The period of 223 months is still mistakenly called a *saros* by some modern writers, using a Greek word that had been used to denote a much longer Babylonian period (3600 years). The modern usage has a long and convoluted history, but goes back only to 1000 AD, or so, and the *Encyclopedia* of Suidas. The ancient Greeks themselves, to get rid of the ⅓ day, sometimes took a period of 669 months, the so-called *exeligmos*. This too was earlier known to the Babylonians.

It is sometimes mistakenly suggested that the 223-month period was primarily used to bring solar and lunar *calendars* into correspondence, but its value to astronomers was clearly much more than that. Those who wish to check its accuracy may use these modern figures for the lengths of the three sorts of month named: synodical, 29.5306 days; anomalistic, 27.5546 days; and draconitic (or 'nodical'), 27.2122 days. For future reference, a fourth type of month is the *sidereal*, defined by reference to the Moon's circuit of the stars as seen from the Earth, its mean value being 27.3217 days. All days here mean solar days.

By comparison with these discoveries, finding the period-icities required by a lunae-solar calendar (one combining months and years) must seem a relatively trivial affair, but it was one that was nevertheless judged to be a matter of great importance in all Near Eastern religions, in Ancient Greece, and later in the Islamic and Christian worlds. For a time in the sixth century the Babylonians used an eight-year period (99 synodic months), and later a twenty-seven-year period (334 months), but the most widely used period equated nineteen years to 235 months. The excellence of this relationship can easily be verified approximately, using the length of the synodic month as quoted above, but in doing so we might very easily miss an important point. What, on this basis, should we take as the length of the year?

If we take 365.25 days, we shall find the discrepancy to be about 0.06 day. As we know, 365.25 is only an approximation, but again we must distinguish between two ways of defining our unit of time, in this case the solar year. Just as we have a sidereal month, so we have a *sidereal year*, measured by referring the Sun to the fixed stars. The more common astronomical definition, at least since Greek times, measures the year (the *tropical year*) differently, namely by the Sun's return to one of the equinoxes or one of the solstices. Reference is now usually made to the Sun's return to spring equinox, or rather to that highly abstract point that will later be used as the origin of astronomical co-ordinates, the 'vernal point' (or 'Head of Aries'). The ecliptic and celestial equator meet there, and the Sun passes through the point in its passage from south to north at the vernal equinox. It must not be thought that its origins were so abstract, for as we have pointed out, the prehistoric origins of astronomy involved the observation of these points in the solar year.

Why should the two definitions give different results? That they did so was discovered effectively only by the Greek astronomer Hipparchus, in the second century BC. To take a longer view, it can be said to be a consequence of the fact that the Earth's axis does not hold a constant orientation in space, with the result that the vernal point is moving with reference to the stars. As a result, the tropical year is (today) 365.2422 mean solar days, and the sidereal year 365.2564 mean solar days.

The difference is slight, but the accuracy of Babylonian astronomy was high, and it is possible to see from several of the cyclical relationships quoted by them that they were derived using the *sidereal year*. Since early Greek astronomers generally used the *tropical year*, we have here one of several fundamental differences in the approaches adopted by these two important groups of astronomers.

Just as with the Sun and Moon, so with the Sun and the planets, there are simple periodic relations that were found at an early date by the Babylonians. We have already discussed Venus, which makes five circuits of the Sun and eight returns to the same place in the stars in eight years. Abbreviating this relationship as (5c, 8r, 8y), we can say that some at least of the following relationships were used long before they were recorded in Seleucid times: Mercury (145c, 46r, 46y); Mars (37c, 42r, 79y) and also (22c, 25r, 47y); Jupiter (76c, 7r, 83y) and also (65c, 6r, 71y); Saturn (57c, 2r, 59y). How these periods were used is explained in a type of text that A. Sachs called 'Goal-Year texts'.

The Babylonians knew that these relationships were not exact, and used additional rules for correcting them. With Venus, for example, in 5 circuits of the Sun Venus was said to complete $2\frac{1}{2}$ degrees less than 8 revolutions. This sort of consideration led to the statement of longer periodicities. In the case of Venus, since the deficit is 1/144 part of a circle in 720 circuits of the Sun (720=144×5), Venus will complete 1151 revolutions (8×144 − 1), and this in 1151 solar years.

The long-period relations often quoted for the other planets were found in a similar way, and a certain delight in the calculation of these long periods of time led to this becoming almost a subject in its own right, finding favour in India to the east and – in more restrained forms – in Greece to the west. According to one account (Seneca), the Babylonian Berossos, who was a priest of Bel and who founded an astronomical school on the Greek island of Kos in the third century B C, taught that when the planets all conjoin in the last degree of Cancer there will be a world conflagration and then a flood. Here is the idea of periodicity. Given the periodicities described, the recurrence of planetary influences easily becomes a part of religious and astrological dogma, and fits well with the idea that history generally, and even human existence, is a recurrent phenomenon. All one has to do to find the scale of the time after which history repeats itself is to find the least common multiple of the long planetary periods. This was perhaps too much to ask. In some accounts of the Bel story, simple round numbers of years are quoted, for instance 2160 000 years (600×3600). The much later Hindu period of time known as the Mahāyuga is just double this amount, showing the clear connection with Babylonian time-reckoning. When the Greeks set out their concept of a Great Year, that too tended to have large multiples of 360 in it.

The figure of 360, which we recognize as the number of degrees in the circle, is almost as good as a Babylonian signature. By the early fifth century B C the Babylonians had the makings of a co-ordinate system, for they had by then begun their division of the zodiac into twelve 'signs' of equal length, naming them after the constellations or important star groups − Aries, Taurus (or the Pleiades), Gemini, Cancer, and so forth. As in later periods, however, this left the risk that constellation and sign would be confused, sharing as they did the same name. The potential confusion became more serious as time went by, and as the precessional drift of the equinoxes moved the stars out of their old signs completely. The Babylonian co-ordinate system, fixed in relation to individual stars rather than the vernal point, was for this reason not ideal, although it was easy to understand − since the stars can be seen and the vernal point cannot.

That the systems differed, but that the differences were often not appreciated, is nowhere better illustrated than in the common Babylonian statement that the vernal point is at 8° Aries. This remark found its way into some second-rank medieval astronomy, and when the context was lost it made no sense whatsoever. (It was later provided with a new meaning in the context of the theory of access and recess, a subject to which we shall return in chapter 8.)

The first Greek text showing the astronomical use of degrees is by Hypsicles, from the mid-second century B C. It was adopted by Hipparchus, the most influential astronomer before Ptolemy. Strabo, however, writing a century later, says that Eratosthenes divided the circle into sixty parts.

THE SELEUCID PERIOD

The establishment of a system of celestial co-ordinates – in this case the division of the zodiac into twelve signs of thirty degrees each – was of the greatest importance for the advance of mathematical astronomy. Accurate planetary periods can be found without it, from observations made over long periods of time, but for an analysis of the finer points of planetary motions such a system is essential. The motives for making that analysis must have been in part intellectual, but they also had much to do with religion and astrological prediction.

The old Mesopotamian stellar religions had encouraged only a crude astrology of simple omens. Various Near Eastern religions, such as Orphism and Mithraism, supported a slightly more developed zodiacal astrology, and some of these were brought to the Latin and Greek worlds with the spread of the Persian empire. Of these oriental religions, one in particular, Zoroastrianism, deserves mention here. This religious doctrine, ascribed to the prophet Zoroaster (Zarathustra), gradually became the dominant religion in Iran, and is still practised by isolated communities there and in India. Its doctrines involve a moral dualism of good and evil spirits, and it readily became intermingled with old Babylonian myth, for instance the myth describing the contest between Marduk and Tiamat. Its relevance to astronomy, however, is less obvious. In its later versions it helped to spread the doctrine that the natural place for the human soul is in the heavens – or more specifically, in its western manifestations, the spheres of the planets. B. L. van der Waerden has argued for a link between Zoroastrian doctrines and the rise of birth horoscopes, especially in Greece. There was an idea that the soul, coming from the heavens where it partakes of the rotation of the stars, when united with a human body continues to be governed by the stars. The notion is to be found, for instance, in the *Phaedrus* of the Athenian philosopher Plato.

This philosophical motive might not entirely explain the phenomenal rise of astrology in the Hellenistic world, but whether or not it does so, that was real enough, and put astronomical prediction at a premium. The Greeks had heard of Zoroaster by the fifth century BC, a century before Plato, but it is interesting to learn that it was one of the greatest of Greek astronomers, a contemporary of Plato, who was largely responsible for introducing Zoroaster's philosophical ideas to the Greeks. The scholar in question was Eudoxus of Cnidos.

Whatever the philosophical impact of Zoroastrianism, it was astronomically naive. There were some routines that would have benefited from astronomical knowledge, such as predicting the harvest from the sign in which the Moon stands in the morning of the first visibility of Sirius, but there is no reason to think that the Persians made any great advance in actually *predicting* such things. And even that astrological doctrine was probably borrowed. The mathematical astronomy necessary for the practice of horoscopic astrology owed almost everything to the Babylonians. The oldest known cuneiform horoscope is datable to 410 BC, and is from a Babylonian temple. The Greeks were already in Plato's century giving due credit to the 'Magi' or the 'Chaldaeans', and throughout the world of classical antiquity these epithets stuck, as synonyms for 'astrologer'.

This fact should not, as it has so often done in the past, obscure the brilliant mathematical contribution made by the Babylonians, by which their astrology was underpinned.

Over three hundred cuneiform tablets survive in this category, many of them damaged, and some even in fragments divided between museums in different countries. They are usually divided into 'procedure texts' (in which the methods of calculation are explained) and 'ephemerides' (in which the results of the calculations are listed over a period of time, rather as in the modern Nautical Almanac). Ephemerides (the Greek word *ephemeris* simply means 'daily') are more than three times as numerous as procedure texts. All come from Babylon (excavated 1870–90) and Uruk (excavated 1910–14), so that even now we may be very ignorant of the total achievement of these people.

It is possible that concentrating on the problem of the Moon led to comparable solutions for the planets. How many days are there in a month? A Babylonian month began with the first visibility, after sunset, of the thin lunar crescent. Days were likewise counted from the evening. On this definition, a month has a whole number of days, and experience taught that the number was either 29 or 30. But how do we decide in advance? Today we have a general picture of the situation, a model to which we can apply standard geometrical procedures to derive an answer. Even now, it is far from easy to do so. The Babylonians, having been provided with no such model by their forebears, were obliged to work in reverse. Let us first try to appreciate the intrinsic difficulties they faced by showing how our own analysis would proceed.

We begin with the crude approximation that a month is just 30 days in length. The Sun covers about 1° of the zodiac per day, so that the Sun in moving from conjunction to conjunction with the Moon will have moved 30°. They are in conjunction, so the more rapidly moving Moon will have moved 390°, or 13° per day over the thirty-day period. (A more accurate average figure is 13.176°, but the Moon's velocity varies appreciably.) Now to predict when the crescent Moon will first be seen, several factors have to be taken into account:

(1) The brightness of the Sun means that they have to be separated by a certain minimum distance.
(2) The relative speed of the Moon and Sun decides how long it takes the Moon to cover this distance. The relative speed is on average roughly 12° per day, but this 'daily elongation' can vary by two or three degrees either way.
(3) The critical distance is affected by the brightness of the background sky, which is in turn affected by the angle between the horizon and the line joining the setting Sun and the crescent Moon (see figure 3.3). This can in turn be resolved into various factors: (a) the season

FIG: 3.3 *First visibility of the lunar crescent*

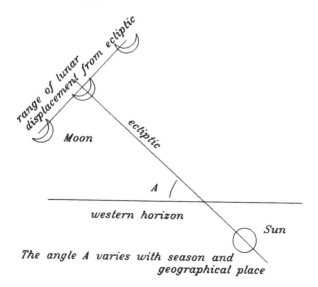

western horizon

The angle A varies with season and
geographical place

of the year, which is another way of talking about the Sun's place on its own path, that is, the 'ecliptic' through the middle of the zodiac; (b) the Moon's divergence from that path, its 'ecliptic latitude', which can exceed 5°; and (c) the geographical place of the observer (latitude), which fixes the angle at which the stars generally cut the horizon in rising and setting.

This summarizes our probable procedure. The astonishing thing is that the Babylonians somehow managed to extract the various factors by analysing their observations of the beginnings of their months. They did so using wholly arithmetical methods, that is, without resort to geometrical models, and if we represent their results here in graphical form, that is only because it is much more economical to do so. To get the full flavour of the originals, one should look at such a work as O. Neugebauer's *Astronomical Cuneiform Texts*, where the material is transcribed and analysed.

Two chief systems for the representation of varying solar, lunar and planetary movement are recognized. The first, known as 'system A', assumes that a velocity (say that of the Sun) is held constant at a particular value over a certain substantial range of the zodiac, then changes to another value, which it holds for an appreciable time before changing again, and so on. If we think of velocity (not position) as being plotted against time, this gives us a 'step function'. Represented graphically it looks like a somewhat irregular battlement. 'System B' is more sophisticated, and assumes that each step, as represented by a new line in the tabulation of positions or whatever, is of constant positive or negative amount, except where a predetermined maximum or minimum would be passed. In these cases, the direction of change (increase or decrease) is reversed.

This change of increment is effected according to strict rules that are most easily explained by reference to figure 3.4. In this particular instance, drawn from an ephemeris for the year 179 of the Seleucid Era (133–132 BC), the hori-

FIG: 3.4 *The Babylonian zig-zag function*

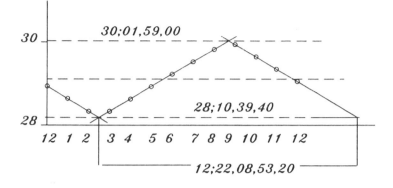

zontal scale is of monthly units, in some sense, these being named in the first column of the tablet. They turn out to be those points in the month at which mean conjunction of the Sun and Moon occur. The vertical scale corresponds to the second column on the tablet, which contains sexages-imal numbers of the order of 28 or 29. (Only after the numbers were analysed by modern scholars could it be seen what it was they represented.) Another column on the tablet, not reproduced on the graph, can be shown to give the longitudes of Sun and Moon at the time of their con-junction. Since it emerges that the second column contains the differences between successive entries in that third column, that second column evidently contains, as we should say, solar velocity (the change of longitude per month). The zig-zag created (in our graphical way of describing purely arithmetical quantities) out of straight lines was their equivalent to something for which we should now use at the very least a sine curve. All told, theirs was an extraordinary achievement.

The period of the zig-zag function in months can be easily found – it is marked on the figure as 12;22,08,53,20. This is the value accepted for the year, measured in synodic months. It was clearly not obtained by direct observation, but from one of those cyclical relationships we referred to earlier, in this case seemingly equating 810 years with 10 019 months. Of course observations must ultimately have been involved, but numerical convenience too must have entered into the calculation. Unfortunately only the end results of the enterprise are known.

Other equivalences are also found, for instance that of 225 years and 2783 months, yielding 12;22,08 months to the year. This number is found in tablets drawn up on both system A and system B. One of the more surprising discoveries of those who have laboured with these cuneiform tablets is that although system A is the older, nevertheless the two systems were in use during the entire period from which they have been preserved, say 250 to 50 BC, and this in both Babylon and Uruk.

Relatively few lunar ephemerides cover more than one year. Most have columns for lunar and solar velocity and position. Some list the length of daylight or night, corresponding to the Sun's position, in an earlier column. For us, this is a matter for calculation using the methods of spherical geometry, but for the Babylonians, only arithmetical methods were in use. There were in some cases columns for the Moon's latitude, and others for the magnitudes of eclipses. A procedure – a formula – for finding eclipse magnitude was applied *every* month, whether or not an eclipse was due. This might be regarded as contrary to the spirit of an empirical science, but it certainly speaks for a high level of abstraction, and a clear grasp of the notion of a mathematical function. Among the procedures followed were some for correcting results on account of a solar velocity that had been accepted as constant at an earlier stage in the calculation, but that was known to be variable. On system B, this correction was bound to be more difficult than on system A, which goes some way to explaining the survival of the latter.

It was realized that lunar and solar eclipses are impossible if the eclipsed object has too great a latitude at the time of new or full Moon. The problem of solar eclipses is much more difficult than that of lunar eclipses, and all that could be said in their case was that an eclipse was impossible. To say in advance that it would be observed, they would have needed much more information about the distances and sizes of the Earth, Sun and Moon. There is no firm evidence that patterns in the recurrence of solar eclipses – another route to their prediction – were known, although some have claimed that they were.

The tablets mentioned so far deal with long periods of solar and lunar movement, but there are others using similar methods for daily changes, and from them, for example, an equivalence of 251 synodic months and 269 anomalistic months (see above for definitions) may be derived. Here 29;31,50,08,20 days was taken to be the length of the synodic month and 27;33,20 days the anomalistic. Lest the high accuracy of these figures is not obvious, let it be noted that the values quoted today are identical to these to within one and four parts in six million, respectively. (There have been much smaller changes in the periods in the meantime, so the comparison is not absolutely strict.) A more interesting historical comparison, however, is between the length of the Babylonian synodical month quoted here and the length used in the high middle ages in the so-called Toledan Tables. The two are to all intents and purposes identical.

When the Babylonians of the Seleucid period turned their attention to the planets, their arithmetical procedures – to use our graphical analogy once more – came one step nearer to the ideal sinusoidal curve. (There are in fact tablets giving the Moon's latitude where the lines of the simple zig-zag had already been modified, bent nearer to the ideal, as it were, as shown in figure 3.5.) Before explaining how the Babylonians proceeded, it will be as well to have an approximate picture of the way the planets are seen to move by an observer on the Earth. The following non-historical aside is offered as background material for

FIG: 3.5   *Babylonian theory of lunar latitude*

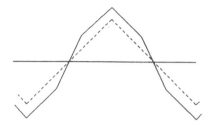

this and the following chapters, dealing with other classical theories of planetary motion.

TWO APPROACHES TO PLANETARY MOVEMENTS

Of the planets recorded before the eighteenth century, all in orbits around the Sun, Mercury is the nearest to the Sun, followed by Venus, and then the Earth. Mars, Jupiter and Saturn have orbits outside the Earth's (figure 3.6). The 'inferior' and 'superior' planets have somewhat different patterns of behaviour, as seen from the Earth. Each orbit is an ellipse, with the Sun at a focal point, as Kepler was able to show, but to a first approximation the orbits may be taken as circular, with the Sun at the centre of all. We imagine a diagram of the system with a pin through the point representing the Sun. If we remove the pin and put it through the point representing the Earth, the relative positions of the planets will be unaltered, and yet it should be clear that now we may regard the planets as moving on circles centred on the moving Sun.

Consider a simple case, that of the inferior planet Mercury, which is now to be thought of as a satellite of the Sun. The Sun's circle may be called by the traditional name of *deferent circle*, literally a 'carrying circle', and the satellite's orbit, the carried circle, will be an *epicycle*. The other inferior planet, Venus, will travel on a larger epicycle. With the

FIG: 3.6 *Planetary orbits*

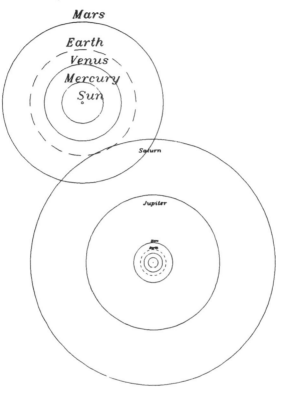

In each drawing the sizes of the orbits are approximately to scale but the upper figure fits into the middle of the lower as indicated.

superior planets, the roles of epicycle and deferent will be reversed, but we need not go into more detail for the time being.

The description of individual planetary motions in terms of epicycles is of great importance in the history of astronomy, although it has to be emphasized that each planet was treated separately, and that it was a long and painful trek to an appreciation of the fact that the Sun is present in the epicycle–deferent system for each planet. Only when this was fully realized, by Copernicus, was it possible to bind the planets into a single system.

This type of explanation is capable of great refinement; for instance, by giving the epicycles a slight tilt to the plane of the Sun's orbit it is possible to account for the fact that the planets are not precisely in that plane, but have movement in latitude. 'Ecliptic longitude' is measured around the ecliptic from a zero-point where the ecliptic, the Sun's path, meets the celestial equator. 'Ecliptic latitude' is then the co-ordinate measured towards the poles of the ecliptic, northern or southern.

The epicyclic description was typical of later Greek astronomy, but not of Babylonian. Once it was appreciated, the astronomer's task was much simplified. The strategy was to assume a geometrical model and then draw consequences from it, for instance about the patterns of rising and setting of the planets. Those theoretical consequences could then be compared with observation. If the model seemed to be worth retaining, then those and further observations would allow the astronomer to determine, or improve, the numerical properties of the model (such parameters as the relative sizes of the circles, the angular speeds in the circles, and so forth). The Babylonians, arriving on the scene first, worked in more or less the reverse order: what was to the Greeks – as it is to us – a derived consequence was to them a starting-point, a datum. Consider, for example, their concern with risings and settings over the horizon, a concern shared by most early cultures. Such observations yield very unpromising information for an astronomer, and the miracle is that powerful theories emerged from them when they did.

We have already mentioned the first and last appearance of the fixed star Sirius, which spends part of the year lost in the Sun's rays, and the fact that the planets too may be lost to view for a time, for similar reasons. The inferior planets Mercury and Venus, which never wander so far from the Sun as to be in opposition to it, have patterns of behaviour touched upon earlier (see figure 3.7 and our earlier figure 3.2). Once again we may consider the matter from a modern perspective. When an inferior planet is on that part of the orbit represented by a broken line, its angular separation from the Sun is so small that it is lost in the glare of the Sun. At the point *FM* it becomes visible, in this case for the first time as a *morning* star. From the Earth the planet will move round the sky in its daily path as though carried by the Sun. The fact that this will be seen in the morning, shortly before the Sun rises, should be obvious from the upper part of the figure, where the orbit is roughly drawn nearly edge-on to the observer. *LM* is the point at which it will be last seen as a morning star, and *FE* and *LE* are the points of first and last appearance as an evening star.

FIG: 3.7 *Morning and evening risings and settings of planets*

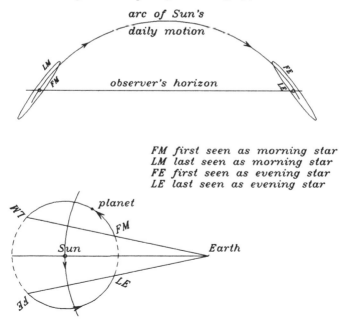

When dealing with the inferior planets, which have to be near the same horizon as the Sun, our fourfold division (*FM, LM, FE, LE*) is unambiguous. *Heliacal rising* is then, as we have seen, the first visible appearance on the eastern horizon before sunrise (*FM*) and *heliacal setting* is the last visible setting just after sunset (*LE*). Mars, Jupiter and Saturn, however, can be seen rising just after the Sun has set, and setting just before the Sun has risen, so that to discuss them we might need an extra qualification, such as 'first morning setting'. The words 'acronychal' (not to be spelled 'achronycal') and 'cosmical' are often used of risings and settings in ways that are ambiguous when they overlook this fact. It is better not to define them at all, but to remember, when reading others who do, that the first adjective concerns *evening* sightings (first or last) and the second *morning* events.

As the Sun goes round the Earth, and any planet goes round the Sun, a spiralling motion will result, as shown in differing degrees of detail in figures 3.8 and 3.9, for Mercury and Venus. The planet is never far from the plane containing the path of the Sun through the stars (the ecliptic), along which path we measure (celestial) longitude, and for an alternative picture of the planet's motion we may plot the longitude against time. This is done in figure 3.10, where the horizontal axis spans approximately one year (the divisions mark intervals of thirty days) and the vertical axis gives the longitude of Mercury (from 0 to 360 degrees).

The diagonal running through the middle of Mercury's graph shows the movement of the Sun, around which Mercury seems to oscillate. Mercury goes round the Sun about four times in a year. (Its sidereal period is 0.24 tropical years.) Marked on the graph are the points of first and last visibility (with the notation of the last paragraph) and the so-called *planetary stations*, where the planet seems to stand still in relation to the background of fixed stars, that is, when its motion changes from direct to retrograde (*S1*), or conversely (*S2*).

FIG: 3.8 *The spiral path of Mercury with respect to the earth*

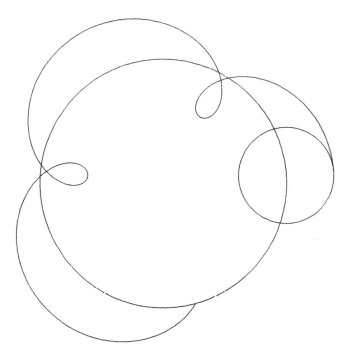

The case of the superior planets may be illustrated by Mars (figures 3.11 and 3.12). The rectangular graph now covers a period of over six years, and since the vertical axis represents longitude, a co-ordinate that is essentially cyclical, the lines representing both the Sun (roughly straight) and Mars have to be repeatedly broken. Some general principles will be obvious, even so. Proximity to the Sun, making the planet difficult to see, takes place in the middle of long-term, moderately regular, direct motion, while during the period of retrograde motion the planet is approximately in opposition to the Sun (180 degrees of longitude distant from it). The 'Sun lines' are of course spaced at yearly intervals, and from them it is evident that

FIG: 3.9 *The spiral path of Venus with respect to the earth*

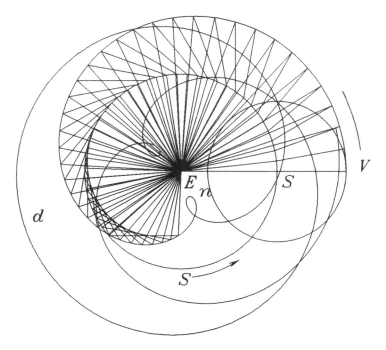

the kinks in Mercury's graph are a little under two years apart. The sidereal period of Mars is in fact 1.88 tropical years.

So much by way of our non-historical aside.

BABYLONIAN PLANETARY THEORY

We have already seen how important to the principles underlying the Venus tablets of Ammizaduga was a concern with horizon phenomena. It is interesting to see how, when the Babylonians analysed the behaviour of first and last appearances, morning and evening, they treated them as *separate phenomena*. It was as though each had its own exist-

FIG: 3.10   *Graph of Mercury's longitude*

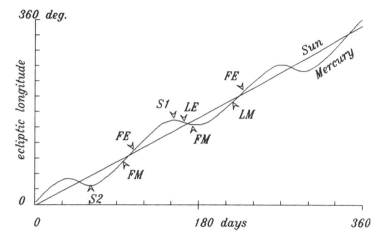

The changing longitudes of Mercury and the Sun over approximately one year.

ence as an object residing in the ecliptic. Consider the points on the Mercury graph (figure 3.10 above) marked as *FM*, for instance. Taking a large number of such points in isolation from the rest of the graph, they fall on a moderately straight line parallel to the Sun's graph, but by no means a perfect one. The Babylonians used the same sort of arithmetical methods as they had used for the Sun and Moon to account for deviations from (as we are representing it) the straight-line graph. The problem, expressed graphically, is one of breaking the line into segments, that is, finding the breaking-points and gradients of suitable component parts. Those gradients (angular speeds) were seemingly expressed in terms of convenient whole-number relationships, such as 'Mercury rises 2673 times in 848 years'.

This was not easily done, and the fact that the Babylonians were tied to a lunar calendar added an additional complication, for of course the planetary periods have virtually

FIG: 3.11   *The spiral path of Mars with respect to the earth*

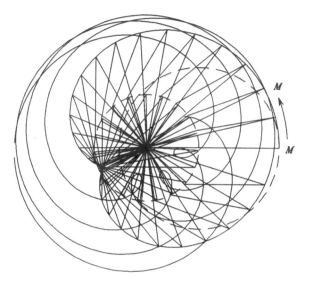

nothing to do with the Moon's motion. In the end, this did not matter greatly, for although they did not express their dates in days, they used a unit of one-thirtieth of a *mean* synodic month. This is now usually called a *tithi*, a word that comes from later Hindu astronomy, where the same unit was in use. The *tithi* is roughly a day in length. The Moon's motion varies in intricate ways, but since the *tithi* is by definition an average value, it can in principle be used in a perfectly workable astronomical system. This is not to say that it is an astronomically well chosen unit, but it had clear advantages for those whose religion bound them to the old lunae-solar calendar.

Having produced a set of rules for the behaviour of phenomena of type *FM*, analogous rules were found for the behaviour of *LM*, *FE* and *LE*. The general principle was that of finding the amounts (in longitudes, and *tithis* of time) to be added to the base values (for *FM*), in subsequent

FIG: 3.12   *Graph of Mars' longitude*

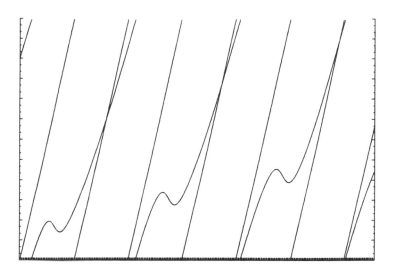

The changing longitudes of Mars (kinked lines) and the Sun over a period of 2400 days (cf. figure 3.10). The time axis is the lower axis.

phenomena. The resulting graph by which we can represent these arithmetical procedures is a plausible approximation to the sinusoidal form of those we have given above.

One highly sophisticated element in the exercise was the introduction of certain 'phenomena' into the scheme of calculation that could not in fact have been observed. The situation can perhaps be best explained by analogy with the case of full Moon. Much of the time, at the precise moment at which this occurs, the Moon will be below the horizon, but the calculation proceeds regardless of whether it is to be seen or not. The other 'phenomena' we have been discussing (*FM, LM, FE* and *LE*) similarly began life as observable events, but were eventually turned into computational ideals – ideal points on our ideal graph, so to speak.

Broadly similar procedures were followed for all the

planets, and for the superior planets attention was paid also to the two *stations*. Much Jupiter material survives. A Mars tablet from Uruk is remarkable for the plausible representation of the highly variable velocity of that planet. All the planets then known are represented by tablets from Babylon and Uruk, and in both places a variety of different methods of arithmetical representation were in use. To say that the general principles were much the same in all cases is rather like saying that all aircraft are more or less alike; but in a work such as this it is not possible to say much more. What should at least be mentioned is that there are a few surviving tablets showing that points on the graph intermediate between the key phenomena (*FM, FE*, first station, and so on) were supplemented by intermediate points that were decided, not by a straightforward interpolation (not by straight-line segments, as we should say), but by schemes based on second-order and even third-order differences. There are sciences that even today have not advanced to this level of sophistication.

4

The Greek and Roman World

ASTRONOMY IN HOMER AND HESIOD

Babylonian astronomical documents contain evidence of two complementary processes, one in which theories capable of representing and predicting observation were created, the other involving the use of those theories to predict phenomena. The second process is of course what is usually encountered in surviving tablets, and the first usually has to be reconstructed from it. That second type of activity required a set of skills that could be exercised by people well drilled in routine procedures of which they need have had little understanding. They occasionally added their names to the tablets they produced, and even the names of their forefathers, together with a date, and the name of the ruler. This all suggests a level of professionalism that might hint at some sort of formal education encompassing the *reasoning* behind the quantitative fundamental theories.

In the case of ancient Greek cultures and civilizations, matters were not very different once the art of processing large sets of observational data had been learned from eastern sources, but this happened at a relatively late date. The most significant influences took effect only in the second century BC, and the man deserving most of the credit for this change was Hipparchus. By then, however, the Greeks had developed a geometrical method of their own that was to assume an extraordinary importance in subsequent history. They modelled the heavens on a sphere, with stars, planets and circle on it, and they learned to explain the simple daily and annual movements in terms of the rotation of the celestial sphere.

Our view of Greek astronomy tends to be formed by Ptolemy, its greatest practitioner, but by his time (the second century of the Christian era) Babylonian arithmetical methods had been very efficiently grafted on to the native geometrical astronomy, a fact that tends to obscure just how cavalier were the first of the great Greek astronomers about observational data. As we shall see, this is true of even the greatest of them, Eudoxus, of the fourth century BC.

The Greeks developed a notable picture of the Universe as a whole, and explained its workings on a rational mathematical and philosophical basis that by the time of Eudoxus was tearing itself free from the creation myths and legends of earlier ages. The Greeks shared in some of the traditions of prehistoric astral religion that we have described already, and there was some cultural interchange with the great neighbouring cultures – for instance, with the Egyptian and the Persian – but very little is known of the state of astronomical knowledge in early Greece, even in Minoan or Mycenaean times. Names of the months occur in the famous Linear B tablets, and worship of the Sun and Moon in some form seems likely, judging from their art. Some four or five centuries after the Mycenaean period came the age we know from the poetry of Homer (perhaps from the mid eighth century) and Hesiod (around 700 BC).

Homer's *Iliad* and *Odyssey* contain only fragments of any relevance to our theme, but they are of great interest. In the first, Achilles' shield is likened to the Earth, which is surrounded by an ocean-river, the source of all water and of the gods. In the *Odyssey*, the starry heaven is said to be of bronze or iron, and to be supported on pillars. Several star groups are named – the Pleiades and Hyades (groups in Taurus), Orion, Boötes, and the Wagon – which is unusual in that it does not rise out of the ocean. (It is too close to the pole to rise and set at all.) There are references to the evening star and the morning star, presumably Venus, possibly not then recognized as a single planet. The 'turnings of the sun' seem to refer to the solstices. The phases of the Moon are often alluded to; the winds and seasons are personified, and Athene is once likened to a shooting star. All told, although this was heroic poetry meant for the courts of princes, the astronomy is very homespun, and that to be found in Hesiod's *Works and Days* is only marginally less so. This is a manual in verse, relating the farmer's year to the changing seasons, as judged by the Sun and stars (heliacal risings, and so forth). There are here no echoes whatsoever of Babylonian expertise.

COSMOLOGY IN THE SIXTH CENTURY

It was Aristotle, the greatest philosopher of antiquity, who in the fourth century B C established a tradition of collecting together the opinions of previous thinkers and subjecting them to criticism, as vigorously as if they were still alive. Some of his material goes back to the late sixth century, but like others who collected early teachings he was largely dependent on unreliable intermediaries. This is particularly true of the four earliest philosophical thinkers of note – Thales, Anaximander, Anaximenes and Pythagoras, all of the sixth century. Thales was considered by Aristotle to be the founder of Ionian natural philosophy. Stories were told of his intense practicality, for example, of his having used his astronomical knowledge to predict a glut in the olive crop. He got a monopoly of the presses, and so made a fortune. On the other hand, he was also presented as a visionary, so engrossed in a study of the heavens that he fell down a well, unable to see what was at his feet – in the words of a Thracian servant girl. (Aristotle retailed the first story, and Plato the second.) He is supposed to have predicted an eclipse of the Sun that took place during a battle between the Lydians and the Persians, and that is now usually set at 28 May 585 B C. This story has been much debated, but can be almost certainly disregarded except as a symbol of myth-making at the time of Aristotle.

It was Aristotle who set his pupils the task of writing a compendious history of human knowledge. Eudemus of Rhodes was assigned astronomy and mathematics, and from him we hear that Thales, after a visit to Egypt, brought these studies to Greece. Other have argued for Thales as a borrower from Babylon. One might ask what we know about, that he might have borrowed. It has been claimed that he introduced the method of geometrical proof, no less; but the evidence is extremely tenuous, and disregards the strong possibility that European traditions of semi-formal geometrical reasoning are much more ancient.

Anaximander and Anaximenes held cosmological views that are almost as similar as their names. The second was possibly the pupil of the first, around the time of the fall of Sardis (546 BC). Like Thales, both came from Miletus, the southernmost of the great Ionian cities of Asia Minor (like Sardis, at the western extreme of modern Turkey), a fact that serves to remind us of the great spread of the civilization that we know as Greece. Among the greatest of ancient Greek astronomers and mathematicians, Eudoxus was from Cnidos, Apollonius was of Perge, Aristarchus of Samos, and Hipparchus of Nicaea and Rhodes, all in, or off the coast of, Asia Minor; Euclid and Ptolemy taught in Alexandria, albeit more than four centuries apart; and Archimedes lived and worked at Syracuse, in Sicily.

Anaximander is said to have made a map of the inhabited world, and to have invented a cosmology that could explain the physical state of the Earth and its inhabitants. The infinite Universe was said to be the source of an infinity of worlds, of which ours was but one, that separated off and gathered its parts together by their rotatory motion. This analogy with vortex motion has perhaps more to do with the observation of cooking vessels than of slings, and yet not entirely dissimilar theories were being advanced in Newton's century. Masses of fire and air were supposedly sent outwards and became the stars. The Earth was some sort of floating circular disc, and the Sun and Moon were ring-shaped bodies, surrounded by air. The Sun acted on water to produce animate beings, and people were descended from fish.

Strange though these ideas now seem, we catch glimpses here of a type of scientific reasoning that is by no means trivial. When Anaximenes elaborates on Anaximander's ideas, and argues that air is the primeval infinite substance, from which bodies are produced by condensation and rarefaction, he produces logical arguments based on everyday experience. (Admittedly they were not always well chosen. He considers breathing through pursed and open lips, breathing into cold air, and so forth.) Again he introduces rotatory motions as the key to understanding how the heavenly bodies may be formed out of air and water. Such attempts at a physics of creation are characteristic of much Greek thought, and the fact that in these early centuries they are combined with a rather weak grasp of the circulation of the heavenly bodies, and of how they proceeded when they passed below the horizon, the rim of the world, does not take away their importance for the later history of cosmological thought.

The name of Pythagoras needs no introduction, although the geometric theorem for which he is remembered certainly had little to do with him, at least in the Euclidean form in which it is – or was until recently – taught. He lived in the late sixth and early fifth centuries. Despite his large religious following, nothing he wrote has come down to us, but it seems that he took the cosmic ideas of Anaximander and Anaximenes one stage further, saying that the Universe was produced by the heaven inhaling (note the metaphor) the infinite so as to form groups of numbers. Why numbers? He maintained that all things were numbers. His greatest claim to fame is his discovery of the arithmetical basis of musical intervals, which discovery gave rise to mystical forms of numerology that even now have their adherents. Pythagoras seems to have been convinced that everything, from opinions, opportunities and injustices to the most distant stars, is rooted in arithmetic and has a corresponding place in the structure of the Universe as a whole. Whether or not this mystical belief can be defended, there has been scarcely any period in history since that time when it has not had important repercussions on scientific thought.

Aristotle tells us of a geometrical model of the Universe proposed by the Pythagoreans, involving a central fire around which the celestial bodies move in circles. The central fire was not the Sun, although the Earth was certainly of the character of a planet to it. To account for lunar eclipses, the Pythagoreans postulated a counter Earth. Slowly but surely a characteristically physical style of Greek thinking was beginning to develop, and was soon to begin to yield important results.

GREEK CALENDAR CYCLES

It was at about this time, or a little before, that the zodiac was introduced into Greece from Babylonian sources, and Babylonian influences seem to have been at work in other ways too. The solstices had long been under observation, but it seems that some attention was now being given to the recording of these seasonal events with a view to improving either the civil calendar or the calendrical scheme into which astronomical observations were fitted. In the century or so before the time of Eudoxus, Plato and Aristotle, there was much Greek concern with the improvement of the civil calendar, but that this was not viewed as a markedly astronomical question is clear from the somewhat random way in which magistrates took corrective action when the solar and lunar cycles slipped out of alignment. Or perhaps it was simply that magistrates missed the point.

Whether or not it is correct to speak of a 'school' of astronomy founded by the Athenian astronomers Meton and Euctemon, both seem to have collaborated in proposing a regular calendar cycle of nineteen years, the so-called

Metonic cycle. Meton, indeed, is said to have set up an instrument for observing the solstices, on the hill of the Pnyx, in Athens. We have already seen that the Babylonians knew the nineteen-year cycle that brings the solar and lunar cycles back into agreement – 235 months being in fact very close to nineteen years. No immediate documentary evidence survives. An observation by Meton and Euctemon, albeit one that had its reliability called into question by Ptolemy long afterwards, had been made in 432 BC. The earliest known use of the nineteen-year cycle in Mesopotamian texts is from about half a century later. Both groups of astronomers made the cycle the basis of rules for intercalation, the insertion of extra days (like our leap-year day) to correct the calendar's slip: 235 months were equated by Meton (according to Ptolemy) with 6940 days. According to a late authority Geminus, out of the 235 months, 110 were 'hollow' months of twenty-nine days and 125 were 'full' months of thirty days. Presumably Meton drew up a scheme for intercalation. There is no evidence that this was used as the Athenian civil calendar, but the history of the calendar is a morass of alternatives, and it would be wise to keep an open mind here. Whether the Babylonians knew the nineteen-year cycle before the Athenians we cannot say, but they do seem to have used other intercalation rules based on the risings of Sirius from much the same time, and very probably pre-empted the Greeks.

Calendar cycles, at all events, became something of a Greek astronomical speciality. A century after Meton and Euctemon, Callippus improved their cycle further by taking four periods (seventy-six years) and removing one day (27 759 days). The Callippic cycle was used yet later by Hipparchus and Ptolemy in a modified form. Again, however, it is clear that Hipparchus' refinement (equating 304 years to 111 035 days and 3760 synodic months) was never practically employed. The simpler cycles of nineteen and seventy-six years were for most purposes enough, and the former eventually became enshrined in the Easter computus of the Christian Church, where it remains in use to this day.

THE GREEKS AND THE CELESTIAL SPHERE

Greek astronomy of the fifth century, like that of the Near East, was intertwined with a study of meteorological phenomena generally – with clouds, winds, thunder and lightning, shooting stars, the rainbow, and so forth. This component remained, with an astrological underpinning, until modern times, but far more important in the long term were the seeds of the geometrical methods that early Greek procedures contained. The discovery that the Earth is a sphere was traditionally assigned to Parmenides of Elea (southern Italy, born around 515 BC), who was also said to have discovered that the Moon is illuminated by the Sun. A generation later, Empedocles and Anaxagoras seem to have given a correct qualitative account of the reason for solar eclipses, namely the obscuration of the Sun's face by the intervening Moon. Astronomy was spread very thinly through the period leading to the first great age of mathematical advance, the fourth century, which began with the remarkable planetary scheme of Eudoxus and ended with the first extant treatises on spherical astronomy, those by Autolycus and Euclid. Small but important developments were taking place, however. We are told that a catalogue of stars was drawn up by Democritus in the fifth century, and many famous examples followed its lead. It is hard to say what they all contained: some were just star-lists, and not until Hipparchus do we have clear evidence of a consistent scheme of co-ordinates on the sphere.

The change that took place in Babylon, from the listing of stars by reference to the zodiacal constellations to a system of numerical ecliptic longitudes, had occurred around 500 BC. Reckoning as we now do from a zero-point where the equator and ecliptic meet did not come for another six centuries. Ptolemy introduced it, for his definition of the (tropical) year. The Babylonians had reckoned from the zero-points of each zodiacal sign, measuring in each from 0 to 30 degrees. That system remains in astrological use. The Babylonian signs were shifted 8° or 10° from where a 'Ptolemaic' astronomer places them (on systems B and A respectively), and we have already noted that traces of this discrepancy are to be found in medieval western sources, where the idea was repeated by scholars who had no idea what it was all about. Unpolished though this system might seem, the Greeks of the fifth century, at all events, had nothing comparable.

The discovery of the sphericity of the Earth, and of the advantages of describing the heavens as spherical, captured the imagination of the Greeks of the time of Plato and Aristotle, in the fourth century, and of one man in particular: Eudoxus of Cnidos (*c.* 400 to 347 BC), produced a very remarkable planetary theory based entirely on spherical motions. In terms of its predictive power, this theory cannot bear comparison with the Babylonian arithmetical schemes, but it was in many ways more important. First, it showed posterity the great power of geometrical methods; and secondly, by an accident of history – its adoption by Aristotle – it was for two thousand years instrumental in shaping philosophical views on the general form of the universe.

THE HOMOCENTRIC SYSTEM OF EUDOXUS

Eudoxus was born in Cnidos, an ancient Spartan city on a peninsula at the south-west corner of Asia Minor. In his youth he studied music, arithmetic and medicine and in geometry he was taught by a notable mathematician, Archytas of Tarentum. On a first visit to Athens he studied with Plato, thirty years his senior. He later visited Egypt, perhaps on a diplomatic mission, and is said to have composed an eight-year calendar cycle (the *octaëteris*) while studying with the priests at Heliopolis. Returning to Asia Minor, he founded a school at Cyzicus, rivalling Plato's Academy in Athens – which he again visited at least once more. (Cyzicus was a Greek city that had been abandoned to the Persians in 387 BC.) The school at Cyzicus is said to have attracted so many pupils that on his return to Athens Eudoxus succeeded in embarrassing Plato. Eudoxus professed the principle that pleasure is the highest good, and it is likely that Plato had him in mind when writing on this subject in his work *Philebos*. Whether or not this is so, Eudoxus' influence in arithmetic, geometry and astronomy was considerable. His pupils included the great Menaechmus, inventor of conic sections. Eudoxus was largely responsible for some of the finest sections – books V, VI and XII – of Euclid's *Elements of Geometry*. The merits of his rigorous definitions of number were not fully appreciated until recent times, when it was realized that they bear a strong resemblance to those of Dedekind and Weierstrass from the nineteenth century. His planetary theory, however, attracted much attention from the beginning.

We have already discussed the motions of the planets from a modern perspective (in the penultimate section of the last chapter). It has often been said – on the basis of a much later authority, Simplicius, through Sosigenes – that it was Plato who set the problem to posterity of explaining how the observed movements of the planets may be explained in terms of 'uniform and orderly' motions in the heavens. While the views of this great philosopher are of much interest, his influence on mathematics and astronomy is easily exaggerated. His contribution was not direct, but stemmed from his concern that both subjects be a part of the education of the ruling class (the Guardians) and the ordinary citizens. As a propagandist his influence is still a force to be reckoned with: he saw these studies as a means of training the soul to look beyond the transitory things of this world to a true reality beyond that only thought can grasp. One might say that as an astronomer Plato was little more than a straw in the wind, were it not that, having eloquently insisted that the Universe operates according to mathematical laws that can only be understood by a properly trained intelligence, he helped to *create* an entire educational climate favourable to the subject.

The then recent discovery of the spherical form of the Earth, extended to that of the heavens, had obviously already taken a hold of Athenian thinking. In the tenth book of one of his finest works, *The Republic*, Plato introduces a myth, told with much use of poetic imagery by Socrates. It is the story of Er, a man killed in battle whose soul visits the land of the dead, to return after his miraculous revival. He tells how Er's soul went first to a certain magic place, described in some detail, and how eventually he saw what amounted to the mechanism of the entire planetary system, with nested whorls ('bowls', according to one of several possible interpretations of the text and 'hoops' following another) turning around a spindle of steel, and each carrying a planet. This rested in turn on the knees of Necessity, the daily rotation as well as the planetary motions being thus taken care of. The whorls were turned with their various characteristic speeds by the Fates (daughters of Necessity), and on each a Siren sang a single note, so that together they made a harmonious sound.

The counter Earth of the Pythagoreans is nowhere to be seen in this account. There is no indication that the zodiac is inclined at an angle to the equator, but it would be foolish to look too deeply into the myth of Er for such niceties. What seems to be suggested very strongly by this myth is that *physical models* of the Universe were being made, and not merely described. To describe such a Universe as Er's, one would surely have introduced *complete* spherical shells; whatever the whorls were, they would have been open-topped, to allow one to see into the workings of the cosmos. In a later work of Plato's, the *Timaeus*, he describes the creation of the Universe by the Demiurge out of the four basic elements in terms that show even more clearly that he has a real model in mind. What he now describes amounts to a simple armillary sphere – a model of the celestial sphere of the astronomers, made with hoops.

There is in Plato's *Laws* an Athenian stranger who says that he was no longer a young man when he realized that each of the planets moves on a single path, and that it is wrong to call them 'wanderers'. Plato had called them erratic in his *Republic*, so this is perhaps an autobiographical statement, and it is tempting to suppose that it was Eudoxus who persuaded him to change his mind.

No writings by Eudoxus survive, but his system can be pieced together from the writings of two others in particular, Aristotle – a late contemporary – and Simplicius. Simplicius was a Platonist who wrote influential commentaries on Aristotle's work, but he was no mathematician. Since he was born around AD 500, and died after AD 533, his testimony, nine centuries after the event, would be dubious were it not for two or three invaluable remarks: he describes the shape resulting from the Eudoxan construction as a 'hippopede', a horse-fetter, a figure-of-eight; and he speaks of an attack on Eudoxus for the breadth he thereby gave to the planet's path. Taken in conjunction with the general form of the theory, on which he and Aristotle are more or less in agreement, this tells us a great deal, as we shall see.

Eudoxus' system is built up from concentric spheres, spheres centred on the Earth. They are inside one another, but this is a mathematician's Universe, in which their differences of size are ignored. The idea that spheres are needed had not always been as obvious as it now seems; but given that spherical models, real or imaginary, were then under discussion, it would have been clear that to describe the Sun one needed at least two spheres, one for the rapid daily rotation, the other for the Sun's annual motion in a contrary direction. The second sphere needs to be pivoted around the poles of the ecliptic circle, of course. The Moon could have been roughly described along the same lines. (In both cases the object is imagined to be situated mid way between the poles of the sphere on which it resides.) In fact for both the Sun and Moon Eudoxus added a third sphere. For the Moon this could well have been meant to take care of the fact that the Moon's sphere is inclined at about five degrees to the ecliptic, which it intersects in points (nodes) that move slowly backwards round the zodiac (circling the sky in about 18.6 years). A rudimentary knowledge of eclipses could have supplied this insight. If so, then Aristotle and Simplicius seem to have the order of the Moon's second and third spheres wrong, but otherwise their accounts make reasonable sense. What is puzzling is the fact that Eudoxus seems to have added a third sphere for the *Sun's* motion too, in the belief that when at the winter and summer solstices the Sun did not always rise at the same point of the horizon. Simplicius says that those who preceded Eudoxus had thought this, and the idea is found repeated by several later writers.

It was in his explanations of the *direct* and *retrograde* motions for the planet that Eudoxus' pivoted spheres came into their own. He proceeded to show how a point could describe a figure-of-eight, which in turn could be carried round the sky with the planet's long-term motion, more or less in the zodiac. To create that figure, the hippopede, he simply took a pair of spheres, one turning in one direction, and the other turning with the *same speed* in the *opposite direction* around an axis that was carried by the first sphere (but that did not coincide with its own axis). Ten specimens of the mathematical curve in question, corresponding to different inclinations of the two axes, are shown for convenience on the same figure (figure 4.1). One is to imagine the planet moving round the figure-of-eight in time. It is easy enough to see how, pivoted around another axis at right angles to the length of the hippopede, it can be carried round the zodiac (or a path near the zodiac) producing occasional retrograde motions. To that third motion one will add the daily rotation of the sky, the 'rotation of the fixed stars'.

FIG: 4.1 *An assortment of hippopedes*

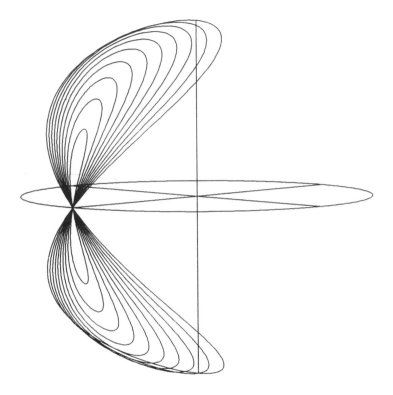

Omitting that daily rotation, the path followed might have the general character of the line drawn in figure 4.2, a figure accurately drawn, but using arbitrary parameters for the speeds and inclinations. We may put aside for the time being the question of accurately representing the observed planetary motions.

In approximating in this way, at least qualitatively, to the motions of the planets, their seemingly erratic motions had been reduced to law. In short, they were not truly erratic at all, and Plato was evidently pleased to discover that fact. But what was Eudoxus' real aim? There is good reason to

FIG: 4.2 *A qualitatively acceptable, but in practice impossible, Eudoxan planetary path*

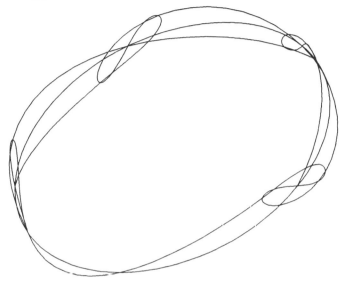

suppose that the delight taken by the Greeks in the explanation he had provided had less to do with its precise predictive power than with its geometrical virtues. To appreciate the real character of Eudoxus' achievement it is necessary to sketch, however briefly, the geometrical reconstruction offered by G. Schiaparelli in the 1870s. Using simple theorems of Greek geometry available in Eudoxus' time, he showed that the hippopede is a curve of intersection of a certain cylinder, a certain double cone and the sphere on which the curve lies (figure 4.3). The cylinder touches the sphere internally at the cross-over point of the hippopede, and the cone has its vertex there.

This beautiful result, at which the descriptions by Aristotle and Simplicius hint only very darkly, is of a sort not altogether unknown from the period. Archytas, Eudoxus' teacher, in solving the problem of the duplication of the

FIG: 4.3 *Hippopede as intersection of sphere and cylinder*

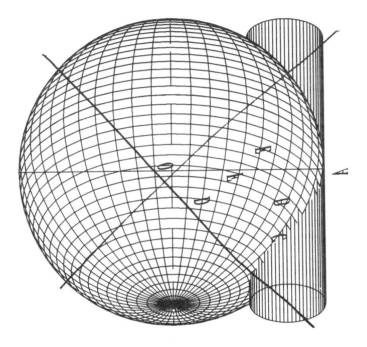

cube, considered the intersections of *three* surfaces of revolution, a torus (anchor ring), cone and cylinder. Those who feel that Eudoxus should not be outdone, but who are reluctant to talk in this context of transcendental curves of the fourth order, might like to add to the sphere and cylinder another simple surface on which the hippopede lies. This is a surface with a parabola as its uniform section. (Imagine a sheet of paper, curved so that two opposite edges are parabolas, and on which the line of the hippopede is drawn.) We have no good reason to suppose that Eudoxus knew of this property of his hippopede, but the same goes in all strictness for the intersecting cylinder and cone. It seems likely that he knew at least of that pair.

This model is so important in the history of geometrical astronomy that a bare outline of a proof of the cylinder property is called for, if only to show the sophistication of an astronomical doctrine now more than twenty-three centuries old. We distinguish between the carrying sphere and the carried sphere. In figure 4.4 we are looking down along the axis of the former, and of the cylinder (on which are points F, E, A) lying parallel to it. (It is instructive to ask why it is not parallel to the other axis; or for example symmetrically placed between them.) A is the starting point of the planet, and the arc AB is its motion around the equator of the carried circle, in a certain time. Looking down on this, it seems to be an ellipse, and angle AOB on the figure is smaller than the angle in three dimensions. This is in fact equal to angle AOC on the figure, where C is a point starting out from A at the same time around the other circle. Clearly B and C will be on the same level (CB perpendicular to OA). Consider now the planet's compound movement at the time in question, as seen against the plane of the diagram (i.e in orthogonal projection on it). It moves up

FIG: 4.4   *The geometry of the hippopede*

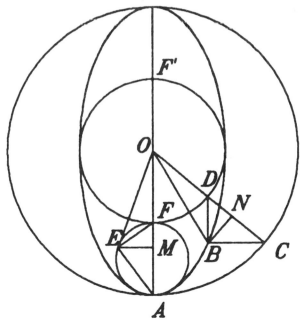

to *B* with the carried motion, and then swings round with the carrier sphere's motion, bringing *OB* to *OE*; and here the angle *BOE* is equal to *AOC*. It has to be proved that *E* lies on a circle, the section of the cylinder. If angle *CBD* is a right angle, with *D* on *OC*, then it is enough to show that *CD* is of constant length; for then all points of type *D* (including *F*) will lie on a circle centred at *O*. The angle *FEA* will be a right angle, so that *E* will lie on a circle with diameter *FA*, the section of the cylinder.

To prove the constant length of *CD* from the properties of the ellipse is more immediate, but considering the relevant part of the diagram in three dimensions, the proof is not difficult on the basis of proportions of sides in similar triangles. This is easier than the initial act of visualization; and considerably easier than proving the theorem of the paraboloidal sheet. I should add that the focus of this is a quarter of the distance from *A* to *F'*.

We have the makings of a powerful geometrical model for the planetary motions, but alas, as it stands it suffers from severe limitations. These are sometimes misrepresented. It is not true to say that all the loops in the planetary retrogradations are identical, as we saw from figure 4.2, nor is it true that the planet's motion in latitude is necessarily great. The retrogradations of Saturn and Jupiter may be fairly plausibly represented without introducing undue latitude (figure 4.5 is for Jupiter). Unfortunately, without

FIG: 4.5 *An accurately drawn Eudoxan-type model for Jupiter*

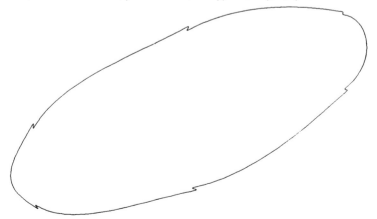

adding more spheres the model has only two main parameters that can be varied, the relative speeds in and of the
hippopede and the size of that hippopede (which depends
on the inclination of the pivoted sphere). These are simply
not enough to accommodate the real motions of Mars,
Venus or Mercury. If the speeds are roughly right, then
the length of the retrograde arc will be wildly wrong, and
conversely.

From a modern standpoint, the relative *speeds* in and of
the hippopede are related to the planet's and our own angular speeds in orbit round the Sun, while the *size* of the
hippopede in relation to the sphere is related to the relative
sizes of the planet's orbit round the Sun and our own. First,
without entering into any details, it should be obvious that
the facts might require the hippopede to move as a whole
so rapidly in relation to the movement of the planet on it
that this can never enter a retrograde phase. This happens
in the cases mentioned. And secondly, if we fix the length
of the retrograde arc in the model on the strength of observation, we are obliged to accept whatever breadth the
hippopede turns out to have as a consequence. It is not
only that in the case of Mars and Venus these happen to
be excessive, but that planetary movement in latitude has
nothing to do with orbital sizes. It is a consequence of the
fact that the planetary orbits, and our own, are all in slightly
different planes.

ARISTOTELIAN COSMOLOGY

There are many unanswered and unanswerable questions
about the Eudoxan schemes, not least about Eudoxus'
motivation in drafting them. Since he taught in a Greek
colony in Asia Minor – Cyzicus is on the southern shore of
the sea of Marmara, across the water and south-west of
modern Istanbul – it is not unlikely that Eudoxus was aware
of astronomy's astrological and religious affiliations, but the
intellectual attitudes of the Greeks at this time did not coincide with those of their Asian neighbours. Star worship may

not have been entirely alien to the Greeks, but it took a secondary place in their religion, as did even worship of the Sun and Moon, divinities though they were (Helios and Selene). The great poet and playwright Aristophanes, who died at about the time of Eudoxus' birth, characterized the difference between the religion of the Greeks and that of foreigners by noting that, while the latter sacrificed to the Sun and Moon, the Greeks made offerings to personal gods, such as Hermes. Hellenic religion had long been moving away from the old and simplistic celestial religions, even though some centuries later the trend was checked, with the advent of eastern astrology.

In Eudoxus' time the philosophers were happy to make a place for the heavenly bodies in their pantheon. For Pythagoras they had been regarded as divine, and Plato shows that he was shocked by Anaxagoras' atheistic claim that the Sun is a burning mass, and the Moon a sort of Earth. For Plato the stars were visible gods in which the supreme and eternal Being had put life. His was no longer a celestial religion for the populace, but a religion for intellectual idealists, and in the hands of his numerous successors, many of them Christian, the Platonic view of the heavens proved highly influential. Even his rival Aristotle defended the divinity of the stars, representing them as eternal substances in unchanging motion. Divinities they may have been, but all this is a far cry from the doctrines of those 'Chaldaeans' (the Babylonians and their neighbours) who claimed to predict from heavenly signs the life and death of nations and individuals, as well as the weather and what depended on it.

We do not know with any certainty Eudoxus' attitudes to these things, but there can be little doubt that in his astronomy he was largely driven by the intellectual pleasures of the geometer – a commodity of which many social historians seem to be entirely unaware. Although Cicero's testimony in the work *On Divination* is relatively late, it is that Eudoxus maintained that 'no credence should be given to the Chaldaeans, who predict and mark out the life of every man according to the day of his birth'. By Cicero's time the Roman world was deeply conscious of these practices, and some have therefore considered this reference to be anachronistic, but there is no reason why it should be so. In fact it might have come from a source that was referring to a non-astronomical way of predicting human life, for there were Babylonian techniques known in Egypt long before Eudoxus that rested only on the calendar. If this is so, however, then it takes away the force of the argument that Eudoxus was deliberately spurning astrology in its better known forms.

Whether or not Eudoxus had motives of a sort we might now be inclined to describe as purely intellectual, we cannot consider his achievement to have been complete, in the sense of later astronomical ambitions. The fact that we can fit the behaviour of Jupiter and Saturn to his model does not mean that Eudoxus did so at all accurately. That *we* find it easy to vary the schemes, for instance by changing the speeds of carried and carrying spheres, does not mean that others did so in ancient times. For the most part this produces unpalatable results. Among the geometrically curious results we may mention that by simply doubling the carrier speed one produces no hippopede at all, but a reflection of the inclined circle. It is this sort of possibility that warns us against speculating on the nature of the next step in its development, at the hands of Callippus of Cyzicus, around 330 BC.

Callippus was, as it happens, the pupil of Polemarchus, who had been a pupil of Eudoxus; and he followed Polemarchus to Athens, where he stayed with Aristotle, 'correcting and completing, with Aristotle's help, Eudoxus' discoveries'. So speaks Simplicius, who tells us that Callippus increased the number of spheres by two each for the Sun and Moon, and by one each for the planets, excepting Jupiter and Saturn. This suggests that there was a measure of satisfaction with those two planets, the very pair that we ourselves can best fit into the Eudoxan model. Under these circumstances it is uncharitable to suggest that the Greeks of this period were concerned only with a *qualitative* model of retrogradation. Simplicius goes so far as to say that Eudemus had listed the phenomena that had induced Callippus to expand the system.

It is customary to note that the total number of spheres in Eudoxus was twenty-six, and in Callippus thirty-three, but quoting totals in this way does not imply a unified scheme, a single system. As far as can be seen, these two men were content to advocate a separate scheme for each planet or luminary. Whatever progress Callippus made, we do know that Aristotle enlarged upon his ideas and turned what had presumably been a set of abstract geometrical theories into a unified mechanical system. As such, the theory held an important place in natural philosophy for two thousand years.

Aristotle was by far the most influential ancient philosopher of the sciences. He was born in Stagira in 384 into a privileged family: his father had served as a personal physician to the grandfather of Alexander the Great, and Alexander was in turn his own pupil. Aristotle studied under Plato in Athens until the latter's death (348 BC), and after moving to Mycia, Lesbos and Macedonia he returned to Athens where he founded his own school of philosophy, the Lyceum. His very extensive writings are highly systematic and coherent, and cover a large part of human knowledge. Since they were written over a long period, there are naturally a few minor inconsistencies. The most important single source for Aristotle's cosmology, his *De caelo* ('On the heavens' – but the Latin title is usually used) was an early treatise, and does not contain all of what in his work was most influential. It does not, for instance, have the theory of the Unmoved Mover, for which we must consult his *Physics*. This, at the outermost part of the Universe, was taken to be the source of all movement of the spheres within it.

Aristotle writes in a semi-historical vein, reviewing the main arguments of his predecessors. The longest chapter in *De caelo* concerns the celestial sphere – a construction by this time generally accepted by the Greeks – and the Earth of similar shape at its centre. He mentions the theories of the Pythagoreans and of an unnamed school according to which the Earth *rotates* at the centre of the Universe. He dismisses the idea, as well as that of an *orbital* motion of the Earth. Both of these we now of course accept. He seems to have been persuaded by Eudoxus' theory that, if accepted, they would imply that the stars are subject to 'deviations and turnings', and that these we do not in fact experience. Eudoxus had unwittingly scored a hit for the fixed-Earth doctrine. Had he been alive, he might have pointed out that if the stars are at great distances, the argument fails.

Aristotle offers various arguments for the spherical nature of Earth and the Universe. The natural movement of earthly matter is from all places downwards, to a centre, around which a sphere of matter will inevitably build up. There is also the observed fact that the line dividing light from dark regions of the Moon's surface during a lunar eclipse is always convex – not a perfect argument by itself, of course. He refers to mathematicians who try to measure the Earth's circumference (Archytas or Eudoxus?) and put it at 400 000 stades, or about 28 700 kilometres – nearly twice the correct figure, but the oldest estimate known to us. The sphere, says Aristotle, is the most perfect solid figure, in the sense that when rotated about any diameter it continues to occupy the same space.

The Universe he conceives to be built layer on layer over a spherical Earth. Only circular motion is capable of endless repetition without a reversal of direction, and rotatory motion is prior to linear because what is eternal, or at least could have always existed, is prior, or potentially prior, to what is not. Circular motions are for Aristotle a distinguishing characteristic of perfection, and in this way the heavens acquired a special place in any discussion of perfection. On the Earth, natural motion was up (for smoke and so forth) or down (for earthy material), whereas in the heavens natural motion was circular, allowing of no essential variation, which would have been a mark of imperfection, incapacity. They are simple and unmixed bodies, made not of the four elements that we know close at hand, namely Earth, air, fire and water, but of a fifth element, the ether. Its purity varies, being least where it borders on the air, which reaches to the sphere of the Moon. (This *fifth*, incorruptible, element or *essence* gives us our word 'quintessential'.)

Aristotle has, then, a heaven that is in sharp contrast with the sublunary world of change and decay. It is unique, ungenerated and eternal – qualities that will provide future Christian Aristotelian apologists with problems. He was here opposing the beliefs of the atomists, Democritus and Leucippus, who had argued for void space (which Aristotle rejected on philosophical grounds) and a multitude of worlds. He was opposing Heraclitus, who had said that the world was periodically destroyed and reborn; and also Plato, who had held that the world was created by the Demiurge.

It is surprising to find that Aristotle even opposes the notion of celestial harmony such as we found in the myth of Er. There is, he says, the absurdity of the claim that we hear no sound because it has been in our ears since our birth. And what of the general principle that the greater the object the greater the sound? Thunder would be as nothing by comparison. But Aristotle did not manage to stem the idea of celestial harmony entirely, and his insistence on the relative perfection of the etherial regions helped to keep in play Platonic belief in the divinity of the heavenly bodies.

For the technicalities of Aristotle's planetary system we must turn to his *Metaphysics*. There he seems to be accepting the theory of Callippus, but he tells us that 'if all the spheres *put together'* are to account for what we see, then for each of the planetary bodies (the carrying and carried spheres, to use my previous words) there must be other 'unrolling' spheres to counteract the effects of the spheres above them that do not belong to the planet in question. For Jupiter, for instance, its own spheres are enough to explain its motion, apart from the star sphere. This being so, since all the spheres for Saturn are outside its own, they must be neutralized, by giving to Jupiter counteracting spheres, with the poles appropriate to Saturn's, but equal and opposed angular velocities. When we come to Mars, we shall have to neutralize Jupiter's spheres, but not Saturn's, which have already been taken care of; and so for the rest. The spheres in Callippus are as follows, with the required numbers of counteracting spheres in parentheses: Saturn four (three); Jupiter four (three); Mars five (four); Venus five (four); Mercury five (four); the Sun five (four); the Moon five (none). The total is thus fifty-five spheres, and Aristotle actually quotes this number. He adds a puzzling remark that has never been convincingly explained, to the effect that omitting the extra movements for the Sun and Moon the total is forty-seven. I suspect that at an earlier stage he had given four counteracting spheres to the Moon, to secure the fixity of the Earth.

Aristotle's is thus a mechanistic view of a Universe of spherical shells with various functions, some carrying planets. Motions were no longer being postulated as though they were mere items in a geometry book, nor were they justified in terms of Platonic intelligences, but rather in terms of a physics of motion, a physics of cause and effect. The first sphere of all, the first heaven, shows perpetual circular movement, which its transmits to all lower spheres; but what moves the first heaven? What moves it must be unmoved and eternal. There is much theological interpretation of this prime mover, whose activity is the highest form of joy with pure contemplation with itself as object, a natural condition for something divine. One might then imagine that as far as ultimate causes are concerned, Aristotle has all that he needs. Certain later commentators speak as though the first mover of the outermost sphere is enough for the system. Aristotle nevertheless speaks as though each planetary motion of Eudoxan type has its own prime mover, so that there will be fifty-five (or forty-seven) of them altogether, and it does seem that in the end he accepts them as gods. Those who in late antiquity and the middle ages found the idea repugnant usually spoke instead of 'intelligences', or of angels.

In Simplicius we read that a system of concentric spheres continued to be taught, and that it was accepted by Autolycus of Pitane (around 300 BC). Autolycus wrote works of 'spherical astronomy', the geometry of the sphere, and these had a certain currency in Arabic, Hebrew and Latin well into the middle ages; but they contain no theory of the Eudoxan type. Autolycus defended the theory, however, against one Aristotherus, who is known to history as the teacher of the astronomer-poet Aratus. The theory was recognized as deficient in failing to account for the changes in brightness of the planets. Simplicius suggests that this was something of which Aristotle was aware.

HERACLEIDES AND ARISTARCHUS

Heracleides, a near-contemporary of Aristotle in Athens, was a man whose fame in the history of astronomy probably far outstrips his achievements. A colourful figure, whose greatly admired literary works have failed to survive, he is said to have died suddenly while being presented with a golden crown in the theatre. The justice in this lay in his having obtained it by subterfuge: he had – so one story goes – persuaded envoys from the Delphic oracle to say that the gods had promised to lift a plague on his city, Heraclea, if he were crowned while alive and given a hero's cult after death.

While he failed in that ambition, he seems to have achieved more success with his audience of historians. He is supposed to have maintained that the orbits of Mercury and Venus have the Sun at their centre, while the Sun is in orbit round the Earth. That this would have been a step in the direction of Copernicanism is what gives it its special interest. At all events, it is reasonably certain that he did believe in the rotation of the Earth on its axis, a doctrine to which Aristotle alluded, and he is the earliest astronomer known to have held to it. Copernicus actually mentions his name in this connection. Perhaps a silver crown is called for.

The idea that the Sun is the centre of the orbits of Venus and Mercury would have been very difficult to work into astronomy during its 'Eudoxan' period, that is, during the lifetime of Heracleides. In an epicyclic theory, the question occurs more naturally. It is mentioned in this context by Theon of Smyrna, but he lived in the early second century A D. A commentary by an even later writer, Chalcidius, on a passage from Plato, and mentioning Heracleides, has been thought to report the doctrine, but when it is said there that Venus is 'sometimes above and sometimes below the sun', it is clear from some numerical data that the meaning is simply 'ahead of in the zodiac' and 'behind in the zodiac'.

The first astronomer to have clearly put forward a true sun-centred theory was Aristarchus of Samos. He was born around 330 BC on the island of Samos, off western Asia Minor, near to Miletus. From this centre of Ionian culture came another astronomer and mathematician, Conon of Samos, in the following century, and he in turn was a friend of Archimedes. It is from Archimedes that we learn of Aristarchus' heliocentric theory, for the only surviving work by Aristarchus himself is his treatise *On the sizes and distances of the Sun and moon*, and this naturally enough takes the Earth at the centre from which distances are measured.

According to Archimedes, near the beginning of his *The Sand Reckoner*, Aristarchus' hypotheses are that the fixed stars and the Sun are stationary, that the Earth is carried in a circular orbit around the Sun, which lies in the middle of its orbit, and that the sphere of fixed stars, having the same centre as the Sun, is so great in extent that the circle on which the Earth is supposedly carried is in the same ratio to the distance of the fixed stars as the centre of the sphere has to its surface.

Archimedes criticized Aristarchus for the last meaningless statement, involving the ratio of a point to a surface, and supposed that he meant rather that the ratio of the diameters of Earth and Sun was equal to the ratio of the sphere in which the Earth revolves to the sphere of fixed stars. Some modern interpreters accept this reading, while others take the ratio of point to surface to mean no more than an extremely great ratio, so great that we should not expect to observe star parallaxes (changes in apparent star positions) as the Earth goes round the Sun.

Whatever his intentions, there is no doubt that Aristarchus believed in the motions that we now generally associate with the name of Copernicus – who certainly knew of his predecessor. The strange thing is that the only astronomer known to have supported these ideas in antiquity was Seleucus of Seleuceia. Seleucus is said to have tried to prove the hypothesis. He lived in the mid-second century BC, flourishing about eighty years after the death of Aristarchus in 230 BC. Seleuceia is on the Tigris, but the fact that Seleucus was later described by Strabo as a Chaldaean probably implies more than merely a Mesopotamian origin: it suggests that he *practised* astronomy in the style of the Babylonians. He was certainly no lightweight, for Strabo says of him that he discovered periodical variations in the tides of the Red Sea, which he realized were related to the Moon's position in the zodiac.

If Aristarchus did indeed believe that the Sun was at the precise centre of the Earth's orbit, then it is very unlikely that he could account for the variation in the Earth's motion throughout the year, that is, for the inequality of the seasons. As we shall see, this inequality was known and explained (on a geocentric hypothesis) by Hipparchus in the following century. Aristarchus' failure in such technical respects as this are unlikely to have been the prime reason for the failure of his heliocentric theory to become popular. Far more important would have been the influence of Aristotle's geocentric cosmology, with its appealing doctrine of the natural movement of bodies towards or away from the centre of the world, identified by Aristotle as the centre of the Earth. There was also a religious dimension to the question, and according to Plutarch, the Stoic philosopher Cleanthes thought that Aristarchus should be put on a charge of impiety for maintaining that the Earth moved. Cleanthes is notorious for the fervour with which he introduced religion into philosophy, but his attitude to heliocentrism is a strange one, bearing in mind his belief that the Universe is a living being, with God its soul and the Sun its heart.

To mention Aristarchus only in connection with his heliocentric theory would be to overlook an important aspect of early Greek astronomy, that is, its practical side. The architect Vitruvius tells us that he invented the *scaphe*, a hemispherical bowl in which stood a gnomon (pointer), the shadow of which marked the time on a network of hour-lines. Very many such dials survive, although not of course all following the same geometrical principles. The need to design them provided an important stimulus to astronomy and geometry, and the art of astronomical projection – comparable with that of map-projection – was soon to pay high dividends, with the design of the plane astrolabe. This we shall discuss in connection with Hipparchus.

While we must not exaggerate the accuracy of Aristarchus' astronomical observations, his work *On Sizes and Distances* gives an indication of the interplay between mathematical and observational methods in Greek astronomy. He presented a sequence of deductions relating the sizes and distances of the Sun, Moon and Earth, notable for the way in which the underlying assumptions are spelled out explicitly, and for techniques that herald trigonometric procedures to come. Among his basic assumptions were that the Moon gets its light from the Sun, that when the Moon is seen by us to be exactly half-illuminated, the observer's eye is on the great circle dividing light from dark regions and that the angle moon–earth–Sun is then 29/30 of a quadrant (87°). (See figure 4.6, not to scale, in which *T* is

FIG: 4.6 *Distances of Sun and Moon according to Aristarchus*

the observer on the Earth, *S* is the Sun, and *M* is the moon.) He assumes the Moon's and Sun's apparent angular diameters to be 1/720 of a circle (0.5°), which is close to the average of quantities that do in fact vary. If we use *m* and *s* to denote the distances of the Moon and Sun, then *we* should write down immediately *m*/cos 87° = *s*. Now 1/cos 87° is approximately 19.1, and Aristarchus quotes 'more than 18 and less than 20', using in his proof a theorem equivalent to what we should write down as

$$\frac{\tan A}{\tan B} > \frac{A}{B} > \frac{\sin A}{\sin B}$$

Whatever this number is taken to be, since the angular sizes of Sun and Moon are about equal, the ratio of their true sizes is – as he saw – the same. Other theorems as to their volumes followed very simply. Unfortunately the entire set of calculations was vitiated by the figure for the basic angle (87°), which should have been approximately 89.8°. The exact moment of the halving of the Moon is of course extremely difficult to judge, and the angle difficult to measure with primitive devices, but the method was sound enough.

APOLLONIUS AND EPICYCLIC ASTRONOMY

Apollonius of Perge (often rendered Perga), an ancient Greek city in southern Asia Minor, lived in the second half

of the third century B C and into the following century. He
visited Alexandria. It is doubtful whether he studied long
with the pupils of Euclid there, as Pappus claimed, but he
was certainly one of the greatest of Greek mathematicians
in antiquity, to be compared perhaps only with Archimedes.
He did for the geometry of conic sections (parabola, hyper-
bola, line pair, circle and ellipse) what Euclid had done for
elementary geometry. He set out his own work, and much
of that done by his predecessors, in a strikingly logical way.
He also showed how to generate the curves in ways strongly
reminiscent of modern algebraic geometry. Those methods
were to prove enormously important to astronomy, in the
century of Kepler, Newton and Halley – who studied Apollonius' text closely.

Apollonius' interest in astronomy is known from various
oblique references. One writer tells us that he was known
as Epsilon, since that Greek letter is shaped like the Moon,
which he studied most intensively. Another source quotes
the distance of the Moon from the Earth according to him
as five million stadcs (about 0.96 million kilometres), which
is about two and a half times too great. Another writer, the
astrologer Vettius Valens (around A D 160) claims to have
used tables of the Sun and Moon drawn up by Apollonius;
but this might be another man of the same name. The most
intriguing statement relating to his astronomical interests,
however, concerns a theorem of his in the theory of planetary motion. According to Ptolemy (around A D 140), Apollonius found a relationship between the velocity of a planet
moving in an epicycle, the velocity of the centre of that
epicycle round the deferent circle (see the penultimate section of the previous chapter, where these ideas were introduced in advance of their historical place), and two distances on the figure representing the situation when the
planet appears stationary (between direct and retrograde
motion).

This situation is shown in figure 4.7, where O is the centre
of the epicycle and P is the planet. It appears stationary to
an observer on the Earth (taken to be at a point T). The

FIG: 4.7   *Apollonius' theorem*

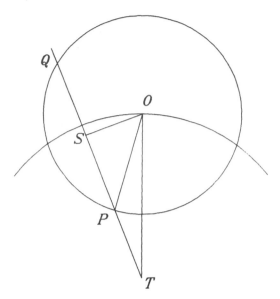

motion of *P* at right angles to the line of sight *TQ* must have two equal and opposite components: one is due to the fact that it is carried with the velocity of *O*, and the other is the result of its velocity around *O* and along the tangent to the epicycle at *P*. Resolving these velocities using elementary modern methods leads easily to the following theorem: the ratio of the angular velocity in the deferent to that in the epicycle relative to *OC* is equal to the ratio of *PS* to *PT*. (*PS* is half the chord *QP*.)

The same result is obtainable by using the methods of limits with classical geometry, as it was derived in Ptolemy's *Almagest* of more than three centuries later. No matter what method Apollonius of Perge used, it is clear that he was well capable of handling motions in two dimensions. This is a matter of some importance, since it seems that he is a key figure in the early development of the idea of epicyclic motion. Ptolemy says that when he proved the relationship explained above, he did so both for an epicyclic arrangement (as here) and for an equivalent arrangement where the planet moves on what is now called a movable eccentric circle.

The equivalence of the two schemes can be readily seen from figure 4.8, where the solid lines show the epicyclic

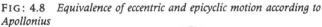

FIG: 4.8 *Equivalence of eccentric and epicyclic motion according to Apollonius*

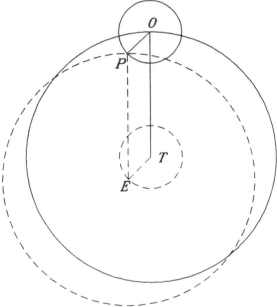

arrangement, and the broken lines the alternative. Ignore the broken circles for the time being. To reach *P* from *T*, one may clearly go from *T* to *O* and thence to *P*, or one may go from *T* to *E*, where *TE* is equal and parallel to *P*, and thence to *P*, where *EP* is equal and parallel to *TO*. The equality of the lengths mentioned here means that *E* and *P* lie on the broken circles, as shown. *E* is usually referred to as an *eccentric* ('off-centre') *point*, and the larger broken circle as an *eccentric circle*, but it should be noted that it is a *movable* circle.

From a strictly geometrical point of view, where the difference between large and small circles is of no consequence, the two constructions are equivalent, and the only compelling reason for attaching different words to them is historical.

The eccentric circles we shall meet with in later constructions are simply fixed circles centred on points away from the Earth. This is a suitable place at which to note that they too can be equivalent to epicyclic motions, albeit of a special sort. Suppose that the large continuous circle in figure 4.8 is fixed, with *T*, the Earth, at its centre. If, as *O* goes round the large circle, *OP* remains always parallel to the fixed line *TE*, then *P* will lie on a fixed eccentric circle (the large broken circle in the figure), with centre at *E*.

HIPPARCHUS AND PRECESSION

Apart from Vettius' tantalizingly ambiguous remark about certain tables drawn up by Apollonius, we know nothing of any attempt he might have made to relate his epicyclic theories of planetary position to observation. His theoretical work, however, led others to do just that, and all the evidence suggests that an infusion of Babylonian methods was essential. The first Greek astronomer who is known to have systematically applied arithmetical methods to geometrical astronomical models was Hipparchus, who flourished between 150 and 125 BC. Born in Nicaea (modern Iznik, in Turkey) in north-west Asia Minor, Hipparchus seems to have worked mainly on the island of Rhodes. The importance of his contribution to astronomy was very great, and his numerical methods are known, often in detail, from the *Almagest* of Ptolemy – who quotes similar work by no other astronomer.

We can better appreciate how important Babylonian influence was to the development of astronomy if we bear in mind that if we were to list all that we know from the Greek world before Ptolemy, we should find scarcely more than twenty reports of precise observations antedating Hipparchus. The earliest is of a summer solstice observation on −431 June 27 (Athens) (i.e. 27 June 432 BC). The others are all Alexandrian, beginning with a series of lunar occultations of stars by Timocharis. This is not to say that there are no reports of other, qualitative, observations. Among the more famous is that of Thales' observation of an eclipse in 584 BC, the eclipse that, without good reason, he is often said to have predicted. The same applies to Helicon of Cyzicus' supposed prediction of a solar eclipse, set by some scholars at 12 May 361 BC. Diodorus reports an eclipse that occurred during a military encounter between Agathocles and the Carthaginians (15 August 310?). Archimedes, who died in 212 BC, is thought to have observed the solstices; but we have no more detailed information concerning this, and even the often-repeated story of his having measured the diameter of the Sun at half a degree (1/720 circle) is not substantiated.

This is a very different picture from the Near Eastern scene. There were of course numerous points of contact between Greek and eastern cultures, even in astronomy. We have already spoken of the calendar and the zodiac, and for example Cicero in one place intimated that he had seen a written statement by Eudoxus, giving an adverse opinion on the astrological predictions of the Chaldaeans. Degree measure and sexagesimal arithmetic first appear in Greek in the *Anaphorics* of Hypsicles, not long before Hipparchus, but Hipparchus clearly had access to Babylonian data and theory of a much more intricate sort than any to be found in other Greek sources. It was F. X. Kugler, at the very end of the nineteenth century, who first realized that Hipparchus had taken fundamental period relations for his lunar theory (so many months equal so many years) from Babylonian lunar theory, on what we call system B. Since then, many other lesser examples of indebtedness have been found, and it seems likely either that an abstract of the Babylonian archives made for local use was translated into Greek, or that a Greek astronomer, perhaps bilingual, himself had access to the archive and made such an abstract. Babylonian methods continued in use in traditional form until after Ptolemy's time, even in Roman Egypt, and Hipparchus himself might have learned them at source.

Essential to any programme that linked geometrical models to observational data was something equivalent to what we should now call trigonometry. Hipparchus played an important part in the foundation of this subject. He wrote a work on chords and drew up a simple table of chords. A chord is the line joining two points of a circle. If we take the radius as our unit, in our terminology it is of course equal to twice the sine of half the angle it subtends at the centre, so that a table of chords will have much the same sort of use as a table of sines. Hipparchus, following Babylonian practice, took his circumference to be divided into 360 degrees, each of 60 minutes, and he took his standard radius to be divided into the same number of units and subdivisions. Ptolemy later set the radius at 60 units, a standard that began to fall into disuse only in the sixteenth century. Indian astronomy long continued to use the Hipparchan norm, however, and also followed Hipparchus in calculating chords for successively halved angles, starting with simple chords such as those for 90° and 60°. This explains why angles of 22°, 15° and 7° are often mentioned in later astronomical texts as fundamental.

As we know from the work of Eudoxus, Greek three-dimensional geometry was highly developed, and it is very probable that Hipparchus broke down problems on the surface of a sphere – for instance problems concerning the risings and settings of the Sun and stars – into problems involving circles and triangles in a plane. He seems more often to have solved such problems arithmetically, however, no doubt enlarging on Babylonian techniques. Another geometrical method is one that requires the three-dimensional celestial sphere, with its appropriate great circles, to be projected on a plane in much the same way as the Earth's surface is projected on terrestrial maps. There is no doubt that Hipparchus exploited this technique successfully, using projections of various sorts.

One in particular is important for its influence on the design of astronomical instruments, even to the present. This is the projection we know as 'stereographic'. It may be easily visualized by supposing the network of circles making up the celestial sphere to be of wire. If the sphere stands on a plane sheet, touching it at one of its poles, and a bright point source of light is placed at the other pole, then the shadows of the wires cast on the sheet will be in stereographic projection. A moment's consideration will show that if the sheet were in the equatorial plane, the same shadow diagram would result, at half the scale. For this reason, a stereographic projection is sometimes described as a projection from any pole on to an equatorial plane. Circles project into circles, and angles on the sphere into equal angles on the plane.

Why these ideas are important can be understood by supposing that we are observing the stars and (to use a little astronomical licence) the equator, ecliptic and other circles on the celestial sphere, all moving round the pole with the daily rotation. We shall want to distinguish these circles from another set, now *fixed*, that allow us to specify the positions of objects in the heavens. One is our local horizon; another is the meridian line; and we might imagine a line drawn right round the sky just one degree above the horizon; and another at an altitude of 5°, and so on up to the zenith overhead, at 90° from the horizon; and we might add lines (compare the meridian line) that serve to locate the bearings of stars in azimuth. This network of fixed lines could also be represented by a wire mesh, so to speak, but it would of course need to be distinguished from the moving mesh. The shadow of the latter would rotate, but that of the mesh of local co-ordinates would not.

Now this is precisely how the instrument known as the (plane) astrolabe is to be understood. In its later form, it was usually made of brass. We shall put aside its uses as an observing instrument, for which it was suspended from the thumb while the altitude of an object was measured using a rule pivoted about its centre. As an instrument for calculation, it had a solid plate of fixed co-ordinate circles, above which rotated a so-called *rete* (net) of pierced metalwork on which were pointers for the brightest stars and parts of the moving circles, the equator and the ecliptic (zodiac). The first corresponds to our fixed set of shadow lines and the second to the moving shadow lines. There was usually a shallow tray, the 'mother', in which the discs were held, but this is not involved in the computing function of the astrolabe. Through the centres of the two discs there is a pin about which the rete rotates, and this represents the north pole. (It could in principle be the south, but that is almost never the case. The projection, in other words, is from the south pole.) Of all the circles on the rete the most obvious is the ecliptic, while the most important on the plate is the horizon.

To take a very simple illustration of the pattern of the most important lines on the instrument, see figure 4.9. If two positions of the Sun are as shown, then the arc of the Sun's rotation around the pole, like that of the stars, is the angle marked as *A*. This would often have been measurable on a scale at the rim, either in degrees or directly in hours.

The evolution of the astrolabe continued for two millennia, and it would require a treatise of its own to explain its many uses. Some brief additional remarks on its history are included in the last section of the present chapter. It does seem probable that we owe its invention to Hipparchus. Our source is Synesius (about AD 400). Certainly Ptolemy

FIG: 4.9   *The main circles on an astrolabe*

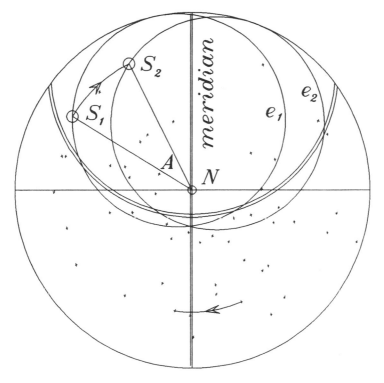

This diagram illustrates the principle of the astrolabe. S_1 and S_2 are two positions of the Sun, as it turns around the poles (N is shown, at the centre). The corresponding positions of the ecliptic, the path of the Sun over the year, are also shown (e_1 and e_2), as are those of the stars. The movement of just one star is marked. A more extensive disc of stars could have been shown, including stars nearer to the south pole, but it was usual to construct a star map, just big enough to include the complete ecliptic, since for northern observers this was likely to include most visible stars. The movement shown, covering the angle A, would take about a couple of hours. Not moving with the Sun and stars are the local horizon and meridian, represented here by double lines. The Sun at S_1 obviously rose above the horizon about half an hour previously. The placing of the circular arc representing the horizon depends on the geographical latitude for which the astrolabe is constructed. For a better idea of the instrument as a whole, see figures 4.15 and 4.16.

knew the theory of stereographic projection, and if Synesius was right, then we can speculate as to how Hipparchus managed to compute so many simultaneous risings and settings of stars, as set down in his various writings, including his sole surviving work, *A Commentary on the Phenomena of Aratus and Eudoxus.*

Aratus, following Eudoxus, had written a poem (*Phenomena*) on which the mathematician Attalus of Rhodes wrote a commentary, shortly before Hipparchus followed suit. This was no new tradition: a Babylonian text from around 700 BC, which lists twenty sets of stars that culminate simultaneously, shows that such matters had long been a human concern. The difference between Hipparchus' work and that of his predecessors was that he listed the points (degrees) of the ecliptic that rose and set at the same time as the stars. (We can call this quantity the *mediation* of the star.) What might at first seem a pointless exercise was in fact quite expressly intended to allow astronomers *to tell the time by night* when making observations. It paid far greater dividends to Hipparchus himself, and he probably made much use of an instrument of the astrolabe type to help him perform the necessary calculations. We do at least know that he had a three-dimensional globe with the constellations indicated on it.

This work of Hipparchus marks the beginning of a system of rigorously applied star co-ordinates. Hipparchus did not have our 'pure' systems either of ecliptic latitude and longitude or of declination and right ascension. These gradually evolved from his, that is, from a system of declination and mediation that was soon to pass into Indian astronomy. Hipparchus built up his own catalogue of stars – not necessarily with co-ordinates for each, however, but in some cases perhaps just giving statements about stars that were in line, and estimates of distances. Aristyllus and Timocharis had listed a few declinations, in the third century BC. According to Pliny, Hipparchus noticed a new star – whatever that may have been. Realizing that it was moving, he asked himself whether others did likewise, and so – if we follow this route to the discovery – he found that indeed *all stars have small motions parallel to the ecliptic.* Their ecliptic longitudes increase.

Until the age of Copernicus, this was regarded as a 'movement of the eighth (star) sphere'. As we should now say, taking a Copernican perspective, it is the *reference system* that is moving. There is a slow conical motion of the Earth's axis that makes it seem that the equinoxes are moving round the ecliptic from east to west. This 'precession of the equinoxes' we know to be a little over 50″ per year, or 1° in 72 years. Hipparchus put the figure as at least one degree in a century – a very remarkable discovery. But was it found purely from star positions?

The movement of the equinoxes obviously affects the relationship between the length of the year, measured as the return of the Sun to a particular star, and as measured by its return to one of the equinoctial (or solsticial) points. The former, the 'tropical year', is shorter than the latter, the 'sidereal year', as we saw in the last chapter. Hipparchus realized as much, and although he certainly tried to find the slow movement by considering star positions as quoted by Timocharis, his most accurate findings fairly certainly came from a comparison of the sidereal with the tropical year. His data for the latter cover equinox observations between 162 and 128 BC, and lunar eclipse observations (which are useful because they give an accurate moon–earth–Sun line). He settled on a very accurate figure for the tropical year of 365 days minus 1/300 day. (We do not know his figure for the sidereal year, and can only make a rough estimate, based on the upper limit he gives to the precessional motion.)

It is instructive to see how ignorant we are of the sequence, and hence the motivation, of so much of this astronomical work. Was it the year or the star positions, or time by night, that set Hipparchus off on the track of precession? There is reason to think that Timocharis was investigating the length of the lunar *month* when he gave his star positions. His lunar observations include no arc measurement – they are simply of occultations of stars, with time in seasonal hours.

There has been much nonsense talked in our own time by the so-called Pan-Babylonians about a Near Eastern discovery of precession. In a certain sense, a 'knowledge of precession' was in the possession of any prehistoric observer who found that the risings and settings of stars were not as marked out by his ancestors. In a sense they were known to the Babylonian astronomers who first realized that there is a difference between the tropical and sidereal mean longitudes of the Sun. But this is not to say that these early observers could rationalize the discrepancy, as did Hipparchus. It is highly significant here that Hipparchus reached an appreciation of the universality of the slow drift of the stars only after a period during which he thought it to be restricted to stars in the zodiac.

HIPPARCHUS: SUN, MOON AND PLANETS

Hipparchus made good use of the two geometrical devices used earlier by Apollonius, the eccentric and the epicycle. The former is enough to account for the Sun's motion quite accurately, and from data for the lengths of the four seasons Hipparchus derived parameters to fit the observational data available to him. He decided that the eccentricity was 1/24 of the eccentric's radius, and that the direction of apogee was Gemini 5°. The latter result is commendable, but the former figure is substantially too great. (In round numbers, the eccentricity is about 1/60.) What was notable here, however, was not Hipparchus' accuracy, but the very fact of his having fitted Babylonian-style observational data to Greek models. He made similar attempts on the Moon's motion, but there he was confronted by far greater problems, even though he was able to draw on Babylonian sources for extremely accurate values of the principal motions of the Moon, the four different sorts of month (the synodic, the sidereal, the draconitic and the anomalistic). Hipparchus was particularly concerned to find eclipse periods, presumably for their own sake but also because they help to give accurate positions – and hence motions – for the Sun and Moon. He was fortunate in being able to compare his own eclipse data with those of the Babylonians. No earlier Greek astronomer is known to have borrowed such materials, but again, more important still is the use to which he put them. He devised a simple, epicyclic, lunar model, notable for the way its motions were made to match the Moon's observed motions. The motion of the epicycle around the Earth he made follow the Moon's known average motion in ecliptic longitude, while the motion of the Moon in the epicycle he made keep time with the Moon's observed 'motion in anomaly'. (The anomalistic month is the period after which the Moon returns to the same velocity, and is to all intents and purposes its time from perigee to perigee.) He found a geometrical procedure that allowed him to derive the relative sizes of the circles and motions around them, based on observations of the times of three lunar eclipses. He applied his method with two different trios of eclipses, once using the epicyclic model as explained, and once using the equivalent eccentric model. (See the earlier section on Apollonius for this equivalence.)

His calculations were flawed, but the method was an excellent one, displaying great originality, and Ptolemy, nearly three centuries later, developed it further. Hipparchus' model can account quite well for the Moon's returns to opposition and conjunction. From Ptolemy's words it seems that Hipparchus himself realized that for intervening positions it is less acceptable, but he does not appear to have improved on it.

Hipparchus did not restrict himself to a model for predicting only the Moon's longitude. Again using his own and Babylonian data, he established the maximum latitude of the Moon from the ecliptic at 5°. He had a clear grasp of the three-dimensional arrangement of Sun, Moon and Earth during eclipses, and he developed geometrical procedures for calculating the actual distances of the Sun and Moon from the Earth that would best account for the observations available to him. His results were very imperfect, and it is much to his credit that he stated them in terms of upper and lower limits. Thus the average lunar distance was set at between 59 and 67⅓ Earth radii. No previous astronomer had come so close to the correct value – a little over 60 Earth radii. For the solar distance he gave a figure that was less than a fiftieth of the true value, but at least he knew how helpless he was: he could not measure the Sun's parallax (figure 4.10), but had to guess at a figure. He took seven minutes of arc, but the figure is in fact close to eight seconds.

According to Ptolemy, Hipparchus did not establish particular models for planetary motion, but did criticize those of his predecessors. Here too, though, he seems to have compiled digests of Babylonian data, mixed perhaps with some of his own, and Ptolemy was able to exploit them to the full. Hipparchus' critical acumen was put to use in another connection. In the middle of the third century BC, Eratosthenes had given a description of the inhabited world, with an estimate of its circumference (252 000 stades, or 48 000 kilometres, about a fifth part too high). Hipparchus was a severe critic of many points in his work. None of Eratosthenes' writings survives, however, and we cannot be at all certain that he ever found either the circumference of the Earth or (as is often stated) the obliquity of the ecliptic on the basis of measurements.

Hipparchus was responsible for changing the direction of Greek astronomy, away from qualitative geometrical description and towards a fully empirical science. He never composed a systematic treatise covering the whole of the science, and his many short works were probably lost because they were too difficult for ordinary readers. His

FIG: 4.10   *Parallax*

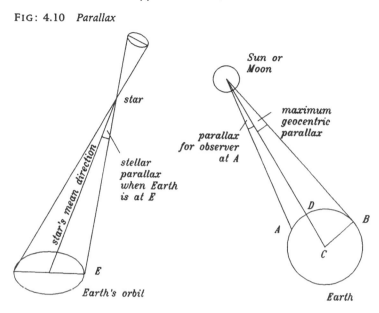

'Parallax' in general refers to the angle through which an object seems to be displaced when viewed from two different positions. For future reference we note that as the Earth moves in its orbit round the Sun, the displacement of a nearby star will so change that the star will gradually describe a tiny ellipse in the sky relative to the background of distant stars (see the left-hand figure). The complete ellipse will be described in a year. Such *stellar parallax* (or 'annual parallax') must be distinguished from the *geocentric parallax* that is so important in correcting predicted solar and lunar positions. These, when calculated from planetary models, are referred to the *centre of the Earth*. Our observations of the Sun and Moon, however, are from a point more than 6350 km from that centre. If we observe either object when it is directly overhead, clearly the parallax is zero, since we (at *D* in the right-hand figure), the centre of the Earth and the object are then in line. The parallactic angle clearly increases to a maximum when the object is near our horizon and we are at *B* in the figure. When we speak loosely of solar or lunar parallax we refer to these *maximum* values. They are obviously directly related to the distances of the bodies in question, and to the Earth's radius, and in quoting accurate figures since the eighteenth century it has been usual to take the *equatorial* radius. This therefore gave rise to the full-fledged notion of 'mean equatorial horizontal parallax'.

reputation in the ancient world was nevertheless considerable. Ptolemy made much use of his writings – although a
modern tradition that Ptolemy was little more than a plagi-
arist of Hipparchus is hardly worth refuting. His influence
is fairly certainly to be found in Indian astronomy, as intimated earlier, and from India it travelled back westwards, as
we shall see, to mingle with other writings produced in a
later (Ptolemaic) tradition. Hipparchus is represented twice
over, therefore, in this curious amalgam.

PTOLEMY AND THE SUN

At much the same time as Babylonian influences were
making themselves felt in the work of Hipparchus, Mesopotamian astrology was beginning to thrive in Egypt. Egypt
was by then at least superficially Hellenized, having been
conquered by Alexander the Great (356–323 BC), and ruled
by his associates and their descendents after his death. Alexander had been tutored by no less a scholar than Aristotle,
but it is the course of his conquests that best explains the
intellectual movements that interest us here. Alexander
was perhaps the greatest military commander of antiquity.
Succeeding to the throne at the age of twenty, he secured
Macedonia, Greece and his northern frontiers before crossing the Hellespont in 334 BC, ostensibly to free the Greek
cities of Asia Minor. Having defeated the Persian armies, he
delayed a thrust eastwards into Mesopotamia until he had
occupied Phoenicia, Palestine and Egypt. He then moved
eastwards, defeated the Persians (under Darius) on their
own territory, and drove on into what is now Turkestan.
From there he went on to India, extending the eastern
bounds of his empire to the lower Indus. After he died – of
a fever, at the age of only thirty-three – his generals divided
up, and fought over, the conquered territories.

We have already spoken of the subsequent rule of the
Seleucids, in Babylonia. The city of Alexandria had been
founded by Alexander himself, perhaps as a future capital.

His friend and general, Ptolemy Soter, became satrap of Egypt, and finally declared himself king in 304. 'Ptolemy' was the name of all the Macedonian kings of Egypt. Under their rule, the old seat of government at Memphis was moved to Alexandria, which grew in importance to become one of the most influential cities in the history of the ancient world.

Alexandria was important as a centre not only of commerce but of learning, and it held its pre-eminent position in the region throughout the period of Roman rule. Under Soter, two great institutions had been founded near his palace, the Museum and the Library. The Museum – there were several less famous examples, that is, of places connected with the Muses – housed a group of salaried scholars under the presidency of a priest. Lectures and symposia were held there, and in these the Ptolemies often took part, even down to the time of the famous Cleopatra, the last of their number. A great fire ravished its Library in 47 BC, during Julius Caesar's siege, but the collection was built up again under Roman rule. The later misfortuncs of the Museum fall mostly after the period of its intellectual importance, namely the second century of the Christian era. It suffered many reversals in the third century, but there were scholars of distinction there until the very end of the fourth century, when Theon, father of the famous scholar Hypatia, was the last member. Both father and daughter were schooled in astronomy and the sciences, and both wrote commentaries on the astronomer Ptolemy's work.

Throughout these centuries, the city had served to channel ideas from its eastern neighbours into a Mediterranean mould. The Arab conquests would eventually make use of Alexandria's eastward-looking intellectual orientation, so that at length it became a largely Islamic centre. Even the ruling Ptolemies had become Egyptianized, and much of the old Egyptian religion reappeared, but with a Greek vocabulary. The native language survived below the veneer of a Greek ruling class, however, especially outside the towns, and this language eventually sprang back into life as Coptic.

Remarkably little is known of the development of Greek astronomy between the time of Hipparchus and that of Ptolemy, and since Ptolemy usually treats Hipparchus as though he were his only significant astronomical predecessor, we can only suppose that little theoretical progress was made over that long period. There is one mathematician we should not overlook, however: Menelaus was active a generation before Ptolemy, and proved a theorem of especial value for calculation in spherical astronomy.

Those who know the theorem of Menelaus for a plane triangle intersected by a transversal may not realize that it is a special case of an analogous theorem, where great circles on a sphere replace the straight lines. Where in the plane theorem we have simple lengths of lines, with the sphere we have chords of arcs.

The astronomer, mathematician, astrologer and geographer Ptolemy was born around AD 100, and died about seventy years later. His name 'Ptolemaeus' shows that he was an Egyptian descended from Greek, or at least Hellenized, ancestors, while his first name, 'Claudius', shows that he held Roman citizenship. His astronomical works are dedicated to an otherwise unknown 'Syrus', and his immediate teachers probably included one Theon, from whom he acknowledges having received records of planetary observations. (This of course was not Hypatia's father. Theon, Ptolemy, even Cleopatra, were common Egyptian names. In the middle ages and in Arabic writings, Ptolemy was often mistakenly represented as a king.) Beyond these simple facts we know virtually nothing about him of a personal sort.

Ptolemy's extensive writings suggest that he was engaged in assembling an encyclopedia of applied mathematics. Of books on mechanics, only the titles are known. Much of his *Optics* and his *Planetary Hypotheses* can be pieced together from Greek or Arabic versions. Some minor works on projection (his *Analemma* and *Planisphere*) as well as the monumental *Geography* survive in Greek, as does his great treatise on astronomy, the *Almagest*.

The title of this, his finest work, is itself an interesting indicator of cultural movements. It began in Greek as 'Mathematical Compilation', and then became 'The Great (or Greatest) Compilation'. When the Arabs translated it in the ninth century, only the word 'Greatest' was kept, but this in an approximation to the Greek word (*megiste*), so that it now became *al-majisti*. From there to the Latin *Almagesti* or *Almagestum*, in the twelfth century, and thence to our *Almagest*, were small steps.

This work in thirteen books begins with a statement of reasons for holding to a largely Aristotelian philosophy – but one that shows the influence of the Stoics, too. We may all attain moral insight in the ordinary course of our affairs, he remarks, but to attain a knowledge of the Universe we must study theoretical astronomy. He follows Aristotle's lead in placing physics on a lower plane, dealing as it does with the changing and corruptible lower world. Astronomy, on the other hand, helps theology, for it draws our attention to the First Cause of celestial motions, the divine Prime Mover. From such relatively brief philosophical beginnings he turns to some rather general cosmological arguments of a qualitative sort, concerning the heavenly sphere and the various motions observed in it. Again he follows Aristotle, more or less, in his physical arguments for the spherical shape, central position and fixity of the Earth. Ptolemy also considers its insignificant size in relation to the heavens. He does not refer to the discussions of the Earth's size by Eratosthenes or Posidonius.

This last point is interesting, because Cleomedes, a near contemporary of Ptolemy's, does inform us of Eratosthenes' measurements, and Cleomedes in the same work writes of the refraction of rays of light passing down through the Earth's atmosphere. It seems that Ptolemy was unaware of this writer. Perhaps Cleomedes was the discoverer of this last highly important phenomenon. In his *Almagest*, Ptolemy considers refraction only as an influence on the sizes of heavenly bodies when seen near the horizon. In the *Optics* he considers atmospheric refraction in more theoretical detail – but that was a later work.

A mathematical introduction now follows, with Menelaus' theorem, a table of chords to three sexagesimal places, and other items that we should classify as 'trigonometry'. His table, for degree intervals, is based on a value of the chord of 1°, which he evaluates by a clever approximation procedure. He is soon at work in books I and II applying his mathematical techniques to astronomical problems, and one item that has repercussions throughout the books that follow is his calculation of the obliquity of the ecliptic.

From the extremes of the Sun's declination he found the value of this fundamental parameter to be between 23;50° and 23;52,30°. Since Eratosthenes and Hipparchus effectively quoted 21;51,20°, which falls in this range, he took that figure, which is a relatively poor one. (A better one would have been 23;40,42°.) His instruments were imperfect, and he probably suspected as much, but one cannot help wondering whether he allowed his admiration for Hipparchus to sway his judgement – or even his instruments.

In book III of the *Almagest*, Ptolemy accepts Hipparchus' solar theory. He compared his own observation of the dates of the equinoxes with those by Hipparchus; and he compared a solstice observation with another by Meton and Euctemon in 432 BC, that is, nearly six centuries earlier. Here he made a calendar error of one day, but even that was enough to throw his figure for the tropical year out of joint, and to persuade him yet again to accept Hipparchus' figure. This was over eleven minutes of time too large, but the theory accounted for most solar phenomena so well that he can have had little incentive to change it.

Ptolemy added tables to allow the rapid calculation of two angles that are needed to settle the Sun's position. The techniques he used were later extended by him to the more complicated motions of the planets, and will serve to give an idea of these theories of heavenly motion generally. Two parameters are needed initially for the Sun, and we shall add a third shortly. Taking the simple eccentric model (but recalling its equivalence with an epicyclic model), these parameters are (1) the mean motion of the Sun on the deferent circle, that is, around its centre, and (2) the eccentricity (*OT* as a fraction of *OS*, in figure 4.11). The angle we

FIG: 4.11 *Eccentric solar model*

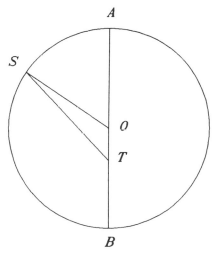

ultimately wish to know is the angle *ATS*. Here *O* is the deferent centre and *T* is the observer on the Earth, the latter being supposedly of insignificant size. The angle *ATS* is the mean motion (angle *AOS*) minus the angle *OST*. Now it should be obvious that the angle *ATS* (the mean motion) is easy to tabulate against the time, say in days, or hours, or both, for it increases at a constant rate. Using trigonometry, the angle *OST* can be easily enough expressed as a function of the mean motion and the eccentricity. (Ptolemy called the angle the *prosthaphairesis*; we should call it an *equation* or an *anomaly*.) He therefore drew up a table allowing it to be found quickly from the mean motion, which had been found from the first table.

One parameter remains, if Ptolemy is to allow us to fix the Sun's position. We need to know the date at which it passed some base point, such as the apogee or perigee; or alternatively we might give its position at a particular date. Ptolemy chose as his standard epoch day 1 of year 1 of the Babylonian king Nabonassar (26 February 747 BC). There is much to be said for such an early date: it means that one does not have to deal with negative years.

Had he been in possession of more accurate data, Ptolemy could have added another parameter, for the line of symmetry (*AB*, the apse line, or line of apsides joining apogee to perigee) does in fact move. He was convinced that the seasons were of the same lengths in his own day as they had been in Hipparchus' time, and so concluded that the line of apsides was fixed.

One subtlety he did not miss was what we now call the *equation of time*. The Sun's daily motion across the sky has been used throughout most of history as a basis for the measurement of short time intervals. This motion, however, is doubly irregular. There is the annual variation in the Sun's speed along the ecliptic, as explained in terms of the eccentric model; but the motion around the poles (i.e. the motion measured with reference to the equator) is variable, owing to the fact that the Sun is moving in a plane (the ecliptic plane) that is inclined to the equator, in fact at more than 23°. Ptolemy explained how to compensate for these irregularities. To this day, the best sundials carry a table to allow for the equation of time, and this correction term is in direct descent from Ptolemy's.

PTOLEMY AND THE MOON

Book IV of the *Almagest* contains a careful discussion of the lunar theory of Hipparchus, accepting a concentric deferent, with new parameters obtained from observation. In book V, when he came to compare it with his own observations, Ptolemy found that it fitted well only when the Sun, Earth and Moon were in line (at conjunctions and oppositions, or *syzygies*, as they are collectively called). This is not surprising, since eclipses had always been the most important factor in settling the details of the simple model. At right angles to these points (at 'quadratures') the error was several lunar diameters – not at all a satisfactory situation. Ptolemy had here found a new variation in the lunar motion, now known as *evection*, and its discovery was a great achievement, but his way of accounting for it was no less remarkable.

The details of his arguments are not something for a short account, but his final model may be briefly explained. As Hipparchus had done, Ptolemy supposed the Moon to move with a retrograde motion on an epicycle, but now he supposed its centre (*C* in figure 4.12) to be eccentric to the Earth, and also to move around a small circle centred on the Earth (*T*). He had to choose, now, velocities that would effectively pull the epicycle nearer to the Earth at quadratures with the Sun. This he did by making the line to the mean Sun (*ms*) the bisector of the angle between *TO* and *TC*. Another refinement was that he reckoned the constantly increasing angle on the epicycle not from the line *TO* but from the line *EO*. This amounts to adding yet another (a third) inequality. It is a mark of Ptolemy's genius that he could add what amounted to new parameters to the model in such ways. Those who dwell on the Greek obsession with circular motions should note the ways in which Ptolemy found it possible to rise above the restrictions they imposed.

This model produced reasonably good results for the Moon's longitude, and certainly better than any before it. The ecliptic has been added to our figure, to show how the key longitudes change. There *mm* is the mean Moon, *A* is the moving apogee of the deferent, and *tl* is the final true longitude of the Moon. As the model stands, however, it has one clear blemish: there is an enormous variation in

FIG: 4.12   Ptolemy's lunar model

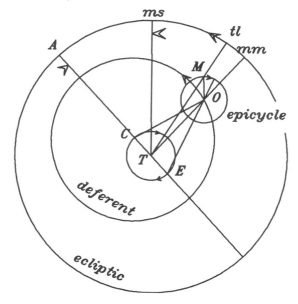

the *distance* of the Moon (*M*) from the Earth, implying that its apparent diameter should vary by a factor of almost 2 during a single revolution. One did not need to be an astronomer to know that this was untrue, and that variations in the Moon's disc are relatively insignificant. Ptolemy kept silent on this point. He had explained longitude well enough, and by placing the deferent and epicycle in a plane at 5° to the ecliptic plane he could explain the Moon's changes in latitude too.

It has often been said that he did not regard his model as describing the motion of real bodies in space, that it was no more than a device for calculating co-ordinates, and that he therefore would not have cared about the variation in the Moon's disc. From his work *Planetary Hypotheses*, however, we know that he cared deeply about creating a planetary system in which *all* the epicyclic apparatus for all the

heavenly bodies was contained. If he noticed the variation –
and he could hardly have failed to do so – it must have
been a great disappointment to him.

Almagest book V ends with a discussion of the distances of
the Sun and Moon, and includes the first known theoretical
discussion of parallax, that is, of the correction it is necessary
to apply to the Moon's apparent position to obtain its position
relative to the Earth's centre. (Refer back to figure 4.10. The
Earth's radius is a significant fraction of the Moon's distance.
Ptolemy's distance for the Sun in Earth diameters was too
small by a factor of about twenty.) From here he could go on
to a geometrical account of eclipses, starting with the theoretically known motions of Sun and Moon and *deriving* the circumstances of eclipses, rather than simply hoping to spot patterns in their recurrences. Ptolemy was not able to chart the
geographical limits within which solar eclipses are visible, a
difficult problem that was not really mastered until studied
by Cassini in the mid seventeenth century.

PTOLEMY AND THE FIXED STARS

Before dealing with the planets, Ptolemy turned to the
longitudes, latitudes and magnitudes (in six classes of
brightness) of the fixed stars. His catalogue of 1022 stars in
forty-eight constellations, and a handful of nebulae, provided the framework for almost all others of importance in
the Islamic and western worlds until the seventeenth century. It was based on materials by Hipparchus that are no
longer extant, and of course took into account his theory
of precession. Where Hipparchus had merely fixed one
degree per century to be a lower limit, Ptolemy took this
as an exact figure. He did not, as is often said, merely add
precession to co-ordinates to update a similar catalogue by
Hipparchus, for his predecessor left his data in a very different form, with descriptions, alignments of stars, co-risings,
and so on. Again, Ptolemy's was a remarkable feat, even
though his stellar longitudes are on the low side.

The reason for this last, and very slight, blemish was the high degree of inter-relatedness of what were superficially different parts of Ptolemy's book. He judged stellar longitudes in many cases with reference to the *moon*, but an error in the *Sun's* motion (which we have just seen enters the lunar model) upset his measurements by small amounts. Most of those who required star positions in later centuries were content to add precession to his longitudes, and so bring his catalogue up to date. The best astronomers incorporated their own observational measurements, but Ptolemy's thoroughness was for long unequalled.

PTOLEMY AND THE PLANETS

Books IX, X and XI of the *Almagest* account for the longitudes of the inferior planets (Mercury and Venus) and the superior (Mars, Jupiter and Saturn). Two different arrangements of the epicycle in relation to the deferent are needed, as we saw in chapter 3, and since Mercury gives rise to difficulties of its own, further refinements are needed in that instance. Again, we shall here give only the end results of Ptolemy's labours. Here he had much less reliable material from his predecessors than for the Sun and Moon. He had the concept of the epicycle, of course, and – through Hipparchus – some Babylonian period relations of the type 'in 59 years Saturn returns twice to the same longitude and 57 times to the same anomaly (e.g. the same stationary point in its retrogradation)'. From such period relations he could construct tables of mean motions, although he needed to trim these later in the light of the models to be developed from them.

We have already seen that the Sun enters into the epicyclic theories. Broadly speaking, for the inferior planets the mean Sun is the centre of the epicycle, while for the superior planets the epicycle radius carrying the planet (*OP* in figure 4.13) is always parallel to the line from the Earth to the mean Sun (*ms*). It will be noticed that in this figure,

FIG: 4.13   *Ptolemy's model for a superior planet*

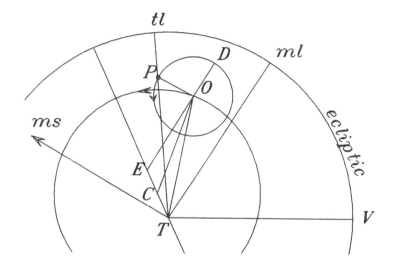

where *C* is the centre of the deferent circle, an extra point *E* has been added to the line joining *T* to *C*, at the same distance but to the other side of *C*. This, the so-called *equant* point, was Ptolemy's device for introducing yet another anomaly. It had always previously been assumed that the epicycle moved uniformly around the centre of the defer-ent. (It is conceivable that Apollonius thought otherwise, but this is a moot point.) Having tried to derive the size of the epicycle, Ptolemy found that it seemed to vary in a way that did not fit well with the simple assumption of an eccentric deferent circle. He adjusted its angular speed, therefore, by the device of making it constant not around *C* but around *E*. (In our figure 4.13, the line *EO* is parallel to the line from *T* to *ml*, the mean longitude.)

   This introduction of the notion of an equant was all the more commendable because it meant breaking with the tra-ditional dogma that all must be explained in terms of uni-form circular motions. Ptolemy introduced an equant *circle* (not shown on our figure) on which a point moved round at constant speed, as a prolongation of the line *EO*. That should have saved him from criticism, but it did not, and fourteen centuries later we find that even Copernicus found the equant distasteful.

When it came to Venus and Mercury, the roles of epicycle and deferent are interchanged, for reasons we have already seen. Venus has a large epicycle, but otherwise a relatively simple motion. The model for Mercury, however, shows Ptolemy at his most ingenious. It embodies all the ideas we have encountered thus far. The equant centre, for instance, is *E* in figure 4.14, and there is an epicycle moving around a deferent circle, but now the centre *C* of the deferent circle is made to move. We have come across a similar device with the model for the Moon's longitude, but here the small circle on which *C* moves is centred not at *T* but at a point

FIG: 4.14 *The oval deferent curve derived from Ptolemy's model for Mercury*

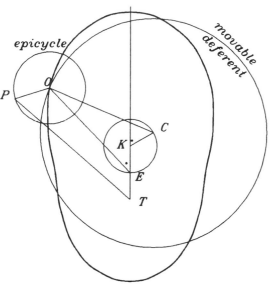

K beyond *E* (so that *KE* equals *TE*). The position of *C* for a particular time is fixed by making the two angles marked with small circles equal. In other words, they move round at a constant rate in opposite senses. Ptolemy arrived at this complex model on the basis of faulty observations that led him to suppose that Mercury had a double perigee, not opposite the apogee, but at points about 120° from where perigee was to be expected. Whatever the merits of his observations, he here effectively provided planetary astronomy with its first oval. For every position of *C* there is a single corresponding position of *O*, and the path followed by *O* is essentially the resulting deferent curve on which the epicycle moves. Its shape is shown by the heavy line in the figure (not to scale). It is an oval pulled in at the waist, so to speak, and for small eccentricities is not far removed from an ellipse. A passage in one of the books assembled at the court of Alfonso X of Castile in the thirteenth century shows that its author had thought it to be of precisely that shape described as a *section* of a pine-cone (compare our use of the phrase 'conic section').

Ptolemy was concerned not only to explain observed planetary motions, but to make it easy to calculate them. Running through all the models here described we have a situation where for every 'mean motion' – that is, an angle increasing at a steady rate – there is another angle, slightly different from it, that has to be used when we come to combine the component angles in the final true longitude. The small differences are called equations, and we have already met with an example, in the case of the Sun. To simplify practical computation Ptolemy drew up tables of mean motions, of course, but also special tables of equations. These can be functions of the mean motions, or of intermediate terms in the calculation. In the end, the astronomer was asked only to add or subtract angles. Even so, the calculation of a full set of planetary positions for a particular time would probably have occupied an hour or two, and more if planetary latitudes were needed.

In book XIII, latitudes were introduced into Ptolemy's scheme, rather as they had been for the Moon, so introducing a third dimension to what otherwise was a two-dimensional treatment. He makes the plane of the planet's deferent inclined to the plane of the ecliptic. In the case of the superior planets the inclination was fixed, but for the inferior planets he made it oscillate. He required the epicycles to lie in still different planes. We can easily see the reason for his difficulties. His system was *earth*-centred, whereas the physical planes of the planetary orbits pass through the *sun*. (The gravitational forces acting on the planets are towards the sun.) It would have been some compensation for his invisible difficulty had he made the planes of the epicycles parallel to the ecliptic plane. To his very great credit, it is clear that he had done precisely this by the time he came to compose a later work, the *Handy Tables*. In them we have only the procedures we are to follow to apply the models, and no justification for the models themselves, so we cannot say how he made his discovery. Here as in so many other ways, however, we have evidence for Ptolemy's high genius in the selection and analysis of astronomical observations for theoretical purposes. Astronomy has many other sides to it, but in this supremely important respect, Ptolemy simply had no equal until Johannes Kepler came to analyse the observations of Tycho Brahe.

It is impossible in a short space to show how the parameters of a particular model can be derived from actual observations, but some very brief general remarks are in order. Using the assumption that motions are on *circles* and are *uniform* (so that angles from the centre are proportional to times), finding the parameters of the model involves at the very least solving the following geometrical problem: *Given three points on a circle, find a point – inside or outside the circle – from which lines drawn to the three points will form given angles* (these, in the astronomical case, will be observed angles). Apollonius is thought to have solved this general geometrical problem. Hipparchus certainly applied it to the cases of the Sun and Moon. Later astronomers realized the advantages of making observations at special times, to give a simpler solution. For instance, if one observes the Sun at the equinoxes and solstices, the angles are multiples of a right angle.

PTOLEMY'S INFLUENCE. ASTROLOGY

So large does the reputation of Ptolemy loom in our view of late antiquity that it is easy to forget that astronomy continued to be practised on a much lower intellectual plane. Of this, by the nature of things, we know relatively little. There is, for example, a papyrus (*P. Heid. Inv. 4144* with *P. Mich. 141*) dating from the century after Ptolemy that gives evidence for the use of a crude scheme for finding the position of Mars. This scheme evidently made use of an epicyclic model, but one that was blended with the so-called 'zones' of the Babylonian 'system A'. There are other indications that Babylonian schemes were known in Roman Egypt. Hellenistic astrology was flourishing, and methods that were easier to apply than Ptolemy's, however inaccurate, were called for. In something approximating to the Greco-Babylonian techniques, astrologers seem to have found what they needed, and there are fragments of texts relating to the Sun and planets and (especially numerous) the Moon. One innovation in these crude schemes is the treatment accorded there to lunar latitude.

This strange blend of arithmetical and geometrical methods makes one thing clear: it is a mistake to imagine that between the time of Hipparchus and Ptolemy there must have been a steady advance in theoretical astronomy. The methods employed even by Hipparchus were a patchwork of elements, geometrical and arithmetical – and under that last heading we should distinguish the 'zig-zag' techniques and the methods of cycles of time after which phenomena repeat. One can say with some confidence that

Ptolemy was *uniquely* responsible for building up astronomy from a coherent set of first principles. With the help of his predecessors' ideas, he was able to conjecture as to how the heavenly bodies moved in space. Having found the parameters of the models by fitting them to observation, he could then predict the phenomena that would be seen, as the consequences of his geometrical assumptions. In short, where others had found patterns of repetition, Ptolemy gave *reasons* for those patterns. With Ptolemy, astronomy had come of age.

The lesser astronomers practised their trade, none the less, and those of a more academic disposition began to write commentaries on the *Almagest* – beginning with Pappus and Theon of Alexandria. It was first translated into Arabic around 800, but improved versions followed quickly. It reached western Europe in two Latin translations, one done from the Greek around 1160, the other – far better known – done from the Arabic by Gerard of Cremona in 1175.

There were two classes of scholar in particular whose needs it did *not* meet, the astrologers and the natural philosophers (or cosmologists, as we might call them). For astrology Ptolemy wrote what again became a standard text, the *Tetrabiblos* ('a work in four parts'), while his *Planetary Hypotheses* went far to providing a more sophisticated version of Aristotelian cosmology. This was based on the assumption that there are no empty spaces in the Universe, but that neither can there be overlapping of matter with matter, so that the outermost point reached by a planet in its epicycle must be equal to the minimum distance reached by the planet next above it. This assumption turned Ptolemy's separate planetary models into a universal *system*. A moment's thought will show that, since the relative sizes of the circles in any planet's geometrical model is laid down by Ptolemaic astronomy, and since the scale of the circles of one planet now fixes the scale of the circles of the planet above it, the entire scale of the Universe is fixed (up to Saturn) in terms of the lowest sphere, the innermost limit of the Moon's possible motion. Since Ptolemy had a distance for the Moon, it became possible to write down distances for all the planets. The answers are plausibly large (they are in millions of miles), but of course they do not correspond to reality. The scheme was seized upon by Islamic writers, and through a summary of the *Almagest* written by al-Farghānī (*fl.* 850), it became a standard part of the curriculum in western universities in the middle ages. It helped to inspire some of the details in Dante's *Divine Comedy*, for example.

The *Tetrabiblos* likewise entered the European consciousness through Islam, but in this case it acquired much extra astrological baggage on voyage. Although its subject matter is certainly not to modern scientific taste, it is nevertheless a masterly book, and in many respects a scientific one. As we have seen, astrology had – among others – Babylonian roots, and we can even trace specific points of astrological contact between the Hellenistic world and Babylon. Most famous is the migration of Berossos, a priest of Bel, from Babylon to Ionia, where he founded an astrological school on the island of Kos, around 280 BC. Greek scholars often claimed to be disciples of the Chaldaeans, and to have received instruction in their schools. Even when they took over what they thought were Egyptian ideas, they were often taking over Babylonian material at second hand. It was possibly at Alexandria, about 150 BC, that treatises were written purporting to be from the pen of the king (mythical, as we know) Nechepso and his priest Petosiris. These books acquired great authority in the Roman world, as did other writings attributed to the god 'Thoth, the 'thrice great Hermes' of the Greeks, Hermes Trismegistos. These works are to *Tetrabiblos* as a crystal ball is to a professional economist: neither is wholly reliable, and both may be wrongly motivated, but there is a world of difference between their techniques.

Where Babylonian and Assyrian divination had mostly concerned public welfare and the life of the ruler, the Greeks applied the art in large measure to the life of the individual.

The activity was unintentionally encouraged by the teachings of Plato and Aristotle on the divinity of the stars, and in
late antiquity many astrologers regarded themselves as interpreting the movements of the gods. With the rise of Christianity this attitude was of course repressed, although it
flourished as a literary device throughout Roman antiquity,
and has been a characteristic of Christian Europe almost until
the present day. Ptolemy's *Tetrabiblos* was thus a handbook
for people of many different persuasions.

It opens with a defence of astrology, and is ostensibly
written around the idea that the influences of the heavenly
bodies are entirely physical. In the end, however, it
amounts to a codification of unjustified superstition, largely
inherited from Ptolemy's predecessors. Book II deals with
cosmic influences on geography and the weather, the latter
a popular and spiritually safe subject in later centuries.
Books III and IV deal with influences on human life, as
deduced from the state of the heavens, but oddly lacking
in any of the mathematics of casting the houses that so
obsessed astrologers in later centuries. (Something more on
this theme will be found in chapter 10.)

In late Roman antiquity, the so-called 'Chaldaei' and
'mathematici' – words we can interpret simply as 'astrologers' – were very numerous, judging by the frequent criticism levelled at them by Roman magistrates and satirists. A
number of expulsions from Rome and Italy are known from
before the first century, and there were edicts in force
against them in the fourth century, when the Christian
emperors added their own religious scruples to old political
objections. In 357, Constantius made divination a capital
offence, and the ban was repeated in 373 and 409.

The ancient tradition of astrological divination had a
marked influence on the practice of medicine. Latin literary
style was also much influenced from an astrological quarter,
for instance by the first-century poem known as the *Astronomica* of the Stoic philosopher Marcus Manilius. The Stoa
was a philosophical sect with a long history, beginning

around 300 BC, and one of its chief doctrines was that the philosopher's aim should be to live in harmony with Nature through the use of reason. As time went on, the sect became increasingly concerned with ethical questions and it is not surprising that the Babylonian idea of a stellar necessity ruling the world found a sympathetic audience among Stoic philosophers generally. Manilius put about the idea that human life is absolutely determined by the stars, but he did so in the course of a work that was no doubt valued more for its astrological technicalities than for the philosophical ideas underpinning it. The philosophers helped to give respectability to ideas that were broadly astrological, however. Around AD 265 Plotinus, the founder of Neoplatonism, proposed a related doctrine that magic, prayers and astrology are all possible because each part of the Universe affects the rest through a kind of mutual sympathy. Such ideas gave much comfort to later generations of scholars anxious to play with fire.

A literary work that helped to counter such influences was the *City of God* by St Augustine (354–430), in which he warned that astrologers could enslave the free human will by predicting a person's life from the stars. If predictions come true, he said, this is due to chance or to demons. He had in fact been a believer in both astrology and sacrifice to demons, and his testimony was well argued and compelling to many medieval churchmen. However, they continued to believe, as did he, in God's foreknowledge and in celestial influence, and so were presented with a dilemma. How could humans be free, if all was fore-ordained, that is, either by God – who can only know what is to come if indeed it is to come – or by the influence of utterly predictable planetary motions? The usual way out was to say that the stars force us in a certain direction but do not compel us to act against our free will. They 'incline but do not compel'. Prayer would help people to resist. Other Church Fathers touched on these questions. Origen, for instance, tried desperately to purge astrology of fatalism.

These facts have an obvious relevance to the practice of *astronomy*. Regardless of any real astrological association, this was regarded with deep suspicion by those classes of people – that is to say, almost everybody – who did not understand its potential independence of astrology. There are some famous astrological names from the Roman world: Vettius Valens from the second century, Palchus, Eutocius and Rhetorius from the fifth, and no doubt many materials have disappeared without trace from the intervening period, but after them we come to a period in which astrological practice was firmly suppressed in the west until something of a revival came in the eighth century. And from late antiquity onwards, such western astrology as we find being practised tends to be thoroughly derivative.

THE ASTROLABE

Astronomy continued to be practised in Byzantium, the eastern Roman empire, which takes its name from the refounding of the old city of Byzantium as 'New Rome' by the emperor Constantine in 330. (From him comes the city's name 'Constantinopolis', our 'Constantinople'.) Born in Byzantium around 410, for example, was the philosopher and mathematician Proclus (*d.* 485), who was familiar with the intricacies of Ptolemy's astronomical theories but was critical of the arbitrary character – as he mistakenly judged it – of Ptolemy's hypotheses. He was not above making a paraphrase of *Tetrabiblos*, where there was surely much more room for scepticism. A lesser scholar, but important to any history of astronomical instrumentation, was Synesius of Cyrene (*d.* between 412 and 415). He had been a pupil of Hypatia in Alexandria. He was a soldier who, having married a Christian woman, was with great reluctance persuaded to accept baptism and the bishopric of Ptolemais in 410. He appears to have found time to make some sort of improvements to the astrolabe, an instrument we mentioned earlier in connection with Hipparchus. He presented a silver instrument of this sort to a friend in Constantinople, and with it a letter describing it and its uses.

His references to the instrument are valuable, and in some ways their rarity is surprising, for the great Greek architect Vitruvius, who died some time after AD 27, had described a water clock capable of showing the seasonal hours of day or night, and judging by his description, this had a sort of astrolabe as its dial. The Vitruvian mechanism as a whole is called an 'anaphoric clock', since it is based on rising times. Fragments of later Roman anaphoric clocks have been found in France, and there are other reasons for supposing that the instrument was not altogether rare in the ancient world, but early literature is surprisingly scarce.

Although there are several different instruments to which the name 'astrolabe' has at various times been attached, there is one sort of common plane astrolabe that far outnumbers all others. This comprises one or more circular discs overlaid with a pierced circular disc of the same material, usually of brass. The pierced disc is pivoted around a pin at the common centre of the discs (see figure 4.9, and plate 6). As we have seen, the fretted disc, or 'rete' (the Latin *rete* simply means a net), is essentially a star map. It is pierced simply to allow certain circles engraved on the disc beneath it to be seen: these are the fixed local coordinate lines (meridian, horizon, lines at altitude 5°, 10°, and so forth above the horizon, etc.) with reference to which the positions of the heavenly bodies are judged. It is the relative motion of the two discs (rete and plate) that is of importance. It is not necessary that the pierced plate should represent the star sphere, and the plate the horizon, meridian, etc. Although this became the almost universal preference, the roles of the rete and plate could be reversed, as indeed they were in Vitruvius' anaphoric clock.

In size, portable versions of the instrument are typically between ten and twenty centimetres across, although smaller and larger examples survive in plenty. As developed in later centuries, the whole astrolabe assembly was meant

FIG: 4.15 *The two main components of an astrolabe*

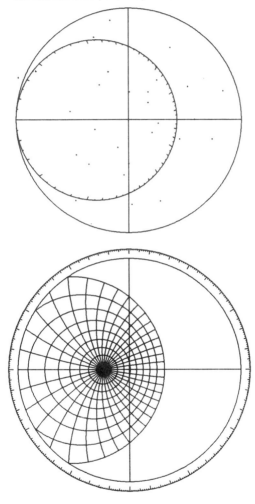

The rete (upper figure) includes the ecliptic and stars, while the plate (lower figure, shown surrounded by an outer scale) shows the horizon, and lines of constant altitude and azimuth (the 'almucantars' and 'azimuths' respectively).

FIG: 4.16   *An exploded view of a typical medieval astrolabe*

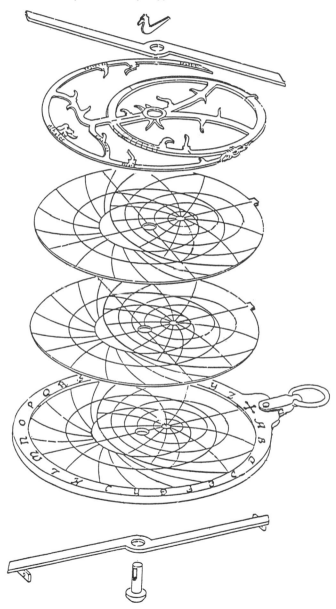

to be used for observation as well as for computation. For the former purpose a ring and shackle are provided, so that the instrument can hang vertically from the thumb of one hand, while observations of objects in the heavens are made with the help of the sighting vanes carried on the centrally pivoted rule. This is usually called the *alidade*, and is on the back of the instrument. It is distinguished from a second rule, without sights, that is to be found overlaying the rete on some – but not all – astrolabes.

The plane astrolabe makes use of certain properties of the projection known as 'stereographic', properties some at least of which must have been appreciated by Hipparchus himself. One may get an approximate idea of its rationale without understanding the intricacies of its geometry. Imagine first two sets of points and circles in the sky around us, one set fixed and one rotating. We may take the *fixed* circles to start with the horizon, above which will come a parallel circle, say five degrees in altitude above it, another at altitude ten degrees, and so on up to the zenith. The *moving* objects will be the stars and other heavenly bodies, the ecliptic, and in fact anything that moves with the daily rotation around the poles.

Astronomers frequently built a three-dimensional model of this double system, an 'armillary sphere' (*armilla* is the Latin for 'ring'). Indeed it was even used for observation, having as it does the merit that if an object on the model can be set in the direction of the corresponding object in the heavens, the entire system will be correctly set for that moment of time. Taking the Sun's place into account – since time is reckoned by relating the Sun to the meridian – this gives us an entry to the general problem of timekeeping. An armillary may of course be used for many other astronomical purposes, for it is no more than a diagram in three dimensions. (Note that it is essentially a representation of angles, not of distances.)

For the time being we are only concerned with the *idea* of the armillary sphere, but it should be noted that actual armillaries were occasionally made with great sophistication. Ptolemy, in his *Almagest*, explains how to make a celestial globe that exhibits precession, and a few designers of armillaries later followed his lead. The armillary was much used by teachers of astronomy throughout the middle ages and after.

Now our imagined armillary sphere may be mapped on to a plane in many different ways, but the simplest ones will be those which project the north (or south) pole into the *centre* of the map, so that the moving part of the map may pivot round it. This is precisely what is done in the case of the astrolabe. The plane on which the circles are projected is any plane parallel to the equator. The moving ecliptic, being inclined to the equator, will not be centred on the poles, but will still be a circle (and not an ellipse, for example). The same is true for the horizon and parallels of altitude (called *almucantars*, from an Arabic word).

One way of imagining the projection would be to suppose an intensely bright point source of light at one of the poles of the armillary, casting shadows of the two sets of rings, fixed and moving, on a screen parallel to the equator. Another way would be to follow an Arab writer who asked us to imagine that a camel had trodden on an armillary sphere, squashing it flat between its poles, but so as to allow the two parts to move relatively to one another still.

The astrolabe has numerous uses – a typical text will explain over forty – but many of them require us to position the moving rete correctly in relation to the fixed plate by reference to the observed altitude of a star. The rete is rotated until the star's marker falls against the correct altitude line (almucantar) on the plate below. The Sun is effectively a star moving along the ecliptic on the rete. To know its position, one needs its longitude. This may be found from a calendar listing the solar longitude for each day of the year, for example, but a simpler way is to consult a so-called 'calendar scale' that is to be found on the back of most astrolabes, correlating longitude with day. (See plate 6 for such a scale.)

Since the Sun's position in relation to the meridian is a measure of the time of day, the astrolabe was always useful as a time-measuring device, by day or night. By night the rete is positioned with the help of a star, but the Sun's place on the rete is known from the calendar scale. By day the rete can be positioned from an observation of the sun.

Although Hipparchus seems to have been its inventor, the oldest surviving treatise with a systematic account of the theory of stereographic projection was Ptolemy's *Planisphere*. The Greek original of this is lost, but the work survives in Arabic translation, and a revision of that was made by an Islamic Spanish astronomer in the tenth century, finally reaching western Europe in a Latin version in 1143.

In the letter mentioned earlier, Synesius claims to have been the first since Ptolemy to have written on the theory of the astrolabe projection, but if so, a superior work by Theon, father of his 'revered teacher' Hypatia, can only have been a few years later. Only the table of contents survives, but it fits closely with a later work by Philoponos (*d.* about 555) and even more closely with a Syriac treatise on the same theme by the bishop Severus Sebokht (*d.* 665).

The identity of the oldest surviving portable astrolabe is a matter of dispute, but from eastern Islam we have an early copy of a ninth-century example, and a dated original from about AD 928. There is a fine Persian example dated at the year 374 of the Hejira (AD 984) that bears witness to a long craft tradition, confirming what we know from eighth-century literary references from Baghdad and Damascus. Within a few centuries we find examples of texts and instruments stemming from every important centre of civilization between India and the Atlantic. Persian craftsmanship remained of consummate quality, and was rivalled in Europe only in the sixteenth and seventeenth centuries.

Ptolemy's was most certainly not the sort of treatise that could ever have made the astrolabe popular, but in the later Islamic and Christian worlds alike many scores of alternative texts were produced. European writings were numerous, but fall into only three main families, all radiating from Muslim Spain. In no centre was there any other scientific instrument of comparable importance, artistically or symbolically. Admittedly, as an instrument for precise observation the astrolabe was of no great value, while for computation it was usually too small to give more than an approximate answer to complex problems. As a teaching device, and for clarifying problems in positional astronomy, it has had few equals.

New types of instrument were developed, especially in the Islamic world. As one consequence of its design, the horizon plate of the standard astrolabe is of use for only a single geographical latitude. Most have a collection of different plates in the body of the astrolabe, and the user chooses the most appropriate.

Apart from much superficial variation in style and artistry, astrolabes changed little over the centuries. There were astrolabes designed for use at any location, with only a single plate; but they were difficult to use, and fewer by far were made. Astrolabes were usually made by astronomers for their own use; but some, made for patrons and men of substance, were richly ornamented and exquisitely engraved. The astrolabe was beyond the reach of most ordinary people, but in the late middle ages was known to very many, since on a grand scale it fronted astronomical mechanical clocks. Indeed, true to its Vitruvian ancestry, it was the prototype clock face. The mechanical clock originated perhaps in England in the late thirteenth century, and was a direct consequence of a desire to represent the moving heavens in a tangible form.

In east and west, well into the seventeenth century, the astrolabe remained both a working tool of the astronomer and a powerful symbol. It symbolized not only the cosmos but astronomy itself, known as it was by sight to many who had no understanding whatsoever of its many intricacies. For all its accretions, it was a tangible reminder of the Greek genius for combining astronomy with geometry.

China and Japan

ANCIENT CHINA:
THE RISE OF STATE ASTRONOMY

Ancient China developed a form of writing – in the form of pictographs – at a very early stage in its history. Chinese is the only great script that has kept an ideographic form to the present day, in fact over a period well in excess of three millennia, and the first great dictionary was produced as long ago as A D 121. Although, therefore, linguistic changes have been continuous over time and space, some historical knowledge of very ancient Chinese affairs is reasonably reliable.

It is as difficult to mark the boundaries between early Chinese religion, astrology and astronomy as in other cultures. Just as in Mesopotamia, several methods of forecasting were developed, based on the interpretations of many sorts of signs. Some were remarkably similar to those used in the west. One example is the technique of interpreting cracks in the heated shoulder blades of animals (scapulomancy), the main difference being that in China, with luck, the cracks would take on the appearance of a standard pictograph. At one level, astrology was simply an additional form of divination. As in the case of lands to the west, it was less concerned with the fate of ordinary individuals than with prince and state, harvest, war and the common weal. There are strong parallels, for instance, between a text from the library of Assurbanipal (seventh century) and the Chinese *Shih Chi* (Historical Record) of Ssuma Chhien (90 B C), both of them interpreting planetary movements through the stars in terms of the destiny of the king and his enemies.

One striking difference between celestial observation in these cultures was that, whereas in countries to the west of China attention was at first generally focused on the *horizon*, in China great importance was attached to the constellations around the *north celestial pole*, and those around the celestial *equator*. (There are of course many exceptions on both sides. Did not the sentries at the siege of Troy change guard with reference to the direction of the tail in the Great Bear?) By concentrating their attention on the stars around the northern celestial pole that never rise and set, the Chinese considered celestial objects in their relation to the Sun, and in particular in opposition to the Sun. They also mentally positioned stars by imaginary lines through circumpolar stars to stars that did indeed rise and set. In the northern hemisphere we all know the method of locating the Pole Star by the same technique of following a line from the stars at the end of the Dipper. The Chinese pinpointed stars by pairs of such lines.

Perhaps as early as 1500 B C, and certainly not later than the sixth century B C, the Chinese divided the heavens into twenty-eight lunar mansions, each regarded as a section of the equator, or as the stars within its limits, in which the Moon happened to be at the time of interest. Historical influence is here difficult to determine. There is an oracle bone older than 1281 B C that mentions stars by name, some but not all of which have been identified. Another oracle bone refers to a solar eclipse identifiable as that of 1281 B C. It is possible that the Chinese in the first millennium B C took from the Mesopotamian culture some of the broad principle of divination using the stars and Moon, but that they interpreted it in accordance with their own patterning of constellations, based on an equatorial rather than an ecliptic system.

The Chinese continued to absorb western astrological ideas, but rarely took them over in quite their original forms. In the fourteenth century, we find horoscopes in Chinese sources (for instance in the *Thu Shu Chi Chhêng*) remarkably similar in all but outward shape to European horoscopes. The reason is simply that both have a common source, namely in Islamic, and still earlier Greek, astrology. The Chinese in such cases tended to increase greatly the number of alternative interpretations. They did so under the influence of other ancient systems of divination already in place, for instance some based on biological theories and others on the calendar. Even here, when we find what appears to be western influence – for instance in the matter of lucky and unlucky days – there is often the possibility of a similar earlier tradition on which the later was grafted.

In ancient China, astronomy was intimately connected with government and civil administration. Perhaps the most famous illustration of this concerns an incident in the eighth century BC, and is reported in the *Shu Ching* (Historical Classic). It concerns a commission sent by a legendary emperor Yao to six astronomers, of whom two are named – the brothers Hsi and Ho. They were instructed to move to various places from which they were to observe the rising and setting of the Sun, to determine the solstices and make other observations of importance to the drafting of a calendar. In a later chapter of the *Shu Ching* there is an account of an expedition led by the prince Yin to punish other astronomers for failing to foresee or prevent an eclipse. These legends, for three millennia the official account of the origin of Chinese astronomy, have for centuries helped to create the image of a venerable branch of learning, but they are now known to stem from an early mythological tradition in which Hsi-Ho is the name of either the mother or the chariot driver of the Sun. The brothers Hsi and Ho are no more.

The all-pervading Chinese view of Nature as animistic, as inhabited by spirits or souls, gave to their astronomy a character not unknown in the west, but at a scholarly level markedly less important. At a concrete level, we have such Chinese doctrines as that there is a cock in the Sun and a hare in the Moon – the hare sitting under a tree, pounding medicines in a mortar, and so forth. At a more abstract level there is the doctrine that, corresponding to the Sun and Moon, there are on Earth two great principles, the *yin* and the *yang*, the first being male, associated with heat and day-time, the second female, associated with cold and night. This spiritual system was extended to the planets too, and *yin-yang* techniques of magical divination continue to assume great importance in modern times.

Early Chinese interest in cosmological matters was not markedly scientific, in the western sense of the word. It did not develop any great deductive system of a character such as we meet in Aristotle or Ptolemy, for example. The great scholar we know as Confucius (551–478 BC) did nothing to help this situation – if help it needed. Primarily a political reformer who wished to ensure that the human world mir-rored the harmony of the natural world, he wrote a chapter on their relation, but it was soon lost, and a number of stories told of him give him a reputation for having no great interest in the heavens as such. It is worth remembering here a definition of Confucianism as 'the worship of the Universe through the worship of its parts', a programme very different from that of the great system-builders in the west.

The Chinese rarely speak in terms of a god beyond the world, a maker of it. Heaven is the highest god, and the Emperor is heaven's son, and head of the state religion. The most important sacrifice to heaven took place on the night of the winter solstice, when the *yang* begins to increase again after reaching its lowest ebb. Of course there were many other and related imperial rites. It is clear that the record of them set down in the book *Chou Li* around the second century BC describes a situation that had long been evolving. The tasks of the imperial astronomer and the imperial astrologer are there distinguished. Planetary obser-vation belonged to the second office, for instance observa-tions of Jupiter's twelve-year cycle that corresponded, it was said, to the cycle of good and evil in the world. As in other cultures, the astrologer observed the weather – here the colours of five types of clouds, the state of the twelve winds, and so on. For more than two millennia these (hereditary) officials headed government departments with a large staff. They were the keepers of time, even of the hours, and the Chinese developed a whole series of water clocks, some of them extremely intricate hydromechanical affairs, that is, driving astronomical and other models by water power. It is said that Chang Hêng, about AD 132, was the first to devise a means of turning an armillary sphere by a water wheel so that it kept pace with the heavens. The general idea was further developed, although sporadically.

By the time of the Northern Sung dynasty (960–1126) there were two separate observatories in the capital, one imperial and the other belonging to the Hanlin Academy. Each was equipped with water clocks and instruments for observation. The two were meant to make independent observations and then compare the results. When Phêng Chhêng became Astronomer Royal in 1070, he discovered that the two sets of astronomers had for years been simply copying each others' reports, and even taking the positions of heavenly bodies from old tables. His successor Shen Kua discovered that the examiners in the state examinations were still no more competent than their predecessors had been.

Since calendars were important symbols of dynastic power, and were often totally revised with changes of dynasty, old calendars could be politically sensitive material, and perhaps this explains why few old astronomical documents survive, in comparison with mathematical materials, for example. Secrecy was enjoined, which explains some of the difficulties encountered by Jesuit missionaries visiting China at the end of the sixteenth century. The political character of calendars meant that they were numerous: no fewer than 102 were produced in the period between 370 BC and 1851, and many of them have star tables and planetary ephemerides in addition to solar and lunar material, making them an excellent historical indicator of the progress of astronomical theory.

From the fourth century B C onwards, there was a steady increase in the number of works devoted to the visible Universe of stars and their grouping. Although there were no great cosmological systems, in the western style, there were simple cosmological pictures. For example, in the *Kai Thien* cosmology, which is of uncertain age but might be earlier than the fourth century B C, the heavens were pictured as a hemispherical bowl placed over a hemispherical Earth (perhaps somehow trimmed square). This arrangement might have come from Babylonian sources. In places it lasted into the sixth century A D, and figures were sometimes quoted for its dimensions, although they seem rather arbitrary.

The oldest surviving Chinese description of the heavens as *completely* spherical is by Chang Hêng, of the late first century A D. Its circumference was divided into 365¼ units, each unit the distance traversed by the Sun in a day, and this corresponds to a length of the year known at least as early as the thirteenth century B C. There was a school of thought, the Hsüan School, older than Chang, according to which the heavens were endlessly extended, and this view found favour with neo-Confucian philosophy well after the twelfth century A D. It fitted well with Buddhist ideas and did not conflict head-on with the spherical picture in the same way as that had done with the *Kai Thien* cosmology. Yet another theory was that when the Sun is on the meridian it is five times more distant than when rising and setting, suggesting that the sky was conceived of as a highly elliptical dome.

There are other 'philosophical' systems of cosmological thought, but it is easy to find ourselves searching for them merely to draw parallels with what was happening at the same time in Greece. Unlike Platonic and Aristotelian thought, Chinese was not overtly philosophical but rather historical. Joseph Needham has suggested that the reason for this is that Chinese religion had no lawgiver in human guise, so that the Chinese did not naturally think in terms of laws of Nature. An important instance of writing in a historical style is the *Shih Chi* (Historical Record) of Ssuma Chhien – a man who had held the highest astronomical position in state service – completed in 90 BC. One chapter of the book provides a survey of current astronomical doctrine, including much on natural meteorological phenomena. Later official histories of dynasties usually had astronomical chapters, including astrological portents with a possible bearing on the future of the state. They typically contained instructions on calculating planetary positions using only moderately accurate (synodic) periods, and instructions on simple lunar eclipse calculation. Astronomical styles changed materially as more use was made of the ecliptic in the first century AD.

To assist with a calendar reform by Chia Khuei around AD 85, instruments were constructed to measure the ecliptic's obliquity. Early in the fourth century AD the astronomer Yü Hsi (*fl.* 307–338) discovered the changing longitudes of the stars, our 'precession of the equinoxes', seemingly independently of western knowledge of it, as first found by Hipparchus. The character of Chinese astronomy continued, however, to be an act of compilation of data rather than of the creation of complex mathematical theories. Many simple regularities were appreciated, but when they broke down in a particular instance, the observation was simply regarded as 'irregular', and labelled as such. (One's thoughts naturally turn here to the role of miracles in western religions.) Many Chinese writers betray a belief that, while broad analogies are to be found in the world, reality is essentially too subtle to be encoded in general principles.

This attitude had at least one very fortunate consequence. In western astronomy, phenomena – such as comets, novae and oddities in the appearance of the Sun – that did not readily lend themselves to treatment in terms of laws were taken far less seriously than those that were. The history-conscious Chinese, on the other hand, kept massive records of *all* such phenomena, and these records were later to prove, as they continue to prove, an important source of astronomical information. Sunspots, in European terms a telescopic discovery of the seventeenth century, were already recorded in China in the time of Liu Hsiang, 28 BC, and perhaps long before that. Between then and AD 1638, there are well over a hundred references to sunspots in official histories, and many more in local records. The Chinese knew the art of looking at the Sun through smoky crystal or jade, and haze and dust storms could also be turned to advantage. It is interesting to discover that when Thomas Harriot first turned his telescope towards sunspots, in seventeenth-century England, he too took advantage of natural haze.

There is a long and continuous tradition of drafting starmaps in China. This is not surprising, bearing in mind that omens in the heavens were thought to have relevance to the region of the stars in which they were sighted, and these regions were associated with regions of the Earth – whether foreign territories, provinces, cities, or even divisions of the Imperial Palace. One of the most famous divisions of the sky is flanked by two chains of stars, seen as walls of the Imperial Palace that enclose the 'Purple Forbidden Enclosure'. Solar haloes, sunspots, coloured clouds, the aurora borealis, or indeed almost any other astronomical or meteorological event would be heeded, especially before battles, which needless to say usually had territorial implications.

Although not the oldest of the various maps on which Chhien Lu-Chih (fifth century) based his own, those by Shih Shen, Kan Tê and Wu Hsien in the fourth century BC, and their combined equivalent by Chhen Cho in the fourth century AD, were all important for the later tradition. They allowed Chhien to create a map on which he assigned colours to the stars, so indicating the astronomer responsible. (Earlier maps are known to have existed, but no details are known.) The combined list, with 1464 stars in all (compare Ptolemy's earlier 1022), was grouped into 284 constellations – a very large number.

Although the co-ordinates were equatorial, eventually ecliptic co-ordinates were recorded, following in the western tradition, with at least one notable discovery in consequence. Around the year 725, the monk I-Hsing, using instruments built by an engineer, Liang Ling-Tsan, found star co-ordinates differing from those in the old lists. Precession could of course account for most discrepancies, but not for ten and more cases of movements that also changed the ecliptic *latitudes*. Now the stars have their own 'proper motions' in addition to precession, as was to be discovered eventually by Halley in the seventeenth century. It is impossible to say that I-Hsing had 'discovered the proper motions' of the stars. His instrumental errors no doubt accounted for many of the observed effects. He offered no systematic statement or argument, and nothing came of his potential discovery, but there can be little doubt that a part of the observed effect was due to non-precessional movements, proper motions.

Many stories are told of I-Hsing that reflect on his reputation for magic of an intellectual turn. The occasion of one, from a collection dating from AD 855, was the imprisonment of a friend on a charge of murder. In the Temple of the Armillary Sphere with his hundreds of assistants, I-Hsing ordered seven pigs to be caught and put into a pot. Subsequently the emperor complained that the head of the Astronomical Bureau had found that the constellation of the Great Bear was missing. I-Hsing said that he could recall only one remotely similar occurrence, when Mars had once been lost. The new mishap he interpreted as a dire warning, perhaps of frost or drought, but this could, he said, be averted if the emperor followed Buddhist preaching and issued a general amnesty. The amnesty did indeed restore the stars to the heavens; and when the pot was opened, the pigs had disappeared. Putting aside the undoubted talents of the astronomer in question, that the story could be told with a fair degree of seriousness goes far to distinguish between Chinese and western styles of astronomizing at this period.

CHINA BETWEEN THE TENTH AND SIXTEENTH CENTURIES

Chinese astronomy had a strong cohesion: bound together as it was by language and script, it was widely regarded as the work of a single national group. It was admittedly enriched from time to time by contact with astronomers from countries to the west, and it was passed on, with an admixture of Korean astronomy, to Japan in the sixth to eighth centuries. Very early contacts with Babylon have been suspected, since by the mid second millennium BC the Chinese were using a count of days by sixties. Persian astronomers certainly visited China in the eighth and ninth centuries of the present era, and carried with them the Babylonian and Greek methods of computation.

Great political changes were then taking place. Aristocratic land-owning families held power, with the emperor at the head of the pyramid of power. The elitist civil service, despite its rigorous system of examination, was really accessible only to the upper echelons of society, and the system did not encourage high enterprise. For three centuries the Thang dynasty gradually declined until it was reduced to a series of rival kingdoms. Something approaching a new 'universal state' came with the Northern Sung (960–1126). Taxation weakened the old families, the centre of economic power moved from the north – the old centre of civilization of the Han people – to the lower reaches of the Yangtze valley, and society became more open to innovation.

This situation is well reflected in astronomy. The second Sung emperor is known to have had a large astronomical library, and one work that has survived from it is of great interest, for it seems to parallel the Greek and Arab tradition of water-driven representations of the cosmos. This 'astronomical clock' was described in the *Hsin I Hsiang Fa Yao* (New Description of an Armillary Clock) by Su Sung (1020–1101). He was first privy councillor during a hectic political period, but managed to guide a vast imperial plan to amalgamate medical writings and print the ancient medical classics. The clock, with an escapement that used a technique of tipping buckets on a chain drive, was built between 1088 and 1095 under his direction.

The description of the mechanism also includes the oldest extant printed star map, based on a new survey of the heavens. Since so much of Chinese learning in general was spread by books printed with wooden blocks, it is not surprising to find evidence for star maps in this medium from at least the eleventh century. (A later example, from AD 1092, is shown in the plate section.)

A scholar of still greater repute who served the same emperor and became the emperor's confidant, only to die in disgrace, was Shen Kua (1031–95). He was a good mathematician who applied mathematics to a number of physical problems – to music and harmonics, for instance – and who made a number of attempts to do the same for astronomy. What is especially interesting is that whereas earlier Chinese astronomers, like the Babylonians, had used primarily arithmetical methods to account for the retrogradations of the planets, Shen Kua proposed a geometrical model. By Greek standards his might now be thought a primitive theory. It serves to remind us that a solution in terms of circular motions alone was not an automatic choice for all people. Shen Kua thought that the planet followed a circle until it came to a 'willow-leaf' part of its path, that might be inside or outside the main orbit, and then it performed a detour before returning to the main path (figure 5.1).

One of Shen's early services to the court was to have improved instruments made – including a gnomon, for

FIG: 5.1   *Shen Kua's willow-leaf model*

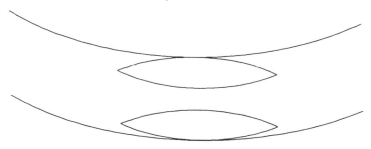

solstice measurements, and an armillary sphere. For the alignment of the polar axis of the latter, a faint star near the north celestial pole was used. Since it was not quite at the pole, it described a small nightly arc of a circle. Shen made a tube of such a size that, looking through it towards the faint star, one saw a circle of open sky of just the size described by the star's path.

The star in question was one of a succession of stars used as a pole star, the drift in and out of acceptability being of course due to precession. It is a remarkable fact that very many stars have Chinese names that indicate their earlier or later use as pole stars, taking us back perhaps as far as 3000 BC. In Shen's day the star was our 4339 Camelopardi. In the fifth century this had been only a little over a degree from the true pole, but now Shen found the distance to be in excess of three Chinese degrees (each $360°/365.25$, as mentioned above).

Another decision Shen made was to discard the ring on the armillary sphere representing the Moon's path, for it could not be made to move so as to reflect the backward motion of the Moon's nodes.

These were not great reforms, but they helped to simplify a system hidebound by tradition. The same would have been true of his suggested reform of the calendar, had it been adopted. He proposed a purely solar calendar in place of the lunae-solar calendar that had been in use from ancient times. He was right in supposing that it would give offence, and radical reform was not to come until the mid nineteenth century. Even that, instituted after the Taiping rebellion, was short-lived. China finally adopted the western (Gregorian) calendar for most public purposes in 1912.

The period of Mongol domination, known in China as the Yuan dynasty, lasted from 1260 to 1368, and saw a revival of standards in astronomy as a consequence of Persian and Arab influence. The greatest Chinese astronomer of the period was Kuo Shou-Ching (1231–1316), a great mathematician, and a designer of sophisticated water-clocks into the bargain. Kuo carried out a sustained programme of observations and wrote an important *Shou Shih Li* (Calendar of Works and Days, 1281). A magnificent armillary made under his direction for the latitude of Phin-Yang in Shansi around AD 1276 still survives, if not in original form, then as a replica dating from 1437. Still also surviving is the Tower of Chou Kung, a Ming renovation of a building set up by Kuo Shou-Ching around the year 1276 for use with a twelve-metre gnomon for measuring the length of the shadow cast by the Sun, and so the solstices. This magnificent instrument had a stone scale, over thirty-six metres long, flanked by parallel troughs carrying water for purposes of levelling. As ever, the collection of data remained a matter of the highest priority, but no striking new theories were forthcoming.

During what in Europe are treated as the late middle ages, when astronomy there was rapidly gathering momentum, in China affairs seem to have been showing signs of decline. That great speciality of Chinese workmanship, the grand armillary sphere cast in bronze, with its supporting lions, dragons and other symbolism, reached a high point with the Northern Sung in the eleventh century. The fall of the capital to the Chin Tartars put an end to this period of practical expertise. Instruments of various sorts were produced for common use, for example sundials, water clocks of various ingenious sorts, and even incense clocks, on which time was judged by the burning of incense pressed into a spiral groove, or maze of grooves. The Ming dynasty (1368–1644), which produced so much fine art, would have been unmemorable from a more serious astronomical point of view had it not been for a remarkable historical episode, the coming to China of the Jesuits – the subject of a later section.

KOREA AND JAPAN

In the ancient mythology of Japan, the Sun-goddess, Amaterasu, has a central role. The Moon-god, her brother, Tsuki-Yomi, is relatively unimportant, except in some stories of the birth of the Japanese archipelago, which make this out to have been born of the union of Sun and Moon. The stars seem to have had an even less significant place. There are early traces of festivals of certain stars, but that idea seems to have come from China, and when such records become numerous, it was as a result of the introduction of Buddhism from China in the sixth century A D. That cultural invasion immediately gave rise to a discussion of the relationship of the old Japanese gods and those of the Buddhist pantheon. The most striking parallel was between the Sun-goddess and the Sun myth used to explain the Buddha's personality. Buddha Vairochana (The Illuminator) was the resulting notion, one that influenced worship in Japan until the nineteenth century, when it was forbidden.

Korea was in many respects a staging-post for Chinese astronomy on its way to Japan. An incident that well illustrates the drift of ideas across Asia begins with no fewer than three schools of Indian experts serving in the Tang national observatory in the eighth century. In the seventh and eighth centuries two Indian astronomers, Qutan Luo and Gautama Siddhārtha, actually rose to the post of Director of the Astronomical Bureau in China. In 729 a new calendar was put into effect, the *Dayan li*, designed by I-Hsing. Three years later there was a dispute at court about its accuracy, led by an Indian astronomer Qutan Zhuan, who felt that he should have been consulted. He accused its author of plagiarizing Indian work, and adding his own mistakes. A competition was therefore held between three different calendars, one Indian (the *Navagrāha*, known in Chinese as *Jiuzhi li*), one the old Chinese version, and one the new. In this competition the *Dayan li* proved to be much the best. The Indian *Jiuzhi li* thus came to exert little influence on official Chinese practice, but it was taken to Korea, and there in fact it was adopted for a long period. It also influenced Korean mathematics to some extent, since it contained trigonometrical tables and explanations of their use.

Not only astronomy but many other technical and scientific professions were instituted in Japan after the immigration of peoples from Korea and China in the sixth, seventh and eighth centuries. Until the first influx of European science (1543), Japanese astronomy was based almost wholly on that of the Chinese and Koreans. In 607, during the Sui dynasty in China (581 to 618), the Japanese emperor sent an embassy to China. Over the next century Korean masters were invited to teach their art in Japan, and as the Chinese view of Nature began to take hold there, new institutional patterns were created on the Chinese model. Following the style of the Astronomical Bureau in China, a 'Board of Yin-yang' was given tasks in astrology, *yin-yang* divination and calendar-making – especially for ordering court ceremonies. In terms of analytical skill, we should now regard the third class of tasks as the most demanding, and yet in China astrology and alchemy were always rated higher in the scale of human wisdom. Even in the Tokugawa period (1600–1867), when a Confucian set of values persuaded the Shogunate (the military government) to value more the mathematical aspects of astronomy, the hereditary family of *yin-yang* diviners were ranked above them.

Long before the military class gained power, the Japanese government was monopolized by a hereditary court aristocracy that eventually destroyed the bureaucratic pattern that had been borrowed from China. A hereditary system of responsibility for astrology and calendar-making – it was largely controlled by two families, the Abe and the Kamo families – did nothing to encourage scientific standards, and even the sober mathematical science of the calendar became increasingly arcane. At least this saved the calendar from repeated revision. A much greater interest was shown in auspicious and inauspicious days than in the calendar's astronomical quality. The Japanese did not make or use many instruments comparable with those of the Chinese astronomers and there were even heavy legal penalties for the private use of time-keeping instruments. As official standards declined, however, an element of competition crept into calendar-making. Unauthorized agricultural calendars were produced in large numbers, and the Buddhist *sukuyō dō* school challenged and competed with court calculators in predicting eclipses. Knowledge in fact remained almost entirely dependent on antiquated Chinese sources until both cultures were stirred into a new kind of activity by the impact of a European tradition in the sixteenth century.

THE JESUIT MISSIONS TO JAPAN

The Jesuits, members of the Society of Jesus, were an order founded by Ignatius Loyola. Converted during convalescence following a wound received in battle, he gathered together a number of companions, and created a religious order that received the Pope's approval in 1540. In style, the order took much of its stern discipline from military models, but many of its members soon began to regard themselves also as an intellectual elite. Almost at once they began to carry the Christian message to all quarters of the known world, first notably with St Francis Xavier's journeys to India and Japan (1541–52).

The next great mission, that concerns astronomy, was that – partly scientific and partly apostolic – of Matteo Ricci and Michael Ruggerius (who, as we shall see later, began in China in 1583) and Ferdinand Verbiest (who followed them there). In both eastern countries the effect of their work was great, even though between 1600 and 1640 every missionary in Japan was put to death or deported. After 1638 the only foreigners who were allowed to remain in Japan were the Chinese and the Dutch, and they were obliged to remain in Nagasaki for purposes of trade. Their Japanese official interpreters continued to read western works, however, and so western learning influenced their culture indirectly. In China the Society of Jesus survived until it was suppressed in 1773. Despite prohibitions, western books reached Japan to some extent through translations done in China, especially after a relaxation of the strict ban in the time of the eighth shogun Yoshimune, in 1720.

Francis Xavier landed in Japan in 1549, and despite initial difficulties of language his message was eventually found acceptable to many Japanese. He found them eager to learn of cosmic phenomena, planetary motions and eclipse calculations, for example, since they recognized the superiority of western methods over those that had come from China. Astronomy became a means of converting the elite classes to their faith. Having converted the elite, the lower orders of society followed suit wholesale.

As early as 1552 Francis Xavier was teaching the sphericity of the Earth and other Aristotelian ideas in Japan. An excellent insight into Japanese attitudes to all this can be had from a point-by-point commentary on a western work published by Mukai Genshō (1609–77) around 1650. He contrasts the views of 'those who write vertically and eat with chopsticks' and 'those who write horizontally and eat with their bare hands'. He thought westerners ingenious in matters of techniques dealing with appearances and utility, but poor on metaphysics, especially in understanding heaven and hell. Indian ideas he thought to have only spiritual meaning, and to be fantastic and incomprehensible. As for Chinese and Japanese traditions, he remained a loyal neo-Confucian in his admiration for them. Against his better judgement it might have been, but he and other similar commentators showed that they had learned much from western purveyors of appearance and utility.

The first official astronomer to the Japanese shogun was a competent writer on mathematical astronomy, Shibukawa Harumi, who was responsible for the first important native calendar reform in Japan. He used chiefly a Chinese calendar (the *Shou-shih*) of 1282, but referred to two others, one due to Chinese Jesuits (the *Shih-hsien* of 1644). He used no new observations, but at least was competent enough to adapt the calendar to a Japanese longitude. After much controversy his *Jōkyō* calendar was accepted (1684), and at last Japan had something it could call its own. Although traditional in many respects, it incorporated some ingenious new mathematical techniques, and had it been presented at a session of the Royal Society in London, some aspects of it would not have been deemed uninteresting – its interpolation techniques, for example.

A man who carried much greater responsibility for turning Japanese astronomy towards European models was Asada Gōryū (1734–99; this was his later pen name – his real name was Ayube Yasuaki). A member of a family of Confucian civil administrators under the Kizuki fief government, he had access to Chinese and Jesuit–Chinese works, and earned something of a reputation when his calculation of the solar eclipse of 1763 was much closer than the official predictions. Employed as physician to a feudal lord who refused him leave to pursue astronomy as he wished, he fled to Osaka and supported himself by practising medicine there among the wealthy merchant class. He also taught astronomy, and with the help of new instruments, many of them – including telescopes – made by Asada himself, his school began to collect new data of an accuracy unprecedented in Japanese science. When he published a theory of planetary motion, based largely on that of the long-outdated system of Tycho Brahe, it was with sound new parameters that he and his pupils had evaluated.

Many of these pupils were of the Samurai class, and Asada was offered preferment by other lords and the shogunate itself, but his shame at his earlier desertion led him to refuse the offers. In later life he assisted the movement, that was by now well established, of translating Dutch scientific works into Japanese. He helped produce a synthesis of astronomy that was curious because it was such a chronological jumble, with elements taken from Newton, Kepler, Copernicus and Ptolemy without regard to the sequence of their discovery. He was at heart an algebraist, and never fully appreciated the advantages of geometrical models. He did not master the Newtonian theory of gravitation. Analysing existing European data, however, he developed a number of useful techniques, and managed to devise quite a passable formula for the length of the tropical year.

The last great traditional astronomer in Japan who shared Asada's ambitions to turn the direction of his subject was Shibukawa Kagesuke (1787–1856). Jesuit influence hardly mattered any more, but foreign relations were not without their sombre side to a man whose brother had been executed for helping a German traveller to smuggle forbidden materials out of the country. Shibukawa, working in the Astronomical Bureau, was privileged to read all the foreign works he could obtain. His struggle to rectify the inferior data that were circulating in his time, much of it forged by astronomers who had no feeling for the need to represent reality, makes almost tragic reading. In the end he carried the day on technical matters. He saw that his compatriots would soon have to make themselves familiar with the doctrines of Copernicus and Newton, and there too he prepared the ground, although with less sympathy for the end result. 'Let us melt down the mathematical principles of the west and recast them in the mould of our own tradition', he wrote, using a Chinese adage.

THE JESUIT MISSIONS TO CHINA

Matteo Ricci (1552–1610) was the son of a pharmacist in Macerata in Italy. He joined the Jesuit order and studied astronomy among other subjects at the Collegio Romano in Rome. He was much influenced by Christoph Clavius (1537–1612) who taught there, a friend of Galileo and one of the most distinguished of European astronomers, although it has to be said that Clavius did not accept the Copernican idea of a sun-centred Universe. Ricci left Rome in 1577, and sailed from Lisbon to Goa and thence to Macao, which he reached in 1582. In 1583 he arrived in China and settled at Ch'ao-ching in Kwantung province. Having established several missions throughout the empire, he finally settled in 1601 in Peking under the protection of the emperor Wan-li, and there he remained until his death.

Having won their way into official Chinese circles by their expertise in calendar computation, Ricci and his companions published a well chosen selection of European material that generally put rival Chinese material in the shade. Ricci's writings in Chinese were largely on theology and ethics, but he also translated or abbreviated Clavius' writings on the astrolabe, the calendar, spherical trigonometry and mathematical subjects, including the first six books of Euclid. He was assisted here by a pupil Hsu Kuang-ch'i. He published a large map of the world (179 cm by 69 cm) in various editions, giving the Chinese their first awareness of the distribution of lands and seas across most of the globe. Among other useful astronomical techniques, the Jesuits introduced those of the new European algebra, and later logarithms and the logarithmic slide-rule for help in calculating.

Needless to say, the impact of all this was considerable, and changed drastically the direction of a Chinese science that during the Ming and early Chhing dynasties had become virtually static. In their letters home, however, the Jesuits presented a rosier picture of Chinese science. They gave Europeans a sense of the great value of the astronomical culture they had discovered. Among other things, it was a rich mine of potentially valuable astronomical data from the distant past, and as such it attracted – as it continues to attract – attention that is only superficially historical.

The first telescope to reach China was brought by Johann Schreck (Father Terrentius) in 1618, and was given to the emperor in 1634. Schreck was a talented friend of Galileo, who corresponded with him and with Kepler. It was during Schreck's time that the Jesuits began to make a monumental compilation of the scientific knowledge of the day, traditional and western, a project of both scientific and historical value that was augmented well into the eighteenth century.

The last of the great Jesuit scientific missionaries to China was Ferdinand Verbiest (1623–88), a Flemish Jesuit, who served at the court of the emperor Kangxi (K'ang Hsi), of the Qing dynasty (Manchu). Verbiest's published correspondence is an impressive collection in Latin, Portuguese, Spanish, Russian, French and Dutch, and the most important of his surviving petitions to the emperor run to more than a thousand pages of Chinese text. The great strength of the Jesuit order was its learning: he had in fact been singled out for missionary service by his predecessor in China, Adam Schall, who had taught for eight years in the Jesuit College in Ghent and appreciated his skill in mathematics and astronomy. Verbiest reached China in 1669. Schall, who had died in 1666, had been working with twelve Chinese assistants on a vast translation project since 1631. With Terrentius (Johann Schreck) and Jacobus Rho, Schall had translated substantial sections of a hundred and fifty western books, and Verbiest continued to write for the Chinese in the same spirit.

Jesuit planetary astronomy in China was at root that of Tycho Brahe, Earth-centred but sharing many of the advantages of the earlier Copernican system. The astronomical instruments Verbiest introduced into his writings were also based on Tycho's — sound, but certainly not at the cutting edge of astronomical research by the 1670s. The calendar is of much interest, since as always it had a political dimension. There were rival factions at the Chinese court, one of them defending Muslim calendar techniques.

Verbiest wrote a highly influential Latin work, *Astronomia Europaea* (European Astronomy), written before 1680 but published in Dillingen only in 1687. Despite its title, it reproduced a series of observations made in China in 1668 and 1689 and already published in Chinese. One of its obvious aims was to demonstrate the superiority of European learning and technology over its Chinese equivalents. Instrument technology, calendar theory, optics, gnomonics, pneumatics, music, horology and meteorology, for instance, there rub shoulders with fundamental astronomy. Running through it there is an amusing undercurrent of boastfulness about the fact that they, the Europeans in China, were casting fine cannon, building clocks, automatic organs, telescopes, sketching in perspective, and so forth, all in the best European tradition. Like most Jesuit writings, it whetted the appetite of a number of western scholars to learn more about a system of astronomy that was respectably ancient. It had some merits that the Jesuits did not perceive, however, bound as they were by the Catholic Church's rejection of a Sun-centred and potentially infinite world. It is ironical that other scholars in Europe were propagating the idea of infinite space at the very time when the Jesuits were turning the Chinese away from the Hsüan Yeh doctrine of heavenly bodies floating in endless space. In fact so different was it, in overall character, that much of Chinese cosmology remained little appreciated until long after the exodus of the Jesuits from China.

Pre-Columbian America

MAYA, AZTEC AND SOUTH AMERICAN ASTRONOMY

In a region bounded by Arizona and New Mexico to the north and Honduras and El Salvador to the south, Central America saw a number of advanced city-cultures rise and fall in the two millennia ending around the time of Columbus' discovery of America. The four most notable are the Olmec, Zatopec, Aztec and Maya, of which at least the Maya developed the ability to analyse astronomical events using mathematical techniques. Somewhere along the way, all of them seem to have shared a theory of a layered Universe, each layer containing only one sort of celestial body. First above the Earth was the Moon's layer and then followed one for the clouds, another for the stars, then those for the Sun, Venus, the comets, and so on until the thirteenth, where the creator-god resided. If nothing else, this scheme gives an answer to those who suppose the Greek system of the spherical Universe to be self-evident or inevitable.

Columbus first met inhabitants of Yucatán in a large canoe on the high seas, and visited their homeland briefly. Others followed, and seizures of gold and reports of stone-built cities soon led to exploration and conquest by Cortes and others. It is one of the great tragedies of the entire episode that — according to Diego de Landa, first bishop of Yucatán — 'a great number' of the books of the Maya of Yucatán were burned on account of the superstitions they were supposed to contain. Before expostulating, however, it is as well to remember that de Landa had been appalled by witnessing the sacrifice of children, some in his own churches by supposedly converted Maya.

De Landa left a substantially correct account of the Maya calendar, and Maya dates can be stated with high accuracy. Apart from books that were destroyed, others have disappeared through neglect. It seems that only five Maya manuscript books or fragments now survive, one of them a congealed block of pages, but they include lunar and solar calendars and a Venus calendar of great interest. The best of them is the *Dresden Codex*, named after its present location, which also has many drawings of the gods of the Maya.

Each of the Maya books is made of a single sheet of barkcloth paper up to 6.7 m long and 20 to 22-cm high, pleated into folds making pages about half as wide as they are high. Of the Maya books, the Dresden example dates from the thirteenth, or possibly fourteenth, century. It was probably sent to the emperor Charles V by Cortes in 1519, that is, soon after his discovery and conquest of Yucatán. The books have glyphs and pictures on both sides (see plate 7) that give evidence of a complex calendar and use of astronomy in religious ritual and divination. There are almanacs of various sorts, including farmers' almanacs, and multiplication tables to help in their use. In some the luck of each day in cycles of 260 and 364 days is marked. Another cycle of fortune was the *katun*, a period of twenty 'approximate' or 'vague' years of 365 days. (The Maya counted in twenties.) There are accurate tables giving the synodic revolution of the planet Venus, again accompanied by glyphs showing the fate of humankind according to the day of the planet's heliacal rising. The Maya had studied Venus to such effect that they knew the importance of a period of 2920 days, eight vague years, after which Venus begins to repeat its movements in relation to the Sun. (Heliacal risings, and so forth, thereafter repeat themselves in a Venus calendar, as we saw in connection with Babylonian astronomy.)

Gods are frequently mentioned, including – in rough order of frequency – gods of rain, Moon, death, creation, maize and the Sun. Their almanacs of given sets of days variously arranged (4×65, 5×52, 10×26, and so on) are for different purposes, such as net-making, fire drilling, maize planting, marriage and child-bearing. Some of the computations deal in millions. There is material on New Year ceremonial, and on the weather. It seems to be by mere chance that the surviving books deal with the subjects that most concern us here, for other materials from the Mexican region cover a much wider range of subjects.

Many of the astronomical practices and beliefs of the Maya are a question of inference from archaeological remains. Those of the Aztecs of central Mexico are better known from their literature, in particular from a work known as the *Codex Mendoza* that was written at the time of the conquest. The Aztecs – and in this their king, the famous Montezuma II, still participated after the Spanish conquest – offered incense to certain stars at appropriate times of the night. Montezuma was supposedly born on the same day of the calendar ('Nine Wind') as the god Quetzalcoatl – the Morning Star, our Venus, but seen as a male deity. There was always some form of Sun worship in all cultures, and examples are too numerous to mention. A previous Aztec king, in the middle of the fifteenth century, had set up a large porphyry pillar half-way up the great temple staircase in the town of Tlaltelolco (now part of Mexico City) covered with solar symbols, which some would have us believe were used in eclipse calculation. Certainly the Aztecs are known to have sacrificed hunchbacks at solar eclipses.

The worship of Venus was no less universal and important than that of the Sun, and gave rise to laudable astronomical techniques of prediction. The planet was regularly and closely observed. There used to be a picture – now destroyed by the tourists' habit of throwing bottles at it – near a sacred underworld lake at Chich'en Itzá, a lake into which sacrificial victims were thrown. This showed a square Sun rising over the horizon, and it carried a date equivalent to 15 December 1145. Modern calculations show that a rare transit of Venus across the face of the Sun was indeed to be seen on that day.

The early Spanish historian Juan de Torquemada, who was writing a century after the conquest, was able to discuss with native people the astronomical ceremonies that were still to some extent taking place. It is said that most divination was then done by other means than observing the heavens, although Atahuallpa's general said that the coming of the Spaniards had been foreseen astrologically. The point here is not that it was true, but that it was not out of character with the practices of the time. Among a number of astronomical omens is one quoted from Atahuallpa himself, the last native ruler of Peru, who attributed the death of a man to a comet in the sword of our constellation Perseus.

Torquemada reported having seen rods placed in holes on the roof of the palace of the Aztec king of Texcoco, the rods having on them balls of cotton or silk to help in measuring celestial motions. In some pictographs it seems that a priest used crossed rods as a sighting device. But to what end? Torquemada's informant said that it was to aid the king, who with his astrologers viewed the heavens and the stars. While one of the chief concerns of these peoples must have been the regulation of the agricultural calendar, no doubt there were also rituals that had taken on a life of their own, their origins having been forgotten, and yet other rituals that were derived by analogy. Venus does not provide an immediate guide to the seasons, although by relating its motion to the Sun's it can be made to do so. The historian Fr Bernardino de Sahagún tells us how the Aztecs sacrificed prisoners to Venus when it made its first appearance in the east, splashing blood towards the star. There is no doubt that Venus was of extraordinary importance to the peoples of Central America, and the surviving

Maya books give us an insight into what was achieved on this score.

Probably more often than not, what we are inclined to identify as 'astronomy' was an art of an informal and descriptive kind, even when entering into the rituals of which most must have been aware. There is no doubt that at this level it was a part of the general consciousness. A good example is in the *Popol Vuh* (Book of the People), a Maya story of what happened to the hero, who was the Morning Star, in the underworld. Captured and decapitated, his skull spat at the daughter of the Lord of Death, who conceived a child that became once again the Morning Star. The *Popol Vuh* is one of several sources from which we know the cosmic nature of a ball game that was played right across the continent. The ball, of rubber, represented the Sun, and victory went to the team that first put the ball through a stone ring, about six metres above the ground. Several courts still survive. Board games in use often had a cosmic meaning. The Mexican game of Patolli, for instance, moves a stone representing a heavenly body through four divisions of the board that represent divisions of the sky.

The stone rings and walls down the side of the courts used for ball games were in some cases clearly aligned on astronomical events, but there are numerous other stone monuments throughout South, Central and North America, with pillars, entrances and windows showing clear alignments on sunrise and sunset at the solstices. Some have hieroglyphic marks that have left ample scope for the imagination of their interpreters. One stone found at Chapultepec as long ago as 1775 had under it three crossed arrows pointed accurately to sunrise at equinox and solstices. Alignments on the cardinal points north, south, east and west are commonplace, especially on the pyramids of Mexico and Central America. There is little written evidence about the astronomical aspects of religion in Peru, but there are monuments there with alignments on the Sun's position at the solstices. In the southern region there are linear constructions on the hills behind the site of Nasca with straight rows of whitish stones that might be astronomically aligned. Across the lines there are very extensive outlines of birds that cannot be seen from the ground, but resemble birds on Nasca textiles of a little before or after the beginning of the Christian era.

In Europe, the alignments of the monuments of prehistory are the only testimony to an astronomical character. In Central America there has been ample living testimony. An early Spanish writer, Fr Toribio Motolinía, telling of the Aztecs, reports a festival that took place in the Aztec capital at the equinox. The chief religious building, a double pyramid now called the Templo Mayor, was 'slightly out of true', so Montezuma wished to pull it down and rebuild it correctly. The pyramid is surmounted by twin temples between which sunrise at the equinoxes would have been observed from an observation tower on the Temple of Quetzalcoatl to the west of it. Many urban centres have an appearance of equally careful planning. The most famous is the temple complex at Teotihuacán, which arose in a relatively short space of time ending around 50 BC. This, the largest and most influential of all Central American cities of the pre-Columbian period, includes a vast Pyramid of the Sun and a lesser Pyramid of the Moon. The city was certainly laid out with great accuracy on an axis a little over 15° east of north (and the same south of east, etc.). This angle was in all probability connected with a sighting of the rising of the Pleiades. The dating of the foundation might be the second century AD, which fits the arrangement well, and there are other buildings in South and Central America that seem to follow a similar convention.

As for the stars, throughout most of the American continent we find attention being given to the Pleiades. As so often in Babylonian and Greek astronomy, these were linked with Venus. They were also linked with harvest and rain, from Peru to the Eskimo, so that in Aztec astronomy they had a name the equivalent of marketplace, elsewhere of maize, doves, or a granary. Seven-fold items were obviously favoured. To the Algonkian the Pleiades were the seven heated stones from a ritual bath. Legends concerning them are legion: in many they are boy or girl dancers. To the Maya as well as the Micmac they were the rattle of a rattlesnake, and the last makes understandable the association with the Way of the Dead at Teotihuacán, for from some of the mounds rattle figures have been excavated.

The Great Bear too has its large collection of stories, but they, and the mystery of the many elements shared with legends from the Euro-Asian land mass, are problems more of anthropology than history.

PRE-COLUMBIAN NORTH AMERICA

The culmination of a Central and South American skill in binding together tribes into kingdoms and empires was reached with the Inca culture of Peru. Gold had been known from the second century BC, and became a symbol of gods and kings and of the Sun, whom the Incas conceived to be their ancestor. Spanish greed for the gold helped the downfall of this remarkable civilization, although it had begun from within, when Atahuallpa rose in revolt. In North America there was no comparable civilization, although many of the simple astronomical rituals of observing rising and settings were shared. There was no native written language, and occasional records discovered by archaeologists engraved on rock are too rudimentary to be interpreted with certainty. The lunar crescent was a favourite symbol, sometimes together with a star, especially in the southern part of North America. According to one ambitious interpretation it symbolizes the supernova of AD 1054, supposedly first seen when the Moon was nearby.

The best known of the structures left by the pre-Columbian peoples are the mounds of Earth, some of them no more than a couple of metres across, others many hectares in extent. They are chiefly due to three different cultures. The Adena culture, from about a millennium BC, made some mounds in the forms of animals. An example is the Great Serpent Mound in Adams County, Ohio. The Hopewell culture that followed preferred geometrical shapes. The Mississippian culture, from about AD 1000, built large mounds with platforms, often pyramidal in shape, on which buildings of some sort stood. These were still in use when sixteenth-century Europeans first explored the Mississippi valley. There is some reason to think that alignments not only on the solstices but on the Moon's extremes of rising and setting are indicated by these monuments.

Stars too were perhaps involved in ritual sightings. The Big Horn Medicine Wheel, so called, in Wyoming, and the Moose Mountain Medicine Wheel in Saskatchewan, Canada, have similar arrangements of cairns round them. They seem to align on summer solstice sunrise and on three bright stars visible at dawn in summer. The uncertainties here are very much greater than in the vaguely similar prehistoric monuments of Europe, and the grand monuments of Central America, for the North American structures were generally both cruder and flimsier than either. There is also the problem that in all the literature reporting conversations with Indians – who were still using some of these as centres of ritual activity – there is almost nothing to suggest that their astronomical associations were known at the time.

Whatever the truth of the matter, only by an act of intellectual charity can we at present say that the pre-Columbian inhabitants of North America ever developed an astronomy with more than a token theoretical element. In Central America, on the other hand, we find peoples who had discovered, surely independently of their contemporaries on the other side of the world, that – to quote Galileo – 'the book of Nature is written in mathematics'. Their use of mathematics makes their astronomy doubly intriguing.

Indian and Persian Astronomy

VEDIC ASTRONOMY

The peoples of the Indian subcontinent, like all early peoples, allied their accounts of divine and supernatural powers with accounts of what they observed in the heavens. The Vedic religion, the source of the modern Hindu religion, is of great historical interest because it is among the earliest of all religions to have been recorded in literary form – in this case in Sanskrit – and it shows us this interaction between the cosmic and the divine at work.

The oldest of the Vedic writings, the Rigveda, gives more than one account of the creation of the world. The main version is that the world was made by the gods, as a building of wood, with heaven and Earth somehow supported by posts. Later it is suggested that the world was created from the body of a primeval giant. This last idea gave rise to the principle, found in later Vedic literature, that the world is inhabited by a world-soul. Various other cosmogonies followed, with the creation of the ocean sometimes being given precedence, and place being made for the creation of Sun and Moon. There is a certain circularity in it all, however, since heaven and Earth are generally regarded as the parents of the gods in general; and water was sometimes introduced into the parentage.

There are numerous myths of astral gods, for instance of the Sun, husband of dawn, and drawn in a chariot by seven horses, and there are simple rules for the times of sacrifice, but the Vedic literature gives no clear indication that mathematical techniques for describing the motions of heavenly bodies were discussed in India before the fifth century BC. There is much earlier evidence for contact with Mesopotamia, however, in the neo-Assyrian period, for instance in the matter of omens. Some statements in the Vedic texts can be traced back to statements in mul-Apin. Contact from this direction eventually proved decisive in forming the character of Indian astronomy.

The Vedic texts make much use of periods of time of various lengths, the so-called *yugas* of two, three, four, five and six years; periods of twelve months of thirty days; periods of half-months with fourteen or fifteen days. No clear evidence of any sophisticated calendar schemes can be found here, but one highly developed aspect of lunar observation during the last few centuries before the Christian era is the scheme of *naksatras*. These are twenty-seven stars (sometimes twenty-eight stars, possibly groups of stars) that mark the passage of the Moon through the sky in a month. Each is associated with a different deity. The system had a long and involved history, and in the middle ages went as far afield as Europe, where it entered into astrology and geomancy. We earlier touched upon a similar doctrine from China.

INFLUENCES FROM MESOPOTAMIA AND GREECE

Mesopotamian astronomy reached India in the late fifth century, with the conquest of north-west India by the Achaemenids. (This was the dynasty that held power in Persia between 558 and 330, and various allusions to it have already been made.) Contact is evidenced by the writer Lagadha's use of Mesopotamian, Greek, Egyptian and Iranian calendar techniques, that is, the use of such 'period-relations' as in the equation of five years and 1860 tithis (a Mesopotamian unit discussed earlier), or of twenty-five years and 310 synodic months. (The Egyptians used a twenty-five-year cycle of 309 months.) Lagadha also took over Babylonian doctrine on the length of daylight, using not only their arithmetical technique of the 'zig-zag function' but the water clock by which time at night could be measured.

Another Mesopotamian instrument taken over by the Indians was the gnomon, the vertical pillar that, by its shadow, could indicate the time. They usually divided the gnomon into twelve units, and this convention was curiously persistent, surviving even in western astronomy through the use of tables that had their origin in the east. Tables correlating time with shadow length are of course dependent on geographical latitude, a fact not always appreciated either in India or elsewhere.

Although there was no doubt some small flow of astronomy into India in the centuries that followed, the next significant stage was in the Seleucid period, when Babylonian methods, modified now by the Greeks, found their way eastwards. In AD 149/150 a long Greek astrological treatise was translated into Sanskrit prose, and part of this was given over to mathematical astronomy. In 269/270 this was in turn versified by Sphujidhvaja under the title *Yavanajātaka*. The length of the solar (tropical) year in it turns out to be the same as that accepted by Hipparchus and Ptolemy.

Sphujidhvaja, whose motivation was plainly astrological, uses 'linear zig-zag' techniques for solar and lunar positions and a Babylonian 'system A' method for the rising times of the signs of the zodiac, known also from Greek texts. The twelve-part gnomon is now introduced into the rules, but here and for the procedures given for finding planetary positions Sphujidhvaja gives what are clearly Greek versions of Babylonian methods.

There are other, parallel, instances of the same tendency, where a Greek intermediary text may be inferred from the style of the procedure being advocated. Periodic times for the motions of the planets are repeatedly found, identical to others found in much earlier cuneiform texts. That the

accompanying procedures are often garbled shows conclusively that the parameters were not found independently.

With the 'Roman' *Siddhānta* (*Romakasiddhānta*) of the
third or fourth century, the doctrine of precession is in evidence, and again the same length for the tropical year is
found as in Hipparchus. This same work has material on
solar eclipse calculation that uses Greek geometrical
models, although now there are many signs of slight
adaptations of them and their parameters. The same is true
of another more or less contemporaneous text of Greek
origin, the *Paulisasiddhānta*, but this uses a length for the
solar year that is said by a later Arab writer (al-Battāni) to
be due to 'the Egyptians and the Babylonians'. The text
is known through a summary included in a work entitled
Pañcassiddhāntikā, into which additional matter has been
introduced by its famous sixth-century author, the astrologer Varāhamihira.

These works are invaluable for the insight they give us
into Greek astronomy in the period before Ptolemy, a
period from which few texts survive. The Indian writers
make valiant attempts at problems in spherical trigonometry, using Greek methods, inclusive of methods involving projection on to a plane surface (compare the astrolabe).
They introduce the sine function, in place of the Greek
chord function, and in a sense we may regard their gnomon
shadow tables as tables of the tangent function. They show
a strong interest in problems relating to solar and lunar
eclipse prediction, and in some cases here there are traces
of Babylonian omen literature. Their working is often blemished in ways that are to be expected, bearing in mind the
crossing of so many cultural barriers, and yet it is clear that
Indian astronomy, for all its faults, was by this time far from
being a purely passive science.

It was actively pursued, and persistently so, over the centuries. In 1825 Lt Col. John Warren published a long and
remarkable study of calendars and astronomy in southern

India. He told of a calendar-maker in Pondicherry who showed him how to compute an eclipse using shells on the ground as his counters, and from tables memorized with the help of artificial words and syllables. His Tamil inform-ant, *who knew nothing of Hindu astronomical theories*, was in this way able to compute a lunar eclipse for 1825 with an accuracy of +4 minutes for its onset, −23 minutes for its middle, and −52 minutes for its end. The tradition in which this man was working went back to the *Pañcassiddhāntikā*, and beyond that to Seleucid Babylonian astronomy, in other words, more than two millennia in all.

It is easy to forget how much learning was done by rote in the past: a versified table of the sines of the planetary equations (by Haridatta) is only an extreme example of a common phenomenon, a human equivalent of electronic means of recall. In Europe the much simpler task of learning the ecclesiastical calendar was achieved in similar ways.

Another small point perhaps worth putting on record is the occasional Indian use of 57;18 parts as the length of the radius of a standard circle, chosen because it ensures a circumference of 360 parts. The analogy with our measure-ment of angles in radians should be obvious. (Today we take the radius as unity, making the radian, the angle sub-tended at the centre by an arc of unit length, equal to 57;18°.) Where Ptolemy used a standard radius of 60 parts, Indian astronomers often preferred 150 parts.

THE PURĀNAS AND THE BRĀHMAPAKṢA

Although influenced by texts going back to Vedic times, and by Iranian sources too, the *Purānas* are writings with cosmological sections dating from the early centuries of the Christian era. The Earth is now represented as a flat circular disc, with a mountain, Meru, at its centre. The mountain is surrounded by alternating rings of sea and land, so that there are seven continents and seven seas. Wheels are con-ceived to carry the celestial bodies, these turning around the star at the north pole by Brahma, using cords made of wind. This cosmology was taken over by Jainism, a monastic religion which like Buddhism denies the authority of the Veda, but from the fifth and sixth centuries onwards it was undermined by the influx of a new form of Greek cosmology, with pre-Ptolemaic roots. In short, Aristotelianism reached India.

The sources of these influences from Greece are never likely to be pinned down, but they coincided with the conquests in northern and western India by the Gupta dynasty. Another factor might be persecution of the Nestorian Christians by the Emperor Zeno, in the fifth century, for this led to an eastward migration, including that of scholars with Greek and Syriac texts. Many settled in Gondeshāpūr. At all events, there was an adoption of simple planetary theory, with its period relations. The Indians translated these into their own system of time-reckoning, with yugas of considerable size now. (They are based on the Babylonian figure of 4320 000 years, called now the Mahāyuga, and certain multiples and submultiples of it, each with its own name. Thus a Kalpa is a thousand Mahāyugas, and a Kaliyuga is 432 000 years.)

This all fitted rather well with the Greek notion of a Great Year, the period after which the planets, having been once all in conjunction, will be so again. Not only is the Indian doctrine reminiscent of Pythagorean and Stoic teaching, but the yugas themselves are almost all divisible by the third power of sixty, that is, 216 000. Commonly found are two, four, six, and eight times this figure, as well as these multiplied by powers of ten. The Indian number system was from its beginnings purely decimal, so the Babylonian–Greek influence is plainly revealed even here. The yuga system might have been developed as early as the third century BC.

The Indians quoted the planetary periods in such a form as '5775 330 sidereal months is one Kaliyuga', and then provided a date at which the current Kaliyuga began. (It was a certain day in our year 3102 BC.) This is enough to give mean positions at later dates.

Indian astronomy did not fight shy of the problem of planetary inequalities, but the Indians made matters difficult by taking very seriously the Aristotelian cosmology they inherited, with its insistence on a purely concentric set of spheres. This, however, they blended with Greek epicyclic astronomy in an extremely interesting way. The basic astronomy was pre-Ptolemaic: there is no sign of Ptolemy's equant, for example. The Sun and Moon are each provided with a single epicycle on their deferent circles, which are like all the other deferents centred on the Earth. As for the planets, each has *two epicycles*, with a *single* centre, this centre moving round with the mean planetary velocity. (The epicycles were called the manda and shīghra epicycles.) Each epicycle is taken to have a point travelling round it, in one case at a speed similar to that in the epicycle of the corresponding Greek model, in the other case so that the radius joining the point to the epicycle centre lies in the direction of the head of Aries, the zero of longitude. These two points are simply auxiliary points in the model, however, and neither is identifiable with the moving planet as such. Where, then, is the planet supposed to be?

Of course it is safe to say that the planet was simply meant to be at the longitude determined by the procedural rules laid down in the text, but there are two quite different ways of regarding the situation. We can simply do what an Indian calculator did, apply the rules found in the texts, and produce a figure for the planetary longitude; or we can begin from the Greek epicyclic model and see how the Indian procedures arise from it, how the one model was turned into the other.

The first, 'unthinking' approach, is not without its interest. The complex of computational rules is applied to yield a first approximation to the longitude, and this is repeatedly re-evaluated until repetition produces no significant change in the result. This 'iterative' procedure was one of great sophistication, and this was a very early example of it, and a powerful one. It is already found in the *Paitāmahasiddhānta*, written in the fifth century. (This work has much other elaborate mathematical material of Greek type concerning trigonometry and plane projection.) In this procedure, the first approximation was made by halving each of two independent corrections (see the following paragraph), and for this reason the Arab followers of the Indians used the name 'the method of halving the equation'.

To start from the Greek model and justify these complicated procedures would take us too far afield, but a very cursory account of the situation may be given. Taking first the Greek scheme with a simple eccentric epicycle (figure 7.1, uppermost), this may be transformed into a scheme with two epicycles, not concentric with one another, the second epicycle as carrier for the main epicycle (lower left of the figure). (The transformation makes use of the theorem enunciated by Apollonius, touched on in chapter 4.) Each epicycle has a radius that subtends at the centre a certain angle (not drawn) that serves as a correction ('anomaly' or 'equation'), and that has to be added to or subtracted from the mean longitude of the planet. The two epicycles in question will not be concentric (lower right of the figure), but as far as the task of calculating those correction angles is concerned, they can be made concentric as long as further rules are introduced for correcting the corrections. Of course it is not possible in a short space to explain how this was done.

This is a far cry from a physical model of the planetary circles. Even though the Indians placed the Earth at the centres of the planetary deferents for purely physical ('philosophical') reasons, and perhaps made the epicycles concentric for the same reason, yet in the last analysis they had followed a bizarre path. We find that a Greek model had been accepted, then geometrically transformed, and then distorted yet again before its 'correctness' was restored by purely *computational* devices.

The *Paitāmahasiddhānta* is a text based on the Brāhmapakṣa, revelations supposedly from Brahma. There

FIG: 7.1   Indian planetary models

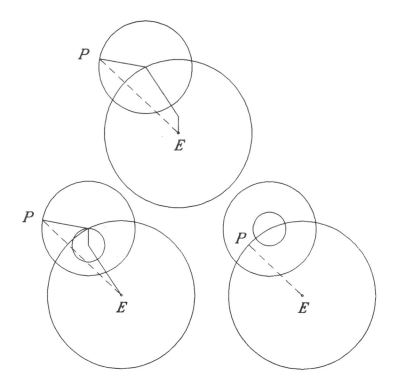

were many other texts in the same tradition, but one influential work in particular followed it closely. It was written (in southern Rājasthān, in Sanskrit still) by Brahmagupta in the year AD 628, and was of considerable historical importance. Known originally as the *Brāhmasphuṭas-iddhānta*, it was introduced to the Arabs in Baghdad in the year AD 771 (or 773) by a member of an Indian embassy there. The Arabs gave it – or a work intimately based on it late in the same century – the name *Zīj al-Sind-hind*, and under this name it had a new and influential career, in the Islamic, Byzantine and western Christian worlds.

Brahmagupta included various short-cuts in the computation of planetary positions. He included new and intricate
rules for planetary longitudes and eclipses. His work made
use of, among other things, a pulsation in the sizes of some
of the epicycles. (He was not the first to make use of the
idea.) Of course if the observer is not at the centre of the
epicycle, as in Ptolemy's models, the epicycle will seem to
fluctuate in size as it moves round the deferent. This way
of adding a correction to the planetary longitude is lost
when the observer is placed at the deferent centre, but
whether Brahmagupta had it in mind to compensate for
what amounts to the loss of a parameter in the Indian
models is a moot point.

Indian astronomers continued to write in this tradition
into the seventeenth century and beyond. In the eleventh
century, extensive changes were made to the planetary
parameters, and material also flowed back to India from
Islamic (and ultimately Ptolemaic) astronomy, but the techniques used were much as before.

ĀRYABHAṬA

We are told by Āryabhaṭa I – who was born in AD 476,
probably near modern Patna in Bihar – that astronomy was
based on a revelation from Brahma. Āryabhaṭa too used
the *Paitāmahasiddhānta* as his main source. He wrote various
astronomical works, one of which had many imitators in
the Arabic-speaking world from the beginning of the ninth
century onwards. (The Arabs called it the *Zīj al-Arjabhar*.)
In his home territory in southern India his methods are said
to be still in use, a millennium and a half after they were
composed. Āryabhaṭa tampered with his inheritance in
ways later unpopular with Brahmagupta. Taking smaller
yugas he effectively required the planets to return to their
original state of conjunction more frequently – after only
1080 000 years, in fact. He introduced pulsating planetary
epicycles, as Brahmagupta was later to do, and that he

knew the Greek models is quite obvious from the fact that he discusses in general terms the relation of the eccentric epicycle to the double epicycle. He established values for the planetary motions and the parameters of the models that were new, at least superficially.

The elder Bhāskara (*fl.* 629), a notable mathematician, was one of the many exponents of Āryabhaṭa's astronomy, and his writings were in turn the subject of much commentary. He wrote much on questions of lunar and planetary visibility, and he introduced a way of accounting for the planets' movements in latitude by supposing that both of the epicycles were tilted. A direct follower in the same tradition was Lāṭadeva (around 500), who helped revise, and contributed to, the *Sūryasiddhānta*. Such revisional activity to this influential work went on for over a thousand years, and to do justice to it one would have to consider not only the numerous contributions themselves, but the local rivalries – between places and even families – to which they bear witness. Astronomers frequently claimed to have corrected the work of their predecessors from observation, but were often doing no more than putting it into different arithmetical dress. For example, the astronomers Keshava (1496) and Gaṇesha (1520) divided time into eleven periods of 4016 days, with some supposed advantages, and the system swept from Gujarāt across northern India, but to no obvious astronomical advantage.

THE CHARACTER OF INDIAN ASTRONOMY

Indian computational activity has few parallels anywhere in the world between the third and ninth century, say, and even after this time was extraordinary in its sheer extent. It is marked not by theoretical originality on the large scale, but rather by a delight in modifying computational technique and supplying revised data. An eighteenth-century work in the genre tends to resemble, superficially, a work of the previous millennium. One is reminded of the Chinese situation, here. Of course in time, especially after the tenth century, new material worked its way back into India from the Arab and Persian astronomers to the west. Ptolemy's *Almagest*, for instance, was translated into Sanskrit in 1732. There was something of a renaissance of astronomy at Jaipur around the same time, and some vast monumental instruments were built there, but they were of a style that Europe had long passed by. Much antiquated material was still following these routes in the nineteenth century, simultaneously with much more advanced European astronomy. This is only surprising if we forget the vast size of India, and its great social complexity.

The reason for the relatively static situation, especially in the earlier centuries, was no doubt that Indian astronomers' motives were largely religious and astrological. They were directed, for instance, towards the preparation of calendars for settling the times of religious observances, and there was no strong drive to link astronomy with other systems of knowledge. The link with physics, for example, was rarely attempted, even long after European contacts with India. I have a copy of Newton's *Principia* that belonged to Reuben Burrow, a British mathematics teacher and surveyor in India in the late eighteenth century, and writer on Indian astronomical history – although he was more famed for his love of the bottle than for scholarship. (On the title page of the book he has scurrilous things to say about Queen Anne.) He was, even so, one of a new succession of astronomers responsible for a two-way traffic in astronomical ideas. Better remembered is Lancelot Wilkinson, who ensured that much European work was translated into Sanskrit. There are many others. Activity of this new sort finally bore fruit, and much work of distinction of a new and original kind was done – for instance in astrophysics at the Observatories of Kodaikanal and Madras. It was there in 1909 that John Evershed discovered the radial motions of material, parallel to the Sun's surface, in sunspots. Since that time, in Subrahmanyan Chandrasekhar, India has produced one of the world's leading authorities on the physics of stars.

There are many other examples, but they would only serve to emphasize that, although they and the traditional work may go under a single name, 'astronomy', we are here dealing with substantially different subjects, different not so much in content as in purpose.

PERSIAN INFLUENCES ON ISLAMIC ASTRONOMY

Indian astronomy has a unity that makes it difficult to ignore. It is tempting, in view of the remarkable Islamic contribution to astronomy that was to come, to forget the astronomical traditions of those other peoples nearer at hand over whom the Arabs gained political domination. At the outset, nevertheless, Arab astronomy was relatively primitive, and most of the civilized peoples in question, conquered between the seventh and ninth centuries – Indians, Persians, Syrians, Copts and Greeks – had reached higher levels of sophistication, whether in mathematical planetary theory, in claims to evaluate the ways in which the stars influence human events, or in calendar systems with astronomical content.

The first phase in the intellectual history of Islam was that of the amassing of superior knowledge, and no short account can do justice to the many astronomical tributaries to that process. Persia (Iran) had perhaps less to offer than India at this time, but it was an intermediary, and Persian culture was ancient and rich. Calendars, for example, are superficially among the dullest of texts, but they do occasionally serve to reveal astronomical knowledge that might not otherwise have been suspected.

This was the case with two Persian calendars that were started from the year 503 BC, the year 19 of the reign of Darius I. It emerges that one of them had the tropical year as its base, the other the sidereal year. An older Persian calendar had been lunae-solar, and did not follow Babylonian rules for intercalation. Of the new calendars, however, the religious one used a year of length so close to the Babylonian sidereal year (system B) that borrowing seems certain. The existence of the two different years in the new calendars shows that the ground had already been prepared for Hipparchus' discovery of precession long before that explicit discovery had taken place – not a trivial historical discovery.

Several pre-Islamic writings in the Persian (Pahlavī) language later became influential in Islamic astronomy. There were translations into that language of the astrologers Teucer of Babylon (probably first-century BC) and Vettius Valens (second century AD). These Persian books were important as the bearers of astrology to the Arabs. In one of the classic Arab texts, by Abū Maʿshar, the writer presents astrology as 'the teaching of the Persians'. There were other translations of 'Indian books' and 'the Roman *megiste*'. All these were available at least as early as AD 250. Around AD 550, the most influential Persian work of all was revised, the *Zik-i Shatro-ayar*. Known after its translation into Arabic (around 790) as the *Zij al-Shāh*, the work was heavily dependent on Indian sources. It was still in use in Spain in the eleventh century for its table of star positions. The work is lost, but much is known about it from others, especially al-Bīrūnī, author of a full-length study of Indian customs and learning. Its zero meridian was taken to be not, as in Indian tables, at Ujjain, but at *Babylon*. This was probably chosen because the capital of the Sassanids – the kings of the new Persian empire between 224 and 636 – was at Ktesiphon nearby.

When Baghdad was founded by al-Manṣūr in 762, a horoscope was cast for the event, the propitious moment having been chosen by two astrologers. One was the converted Jew Māshāʾallāh, the other a Persian whose family did much translation into Arabic, Naubakht. During the following decades, and indeed centuries, translation into Arabic was pursued feverishly, without much attention to the quality of the originals, so that astrology in particular, where inconsistencies are less obvious than in mathematical astronomy, ran wild. Side by side with the written authorities, the oral traditions of the conquered peoples found their way into astrological writings. On the surface, Christianity had been very effective in suppressing astrology for a time. It would be easy to suppose that this was so in Syria, for instance, if one did not take into account the rampant pagan astrology at Ḥarrān. Among the Jews in the region, the story is the same. While frowned upon by the orthodox, astrology was not only practised by many of them in the early years of Islam, but it provided Islam with some of its principal astrological writers – Māshā'allāh, Sahl ibn Bishr, Sanad ibn ʿAlī and Rabban al-Ṭabarī. The Arabs had begun a political conquest, but intellectual domination was largely of them, by others.

Eastern Islam

THE RISE OF ISLAM

The Islamic faith stems from the teaching of Muhammad. Persecuted in his native town of Mecca, his fortunes changed dramatically after he fled to Medina with two hundred followers in 622. Through preaching and warfare, the new faith spread rapidly throughout the Arabian peninsula. Within a decade of his death in 632, Armenia, Mesopotamia, much of Persia to the east, and Egypt to the west were in the hands of his followers. Within a century of his death the conquest of the north African coast was complete, much of Spain was taken, and attacks were being made on the Mediterranean coast. A Muslim advance on what is now France was stemmed by the victory of Charles Martel near Poitiers (732). To the east by this time all of Persia had been taken, as had much of Kashmir and the Punjab, and conquests beyond those limits continued. The faith of Islam spread to northern India, with the establishment of the Mogul empire. Not since the days of Rome had anything comparable been seen.

At first Islam might have seemed irresistible, but in east and west, ground was lost by degrees. There had been a fundamental split in Islam thirty years after Muhammad's death, between the Sunni and Shi'i, followers of the third and fourth of his successors (caliphs). The Syrian-based caliphate, the *Umayyads*, ruled an empire from the Atlantic to China. Overthrown by the *'Ābbāsids* in 750, the seat of government was moved to Baghdad. Gradually these Arabs were forced to share power with dynasties elsewhere, for instance in Persia and Turkey (the *Seljuks*). The first western places to gain independence of Baghdad were in Spain, where an Umayyad prince fleeing from the east set up an emirate in 756. The Umayyad caliphate of Córdoba lasted until 1031. The Muslims were finally expelled from Spain in 1492, but their contribution to European astronomical learning, especially in earlier centuries, had been very great indeed, as we shall see.

Between the tenth and thirteenth centuries a succession of invasions by heathen Mongols took away much of the conquered territory for a new empire that had its capital first in Mongolia, later in Peking. They went as far in their conquests as Anatolia, and were eventually converted to Islam, founding various Mongol-Turkish states. The various Crusades of the Christian Church directly affected a region small in area but not in spiritual importance. The Latin kingdoms of Jerusalem, however, were no more than a minor check on the influence of Islam on the entire area, which was completed by the fall of Constantinople in 1453, and the end of the old Byzantine empire.

Islam was, from its beginnings, a remarkable movement in spiritual imperialism, and during its first five or six centuries the arts and sciences of the ancient world were fostered and then developed under its protection to an extraordinary degree. The peoples who adopted Islam, or who had it thrust upon them, spoke a great variety of languages. Greek was of course the language of most of that inherited learning which concerns us, but many Greek texts had been translated into Syriac, for example, in the fifth and sixth centuries. The Arabs, like most other peoples, had their own native astronomical lore, but it was as yet relatively primitive. With the conquests of the first two centuries of Islam, astronomical techniques were made to serve the needs of religion – in determining more precisely the hours of prayer, the direction (that of Mecca) in which to pray, and the periods of fasting. In the new and most powerful administrative centres the new sciences were co-ordinated as never before. We have already seen examples of the way astronomy was brought from Alexandria and India, from Syria and Ḥarrān, from Persia and the fringes of Byzantium. Of course astronomical activity was only one aspect of a much wider intellectual movement. Astronomy was peculiar in that it was regarded as useful in medicine, and every ruler had his physician. At Baghdad a hospital was built with great pomp, the prototype of many others, modelled on one at Jundīshāpūr. The market for astronomy was growing apace.

Not only were technical astronomy and the mathematics by which it was supported cultivated, but also Greek cosmology, as a special aspect of Greek philosophy. A Syriac Neoplatonic philosophy was developed that spoke much of celestial influence, and that took over Harranian theological ideas. Jewish and Christian learning was imbibed just as avidly. Translations of ancient texts were at first often done at second and third hand, since so much had been taken from the Greeks at an early date by Persians, Indians, Jews and Syrians. Notable patrons of the sciences included the 'Abbāsid caliphs at Baghdad, in particular Abū Ja'far al-Manṣūr, Hārūn al-Rashīd, and 'Abdallāh al-Ma'mūn. Although the description 'Islamic' is often applied to the resulting learning, this obscures its diverse origins. Within a century or two, however, research centres came into being throughout the Islamic world that advanced astronomy and many other sciences in new and remarkable ways.

It was in the reign of al-Ma'mūn that a government-supported library at Baghdad, already established for the translation of scientific works into Arabic, reached the pinnacle of its importance. He is said to have acquired Greek manuscripts from Byzantium and Cyprus. Translators worked in teams, comparing different manuscript versions and checking where possible against earlier Syriac translations. This sort of work continued for about two centuries, after which time it became redundant, and died away.

Some of the translations have already been mentioned – those of Ptolemy's *Almagest*, of the *Sindhind*, and the *Zij al-Shāh*, for example. Thābit ibn Qurra was a distinguished mathematician and astronomer whom we shall meet later, who also acted as translator. What mattered here was not the simple act of making a few important works available to scholars who would otherwise have had no access to them, but the creation of a huge linguistic empire, comparable with those of Greek, Latin, Sanskrit, Chinese and later English, capable of binding together intellectually large numbers of people of different outlooks and beliefs, so that it became scarcely necessary for them to learn more than a single language to practise their science. 'Arabic science', in other words, was far more than the science of the Arabs.

THE ZĪJ

We have come across several works with the name *zīj* in the title without explaining the word, but we encountered the broad principle in connection with Ptolemy.

The *Almagest* of Ptolemy contained a full complement of tables that allowed the practising astronomer – whose ultimate concern was in all probability astrological – to perform more or less all the ordinary calculations needed in his day-to-day work. These tables were interleaved with their theoretical underpinnings, and were hardly convenient for regular computation. In the *Handy Tables* he therefore issued a new version, somewhat changed, with an introduction explaining their use. (Apart from his introduction, which survives in the original Greek, and in some muddled Latin fragments assembled from a Greek original no later than the eleventh century, this work is now only available to us in a revised form due to Theon of Alexandria, who lived about two centuries after Ptolemy.)

In the Orient, the usual word for a single table of this sort is *zīj*, which passed into Arabic from Persian, and which was handed on to Latin and its vernaculars in such forms as *azig*, and *açig*. At an early date the word was applied to a complete set of tables, and this soon became its standard meaning. Another word with this meaning is the Latin *canon*, which comes from the Greek, often through the Arabic intermediary *qanun*. All these words have at least two meanings, first that of a *thread*, or the woof of a fabric – observe the analogy with the parallel rules marking out the columns of a written table – and second that of a *model*, something to guide one's actions. In the Latin languages, therefore, *canones* was the name given to introductory instructions as to how the tables were to be used.

Some of the tables in a zīj were purely arithmetical or trigonometrical aids. Some of them were required for calendar computation – often involving the conversion of dates from one calendar to another. Others had to do with the time of day, and the related problems of risings and settings of Sun, Moon and planets. There were tables to evaluate the daily or even hourly changes in position of these bodies; more specifically, they covered mean motions, planetary equations (correction terms based on the geometrical models then accepted), the stationary points in the planets' paths as they move forwards and backwards in the zodiac, and planetary latitudes. Other tables were added for lunar parallax – a subsidiary calculation – and for calculating the circumstances of eclipses of the Sun and Moon. Since in most eastern cultures the date of first visibility of the crescent Moon was of prime religious importance, tables were also added for the solution of this difficult problem.

For many forms of time-keeping, for instance using the astrolabe, star co-ordinates were listed, usually in the form of a revised version of the catalogue of 1022 stars found in the *Almagest* of Ptolemy. Tables were usually included to allow for precessional change, whether according to the simple theory of Hipparchus and Ptolemy – which added one degree to the longitudes every century – or the more complex theory of what was known as 'access and recess', later known as the theory of 'trepidation'.

Collections of tables were often included for drawing up horoscopes, and others to help in applying such esoteric astrological doctrines as those of the 'projection of rays', of 'aspects' and of '[excess of] the revolution of the year'. Others were included for deriving the length and quality of a person's life, on astrological principles. Many of the tables are of necessity appropriate to one particular geographical latitude and longitude. Geographical tables were therefore often found with the rest, that is, lists of cities with their co-ordinates. These were needed for both astrological and astronomical purposes whenever the calculation concerned a place other than that for which the tables were set up in the first instance.

Instructions for their use were almost invariably added to tables when they were first compiled, although of course they have often been lost over the centuries. They are rarely very long, but they were certainly important, to the extent that they provided those who used them with a smattering of the basic principles of astronomy, something they could not always easily obtain elsewhere. Zijes often therefore became detached from tables and circulated as texts in their own right.

From the middle of the eighth century to the end of the fifteenth, *well over two hundred* recognizably distinct zijes were produced, with perhaps more than twenty of them incorporating new parameters, recalculated on the basis of original observations. If nothing else, this figure should give an idea of the high importance accorded to astronomy in the Islamic world. The basic *theory* in most instances was that of the *Almagest*, although in some influential cases Hindu or Iranian theories were used. A notable surviving example of this eastern influence is the zīj of al-Khwārizmī (*c.* 840), of which we shall have much more to say. Baghdad was for long the chief new centre of activity, the first true successor to Alexandria in matters of science. From the mid tenth century onwards, Iran took over the lead in zīj production in the east. The Jews later played an important part, especially in Muslim Spain. Wherever astronomy was practised, however, zijes were composed, working tools that the average astronomer no doubt valued far more than the underlying theory.

ABŪ MA'SHAR

Abū Ma'shar (787–886) was born in or near Balkh in Khurasan, and although he eventually entered the service of the Abbasids in Baghdad who had conquered his people, he was intellectually inclined towards Iran and the Shi'a sect. Persuaded in his forty-seventh year by a renowned Neoplatonist philosopher, al-Kindī, that he must study mathematics in order to understand philosophy, he turned to astrology. In this subject he acquired a reputation that has lasted in some circles to the present day. In a sense it is a cause for regret that this man stood at the cross-roads of so many different traditions, and that he had the means of comprehending them – the Greek, the Indian, the Iranian, the Syrian, and their various composite forms, for as a consequence of his breadth of knowledge, he made a synthesis with little regard to consistency – the besetting sin of most practising astrologers. His philosophical awareness was unusual, even so, and his use of Aristotelian and Platonic writers in his attempts to justify astrology became an influential document in later debate, in the Islamic and Christian worlds. His debased Aristotelianism entered Europe, strangely enough, before much of Aristotle's own work was available there.

Abū Ma'shar's earlier training in religious exegesis had made him an expert in calendar work and chronology, and it is not surprising to find him later arguing for the idea that scientific knowledge has been handed down, imperfectly, from a divine source which we can come to know from revelation. He wrote his *Zīj al-hazārāt* to recover the lost knowledge of the true astronomy, and in this zīj he made use of Indian planetary parameters and mean motions (using the yugas), but with a Ptolemaic model. If ever proof were needed of the progressive debasement of knowledge, here it is; and yet the zij was supposedly based on a manuscript buried at Isfahan before the Flood.

Abū Ma'shar's historical approach to science is clearly visible in a doctrine that struck a chord in many a later writer, east and west. The idea was that human institutions — for example religious sects and secular powers like his masters in the caliphate — rise and fall according to a timetable set by certain types of conjunction of the planets Saturn, Jupiter and Mars. This was a doctrine of hope for those who looked forward to an Iranian revival, and of apprehension in later centuries for those who awaited the end of the world or the coming of the Antichrist.

The more discerning astronomers of Islam were severe critics of Abū Ma'shar. One of the greatest of them was al-Bīrūnī, and it is a symptom of the real motivation of later 'astronomers' that Bīrūnī's work is relatively rare, and indeed unknown in medieval Europe, while copies of Abū Ma'shar's writings are legion.

AL-KHWĀRĪZMI

Another highly influential astronomer who worked under the patronage of the caliphs of Baghdad was al-Khwārizmī (*d.* before 850), renowned for his zij, a mixture of largely inconsistent Hindu, Persian, and Hellenistic elements. The fundamental era of the zij (the base date from which calculations were done) was the Yazdajird Era and the calendar was Persian. In due course it was revised by the Muslim astronomer al-Majrīṭī from Córdoba (Spain), who not unnaturally changed the Era to the Islamic Hijra and recalculated the tables to his own meridian. The sheer length of this intellectual chain is remarkable: pre-Ptolemaic Hellenistic astronomy passes to India, to Persia, through intermediaries (perhaps al-Fazārī) to al-Khwārizmī, and then along the length of the Mediterranean to al-Andalus, Muslim Spain. The long journey of the somewhat superannuated knowledge thus transmitted was even then far from complete. Although al-Khwārizmī's zīj is the earliest Arabic astronomical treatise of any size to survive, today it is known only through a Latin translation made of it by the English astronomer Adelard of Bath in the early twelfth century – possibly the translator into Latin of an Arabic zīj. Whether or not Adelard was working in Spain is a question for debate, but he was certainly using at best only al-Majrīṭī's revision, and perhaps even a revision of that.

Despite this long journey, the Hindu connection can today be proved from the parameters underlying the tables. Some evidently come from the Persian *Zīj al-shāh*. Many of the procedural rules laid down in the explanatory canons, for example the method of 'halving the equation', are also Indian in character. There are reasons for suspecting that al-Khwārizmī might have made use of the equant in his working, which has led some to suspect that this was a pre-Ptolemaic invention; but the possibility remains that it might have arrived from a Ptolemaic source independently of the Indian material.

Al-Khwārizmī's zīj did not go uncriticized. From the very first, its shortcomings were noted by al-Farghānī, a young contemporary – a fact noted by al-Bīrūnī. This adverse publicity seems to have found no echo in Spain, where al-Khwārizmī's zīj had an enthusiastic reception, and from whence it was launched into a successful European career. One of the strangest proofs of al-Khwārizmī's capacity for survival is the fact that his zīj was still in use in Samaria in the eighteenth century and in the Jewish Geniza, in Cairo, in the nineteenth.

Al-Khwārizmī wrote an influential work on algebra, and indeed our word 'algorithm' comes from his name, but he also wrote what seems to be the oldest extant treatise on the astrolabe in the Arabic–Islamic tradition. It is at present known from only a single manuscript.

AL-BATTĀNĪ

During the ninth century, the *Almagest* and the *Handy Tables* became available in Arabic translation, and the general quality of astronomical work improved greatly as the superiority of Ptolemy's system became recognized. The two centuries following the death of al-Khwārizmī saw five great Islamic astronomers: al-Battānī, al-Ṣūfī, Abū'l Wafā, Ibn Yūnis and al-Bīrūnī. Far from being the products of a single centre of activity, they worked in places as far afield as al-Raqqa, Baghdad, Cairo and Afghanistan. The Islamic world was beginning to dissolve into separate states, in a movement that we sketched at the beginning of this chapter.

Al-Raqqa is on the left bank of the Euphrates (in the northern part of modern Syria). In other words, although he was a Muslim, al-Battānī (*c.* 858–929) came from the region around the city of Ḥarrān where an astral religion was still practised and even tolerated by its Muslim rulers. Thābit ibn Qurra, a generation before him, had adhered to it.

Here al-Battānī composed his zīj – a name that does not do justice to such a solid text – basing it essentially on Ptolemy's superior methods. Despite its fame, it is now known in Arabic from only a single manuscript. There were translations of the text into Latin in the middle of the twelfth century, and later into Spanish and Hebrew, here based on the meridian of Jerusalem.

Between Ptolemy and the end of the eighth century very few astronomers had a clear conception of their science as one that required observation as a test of theory. In the preface to his zīj, al-Battānī made it clear that he at least had understood this precept, implicit in the *Almagest*, and he set something of a fashion for observation. Admittedly he was not the first to recognize the urgency of the matter. Thus during the reign of al-Ma'mūn (*d.* 833), a group of astronomers had set up a new zīj based on new observations made at Baghdad and Damascus, and had given it the name of *Mumtaḥan* (tested) *zīj*. (It became known in the west as the *Tabulae probatae*, the 'tested tables'.) Also under the 'Abbāsids, Ḥabash al-Hāsib (*d.* 862) had made use of extensive observations of planetary positions, solar and lunar eclipses, observed from the same two places and Samarra.

We shall come across more of this kind of activity, essential if astronomy were not to stagnate. The zīj of al-Battānī, however, was written in a new and refreshing style, not slavishly repeating everything that had been set down in earlier works, but concentrating on recent developments, such as his newly derived figure for the obliquity (23;35° as against Ptolemy's poor 23;51,20°), or a new direction for the Sun's apogee, or new formulae in spherical trigonometry. He introduced material on instruments − a sundial with seasonal hours, a new type of armillary, a mural quadrant (that is, for wall mounting) and a triquetrum (Ptolemy's observing instrument of three hinged rods). There is much that is new implicit in his extensive tables, and yet at the same time some of his explanations of planetary theory are hastily thrown together, and even erroneous. We may excuse them as no more than the marks of a talented astronomer in a hurry.

He was not the first with a new figure for the obliquity: a century earlier, al-Ma'mūn's astronomers had found 23;31°, and others 23;33°. Nor was he the first to detect changes in the solar apogee − Thābit ibn Qurra (or perhaps the Banū Mūsā) had by good fortune previously found what we now know to be a marginally better figure. What characterizes al-Battānī's work is his meticulous description of his substantially new methods that allows the reader to assess the *quality* of the result, given the accuracy of the observation. As for this, it was generally good, so that in deriving a new figure for the eccentricity of the Sun's orbit (2;04,45 parts, about three per cent too high for the time) he improved considerably on Ptolemy's excessively high value. He improved greatly on the figures for precession and the tropical year.

This zīj was influential in select circles in medieval Europe. It was translated in the middle of the twelfth century by Robert of Chester – who also had the distinction of being the first man to translate the Koran into Latin. Another version of the zīj, the only one presently known to survive, was done by Plato of Tivoli at about the same time. It was not circulated in great numbers, but those astronomers who, like al-Bīrūnī, praised it, were often astronomers of the first rank, scholars such as Abraham ibn Ezra, Richard of Wallingford, Levi ben Gerson, Regiomontanus, Peurbach and Copernicus.

FOUR ASPECTS OF ISLAMIC ASTRONOMY

If evidence is needed of the new-found vitality of astronomy, it appears plainly enough in the unprecedented numbers of astronomers with well-deserved and lasting reputations. We may take four, whose work illustrates different aspects of the subject, namely Al-Ṣūfī (903–986), Abū'l Wafā (940–997/8), Ibn Yūnus (d. 1009) and Ibn al-Haytham (965–c. 1040).

Al-Ṣūfī and Abū'l Wafā were contemporaries who worked in Baghdad, but their contributions were very different. Abū'l Wafā's achievements were mostly mathematical, and it is impossible to do justice to them here, but in brief one may say that he cut down reliance on the theorem of Menelaus and introduced a number of new theorems of his own. Only the most discriminating of writers took note of his reforms, however, and his reputation was generally achieved at second and third hand. Al-Ṣūfī struck a more easily appreciated note. In his *Book of the Fixed Stars*, he made it his mission to integrate Ptolemy's star catalogue with Arab star tradition and terminology, and to define boundaries for the constellations. The constellation drawings associated with his catalogue soon became canonical, in Europe too. So well established were the Latinized Arabic star names in Europe by the time Johann Bayer wrote his *Uranometria* (1603) that it would have been a foolhardy act to try to reform them. He did not, and even though there have been very many revisions in terminology since that time, many of the common names for the stars still in use today come indirectly from al-Ṣūfī's work.

Ibn Yūnus had talents of a very different order. Although he earned an enviable reputation as a poet, and was certainly concerned with astrology, in many respects his astronomical work has a modern appearance. In his youth he saw the Fatimid conquest of Egypt, and he served two caliphs of the dynasty, making astronomical observations for them between 977 and 1003. To the second, al-Ḥākim, he dedicated his zīj, which is unusual in that it records large numbers of observations, many from previous observers. He has been associated with instruments in Cairo, instruments of very large dimensions – for instance with an armillary with rings large enough for a horseman to pass through, and an astrolabe three cubits across, but both associations are uncertain. What is certain is that many of the parameters he used in his zīj – on the basis of observations about which he is very vague – were much superior to those of his predecessors. His figure of 23;35° for the obliquity was much quoted. In the nineteenth century Simon Newcomb used some of his eclipses to determine the secular acceleration of the Moon.

Like many Islamic astronomers, he gave much time to a far from simple problem in spherical trigonometry, that of determining the *qibla* – the direction of Mecca, towards which Muslims face to pray – say from the Sun's altitude. He also tabulated the times of prayer in relation to the Sun's daily motion, and here his meticulous calculation took into account the atmospheric refraction of the solar ray at the horizon. His figure of forty minutes of arc for the angle between the observed and 'true' (level) horizon is perhaps the earliest specific figure recorded for this quantity.

An important set of tables that Ibn Yūnus based on the parameters of the *Ḥākimʿīzīj* give the 'equations' of the Sun and Moon, but in a new way, cutting out one stage in the computation of the Moon's longitude, but so requiring the table to be very extensive. (They are 'double-entry tables'.) Ibn Yūnus died in 1009, having – it is said – while still in good health predicted his own death in seven days time. He saw to his affairs, locked himself away and recited the Koran until he died on the appointed day.

The caliph al-Ḥākim was patron not only of Ibn Yūnus but of Ibn al-Haytham, one of the greatest of scientific writers of the middle ages. (He is often known in the west as Alhazen.) Ibn al-Haytham seems to have come to Egypt from Baṣra (Iraq). His most renowned contribution to science was his treatise on optics, which as it happens was to be of much greater influence in the west than in Islam. (Its influence held until well into the seventeenth century.) He wrote a score of short works on astronomy, mostly dealing with specific technical problems, and often touching on optical matters, and one of these in particular merits attention: *On the Configuration of the World*. Eventually translated into Castilian, Latin and Hebrew, this work became influential in the west – the only complete astronomical work of his that was known in Europe.

The work was a critique of Ptolemy's *Almagest*, which Ibn al-Haytham viewed as a piece of abstract geometry, with its imaginary geometrical points, lines and circles, for explaining movements in the heavens. This theory, he thought, needed to be clothed in physical reality – even Ptolemy, in his *Planetary Hypotheses*, had said as much. (It is possible that Ibn al-Haytham did not know this work directly when he wrote the *Configuration*.) The principles by which he said this was to be done were all traditional: there is to be no empty space in the Universe; celestial bodies move with uniform, constant and circular motions; a natural body *can only have one natural motion*; and to each motion introduced into the *Almagest*, some *single spherical body must correspond.* These last ideas had repercussions on European criticism of Ptolemy in the fourteenth century, but Ibn al-Haytham's *Configuration* was not particularly iconoclastic.

He later became more critical. A work by him that has not survived, but that is known through a defence he made of it that survives in Arabic, shows the extent to which he became a critic of Ptolemy's model as given in the *Planetary Hypotheses*. A serious problem with any physical representation of the Ptolemaic planetary models, as he saw, is that since the epicycle revolves around the equant point rather than the axis of the deferent circle, the motion around the axis of the principal planetary sphere will not be uniform. The accepted physical ('philosophical') guidelines for natural heavenly motion seemed thus to have been broken. He needed a new model in place of the old, and the one he found had an odd resemblance to that of Eudoxus.

In the theory of planetary latitudes as given in the *Almagest*, the epicycles were taken to be inclined to the plane of the ecliptic. Following the physical principles already set forth, each epicycle had to be associated with a *single* sphere. Modifying Ptolemy's physical model in the *Planetary Hypotheses*, with its earth-centred shells inside the thickness of which are found the eccentrics, equants and epicycles, Ibn al-Haytham found a way of representing the inclined epicycle physically.

The poles of the epicyclic sphere he placed at a distance from the furthest and nearest points of the epicycle (apogee and perigee) equal to Ptolemy's maximum inclination of the appropriate epicycle diameter, and he gave the sphere a rotary motion equal to that of the rotating diameter of the epicycle in *Almagest*. The epicyclic sphere is of course carried round the planet's main (deferent) sphere, but Ibn al-Haytham now suggested adding another, between the main sphere and the epicycle, with a motion equal to the first but in the opposite direction. The resulting motion resembles that produced by the homocentric theory of

Eudoxus. Without describing the arrangement in detail we can say that, as seen from the Earth, the end of the rotating epicycle diameter will trace out a hippopede, a figure-of-eight that is carried round the sky and that accounts for the planet's motion in latitude.

Ibn al-Haytham wrote much more along the same lines, not so much changing the predictive consequences of conventional theory as giving his audience the impression that he had found the true physical configuration of the Universe, whereas Ptolemy had not. Where he objected to Ptolemy's lunar theory, the pattern of his argument is instructive. He claimed, first, to have found the only two possible models equivalent to the lunar theory; and then it was a short step to dismissing them with the remark that it was not possible to assume a physical body with the properties of either. Oddly enough he makes no mention of the great variation in the Moon's distance on Ptolemy's theory, a point on which the lunar model seems blatantly unacceptable.

This type of dismissive physical argument was repeated in the west, at first without astronomical repercussions. In eastern Islam, however, in the thirteenth century the great Naṣīr al-Dīn al-Ṭūsī was stimulated by what he read in Ibn al-Haytham to make yet further criticism of Ptolemy and to offer an alternative theory of planetary motion.

NAṢĪR AL-DĪN AL-ṬŪSĪ AND HIS FOLLOWERS

Naṣīr al-Dīn al-Ṭūsī was one of those figures in history who combined an acute intellect, an abundance of energy, and the good fortune to occupy a key social position at the centre of the Islamic stage. This made it easy for others to appreciate his worth, and his intellectual influence was probably greater than that of any other single medieval astronomer. Born into a warring continent in Ṭūs in Persia – hence his name – he was educated first at home by his father, one of a long line of Shīʾite scholars, and in several institutions – notably in Nīshāpūr, an important centre of learning. He was well trained in virtually all branches of Islamic learning, and he rightly considered himself the heir to Hellenistic science and philosophy. He eventually found security in the service of the Ismāʾīlī rulers, in particular al-Alamūt, Grand Master of the Assassins, 'the Old Man of the Mountains'. With their courts he moved his residence between mountain strongholds, until in 1256 the Īlkhānid conqueror Hūlāgū, grandson of Genghis Khān, ended Ismāʾīlī rule in northern Persia. The astronomer's fame now ensured him a place in Hūlāgū's entourage. He was at the conquest of Baghdad in 1258 and a year later persuaded Hūlāgū to begin the construction of an observatory at Marāgha in the north-west corner of modern Persia (80 kilometres south of Tabriz). Hūlāgū's brother Moengke, who ruled over a vast area of China, had set in motion plans to build an observatory in Beijing, but this was not completed in his lifetime.

The Marāgha observatory was in many ways the first research institution on a large scale with a recognizably modern administrative structure. It had an extensive scientific library with a permanent librarian, and an astronomical staff of at least ten, among whom there was at least one Chinese scholar, Fao Mun-ji, and very probably more. It was equipped with numerous expensive instruments – a large mural quadrant, parallactic rules, an armillary and quadrants adjustable in azimuth. The Īlkhānī astronomical tables were completed there in 1272 under the rule of Hūlāgū's successor Abāqā.

Naṣir al-Dīn al-Ṭūsī wrote a long series of important works on logic, philosophy, mathematics and theology. He was responsible for a revival of the doctrines of Ibn Sīnā (Avicenna, 980–1037), with their Aristotelian cast. He was an exceptional geometer and it is not surprising to find him applying his geometrical skills to the problems in natural philosophy that Ibn al-Haytham had brought to his notice. Composing his own criticism of Ptolemaic astronomy in a work called 'The Treasury of Astronomy' (*Tadhkira*), he made it quite clear that the work was to be seen as a summary for non-specialists, without difficult mathematical proofs. It dealt with the external aspects of earthly as well as celestial bodies. He added to his critique of Ptolemy a positive contribution in the form of some new planetary models. One of the most interesting of these relies on the following theorem:

> If one circle rolls inside the circumference of another with radius twice as great, then any point on the first describes a straight line (a diameter of the fixed circle).

This arrangement, today often called a 'Ṭūsī couple', is illustrated in figure 8.1, where the broken lines join the points of contact of the circles to the point fixed on the circumference of the rolling circle.

FIG: 8.1   *The Ṭūsī couple*

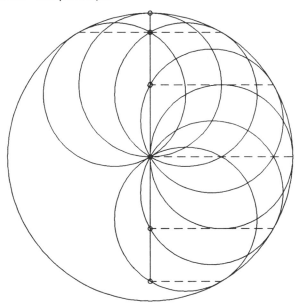

The theorem is very easily proved: the lengths along the circumferences of the two circles that have been in contact have to be equal, and each is the product of a radius and an angle. For the fixed circle, the angle is only half as great, but the radius is twice as great, as for the rolling circle.

To produce a straight-line motion from a double circular motion was an intriguing thing in itself, but we notice how the rolling circle (or sphere) lends itself to a *physical* interpretation of the sort Ptolemy, Ibn al-Haytham and others had been seeking. Providing the theorem with the clothing of geometry is all very well as long as we do not lose sight of this fact. Several Marāgha astronomers in addition to al-Ṭūsī were eager to make use of his ideas and provide physically viable models more or less equivalent to Ptolemy's, and so answer the criticisms of Ibn al-Haytham, al-Ṭūsī and others. They included his colleague al-'Urḍī (d. 1266), who built the observatory, his student al-Shīrāzī (1236–1311), and an astronomer who lived a century later, Ibn al-Shāṭir (1304–75).

Al-Ṭūsī generalized the model to three dimensions. Taking the planes of the two circles to be inclined at a small angle, he found that the oscillatory motion approximated now to an arc of a great circle. This notion he used in the theory of planetary latitude. What he did is doubly interesting by virtue of the fact that Copernicus made repeated use of precisely the same device, as well as other principles from al-Ṭūsī and his followers, so that it is hardly possible to doubt that he was aware of some text or other in which they were to be found. Greek and Latin materials that made use of al-Ṭūsī's device were circulating in Italy at about the time Copernicus studied there.

In his *De revolutionibus*, Copernicus used the Ṭūsī device in his model for the variable rate of precession and variation in the obliquity of the ecliptic, while in that book as well as in his *Commentariolus* he used it to achieve an oscillation in the orbital planes in his theory of planetary latitudes. In the *Commentariolus* he used the simpler plane model to achieve a variation in the radius of Mercury's orbit. The same he did tacitly in his *De revolutionibus*. In the *Commentariolus* he based his models for planetary longitude on the models developed by al-'Urḍī and Ibn al-Shāṭir, although erroneously in the case of the inferior planets, while in *De revolutionibus* his models were related to those, and to others by al-'Urḍī and al-Shīrāzī. In both works, the lunar model is the same as Ibn al-Shāṭir's.

Rather than enter into the details of the various new planetary models, we take as an example the way one of them functions. The problem to be solved is that of replacing a motion on an eccentric (say a deferent circle) with a combination of circular motions, the main centre being the Earth (centre of the universe). Although we shall use epicycles (the two that make up a Ṭūsī couple), it should be stressed that to get a model of Ptolemaic type we need to add yet another epicycle. Here we concern ourselves only with the replacement of the eccentric.

If we take a Ṭūsī couple of carefully chosen dimensions, and attach it (as it were rigidly) at the end of a rotating radius passing through the centre (*C*) of the deferent circle, then by the properties of the Ṭūsī couple, the oscillating point on the small (rolling) circle will always lie on the rotating radial line through *C* (figure 8.2). Several representative positions are shown in the figure, on the assumption that the speed of the 'rolling' motion is chosen so that the point on the smaller circle holds the rolling circle's radius in a constant direction. This makes the places where the point is at its greatest and least separations from *C* exactly opposite one another (apogee and perigee, respectively, at the top and bottom of our figure). It will be seen that an eccentric circle through those opposite points (the broken line in the figure) is almost perfectly accounted for by the three-circle model.

Such models obviously allow great scope for generalization. Inspired by al-Ṭūsī's technique, Ibn al-Shāṭir went further in removing eccentric deferent and equant, using yet

FIG: 8.2   *Replacement of an eccentric by Ṭūsi-style devices*

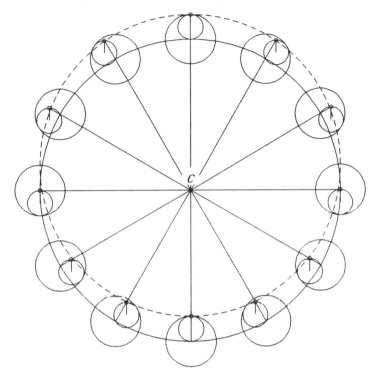

additional epicycles. To the Sun he assigned an epicycle riding on a standard epicycle on the deferent, the latter being earth-centred now. The Moon likewise had a double epicycle, but the proportions and motions were of course different. The lunar model went some way towards correcting the chief blemish on Ptolemy's model, namely the enormous variation in the lunar distance – which the simplest observations show to be an illusion. The planets had *triple* epicycles. The bare essentials of the construction for a superior planet are shown in figure 8.3. *C* is there the centre of the Earth, and *P* is the planet, the position of

FIG: 8.3 *Ibn al-Shāṭir's removal of deferent and equant*

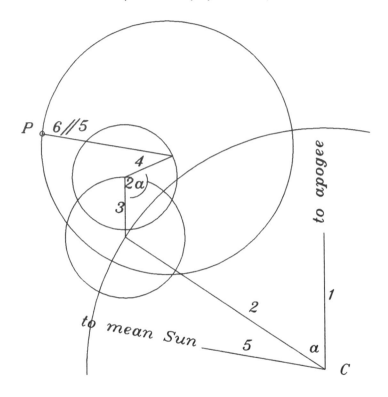

which is determined by a configuration that is arrived at as shown. (The numerals are added to show the sequence of steps to be followed. Lines 3 and 6 are parallel to lines 1 and 5 respectively. All four basic circular motions on the figure are anti-clockwise.)

After Ibn al-Shāṭir, the Muslim fashion for designing 'philosophically acceptable' non-Ptolemaic schemes seems to have died away, although it was to resurface in medieval Europe. It was a strange and long-lived prejudice, but we should be careful to distinguish its varieties. Suppose that someone were to observe an oscillatory motion on a straight line. Using some mathematical formula for the variation in position with time, one might succeed in explaining it – simply as a motion on a straight line. A 'philosopher', however, might maintain that it *should be explained* in terms of a Ṭūsī couple and *circular* motions. Another might go further, and say that whether it is explained or not, it *really is* a pair of circular motions. This last person needs an independent argument for 'what really is'. Here, the astronomers called on Aristotle, whose philosophical arguments had begun from very simple observations.

What might seem a less controversial but weaker philosophical position is when one insists that one is giving only a *working explanation* in terms of circles. In similar cases today one typically appeals, not to an insight into what really exists, but to the aesthetics of the resulting theory, or its simplicity. In the past, the less controversial alternative was usually taken by those who themselves accepted the Aristotelian view of what exists, and wished to avoid conflict with it.

THE OBSERVATORY IN ISLAM

Such a remarkable observatory as that at Marāgha, founded in the thirteenth century, could not have arisen without the prior development of a tradition of observation. As for that, we might list a hundred instances. There were, for example, the solstice observations made by Yaḥya ibn Abī Mansūr in 829; and the solar and lunar observations made in Dayr Murrān with the special instruments that al-Ma'mūn ordered to be made in Damascus – an enormous marble mural quadrant and an iron gnomon. In the late tenth century, al-Khujandī built a colossal meridian sextant of eighty cubits, two parallel walls in the meridian plane carrying a brass scale on which fell the solar image, as cast by some sort of aperture. Over the scale a disc crossed by two diameters was used to mark exactly the Sun's distance from the zenith. Grand as was the conception, al-Bīrūnī was rightly sceptical about its limitations. We have mentioned the colossal instruments seemingly available to Ibn Yūnus. It is worth noticing here how Islamic astronomy placed so much emphasis on *scale*, as a prerequisite of accuracy, overlooking those mechanical factors that so often annulled the advantages of scale.

Large observatories continued to be built in the Islamic world after the main initiative in theoretical astronomy had passed to Europe. Important instances are those of Samarqand (1420/1) and Istanbul (1574/5). The former, housed in a great three-storey building, was founded as part of an important research institute by Ulugh Beg (1394–1449), grandson of the famous Tamerlane. The chief instrument was an enormous stone sextant faced in marble. It lay in the plane of the meridian, bedded in a vast trench (radius forty metres) cut into the hillside. Its remains were located in 1908; and in 1941 Ulugh Beg's tomb was located in the mausoleum of Tamerlane in Samarqand: the skeleton gave clear evidence of a violent death.

The staff of the observatory included al-Kāshī, chiefly remembered as the author of the finest eastern treatise on arithmetic written in the middle ages (1427), a work that presented the theory of decimal fractions. Ulugh Beg's astronomers produced under his name an important zij, which included some excellent sine and tangent tables as well as improved planetary parameters and star positions. An unusually large number of these were based on original observations rather than on a mere updating of Ptolemy or al-Ṣūfī. The star catalogue later aroused much interest in Europe, especially in the early days of serious Arabic studies, in the early seventeenth century.

Al-Kāshī's contributions to astronomy included the design of a new type of equatorium, an instrument that in many ways might offer an alternative to the zij. In the simplest kinds of equatorium, the geometrical models for computing planetary positions were simulated by graduated discs of metal, and these discs, when correctly positioned (usually with the help of simpler ancillary tables), could yield results in a fraction of the time required to grind through ordinary planetary tables, in zijes. The accuracy of the equatorium could of course never compare with that of the zīj, and although the accuracy of the latter was often quite spurious, the true professional continued to use tables for serious computation.

The Istanbul Observatory is interesting because it was so close in time to the great observatory set up by Tycho Brahe at Uraniborg. Like Samarqand and the observatories set up at Delhi, Jaipur and Madras by Jai Singh in the eighteenth century, it made much use of large-scale masonry instruments. Much attention was given to the foundations of these truly monumental instruments, and their graduation, but their usefulness was limited to a very few types of observation, chiefly with the Sun. Even there, the confusion in the solar image cast on the scale (with the umbra and penumbra of whatever aperture was used to limit the image) introduced sizeable uncertainties into the angles measured.

As precision instruments, armillaries can be discounted, since they are so difficult to make mechanically perfect. A well known illustration from a sixteenth-century Ottoman Turkish manuscript, now in Istanbul, shows a large bronze armillary instrument of observation with a supporting frame constructed entirely in wood. As drawn the complete instrument would have been nearly five times the height of the men shown using it.

Instruments of research were of course always relatively rare. Astrolabes, armillaries and globes became the symbols of the astronomer, and standard parts of the instrument-maker's repertoire. As in the manufacture of other instruments, Ḥarrān became an important centre. Where mathematical skills were called for in making an astrolabe, skill in casting a sphere was perhaps as difficult to come by. Those who like al-Sūfī expected accuracy of a globe, did not find it. Not all globes were of metal, but those that were of metal were usually an expensive luxury for the courts of princes and the rich. For more accurate calculation the planispheric astrolabe, or some derivative of it, was used, but again, the highest accuracy was vouchsafed only to those able to work with a zij.

A typical celestial globe had only the main astronomical circles engraved, with perhaps twenty or thirty bright stars. The richest globes would have on them a substantial fraction of the 1022 stars that were to be found, plotted by their co-ordinates, as taken from Ptolemy's star catalogue, al-Ṣūfī's, or their derivatives. In these cases, the constellation images are usually added, with richly decorative results reflecting styles in star catalogues. Together they helped to hand down the tradition of picturing the constellations, a tradition that entered Europe, and one that has left many traces in the modern world.

Western Islam and Christian Spain

ASTRONOMY'S ARRIVAL IN AL-ANDALUS

Before the end of the tenth century a new impetus was given to astronomy, at the western extreme of the civilized world. From the eighth to the fifteenth centuries, Spain was to a greater or lesser extent under Muslim control, and so it became one of the two main channels for the transmission of Arab science in general to Christian Europe. The other principal route was through Sicily, but from an astronomical point of view this was of lesser importance.

Spain had a tradition of learning long before the arrival of the Muslim conquerors. Few scholars had greater general influence in early medieval Europe than Isidore of Seville (c. 560–636), an encyclopedic writer who drew on mainly Roman texts. He wrote on broadly cosmological matters, such as atomism, the disposition of the four elements, and so forth, but his knowledge of Greek ideas was almost entirely indirect. The 'Isidorian' period in Spanish learning survived the first conquests (the end of the Visigothic kingdom in Spain came in 711), but even then astronomy was still little more than the manipulation of the cycles of the solar and lunar calendars, supplemented later by simple rules for prayer towards Mecca and the orientation of mosques.

In Muslim Spain – al-Andalus – original Greek works, and Arabic and Hebrew commentaries on them, began to arrive on the scene in increasing numbers at a time when a measure of anarchy was descending on Europe generally, in the wake of a decline in the fortunes of the descendants of Charlemagne. With 'Abd al-Raḥmān III (912–961) there began the emirate in Córdoba that, culturally speaking, was to outshine the caliphate of the Abbasids. The second Emir had agents sending him books from as far afield as Baghdad, Damascus and Cairo. At Córdoba, in the second half of the tenth century, schools of mathematics, astronomy and other sciences grew up that acquired, commented on and expanded the material coming to them from the east. It was in Córdoba that al-Majrīti worked (d. around 1007), and among his many achievements was an adaptation of the tables of al-Khwārizmī to the meridian of Córdoba and the Muslim calendar (that is, with dates following the era of the Hijra).

The example set in Córdoba was later copied by Arab rulers in other places in the peninsula – for example in Seville, Valencia, Saragossa and Toledo. Astronomy was no mere literary activity: while it would be wrong to exaggerate their importance, some useful new observations were made, and important new instruments of calculation and observation were developed.

One of the reasons for al-Majrīti's historical importance is that he had a number of pupils who spread his knowledge of astronomy over the whole of al-Andalus and beyond. The more renowned of them were al-Kirmāni (d. 1066), who worked in Zaragoza, Abū'l-Qāsim Aḥmad (d. 1034; known also as Ibn al-Ṣaffār), Abū Muslim ibn Khaldūn and Ibn al-Khayyāṭ (d. 1055). Another, Ibn al-Samḥ of Granada (d. 1035; one of several writers known in Latin texts under the name of Abulcasim), wrote extensive treatises on the astrolabe and equatorium and a zīj that again made use of al-Khwārizmī. If the impression has been given that such things were commonplace, we should not forget that in the rest of Europe they were as yet an almost total mystery.

CÓRDOBA AND EUROPE

Córdoba's cultural influences quickly spread to the Christian states in the north of the peninsula, however, and Latin translations of many scientific works were obtained. Within

Islamic Spain there were numerous contacts between religious groups, and potential translators included, for example, the many bilingual Jews and Mozarabs – that is, Christians whose culture was in many respects the same as that of the Muslims. Even in the tenth century, and long before the general movement to translate scientific material from the Arabic into Latin, a collection of treatises on arithmetic and geometry, astronomy and calendar computation was produced at the Christian monastery of Santa Maria di Ripoll, a Benedictine house at the foot of the Pyrenees. Such translations spread rapidly throughout Europe, helped by the network of Christian monasteries and the movement of scholars between them. For long it was a mystery how Hermann the Lame (1013–54), a monk in the monastery of Reichenau, in a remote Austrian valley, could have composed a treatise on the construction and use of the astrolabe. The answer is simply that he had somehow seen a copy of one of the Ripoll texts.

One influential scholar who helped to disseminate Spanish science in the tenth century was Gerbert of Aurillac. While still a novice, he had been put in the charge of the Count of Barcelona to get instruction in the liberal arts. Taught mathematics, arithmetic and music by the Bishop of Vich, he was perhaps the first scholar of note to carry the new learning across the Pyrenees. Taken to Rome, where he astonished the Pope with his learning, he ascended the ecclesiastical hierarchy rapidly: he became abbot of Bobbio and archbishop first of Rheims and then of Ravenna, before becoming the first French Pope (Sylvester II, *d.* 1003). He did not so much advance scientific knowledge as spread the feeling that it was of importance. His influence was mainly through cathedral and monastic schools, first in Lorraine. European knowledge of the astrolabe owed much to him, and he in turn owed it to a Ripoll text. The west took over some of the Arabic terminology of the astrolabe at this time, although much of it was gradually replaced by Latin equivalents. Other early works on the instrument stem directly or indirectly from his source. They were works written by scholars with independent reputations for learning: Fulbert of Chartres, Herman the Lame and Walcher of Malvern.

This was a time when Hindu–Arabic numerals were slowly finding favour among astronomers in Europe. The change was far from sudden. For various reasons, European commerce retained Roman numeration until the sixteenth century and even after, and yet by the late thirteenth century the best astronomers had begun in earnest to use the new numerals and their associated arithmetical techniques – on which al-Khwārizmī himself had written a standard treatise. Here again it was Spain that provided the stimulus.

Scholars from other parts of Europe, wanting to learn more of these matters, soon began to travel to the source. The movement was helped along by publicity offered by Pedro Alfonsi, a Spanish Jew (Moshe Sefardi) who had converted to Christianity, and who visited the court of Henry I of England around 1110. There he found an interest in astronomical matters, and among others met the Lorraine scholar Walcher, who after travelling in Italy had been made Prior of Malvern. One can judge the sense of intellectual excitement in this new movement from a work adapted by Walcher from a text by Pedro on eclipse calculation. The work has many weaknesses, and Walcher was often out of his depth, but he wanted desperately to master the new astronomy. Pedro's message was simple and explicit: abandon your old and primitive methods of calculation and learn the new techniques from the east. The techniques he was introducing, however, were those of al-Khwārizmī. When Adelard of Bath translated and partially adapted eastern tables, it was al-Khwārizmī's he chose, helping to introduce a consequent confusion into European astronomy that lingered for centuries, for as already explained, al-Khwārizmī's techniques were not purely Ptolemaic, but were mingled with incompatible Hindu theories.

Another zīj reported as having been compiled about this time, but again evidently now lost, was by Ibn Mu'adh al-Jayyānī of the town of Jaén (*c.* 1020; in Castilian and Latin he was sometimes called Abenmoat). It was put into Latin by the greatest of Latin translators, Gerard of Cremona (*c.* 1114–87), who worked for over forty years in Toledo. The influence of al-Khwārizmī is again clear, although his rules are occasionally found to have been altered – for example that on first visibility of the lunar crescent. Not the least interesting section of the entire work concerns the division of the astrological 'houses', that is, the casting of a horoscope. There are several mathematical methods of doing this known to history, and here we find a method in use that is traditionally – but mistakenly – ascribed to the great Renaissance scholar Regiomontanus (1436–76), who once seemingly owned the manuscript from which the Jaén treatise was later printed. (This question will be touched on again in chapter 10.)

AL-ZARQELLU AND THE TOLEDAN TABLES

In the second half of the eleventh century – a golden age in Andalusian learning – an important group of astronomers formed something approximating to a school in Toledo. They included Ibn Sā'id and Ibn al-Zarqellu. (The name of the latter scholar occurs in many different forms, for instance al-Zarqālī, Ibn al-Zarqīyāl or Zarqāllu and – in Latin texts – Arzachel.) 'Arzachel' (*d.* 1100) was to become before long one of the most frequently cited of all astronomical authorities, although in truth the reference to his name was usually to a set of tables for which he had only a secondary responsibility. This is not to deny his importance. He was a trained artisan who entered the service of the Qadi of Toledo as a maker of instruments and water clocks. He remained in that city until some time between 1078 and 1080, when the discomfiture of repeated invasion by the Castilians persuaded him to move to Córdoba, and there he lived until his death in 1100.

Ibn al-Zarqellu's true intellectual qualities are manifest in a series of writings. He is said to have found the movement of the solar apogee, which he held to be 1° in 299 years, on the basis of twenty-five years of observations. This problem relates to the problem of precession, or the motion of the stars of the eighth sphere, as the problem was then presented. Does the Sun's apogee move with them?

Ibn al-Zarqellu wrote a work, now lost, on this theme, and another that contained his theory of precession. The second describes a model for an oscillatory movement of the eighth sphere that has gone down in history as having originated with Thābit ibn Qurra, although there is no Arabic source attributing the model to him. An early Latin version of the model is quoted at length by John of Spain, who attributed it to Ibn al-Zarqellu. There was a Latin text on the problem, supposedly by Thābit, that might date from only about 1080, a year when observations of the Sun were being made in Toledo and Córdoba. The question is admittedly not closed, however, for Thabit's grandson Ibrāhim is known to have proposed a complex model of much the same sort, and this was supposed to explain changing obliquity as well as the increasing longitude of the solar apogee.

The model in question was easier to draw than to develop mathematically. We shall here do neither, but give only a description in words to convey a rough idea of how complex was the procedure of determining the movement of the equinoxes, on which the apparent places of the stars depend. The model involves two small circles (radius 10.75°), at opposite sides of the celestial sphere, each of them centred on a point of the equator – the mean equinoctial points. Around each of these small circles moves a point, the two points being diametrically opposed. The points carry between them a moving ecliptic. Yet other points, where that ecliptic meets the equator, are the equinoxes for the time in question. It should be clear that the equinoxes so arrived at shuttle backwards and forwards along the equator, but since the circuit of the small circles was supposed to take more than four thousand years to complete, the variation in the precessional drift was not marked.

This model of 'access and recess', as we know it from its Latin name, is often dismissed today as lamentable nonsense, as something akin to the idea that snakes hatch out of stones. Any assessment of precessional drift, however, required data from the distant past, and since at this time the evidence of earlier authorities seemed to point unequivocally to a *variable* motion, it would seem more just to regard the model as a triumph of ingenuity. At all events, it became a standard part of astronomical dogma. Pedro Alfonsi reported an early variant: he had tables for an epoch in 1116 in which different parameters were given; and various alternatives appeared over the following centuries, notably one developed by Copernicus.

Ibn al-Zarqellu wrote original treatises on an equatorium and on the universal astrolabe. This last type of instrument, known in the west as the *saphea*, was alluded to at the end of chapter 4. It was easier to make than to understand, and has a complicated history. Robert of Chester, for example, writing in London in 1147, claimed to be translating from the Arabic when he produced a treatise by 'Ptolemy' on the universal astrolabe, but the work is almost certainly a purely western fabrication. It is important to remind ourselves here of how thinly spread this sort of knowledge was, and continued to be. A scholar who, like William the Englishman in 1231, tried to reproduce the universal astrolabe from the written source only, could hardly avoid mistakes. The saphea was not well understood in the west until long after Ibn Tibbon's translation of Ibn al-Zarqellu's text in 1263.

More influential by far than these writings by Ibn al-Zarqellu were his canons to the Toledan tables, which were destined to eclipse all others, in Spain and Europe generally, until the early fourteenth century. It is as well to speak of the evolution of the Toledan tables, rather than of their composition, for, like most zījes before and after them, they were a compilation of often inconsistent elements. Roughly speaking, the principal honours go to al-Khwārizmī and al-Battānī, but it is quite possible that some effort was put into a programme of observation, to verify such older material.

The original Arabic version even of the canons to the Toledan tables is lost, but two versions of that introductory material have survived in Latin, together with many variants on the assortment of tables that follow. Those for the solar, lunar, and planetary inequalities (equations) are largely from al-Battānī, with one from al-Khwārizmī. Tables for planetary latitude come from al-Khwārizmī – although a few manuscripts have Ptolemy's added. Others for planetary visibility, stations and retrogradations are from al-Battānī and Ptolemy. And so the catalogue continues, with very little that could not have been found in eastern Islam two or three centuries earlier. There is some astrological material that is possibly new as far as its appearance in table form is concerned, but that is not new in any other sense. Star co-ordinates were updated, of course – and in Latin recensions they were often updated for the time of copying. One of the more fundamental changes to the tables on which the collection is based concerns the mean planetary motions. The basic epoch is that of the Islamic Hijra, and the meridian of Toledo was used. This meant that all mean motions quoted at epoch had to be recalculated for the time difference between Toledo and al-Khwārizmī's meridian. Although the difference was not accurately known, the error was not great.

The Toledan tables influenced western Islam in at least one of three compilations by Ibn al-Kammād (*c.* 1130). Long afterwards an astronomer from Seville, Ibn al-Hā'im, issued a zij (*c.* 1205) in which he claimed to be correcting

Ibn al-Kammād's errors. Ibn al-Hā'im's was one of the last distinctive zījes to have been produced in al-Andalus, but in European terms it left no mark. It could not but be overshadowed by the most famous Spanish zīj of all, namely the Alfonsine tables of the 1270s. These tables, produced under the patronage of the Christian King Alfonso X of Leon and Castile, will be the subject of a later section.

TOLEDAN VERSUS AL-KHWĀRIZMĪ'S TABLES

Although the old Toledan tables, in Latin translation, ultimately carried Islamic astronomy well and truly to the heart of Europe, they did so alongside the al-Khwārizmī zīj, and it was long before they became the dominant force. They then remained so until the Alfonsine tables took their place in the early fourteenth century – and in some places even later.

Translation from Arabic into Latin and the Castilian vernacular was an activity with a long history in Spain, and we have already seen examples of it. There were at various times important schools of translators in the Pyrenees, the Ebro Valley and Toledo, for instance. To understand the reason for this activity one must appreciate Christian scholars' realization that they and their schools were seriously deficient in astronomical learning. We have already mentioned the classic example of Walcher, Prior of Malvern Abbey in England, who had been avid to learn from Pedro Alfonsi. The zīj to which Pedro introduced him (*c.* 1116) was a rather untidy patchwork of materials, closely related to that of al-Khwārizmī. The planetary mean motions were adapted to the Christian calendar, for instance, in a singularly awkward way. When Adelard of Bath chose to translate a zīj, perhaps with Pedro's help, it was a version of the al-Khwārizmī tables, not the Toledan.

The Toledan tables undoubtedly helped to satisfy a growing European need, but it is hard to see by what criteria they were at length preferred to al-Khwārizmī's. It would be pleasant to think that they were judged on the grounds of predictive accuracy. At all events, tables for Marseilles (c. 1140) by Raymond of Marseilles were based on them, and had a certain currency in France. Robert of Chester, who went to Spain in 1141 and even became Archdeacon of Pamplona before moving to London in 1147, adapted not them but al-Khwārizmī's tables to the meridian of London; but it is significant that he introduced some Toledan material.

Spanish learning was carried abroad in other texts, and very often by Jewish scholars. A notable example is Abraham bar Ḥiyya (known as Savasorda, fl. 1110–35) from Barcelona, who composed astronomical Tables of the Prince which mix Jewish, Islamic and Ptolemaic material. At least one Latin manuscript survives. The epoch is the equivalent of 21 September 1104, and the year is not Jewish but the old Egyptian–Ptolemaic–Battānī year of 365 days. Abraham reveals his source in that he uses al-Battānī's meridian (al-Raqqa) – a strange convention for a Spaniard. Another Spanish Jew was Abraham ibn Ezra of Tudela (c. 1090–c. 1164). He wrote in Latin as well as Hebrew, and travelled widely through Italy, France and England. Abraham wrote tables for Pisa around 1143, following, as he tells us, the zīj of al-Ṣūfī. A London zīj of about 1150 seems to be related to the Pisan tables, and might also be due to Abraham.

Such intellectual commerce continued throughout most of the middle ages. Links with Jewish communities in southern France were strong, but influences spread further afield. This process of exchange lasted to the very time of the expulsions from Spain. Abraham Zacut, for example, born around 1452 in Salamanca, who compiled a perpetual planetary almanac and tables that were used by Columbus and Vasco da Gama, was himself expelled from the country, and died around 1522 in Damascus.

Versions of the Toledan tables soon came to hold the field alone for a time. Well over a hundred manuscripts of them survive to this day, an unusually large number for a genre of text that was of virtually no interest once it was superseded. Scholars in many towns adapted them to their local meridians. The Toulouse tables were in addition referred to the Christian calendar. One fourteenth-century (Latin) version of the Toledan work was even translated into Greek. The cultural circle was complete.

THE ASTROLABE AND ASTROLABE-QUADRANT

There are at least forty or fifty western treatises on the traditional form of astrolabe written before the end of the sixteenth century, and yet, including the group already mentioned, they fall into only three families. All stem from the Arabic culture in Spain. The general drift of western Europe's first knowledge of the astrolabe was evidently northwards and eastwards from Catalonia, in the late tenth century, as already explained. This first move into monastic Europe seems to have petered out by the middle of the eleventh century.

Next came far more powerful intellectual influences, based on more complete texts, that were better argued and understood. Of central importance was one by the eleventh-century Spanish–Arabic astronomer Ibn al-Ṣaffār, and this was translated into Latin twice (by John of Spain and Plato of Tivoli). The oldest dated western astrolabe – although not the oldest – carries the date 417 (of the Hijra, that is, AD 1026/7) and it is a curious fact that the maker's name shows it to have been made by Ibn al-Ṣaffār's brother.

The third family of texts stems from the school of writers in Castilian at the court of Alfonso X, in the late thirteenth century.

The most influential of the three was undoubtedly the second. John of Spain's version was used by Raymond of Marseilles and by a writer whose real identity is unknown, but whose work was ascribed to the Jew from Basra and

Baghdad, Māshā'allāh. This second family introduced further Arabic terminolgy (including the words 'zenith' and 'azimuth') to European astronomy.

As we enter the thirteenth and fourteenth centuries, the Iberian influence in Europe remained just as real, but became less obvious, as astronomers began to blend their sources. Some of the better known European writers were Raymond of Marseilles in the twelfth century, and Sacrobosco and Pierre de Maricourt (Petrus Peregrinus) in the thirteenth. With the fourteenth century we enter an age of writing in the vernacular. In French, Pèlerin de Prusse wrote a slender astrolabe treatise for the dauphin, the future Charles V, who was to be crowned king of France in 1364. The English poet Geoffrey Chaucer made a notable contribution with his treatise on the astrolabe – subtitled 'Bread and Milk for Children' in one manuscript, presumably by an ironical scribe. This remained the only satisfactory work in English on the instrument before modern times, but it, too, derived from the work of the pseudo-Māshā'allāh.

In the late thirteenth century, a new and inexpensive alternative to the simple astrolabe was found. Imagine an astrolabe with its rete *fixed* in position. (To follow this brief account it might be helpful to consider figure 4.9 or the astrolabe illustrated in the plate section.) A thread through the centre carrying a small sliding bead as marker (a seed-pearl was the standard marker in the middle ages) would serve to locate any point on the rete, such as a star: stretching the thread over the star, one would simply mark its distance from the centre and note the angle registered on the outer rim. Rotating the thread through a given angle would move the marker to a new position, *exactly as if one had rotated the rete*. The arrangement would be less intuitive, but mechanically simpler, and easier and cheaper to construct than a complete astrolabe. Suppose now that we take matters one step further, and fold the compound diagram (of lines appropriate to the rete and the plate of an astrolabe) along a main axis, and then fold it again along the axis at right angles to the first. We shall have a mass of lines, points, and graduated scales, many of them double; but given a good grasp of how an astrolabe functions, exactly the same problems can be solved on this 'astrolabe-quadrant' as on the astrolabe itself.

To measure altitudes, small sighting vanes with pinholes can be fitted to one edge of the quadrant, and the quadrant tilted until a star or whatever is seen through the two holes. The vertical is settled by means of a hanging plumb-bob on the thread, and the altitude is read off the peripheral scale.

The earliest known description of such an instrument was written between 1288 and 1293 in Hebrew by Jacob ben Machir ibn Tibbon, a writer known in Latin as Profatius Judaeus. Although he lived in Provence, he was born in Marseilles (*c.* 1236) and he died in Montpellier (1305). His family came from Granada, a place his grandfather left because of civil unrest. Both his father and grandfather made reputations as translators from Arabic into Hebrew, and it has often been assumed, although without any proof, that Ibn Tibbon's so-called 'new quadrant' must have been based on an Islamic prototype. The fact that he called it the 'Israeli quadrant' casts some little doubt on that idea. The work was quickly turned into Latin (1299, by Armengaud) and the Latin was expanded by the Danish astronomer Peter Nightingale, through whose text the instrument became well known to Latin astronomers. Its popularity was due to the extreme ease with which it could be made: a careful drawing on parchment or paper glued to a wooden quadrant was within every scholar's means, although sorting out the mixture of scales and overlapping star markers might not have been so. For the former reason, no doubt, it was very popular in the Ottoman empire – usually in the form of a lacquered wooden quadrant – from the fifteenth century onwards. It survived in Turkey even into the twentieth century, valued by those who used traditional astronomical methods for ordering religious life.

MACHINERY OF THE HEAVENS

We have seen that in late antiquity water was used to drive automata, in particular time-measuring devices. Hero of Alexandria (*fl.* AD 62), who is usually described as an ingenious technician but who also has several important mathematical results to his credit, devised numerous machines, including many that were worked by water, steam and air pressure. A work that he wrote on water clocks is now lost. Like his predecessor Archimedes, Hero seems to have influenced the Arabs, and in the tenth century al-Khwārizmī hinted strongly that the old techniques had not been lost.

There are clear signs of Greek influence in the early thirteenth-century writer al-Jazarī. He worked in the service of Naṣīr al-Dīn, the Turcoman king of Diyār Bakr, for whom he made a variety of ingenious machines and gadgets. He acknowledged that the idea for a water clock he made came from another work, one that he ascribed to Archimedes. Other writers (including the great al-Bīrūnī) mention clocks driven by water or sand, and it is clear that there was a continuing interest in the problem, not simply of turning a wheel to keep time in some abstract way, but of turning *a representation of the heavens*, be it a plane image or a globe, with a twenty-four hour motion.

Such devices appear in Spain in the eleventh and early twelfth centuries. Indeed there is a much earlier reference to a water-clock in a poem written around 887 for the emir 'Abd al-Raḥmān II. One eleventh-century work describes in detail an elaborate mechanism for operating automata, so continuing a practice that had continued without break from ancient times, an important source of ideas, although here not overtly astronomical. Another author describes a celestial globe turning with the daily motion, the motive power being provided by a falling weight floating on a bed of sand, the level of which fell as sand escaped through an orifice.

Water clocks from fourteenth-century Fez (Morocco) included astrolabe dials. Whether by that time western European knowledge was being conducted back into the Islamic world of northern Africa is hard to decide, and is best left an open question, but certainly, by the time the Fez clock was set up in the Qarawiyyin Mosque there, the European astrolabe clock with a *mechanical* drive was not uncommon. And that supremely important invention was a direct product of the desire to represent the moving cosmos, as we shall see.

THE ALFONSINE TABLES IN SPAIN AND PARIS

Paris, beginning in the 1320s, was the most important single point of diffusion of the 'Alfonsine tables', which soon thereafter replaced the old Toledan tables. There are those who would argue that these tables are not truly Alfonsine at all, but an essentially Parisian creation. No copy of the original tables is known, although the explanatory canons in Castilian survive, and we can reconstruct some of the parameters in the tables with a fair degree of confidence. The origins of much of the new material are to be traced to the various leading zijes we have already discussed in general terms. However, origins and diffusion notwithstanding, it seems reasonable to say that without Alfonso X of Leon and Castile, 'Alfonso the Wise', they would never have existed.

The Alfonsine tables are only one aspect of the important scholarly activity of this thirteenth-century ruler, who encouraged the translation from Arabic into Castilian of many philosophical and scientific writings, a task already begun under the patronage of his father San Fernando, but one that reminds us strongly of the great patrons of astronomy in eastern and western Islam. The very introduction to the canons uses phrases that present the king in this light. The tables make use of an Alfonsine epoch of noon, 31 May 1252, the eve of his coronation, although they were

assembled between 1263 and 1272, according to the pro-
logue. This epoch of a Christian king is then related to the
old Spanish era, the Islamic Hijra and the (Persian) era of
Yazdijïrd. The cultural continuity symbolizes the astronomical continuity.

Translation from Arabic into Latin was well established at
Toledo at the courts of the archbishops. (Toledo was the
ancient Visigothic capital.) Alfonso established a school that
included Christian and Jewish savants, as well as a Muslim
convert to Christianity. He presided over this group in some
sense, revising their work and writing parts of the introductions to it. The names of fifteen collaborators are known from
the complete collection of Alfonsine books, which includes
important treatises on precession, the universal astrolabe in
various guises, the spherical astrolabe, a water clock and a
mercury clock, the simple quadrant ('old', as it is now called),
sundials and equatoria. This rich and encyclopedic collection
also includes an astrological text, *The Book of the Crosses*. Most
of the Arabic sources used were Spanish–Arabic, and it must
not be thought that the drift of ideas was only in the direction
of Europe. One curious fact that probably has something to
do with contacts with Muslim astronomers working under
Mongol patronage is the emergence, in a *Chinese* manuscript
of the following century, of exactly the same value of the obliquity of the ecliptic as appears in two of the Alfonsine books
(23;32,30°).

In view of discrepancies between new observations and
predictions based on the old Toledan tables, Alfonso com-
manded instruments to be constructed and observations
made at Toledo. Two Jewish scholars are given as the compilers of the new tables, namely Jehuda ben Moses Cohen
and Isaac ben Sid. They made observations at Toledo for
more than a year, but the king moved his court often, and
much work was no doubt done at Burgos and Seville too.

In the intricate history of the links between Spain and
the Parisian tables that went under the name 'Alfonsine',
there are various key documents. Historically the most

significant is one by John of Murs, written in 1321, his *Exposition of the meaning of king Alfonso in regard to his tables.* In 1322, John of Lignères – whose pupils John of Murs and John of Saxony both were – wrote a work heavily dependent on the canons to the old Toledan tables but with clear hints of the 'Alfonsine' tables to come: he used the solar eccentricity that characterizes at least the Parisian 'Alfonsine' tables – and I believe the originals too. He also combined precession (with access and recess) into the positions of the apogees, used twelve signs of 30° (Aries, Taurus, Gemini, etc.) rather than six of 60°, and took other steps that seem to reflect the usage of the Spanish canons. (Later the Parisians preferred signs of 60°, which have certain arithmetical advantages, in preference to the more familiar signs of the zodiac.)

At some time between then and 1327, he and his pupils – it is quite certain that they were aware of each other's work – assembled the essential ingredients of what was to become the most popular version of the tables, that is, an edition composed by John of Saxony in 1327. John of Lignères wrote his own canons for the tables between 1322 and 1327, and some time after 1320 he improved them greatly by incorporating tables of *combined planetary equations.* This was an important step. We recall that planetary equations are terms that are to be combined with the steadily increasing and easily calculated mean motions to give the final position of a planet at a particular time. There are two main equations to be calculated on each occasion, one of them dependent on the other. With the new arrangement, for each planet there was only *one* table, giving one combined equation. It depended on two parameters, the 'mean centre' and 'mean argument', with which one entered the table. The resulting tables were well described as *The Large Tables.* In fact a similar step had been taken, probably unbeknown to the Parisians, by Ibn Yūnus three centuries earlier.

There is no evidence for any awareness of the original Alfonsine tables east of the Pyrenees at a much earlier date than 1321. A remark by Andaló di Negro, made in 1323, suggests that he knew the original canons explaining their use, although John of Murs evidently did not. His work of 1321 was essentially a reconstruction. Andaló is well known to history as an astronomer on whom the great Italian story-teller Boccaccio pours excessive praise. As it happens, three Italian scholars are known to have been engaged in the Alfonsine enterprise, namely John of Messina, John of Cremona and Egidio de Tebaldis of Parma. (The involvement of Italians is not surprising, since until 1275 Alfonso regarded himself as a candidate for the crown of Holy Roman Emperor, and regularly exchanged embassies with the Italian states.) Assuming that the tables were brought from Spain to Paris, we cannot say exactly how this happened. There are several possibilities. John of Murs tells us that he knew someone with knowledge of them, but who was keeping the knowledge to himself. This suggests contact with someone schooled in Spain. A Parisian scholar in the 1340s mentions a *Spanish* version of the Alfonsine book of the fixed stars, 'taken from the king's bookcase'. The writer also said that he had seen a globe that had been made for Alfonso, with stars appropriately marked. Whether these things were seen in Spain or after being transported to Paris is not clear. Perhaps one day documentation will emerge that serves to fill in the gap in our knowledge of the missing fifty years, and the route of transmission.

John of Murs showed considerable determination and skill in extracting several parameters from his material. There was much hard calculating to be done, and it is not without significance that he addressed a sexagesimal multiplication table to a friend in the very year of his *Exposition*. He does not claim to have done anything original, but he thought his account to be right because the tables then fitted with an observation he had made. His loyalty to the Castilian tables seems clear enough from the tone of his writing. A figure for the maximum equation of the Sun

(2;10°) seems to be expressly calculated from the Spanish tables. The Parisian planetary parameters would have been much harder to extract, but they surely came from the same source.

The hardest part of all would have concerned the Alfonsine theory of precession. This, briefly, combined *two* motions, a *steady* (secular) motion like that proposed by Hipparchus and most early followers, and a long-term *oscillatory* motion in the style of the theory of access and recess dubiously ascribed to Thābit ibn Qurra.

The steady component in Alfonsine precession is at the rate of 360° per 49 000 Julian years. This makes use of a very peculiar idea indeed, for it makes a movement of the stars dependent on a purely arbitrary decision as to how many days we should take for our calendar year. (To take 365¼ days is an arbitrary decision – nobody would then have claimed that it was exact. The tropical year, according to these astronomers, was about 10;44 minutes less than 365¼ days. In 49 000 years the discrepancy amounts to 365.23 days, hence the choice of that large period.)

In tracing the history of mathematical astronomy it is tempting to concentrate on the grand issues, such as those that converge on Copernicus and Galileo, at the cost of subtler themes. No blockbuster film on Copernicus is ever likely to mention access and recess, for example, even though he flirted with the idea, and even though it well illustrates an old weakness for the veneration of one's ancestors. Historical clues are often concealed in apparently innocuous, even trivial, data. Consider for instance the star catalogue that occurs in the Alfonsine *Books of the Eighth Sphere*, as well as in some printed editions of the Alfonsine tables. This has Ptolemy's star longitudes increased systematically by 17;08°. The first (1256) version of this work, essentially based on the Arabic al-Ṣūfī intermediary star-list (AD 964), was edited by two Jews, Judah ben Moses and Samuel ha-Levi, and two Christians, John of Messina and John of Cremona, in 1276. The figure of 17;08° does not fit with the double theory of precession, calculating across the time interval between Ptolemy and 1276. It does, however, fit with a calculation for the interval from AD 16 (which happens to be the starting point for the Alfonsine oscillatory model of precession) to 1252 (the coronation of Alfonso). We know that some Arab astronomers thought that Ptolemy was drawing on a lost work by Menelaus, from the late first century. We now also know that Ptolemy's longitudes were on average rather more than a degree too small in his own day. It is therefore at least conceivable that Alfonso's astronomers, who knew Ptolemy's approximate dates, knew too that his longitudes were systematically in error for his own time, and ascribed them to an astronomer working over a century earlier, in year AD 16.

In the following chapter we return to the later history of the Alfonsine tables, which helped to mould European astronomy for well over two centuries. After their first Parisian decade they continued to be developed in important ways in England. J. L. E. Dreyer thought that he had found the 'lost' Spanish tables in their original form in England, in the tables of William Rede, but in fact they can be explained entirely in terms of Parisian intermediaries.

The Spanish role in the revival of western astronomy was far more important than the impetus coming more directly from the direction of Byzantium. The latter was in some respects potentially stronger, for it had access to more polished eastern Islamic material, and in Byzantium there had never been a complete break with a tradition that made use of the works of Ptolemy and Theon. There were even periods there when astronomy became fashionable at the court. The seventh-century emperor Heraclius, for instance, practised astronomy, as did his successor Manuel Comnenus half a millennium later. If we judge their work to be inward-looking, the fault is perhaps ours, for this is largely unexplored historical territory. In the eleventh century new astronomical writings were being produced, but now derived from Arabic zijes. There was again something of a Byzantine astronomical revival in the fourteenth century, but it was mainly directed towards commentary on works long outdated, in particular Ptolemy's. (Three astronomers worth naming here are Theodore Metochites, Nicephorus Gregoras, his pupil, and Nicolaus Cabasilas.) At the same time, new collections of Persian tables became available in translations by Gregory Chioniades. This new activity has its own history, but it came too late to be of much influence in the world outside Byzantium, except in a small way at a literary level. In this last respect, like all things Greek during the Renaissance, it had a measure of prestige, but it was ultimately powerless to affect the course of astronomy in the way the astronomers of al-Andalus had done.

Medieval and Early Renaissance Europe

THE NATIVE TRADITIONS

Since a deep concern with the patterns of movement in the heavens is evident in Europe as far back as neolithic times, it is not surprising to find traces of it in the earliest historical documents available to us. In his account of his wars in Gaul, Julius Caesar ascribes astronomical knowledge to the Druids, and the Roman historian Pliny the Elder (AD 23–79) adds a few details. It seems that the Celts began each month, year and cycle of thirty years with a new moon. Days, in the sense of day-with-night, were counted beginning with the night. (Note the English 'fortnight'.) Fragments of a bronze calendar were found in 1897 at Coligny, near Lyons in France, with fragments of a statue of a god. Slighter fragments of another had been found in the Lake of Antre near Moirans (Jura), in 1802. They seem to date from the end of the second century, and to show that the lunae-solar calendar then in use had months of twenty-nine and thirty days, and that two months of thirty days each were intercalated every five years. The Coligny calendar shows a complex pattern of marking months and days, presumably distinguishing them as favourable and unfavourable. A similar distinction – if this is indeed what was intended – was common in calendars throughout the middle ages, especially using a system known as 'Egyptian days'. Early Irish literature shows that births were sometimes delayed to ensure that they took place on a lucky day. (Note that the Celts grouped their days in threes and nines. The Jewish week of seven days was introduced only with Christianity.)

An earlier division of time, by seasons, as marked in particular by the winter and summer solstices, has left traces in folk custom throughout the Celtic world – especially in Brittany, Ireland, Wales, the Isle of Man and the Highlands of Scotland. It has often been said that the festivals of the Christian Church were cleverly made to coincide with the older pagan festivals. The feast of St John the Baptist on midsummer day is often said to have replaced the feast of Beltane, and this under the influence of St Patrick himself. (Patrick does once allude to worship of the Sun in Ireland.) There are pagan elements even in the Christian festivals, however, and the near-universality of festivals of the Sun – most of them by definition pagan – makes it difficult to say whether the imported festival was markedly less pagan than the native versions.

Ancient Germanic and Scandinavian traditions of a similar kind are if anything harder to discern. There are rock markings from Sweden suggesting Sun cults, and the Sun and Moon, day and night, summer and winter are all personified in the poems of the older Edda. We know from the Latin writer Procopius that when in places north of the Arctic circle the Sun disappeared for (as he says) forty days in the winter, the days were counted off until it was time to send observers into the mountains to watch for the rising Sun, and so give five days warning of its return to the people below, who so prepared for their greatest festival.

There are comparable Sun cults among all ancient Baltic peoples, and naturally also in Iceland, but whenever among them are found elements resembling the more systematic parts of astronomy from elsewhere, cultural influences are almost invariably traceable. A good example is in the so-called 'Golden Horns of Gallehus', two large horn-shaped artefacts of gold found near the Danish village of Gallehus, but at different times. They were stolen from the royal treasury in Copenhagen in 1802, and so lost forever, but minute drawings of them show that they were covered in human and animal forms that W. Hartner interpreted as stylized inscriptions (written in runes) relating to an eclipse of the Sun on 16 April 413. Hartner finds in the symbolism traces of Hellenistic and Oriental astronomy.

COSMOLOGY OF THE EARLY CHRISTIAN CHURCH

The Christian scriptures have by some been considered to have favoured the pursuit of astronomy as a science, and by others to have been hostile to it. They certainly make use of many primitive analogies between the Universe and the furniture of the everyday world. The tabernacle that Moses constructed in the wilderness was the world, the seven-branched lamp the Sun, Moon and planets, the six-winged golden figures the Greater and Lesser Bears, perhaps. This sort of thing was not antagonistic to science until it gradually gave rise to a large body of mystical commentary that – or so it was often felt – had to be defended at all costs for the good of the Faith. In some cases this mysticism was blended with a crude and untutored common sense, as in the extreme case of Lactantius (*c.* 240–*c.* 320), who became tutor to the Roman emperor's son. He preached against Aristotle and in favour of the flatness of the Earth.

It is a common myth – perpetuated, as it seems, by most teachers of young children – that Columbus discovered that the Earth is round. Of course different people believed different things at different times. The teaching of the Greeks and their intellectual successors was plain enough, but there are many echoes of Lactantius' hostility to the round Earth and the cosmic spheres in late antiquity and the early middle ages. The question of the possible existence of people at the Antipodes continued to worry such writers for centuries to come. These forerunners of the cast of *Neighbours* were supposedly not descended from Adam, were beyond redemption, and by walking with their heads below their feet were incapable of rational thought.

Some of the Church Fathers did their utmost to reconcile the scriptures to Greek philosophy: here Ambrosius, bishop of Milan (*c.* 339–*c.* 397), deserves a prize for his remark that a house may be spherical inside and square when seen from the outside. Always the book of Genesis gave problems, for example over the place of the waters suspended somehow above the firmament. St Augustine (354–430) transformed Latin Christianity – to which he was a late convert – by his Neoplatonism and even his paganism, but he was capable of discussing the question of sphericity without daring to express a final opinion on it. By comparison with these authorities, John Philoponus (*c.* 490–*c.* 570) makes a refreshing change. It was he above all others who turned the Alexandrian school into a Christian one, and his commentaries on Aristotle, including criticism of Aristotle on the eternity and the substance of the heavens ('the fifth essence'), are of a high intellectual standard. He would have deserved a place in history for his *On the Construction of the World* had he written nothing else, for this is a devastating attack on Theodore of Mopsuestia's use of scripture as a scientific text proving that the heavens are not spherical and that the stars are moved by angels.

Such early Christian debate at length had repercussions on Europe as a whole. Great changes came about with the collapse of the western Roman empire. Pressures on Rome from the north had been growing for a century when in 476 Odoacer became the first barbarian king of Italy. The other great change in northern Europe came with the conversion of many of its peoples to Christianity. The Church brought with it books and learning. The seventh-century scholar Isidore has already been mentioned in connection with events in Spain. His technique for staying out of difficulties was to quote Greek authorities without any major criticism of them.

Isidore's case does not well illustrate the dramatic mixing of cultures that was taking place at the outposts of Europe. Here the Venerable Bede, his near-contemporary (672–735), is a better example, brought up as he was in the monasteries of Wearside and Jarrow (near Newcastle) that had been founded by his abbot Benedict Biscop a few years earlier. (Monasticism in Britain had begun around the year 430.) Benedict had come far: he had studied at the very famous school of Lérins (an island near Cannes) and at Rome, and he brought with him from Rome two scholars from even farther afield. Archbishop Theodore was ultimately from Tarsus in Asia Minor, and Abbot Hadrian from North Africa and Naples. Later Bede brought other scholars from Ireland and the continent of Europe to join him.

Bede was fortunate, and the dozen volumes of his published works show a mastery of most branches of conventional Christian learning. Astronomy was not strongly represented, but time-keeping, by the hour and the calendar, was to the rituals of monastic life what grammar was to the study of the Bible. Bede's works on the subject, the *computus*, became classics to which reference was made for a thousand years. He made use of the nineteen-year lunae-solar cycle to construct a perpetual Easter cycle of 532 years, so reconciling earlier cycles over which Church opinion had been divided. As a historian, Bede introduced the habit of dating events by reference to the Christian era. He wrote on the tides largely from first-hand experience, but in cosmological belief – for instance in his book *On the Nature of Things* – he inherited a motley collection of ideas on the stars, thunder, earthquakes, the parts of the Earth, and so on, drawing on scripture, Isidore and Pliny. Bede, be it noted, did teach the sphericity of the Earth, and such basic astronomy as was needed to explain the inequalities of day and night, the variation of both with latitude, and the rough general pattern of observed planetary motions.

Bede's scholastic texts for long enjoyed an extraordinary reputation, and they kept alive simple general principles of astronomy at a time when many in the Church reviled it. During the eleventh century his works were largely put aside in favour of others by Boethius. Boethius (480–524), a Roman aristocrat now best known for his *Consolation of Philosophy*, had written on almost every aspect of scholastic learning except astronomy. (This fact probably reflects the scarcity of astronomical texts in Rome in his day.) It is noteworthy, even so, that Theodoric – the Ostrogothic king of Italy who was later to order Boethius' execution – asked his advice when the king of Burgundy asked for a water clock and sundial.

What Boethius did in the long term to western learning, through his wish to reconcile the writings of Plato and Aristotle, was to propagate a broad cosmology that was more or less Aristotle's. He strengthened respect for a Universe governed by chains of cause and effect. He prepared the way for a more thorough-going Aristotelian invasion in the thirteenth century, and this bore fruit – but very slowly – in introducing a more physical approach to cosmological matters.

Two ancient writers with a greater impact than Boethius on medieval astronomy were Martianus Capella and Macrobius. Both Latin authors were seemingly from North Africa. Martianus composed what was unquestionably one of the most popular of all Latin textbooks of the middle ages, *On the Marriage of Philology and Mercury*. In it, each of seven bridesmaids presents one of the seven liberal arts – the 'trivial' grammar, rhetoric and dialectic, and the more advanced 'quadrivium' of the sciences, geometry, arithmetic, astronomy and music. The astronomy was very basic, and after the arrival of Islamic texts might have been judged insufferably vague by those who wanted to get to grips with them, but it had the makings of good cinema. Astronomy's bridesmaid arrives at the wedding party in a hollow sphere of heavenly light, filled with a transparent fire, and gently rotating. She carries a pair of dividers and a globe. These objects were for long used to symbolize astronomy.

Macrobius was a fifth-century commentator on Cicero's *The Dream of Scipio*, in which the dreamer had been portrayed as undertaking a journey through the spheres, allowing him at length to look down on the entire Universe. The setting allowed Macrobius to expound astronomy in a literary vein, less colourful than Martianus', and owing more to the style of Cicero and Plato. Both writers unfortunately owed more to minor handbooks on astronomy than to the leading authorities of antiquity. Martianus' commentary too became very popular, and with Macrobius he provided the poets Dante and Chaucer with models for heavenly journeys of different sorts. In fact Dante's *Divine Comedy* certainly helped to promote the popularity of the sources themselves, as well as that of Aristotle.

It is no accident that in that, the greatest of all medieval allegories, Dante provided a moral theme within a framework of Aristotelian cosmology. His *Divine Comedy* may be read at various levels, but taken most literally it is an account of his vision of a journey to Hell, Purgatory and Paradise. Hell is described as a stepped conical pit, to successive circles of which various classes of sinners are assigned. Purgatory was seen as a mountain rising in a succession of circular ledges on which are the several classes of repentant sinners. In Hell and Purgatory his guide is the poet Virgil, and in both places he sees, and converses with, his former friends and enemies. Paradise, by contrast, is a region of light and beauty, in which his guide is Beatrice, now an angel. This is a vision of what he conceived to be the true state of the world beyond the experience of the living. Allegorically, he tells us himself, its subject is humankind's liability to just rewards or just punishment in accordance with the exercise of free choice. Literally understood – and it must not be supposed that this is more than a surface to the poem – his descriptions are more or less those he had been taught by the classical astronomical authorities in use in his day. His Universe is a modified Aristotelian world with harmonies of various sorts built into it, and running through his verse is the view that human fortune is intimately linked with those harmonies. Beatrice acts as a tutor to him, explaining to him how the innate *forms* of things work, how fire is drawn up to the Moon, how the Earth is bound as one, how the Primum Mobile, the First Mover, moves creatures irrational and rational alike. She explains numerous astronomical technicalities in general terms, the spots on the Earth, the eclipse of the Sun, epicycles and their effects on retrogradation, the pointing of the magnetic needle to the pole star, and so forth. How does it come about, Dante asks her, that he can fly through the spheres? If you, Dante, being free from constraint, were to stay on the ground in this Aristotelian world, she implies, that really would be a cause for surprise.

In the twenty-eighth canto of the *Paradiso*, Dante looks into Beatrice's eyes and sees God as an infinitely small but brilliant point of light encircled by nine radiant rings which Beatrice relates to the movements of the heavens – the Aristotelian or simplified Ptolemaic system, of course – but also to the three hierarchies, or nine orders of angels as described by Dionysius. The author of a popular Neoplatonic work called the *Celestial Hierarchy* was not, as Dante thought, the eminent Athenian Dionysius the Areopagite whom St Paul converted to Christianity, but no matter. The pedantic Beatrice could not refrain from pointing out that another opinion had been held by Gregory. Elsewhere Dante even adopts a third system. The details need not concern us, but there are two points of great interest in all this combination of crude astronomical and theological elements. The ultimate sources are apocalyptic writings of Jewish origin, and earlier still Persian and Babylonian literature that had an astronomical origin of sorts. History had indeed come full circle. And allegorically, the angels represent the operations of divine providence, through which God's love sustains the spiritual order of the Universe. There is here just the slightest hint of astrological determinism, but it is so Christianized as to have been above reproach. It is not without interest that the souls of the blest encountered in the different heavens are placed precisely in accordance with astrological principles.

There is another sort of cosmic symmetry in Dante's poem, not unrelated to the first, a form of number mysticism involving especially the number nine. It is found in the structure of hell, for instance, which mirrors that of paradise in a peculiar way. Dante manages to work the numbers of lines and cantos into the symmetries of his story, and those of the calendar too. The whole story was related to a very specific Easter – it is generally agreed that this was Easter 1300 – and some have argued that he wove planetary positions into it. He certainly seems to have added a personal astrological touch: ascending from the heaven of Saturn to the heaven of fixed stars he finds that he has entered his native sign of Gemini, the Twins. He speaks as though the stars in Gemini 'poured forth virtue' at his birth. 'To you', he says, addressing the stars, 'I owe such genius as does in me lie.' Finally, he looks back for a last tutorial on the heavenly spheres and the characters of the planets. This is a story of divine love, but we are reminded of some of the most striking lines in the whole poem, in which Dante describes Venus in the opening canto of *Purgatorio*:

> The radiant planet, love's own comforter,
> was setting the eastern sky a-smile,
> veiling the fishes that escorted her.

How real was all this meant to be? Was Venus rising with the stars of Pisces at dawn as he set out? At Easter 1301 it was, while at Easter 1300 it was not. Those desperate to prove a point have produced a faulty almanac that he might have been using. True or not, we can say with some confidence that Dante was far more interested in the formal, architectonic, symmetries of his art than in specific astronomical detail.

Chaucer, towards the end of the same century, took a very different approach, when he wove astronomical allegory into his art. As the author of treatises on the astrolabe and equatorium, he was not only a competent but an expert calculator, in possession of the best available edition of the Alfonsine tables, and capable of working with it to a high level of accuracy even in his verse. Many of his *Canterbury Tales* have beneath the surface a strain of astronomical allegory based on quite specific events in the heavens in his own lifetime, indeed almost certainly around the time of composition. There are so many instances, and they are of such different types, that it is hard to know where to begin, but since it is relatively well known, we may take briefly some of the elements in *The Nun's Priest's Tale*.

This is the story of a cock, Chauntecleer, and his narrow escape from a fox. By his nature, the cock knew that when the altitude of the Sun was 41° and the solar longitude a little over 21° of Taurus, the time was nine in the morning. (The poet does not set out these facts so baldly, but plainly enough.) The data we are given fit perfectly with a specific date, Friday 3 May 1392, following astronomical tables by the friar Nicholas of Lynn, whom Chaucer knew and mentioned elsewhere. The characters in the poem have simple analogues in the heavens – Chauntecleer being the Sun, his wives being the stars in the Pleiades, which we can show the Sun to have been actually passing on the day of the story. The fox is Saturn. The story hinges on four different arrangements of the heavens during the fateful day, and it is quite conceivable that these were explained subsequent to the telling of the story with the help of the astrolabe on which Chaucer was so expert. And as a check to scepticism, if not a complete cure for it: it turns out that medieval star lore in many European countries made out the Pleiades to be seven chickens. The Sun was indeed passing the Pleiades on the day in question, and when the action of the tale begins, Chaucer tells us explicitly that Chauntecleer was walking by the side of his seven wives. Astronomy has much to answer for.

THE UNIVERSITIES AND PARISIAN ASTRONOMY

The framework of formal medieval western education was based on the seven liberal arts. They provided the staple curriculum of the universities, and were a great source of strength for that characteristically European institution. The universities derived some of their patterns of organization from Islamic schools of learning, but what was new was their trans-continental recognition, protected and given privileges as they were by local rulers and by the Pope. The ultimate purpose of the universities was in fact to provide the Church with an educated clergy, but this is not to say that they had the same religious preoccupations as the older cathedral schools. The faculty of 'Arts' was basic, while the higher university faculties of Law and Medicine, and the highest of all, Theology, were reached by a very small fraction of the university population. The vast majority of students were obliged to know the 'four sciences', of which astronomy was one. Medicine in any case required a knowledge of astronomy, for not only were such practices as blood-letting related to the phases of the Moon, but astrological prognostication was an important part of the physician's repertoire.

The universities provided an elite with the knowledge needed to serve Church and State. The first meriting the name, in order of foundation, were those set up in Bologna, Paris and Oxford. Dates of foundation were already a matter of dispute in the middle ages, and do not concern us here, but no one would deny that the beginning of the thirteenth century saw a very marked rise in their social and intellectual importance. New introductory texts were needed, and so it was that in the early decades of the century John of Sacrobosco wrote what was to become one of the most widely studied astronomical books of all time, *On the Sphere*.

This work, by a man who was perhaps an Oxford master and who was certainly a teacher in Paris, was much ornamented with quotation from classical poets, and it dealt with only elementary spherical astronomy and geography, and hardly anything on planetary theory; but a start had been made. The same writer produced other popular texts on arithmetic and the computus (calendar calculation).

Other works at the same level followed – for instance by Robert Grosseteste, Oxford's first Chancellor, and an early enthusiast for Aristotelian science – and they were supplemented on planetary theory by a type of book known by the generic title *Theorica planetarum* (Theory of the Planets). One good example of this type of work, used in the schools from the twelfth century onwards, was John of Seville's translation of al-Farghānī's treatise; and Roger of Hereford wrote another. The most famous western example, however, was by an unknown author, and today as in the middle ages is referred to by its opening Latin words ('Circulus eccentricus vel egresse cuspidis . . . '). Its author was guilty of technical misunderstandings, and it was sadly uninformative on the question of planetary parameters, but it greatly helped to stabilize astronomical vocabulary.

With the help of diagrams this book taught the student the essentials of the planetary models, unlike the canons to tables, which usually taught only the rules of procedure, without justification. It gave no understanding of how planetary theories had been arrived at in the first place. Ptolemy's *Almagest*, which of course lay behind this type of treatise, could have done that, for it was twice translated into Latin in the twelfth century, once from Greek and once from Arabic. (The humanist cult of the pure Greek text led to another translation being done in 1451.) The *Almagest*, however, was much too long and sophisticated – and costly – for general use, and had been replaced by astronomical digests even in Islam. One of the best known of these, by al-Farghānī, was mentioned earlier. It introduced to Europe the idea that the spheres are nested in such a way as to leave no empty spaces. We have already alluded to the way in which the dimensions of the entire Universe may be related to each other, and ultimately to the lunar distance, using this essentially Ptolemaic model.

Aristotle's *On the Heavens* was studied for its cosmological content, and many commentaries were written on it. It was kept alive by the fact that it was bound up with the rest of natural philosophy and metaphysics. Its dual physics, with a celestial region where natural motions are circular, and a terrestrial region where they are straight up or down, was rarely challenged. Where astronomy prospered, however, especially at Paris and Oxford, there was a tendency to give ever less attention to such commentary, and so implicitly to disengage from a type of homocentric planetary astronomy that had been superseded long before Ptolemy was born.

At first sight, the very character of the medieval university might seem to have been unfavourable to the growth of astronomy, regarded as a science related to the *observed* world. Medieval attitudes to knowledge were strongly influenced by the techniques used in discussing Holy Scripture, that is, as an inheritance, to be purified and restored to its original form, then analysed and commented upon before being transmitted to later generations. Fortunately, as better material became available and as scholars learned something of the intellectual pleasures their subject could give, not to mention the promise it seemed to offer of astrological prognostication, a new type of European astronomer emerged. In this way the tempo of intellectual life increased, only to be checked occasionally by political threat, for example, and plague (especially the Black Death of 1348–9). These very dangers, however, helped the cause of learning in one important respect, for they encouraged the movement of scholars. With the rapidly changing social and intellectual order, many more universities were founded, especially in the fourteenth century and after.

There was no one moment of enlightenment. Roger Bacon (*c.* 1219–*c.* 1292) introduces a mildly empirical note into his writings, but he was no astronomer. Thomas Aquinas (1225–74) lent his reputation to the idea that revelation must be supplemented by reason – and by Aristotle – in the pursuit of truth; but neither was he an astronomer. For the real signs of change we should look rather to a more modest scholar like William of Saint-Cloud, who flourished at the end of the thirteenth century. About his life we know little beyond the fact that he was somehow connected with the French court. In 1285 he recorded an observation of a conjunction of Saturn and Jupiter. He compiled an accurate 'almanach', giving the calculated positions of the Sun, Moon and planets at regular intervals between 1292 and 1312, and introduced it with an account of the observations and planetary tables (of Toledo and Toulouse) on which the almanac was based, as well as the corrections he found it necessary to make to them.

In connection with his work on almanacs, with their references to solar and lunar eclipses, William considered the projection of the Sun's image on a screen through a pinhole aperture. This would avoid damaging the eyes, he said, as had happened in so many cases at the eclipse of 4 June 1285. Roger of Hereford had mentioned the same technique in the twelfth century, and following a suggestion by William, Levi ben Gerson actually used pinhole images in 1334 to yield a figure for the eccentricity of the Sun's orbit.

Levi observed the Sun at summer and winter solstices, using a combination of the camera obscura and the 'Jacob's staff', an instrument of his own invention. Kepler observed a solar eclipse in 1600 in almost the same way. The derivation of the eccentricity depends on the fact that the diameter of the image is in inverse proportion to the Sun's distance, so that the connection between the quantities observed and the geometry of the eccentric circular orbit is very direct.

The practical side of astronomy had begun to grow rapidly at the end of the thirteenth century, and it continued to do so thereafter in Europe without any real break. William of Saint-Cloud also wrote an unusual work on a dial ('directorium') fitted with a magnetic compass, by which it was to be set. But as illustrating another very practical matter, he composed a new ecclesiastical calendar, commencing from 1292, and notable for the care bestowed on its astronomical basis. Grosseteste and Bacon had complained about the inadequacies of the existing calendar long before, and others regularly did so until the Gregorian calendar reform of 1582. William of Saint-Cloud had perhaps been naive in trying to present both ideal and workable alternatives together. That the reform was nearly four centuries in the making, however, was not so much a question of Church conservatism as of the fact that Church councils had more pressing political business.

Most of Italy, Spain, Portugal, Poland, France and the Catholic Netherlands followed suit almost immediately before the end of 1582. For reasons of religious pride, Protestant countries were often very slow to follow the lead of the Catholic Church, and even in England, where the reform had been so long advocated, this was so. There, a number of learned astronomers, including John Dee, Thomas Digges and Henry Savile, reported favourably, but the English bishops remembered the excommunication of the Queen, Elizabeth, by the Pope's predecessor, and stymied the proposed reform. German protestants spoke in much less temperate language: the reform, according to one German source, was the work of the devil. Scotland adopted the change in 1600, but not until 1752 did England follow suit, by which time most European countries had long fallen in line. By Act of Parliament eleven days were omitted from the English calendar in October of that year, five centuries after the death of the would-be reformer Grosseteste. There was much unrest as a consequence – 'Give us back our eleven days!' became an election campaign slogan, as one of Hogarth's prints reminds us – but in the end reason prevailed. In the words of a sermon preached by a certain Rev. Peirson Lloyd, had England stayed with the old system, 'in Process of Time, the two Festivals of *Christmas* and *Easter* would have been observed on one and the same Day'. He refrained from telling his flock how many thousands of years would pass before this happened; but then, how many who advocate adopting the Treaty of Maastricht know its contents?

To return to the end of the thirteenth century: by observation, William of Saint-Cloud found that the positions of the stars implied that the theory ascribed to Thābit was about a degree in error. For this reason he favoured a *steady* precessional motion. All told, his approach to astronomy was unusually critical and constructive, but in his use of fresh observation he set an example that few were yet ready to follow. It seems likely – although we cannot prove as much – that by his example he encouraged John of Lignères and his pupils in the work that culminated with their various editions of the Alfonsine tables. These, as we saw in the last chapter, date from the 1320s.

Another astronomer of this same generation, with a practical bent and strong Parisian associations, was Peter Nightingale, for some time a canon in Roskilde cathedral in Denmark. He had been teaching astronomy and astrology in Bologna when in 1292 he moved to Paris, and there he stayed for perhaps a decade before returning to Roskilde. Like William, he too composed a calendar, more orthodox in this case, and one that was to become extremely popular. In fact, by his example, Peter Nightingale shows us another highly consequential aspect of medieval astronomy, the invention and improvement of instruments for *calculation*. In his case he developed a simple equatorium (Paris, 1293) and other devices for calculating eclipses. As in the case of the equatorium devised by Campanus of Novara, another scholar who had studied in Paris, now in the 1260s, these instruments may be described as moving Ptolemaic diagrams made in metal. The poor student would have made them in wood or parchment.

Paris was at this time the most important European centre of astronomical activity, and one might be forgiven for thinking that a condition of entry to the study of astronomy at the university there was to have the name of 'John'. There was a John of Sicily, of Lignères, of Murs, of Saxony, of Speyer and of Montfort, all in the space of two or three decades. Each left his mark on astronomy, and each took a strong interest in the refinement and reorganization of the Alfonsine and related tables. Here it is worth noting the emphasis placed on ease and speed of calculation. As one of many examples we have John of Murs' tables for the conjunctions and oppositions of the Sun and Moon (for 1321–96), an aspect of ecclesiastical calendar-reckoning. Not for nothing did the Pope (Clement VI) invite him to Rome, with Firmin de Bellaval, to advise on calendar reform (1344–5).

All of these scholars, furthermore, interested themselves in the design of instruments of observation and calculation. In the first category we may mention the use of a quadrant fixed in the meridian (the mural quadrant), and the parallactic rules used by Ptolemy – there is something to be said for an instrument requiring no circular scales, if workshop practice is not equal to the task of making them accurately. Records of observations are ephemeral things, and the fact that we have so few does not mean that these Parisians were only calculators. We know that this was not so. A manuscript by John of Murs includes records of observations made at five different places between 1321 and 1324.

Falling into the category of instrument of calculation, however, we have many more examples. To take only John of Lignères, for example, we have treatises by him on a new type of armillary, the saphea, the Campanus-type equatorium, and a 'directorium' – a calculating instrument related to the astrolabe, but specifically for applying an *astrological* doctrine, the doctrine of 'directions'. Most astronomers had their price. The few biographical details we have of many of these Parisians are known to us only because they were in the service of princes and high ecclesiastics, the very classes that provided the richest market for astrology.

RICHARD OF WALLINGFORD

Oxford astronomy during the thirteenth century had been largely concerned with the production of simple teaching tracts for the sphere and the calendar, and with cosmological questions arising out of Aristotle's *On the Heavens*. The Toledan tables were then in use, and the canons to them helped to consolidate astronomical learning and to fix vocabulary, but not until the early years of the fourteenth century were there signs of much originality. A set of tables drawn up by the Merton college astronomer John Maudith during the period 1310–16 focused the attentions of various scholars on the trigonometry underlying spherical astronomy, and provided the first exercises in this subject for one of the most remarkable astronomers of the middle ages, Richard of Wallingford (*c.* 1292–1336).

It is customary to rank early astronomers according to their originality in devising new planetary systems, but this is to lose sight of what was felt to be most needed in the fourteenth century, namely, methods of rapid calculation and representation. In his efforts to provide these, Richard of Wallingford provided a number of original ideas that reveal great qualities of mind, and that had significant, but largely hidden, repercussions.

Richard of Wallingford was a Benedictine monk educated at Oxford, where he also taught until 1327. In that year he returned as abbot to his monastery at St Albans – England's premier monastery. He visited Avignon – then the seat of the papacy – to obtain papal confirmation in his new office, but when he arrived back in England he discovered that he had contracted leprosy. Far from shunning him, his monks were so proud of his achievements that they kept him as their abbot until his death. His *Quadripartitum* was the first comprehensive treatise on spherical trigonometry to be written in Christian Europe. It was developed on the basis of the *Almagest*, the Toledan canons, and a short treatise perhaps by Campanus of Novara. While he was abbot he found time to revise it, taking into account a work by the twelfth-century Sevillian astronomer Jābir ibn Aflaḥ (one of two scholars known in the west as Geber, the other more famous as an alchemist).

Before leaving Oxford, Richard wrote three other works and some minor pieces. His *Exafrenon* was a treatise on astrological meteorology, a restrained but unoriginal work. He wrote also on an instrument he had designed, the 'rectangulus'. The third dealt with his equatorium, which he called 'albion'.

The *Quadripartitum* offered exact solutions of problems in the geometry of the sphere, for instance involving spherical triangles, but such calculation was tedious. The armillary sphere could give approximate answers, but was difficult to construct accurately. The rectangulus incorporated a system of seven straight pivoted rods to solve the same sort of problems, and like the armillary was in principle usable for observation, giving co-ordinates directly. The fact that the rods were straight meant that great accuracy of construction and graduation was possible. There were some compensating disadvantages, but the design shows great powers of intuition. The geometrical problem is that of combining vectors in three dimensions. The mechanical problems to be solved arise from the fact that one cannot pivot seven rods in different planes around a single point. One is reminded here of the Rubik cube, or its spherical counterpart.

Richard of Wallingford's most important finished work was his *Treatise on the Albion*. The albion ('all by one') is in many respects the most notable of the entire genre of planetary equatoria. It did not directly simulate the motions of the planetary circles, giving each a metal equivalent, as did most of its predecessors. Instead, mirroring the use of tables, discs were used to calculate the planetary equations, which were then added to (or subtracted from) the mean motions by rotating the discs through the appropriate angles. This entailed the use of non-uniformly graduated scales; and to lengthen the scales, Richard made some of them in spiral form. Any number of turns to the spiral was in principle possible, but thirty was not unusual. As a cursor, a thread was drawn out through the centre. The whole thing is very reminiscent of the circular slide rules that passed into history, more or less, with the advent of the electronic calculator.

The albion had more than sixty scales in all, some of them ovals. Their complexity was not entirely apparent, but was concealed in the various methods of graduation. The instrument incorporated two different types of astrolabe, one a 'saphea', but they were not essential to it. There was almost no problem of classical astronomy that could not be solved on the albion. As subsidiary instruments over and above those for planetary positions, it gave parallax, velocities, conjunctions, oppositions and eclipses of the Sun and Moon.

The other type of equatorium, the simple analogue of the planetary models, was easier to understand and remained more popular, but was much more limited in its possibilities. The versatility of the albion earned it great esteem, first in England and later in southern Europe, and it remained in vogue, in various anonymous forms, until the sixteenth century. At least seven treatises were derived from it, and astronomers began to abstract its subsidiary instruments, especially its parallax and eclipse instruments. The Viennese John of Gmunden produced what was perhaps the most often copied (*c.* 1430). Regiomontanus (1436–76) drew from it, and produced a rather careless edition, and John Schöner abstracted its eclipse instrument. The most striking printed work to use it was the *Astronomicum Caesareum* (1540) of Peter Apian of Ingolstadt, a work dedicated to the emperor and his brother. Replete with moving discs, it was one of the most lavishly illustrated and colourful scientific works of the first four centuries of printing.

These writers took certain underlying principles from Richard (they cannot be explained here, but they concern the graphing of functional dependences) and extended them in ways that had significant repercussions in later history. In France, from 1526 onwards, the mathematician and cosmographer Oronce Fine (1494–1555) wrote several treatises on equatoria that made use of similar principles; at about the same time in Aragon Francisco Sarzosa did the same; and there were several other instances of similarly sophisticated techniques that were in later centuries to be given the name of 'nomography'.

Apian's painstaking print work came in for passing criticism by Kepler, who called it a waste of time and ingenuity, but its purpose was not to please the likes of the serious Kepler – who was in any case not above seeking easy routes to calculation, or, for that matter, princely patronage.

THE CLOCK AND THE UNIVERSE

At St Albans, a wealthy monastery, Richard of Wallingford was able to dispose of extremely large sums of money on the building of a mechanical clock. We know from a reference in a commentary on *The Sphere* of Sacrobosco, written by Robert the Englishman in 1271, that astronomers were then working – as yet unsuccessfully – on the problem of controlling a wheel in its rotation, so as to provide it with the daily motion. From many references to the building of expensive church clocks in the 1280s and after, we know that the key invention, the mechanical escapement, had arrived – that is, more than forty years before Richard began his work. Despite this fact, the disordered pile of documents he left at his death, with several drafts for the design, contains the oldest surviving description of any mechanical clock, and also – this is one of the paradoxes of history – what was mechanically the most sophisticated of the middle ages.

The clock was lost to history after the dissolution of the monasteries in the time of Henry VIII. The antiquary John Leland reported that it showed planetary movements and the changing tides (these would have been at London Bridge, calculated automatically from Moon positions in a standard medieval way). The whole thing was built in iron – Richard's own father had been a blacksmith. It was built on an enormous scale, with frame about three metres across, on the wall of the southern transept inside his abbey church.

This mechanism is highly relevant to the history of astronomy. Not only was it a moving replica of the Universe as it was known to the medieval astronomer, but almost every aspect of the design was inspired by astronomical practice, even down to the method of calculating gear ratios and tabulating the spacing of gear-teeth. It had hour-striking on a twenty-four-hour system – seventeen strokes of the bell at seventeen o'clock, and so on – and it struck the equal hours of the astronomer, rather than the common man's seasonal hours. It had spiral gears, and an oval wheel to give a carefully calculated variable velocity for the Moon's motion around an astrolabe dial. (This left a theoretical error of only seven parts in a million.) It had differential gears for a lunar phase and eclipse mechanism.

The St Albans clock, completed after Richard's premature death, had a single face, and in this was totally unlike the somewhat later astronomical clock, or 'astrarium', built by Giovanni de' Dondi (1318–89) between 1348 and 1364. The son of an astronomer-physician who had designed a clock for Padua in 1344, Dondi became physician to the Emperor Charles IV. His astrarium was in 1381 acquired by one of the Visconti, the dukes of Pavia. It was seen by Regiomontanus in 1463, for whom it was copied, but was beyond repair by 1530, when the Emperor Charles V had it copied.

The Dondi mechanism had a seven-sided frame, with a dial for each of the planets, the Sun and the Moon. There was a clever digital calendar mechanism. Each planetary mechanism was essentially a geared Ptolemaic diagram. Its mechanical virtuosity was less than Richard of Wallingford's (with sliding rods, and eccentric wheels that were oval only to allow them to mesh at varying distances) but it was doubtless more appealing close at hand. Its brass frame was certainly elegant. Like the simpler type of equatorium that simulated the Ptolemaic models, it failed to represent the Universe as a single system. One might say that in this respect it was truer to Ptolemy's *Almagest* than to his *Planetary Hypotheses*. Richard of Wallingford's clock extended the ancient tradition of the anaphoric astronomical clocks of antiquity, depicting the Universe in a single display.

Most early church clocks had no display, and merely sounded a bell on the hour, but in the course of time most great cathedrals and monastic churches had clocks built with some sort of astronomical symbolism in a *single* display – often supplemented by moving human figures and other automata. Here the water clocks of Islam had helped to show the way.

OXFORD AND THE ALFONSINE TABLES

Richard of Wallingford was at the centre of Oxford astronomy when he wrote his *Albion* in 1327, and yet he made no mention of the Alfonsine tables in that work. He used a version of them around 1330. Around 1340, William Rede of Merton College (Oxford) took a Parisian version with sexagesimal time divisions and converted them to tables in the more familiar older ('Toledan') form, for the Oxford meridian. The fact that there are surviving versions for other English towns, some even from the 1320s (for Leicester and Northampton), and others for Colchester, Cambridge, York and London, reminds us that much astronomy was done in religious institutions outside the universities, although no doubt by men with a university education.

These were all relatively simple adaptations, however, and two much more radical revisions were made in Oxford, the first by an unknown man around 1348, perhaps William Batecombe, the second by John Killingworth in the following century.

The 1348 tables go much further than John of Lignères' labour-saving *Large tables*, double-entry tables that yielded a single equation. The 1348 tables allowed the planetary longitudes to be extracted more or less directly, apart from a small adjustment for precession. These tables were extensive, and could be used to carry information about the direct motions, stations, retrogradations of the planets, and other matters that were of an astrological importance. Yet again we have evidence that astrology provided an important motive for the intensive study of astronomy. Of course 1348 was the year of the onset of the Black Death in Oxford, and as scholars of the time noted, this turned their thoughts to God. No doubt it occasionally turned them to astrology too.

The 1348 tables were seized upon by scholars in other parts of Europe. There are early manuscripts from Silesia and Prague with other versions. Henry Arnaut of Zwolle (in the northern Netherlands) used them, referring to them as 'the English tables'. A fifteenth-century Italian Hebrew translation was done by M. Finzi, assisted by an anonymous Christian from Mantua. Giovanni Bianchini, the most notable Italian astronomer of the mid-fifteenth century, was influenced by them in producing a similar set of tables, much used by such leading contemporaries as Peurbach and Regiomontanus. Through this intermediary, and a fourteenth-century eastern European version known as the *Tabulae resolutae* – which had been studied by Copernicus when a student at Cracow – they provided the core of John Schöner's tables of the same name (printed 1536, 1542). These in turn were widely used for many decades, but their origins had been long forgotten. It was the story of the albion repeating itself; but then, in the middle ages the truth was God's truth, and authorship was not contested like territory.

After the 1348 tables, one further set of Oxford 'Alfonsine' tables of great merit was produced, by the Merton College astronomer John Killingworth (c. 1410–45). These were meant to be used in calculating a full planetary almanac (ephemeris). A copy made for Humphrey, Duke of Gloucester, was of great beauty, and heavily interlined with gold leaf. It is not clear whether or not this was meant to reflect the genius of their author, but this was certainly of a very unusual order. Implicit in the tables is a piece of theorizing that he does not write down for us, but that few could reproduce today without recourse to the differential calculus.

PHILOSOPHERS AND THE COSMOS

By 1380, Europe had about thirty active universities, most of them small and recent, but actively competing with one another. Apart from Prague and Vienna, there was nothing that was both east of Paris and north of the Alps. By 1500, there are nearly fifty more foundations to be added to the list. Many of these were of minor importance, but more than a dozen of the new foundations were in German-speaking areas. From there, and places still further east, came a new wave of enthusiasm for astronomy, which for a century and more outstripped that from the older centres.

This movement was not unrelated to the religious changes taking place at the same time, although the relationship of the two is not a simple one. Although Vienna, for example, was an important centre of learning long before the university was founded there in 1365, its rise to importance came only with the schism in the Catholic Church. After that, it provided a natural home to those central European masters and students whose life in Paris had been made difficult – they were unable to endorse the pro-French pope. Later the university was swelled by those Bohemians driven from Prague, after clashing with the ever more assertive German majority there.

Leipzig university was to be founded by secession from Prague in the stormy days of John Hus, the follower of the English reformer, John Wycliffe, and there are many similar stories that relate to the growing sense of nationhood, in scholarship and religion alike. The tragic history of John Hus and Jerome of Prague, both executed as heretics (1415, 1416) in the course of the struggle for ecclesiastical reform, is well known. There followed systematic harassment and execution of Jews – 240 persons were to be burned at the stake in Vienna on a single day in 1421. We should not forget that many of the scholars whose astronomical ideas we are considering were in some way bound up in the academic discussions that justified such actions. They were contributing, however innocently, to an overall Christian philosophical and theological view of the world that could make such things possible. Their lives as astronomers were not compartmentalized as are those of many of their successors today. They were usually theologians, concerned with what they saw as deeper questions of truth, and of who had the right to decide it; and the roots of many of those questions were emphatically bedded in medieval astronomy. After Copernicus, as we shall see in a later chapter, this discussion was transformed into a less dangerous one concerning the nature of the knowledge provided by a scientific theory.

One of the most notable of the masters displaced from Paris to Vienna was Henry of Langenstein (*c.* 1325–97), who was at the same time one of the most notorious of Ptolemy's critics. (He is often called Henry of Hesse.) Like several other western astronomers of the later middle ages he followed the example set by their eastern predecessors, in using physical arguments to criticize the planetary schemes found in the *Almagest*. After leaving Paris, Henry spent the last years of his life in Vienna, and played an important part in the reorganization of the university there. He was at heart, however, a man of the schoolroom, whose *Treatise refuting eccentrics and epicycles* (written in 1364 in Paris) is a rather crabby book, much of it given over to a petty academic criticism of the standard text, *Theorica planetarum*. His more serious aim was that of showing that the circles of Ptolemaic astronomy cannot be considered as physically real mechanisms existing in the heavens.

Richard of Wallingford had made the same point earlier in the century, but in such a casual way that it must have been a commonplace, no doubt under the influence of the Latin version of Ibn al-Haytham. For Henry of Hesse, the circles of astronomy were mere mathematical constructions, justified only by the predictions based upon them. Henry was dissatisfied with Ptolemy's account of planetary distances and sizes, he disliked the equant, and he disliked the irregularities introduced into the theory of planetary longitude by the theory of planetary latitude. In a word, he wanted a Universe that ran on simpler lines than Ptolemy's. Alas, few philosophers have ever understood how complex is the notion of simplicity.

His arguments are at first sight complicated, but they rest on a number of assumptions about the nature of motion that would not now be regarded as acceptable. As a specimen argument: 'If epicycles existed, the same simple body would be moving at one and the same time with different motions', and this cannot be, 'since one and the same cause cannot produce different effects on the same body simultaneously'. Although some scholars were beginning to grasp the point, most shared the difficulty of understanding how two different motions can be combined into one – as when a ship is blown off course by the wind – and conversely, of resolving one motion into two. And why? Because they thought of a motion as a real, unique and ultimate thing. This is ironical, since the aim was to show that the Ptolemaic circles were not at all real, but purely hypothetical.

They had been taught to think in this way through reading Aristotelian physics, and Ibn al-Haytham's influential work *On the Configuration of the World.*

A certain Master Julmann wrote in the same style in 1377, drawing much from Henry, and when he added material of his own, again he revealed the difficulty he had with the idea of a sharing of motions in one body. This problem seemed particularly pressing when it came to visualizing a 'Ptolemaic' model of the spheres, of the kind found in Ptolemy's *Planetary Hypotheses* and al-Farghānī's *Theory of the Planets.* The model for longitudes was complicated enough, but add to it the idea of real epicyclic spheres that explain also the movements of the planets in latitude, and one can soon develop a certain sympathy with the writers in question.

An academic politician of much greater genius was Nicholas of Cusa (*c.* 1401–64), who had been educated first in the Netherlands by members of a devotional sect and then at the universities of Heidelberg and Padua. He is best remembered as a Platonist philosopher, but his interests took in astronomy too, with some curious results. At Padua, with his friend – later a famous geographer – Paolo Toscanelli, he attended lectures by Prosdocimo Beldomandi on astrology; after his ordination as priest he became a friend of the humanist Piccolomini, who was, as it happens, responsible for a simple star atlas; and after performing a number of diplomatic services for the papal court he was made cardinal (1446). Nicholas of Cusa's wealth was such that he could buy very fine astronomical instruments. Fortunately they still survive. In 1458 Piccolomini was elected Pope Pius II, but he was unable to protect his irascible friend, whose wish for reform in Germany caused him much difficulty. Nicholas of Cusa was one of those whose wish for reform extended to the calendar, but he made no real progress there. Here, however, we have an instance of a man whose imagination, unrestrained by much more than broad analogies, led him to ideas about the place of the

Earth in the Universe that would later take on a prophetic appearance.

The most influential of his works was one he completed in 1440, *De docta ignorantia* (On Learned Ignorance). In this he made much use of a 'principle of the coincidence of opposites', a principle at the shrine of which Hegelians and Marxists have since been happy to worship. The general idea was that all problems can be resolved by its use. Apparent contradictions are united at infinity. Each entity is present in every other; the largest number coincides with the smallest ('the maximum of smallness'); the point coincides with the infinite sphere in the same way; and so on.

Now this would hardly be worthy of mention were it not that from the last principle Nicholas drew the conclusion that since a point includes (or mirrors) the whole Universe, there can be neither fixed centre nor fixed periphery to the Universe. In particular, the Earth cannot be said to occupy the centre of the Universe. So much for its *place*. As for its *motion*, this hinges on a principle of relativity: the position of anything is relative to the observer. From this it follows that the Earth may be said to move. And having seemingly dislodged it from its traditional place, he was led to speculate that it might not be the only body on which there are living creatures.

Some of these ideas reflect ancient ideas, held for instance by Hermetic philosophers (followers of the mythical Hermes Trismegistus). They had little immediate impact, and that was only after Copernicus, and even after Giordano Bruno – that is, after the end of the sixteenth century. Then it was that Nicholas of Cusa began to be cited as though he had been a precursor of Copernicus. Descartes quoted him as having proposed the infinity of the world, and his reputation for cosmological sagacity has grown with the centuries. This is hardly merited. When Nicholas wrote on astronomical matters he took a traditional line on the centrality of the Earth and the ordering of the planetary spheres. History, though, is made of reputations, not of merit.

It is true that Nicholas played down the inferiority of the Earth with respect to the regions above the Moon, and in fact he took a most unusual step in detracting from the relative perfection of the Sun. He speculated on the possibility that within its bright envelope there may be a layer of watery vapour and pure air, within which in turn there might be a central Earth. It may be hard to believe that this view could have been entertained in the fifteenth century, and yet in the eighteenth and nineteenth centuries such excellent astronomers as Wilson and the Herschels were not above making very similar conjectures. The main difference was that they avoided the charge of heresy. Nicholas' political rivals accused him of pantheism, and against this charge he defended himself in a book *Apologia doctae ignorantiae* (1449) in which he quoted the Church Fathers and Christian Neoplatonist philosophers – from whom he had taken some of his ideas.

Oddly enough, in view of the Platonist representation of the world as mathematical, he seems to have been led to a very different idea. From the observation that nothing in our experience is mathematically *exact* (no object is truly straight, the Earth is not truly spherical, and so on) he concluded that a mathematical treatment of Nature is impossible. It is not easy to make a scientific hero out of such a philosopher, although many have tried to do so.

PEURBACH AND REGIOMONTANUS

Astronomy is frequently represented as having undergone a sudden revival in the middle of the fifteenth century, as though only then, with the recovery of large numbers of Greek texts, were astronomers capable of enlarging their Alexandrian inheritance. This is an illusion created by the invention of printing, which suddenly made the multiplication of books a relatively easy operation. As a direct consequence of this, the reputations of two men in particular rapidly eclipsed those of the scholars whose textbooks they rewrote. Georg Peurbach (1423–61) and Johann Müller (1436–76) were both of them influential scholars in the new literary movement, but their own astronomical writings continued, rather than overturned, the medieval tradition.

Peurbach was an Austrian scholar who inherited the mantle of an almost equally influential astronomer, the first professor in the subject at the university of Vienna, John of Gmunden. John had died (1442) before his arrival, but had assembled a large number of invaluable manuscripts and instruments, which he had bequeathed to the university. (He edited the albion text himself, for instance, and owned an example of the instrument.) Peurbach received the master's degree at Vienna in 1453, but both before and after this time he travelled throughout France, Germany and Italy. He became court astrologer first to Ladislaus V, king of Hungary, and then to the king's uncle, the emperor Frederick III. At Vienna he taught the classics in the new humanist style, but there he also completed his famous textbook, *Theoricae novae planetarum* (New Theories of the Planets, 1454).

Johann Müller is better known as Regiomontanus, the Latin name for his native Königsberg in Franconia. He took his bachelor's degree at Vienna in January 1454, at the age of only fifteen. A student of Peurbach's, within less than two years he had joined his master in a programme of observation – of the planets, eclipses and comets. They did not allow the astrological implications of what they saw to go unnoticed. In 1460 their careers took a new direction with the arrival in Vienna of Cardinal Bessarion, papal legate to the Holy Roman Empire. Greek by birth, his mission was to placate the emperor in a dispute with his brother, and enlist support for the recovery of Constantinople, captured by the Turks. He was also anxious to accelerate the western intellectual movement to master the Greek classics, and he persuaded Peurbach – who knew no Greek – to produce an improved abridgement of the *Almagest*. George of Trebizond had translated it out of the Greek in 1451, but his was inferior to the twelfth-century version done from the Arabic by Gerard of Cremona. Peurbach relied on the latter, which Regiomontanus said he knew 'almost by heart', and he was about half way through the work when he died in 1461.

Within two years, Regiomontanus completed the task, but the resulting *Epitome of the Almagest* was destined to be printed only in 1496, twenty years after his own early death. The work depends heavily on a widely used medieval *Abbreviated Almagest*, but was more comprehensive. In its finished form it was simply the best available commentary on Ptolemy, and remained so until modern times. It is amusing to recall with what joy the humanists vilified earlier medieval digests, which had supposedly lost the purity of the original.

Peurbach's works included minor treatises on earlier instruments, but more influential by far was his *New Theories of the Planets*, a revised version of the standard text of the middle ages. It was first printed by Regiomontanus in 1474, and went into nearly sixty editions before it fell into disuse in the seventeenth century. Such was the power of the press. The book made popular the solid spheres of Ptolemy's *Planetary Hypotheses*, found in al-Farghāni, Ibn al-Haytham and others. It contained material on the theories of precession attributed to Thābit ibn Qurra, and Alfonsine theory. What it clearly reveals is how strong was the influence from the direction of the various sorts of Alfonsine tables, in particular those (the 1348 tables, and Bianchini's) which were making the calculation of planetary positions so much easier, and therefore discrepancies with observation more readily apparent.

In the late 1450s, Peurbach completed what must have been his most laborious work, *Tables of Eclipses* (printed first in 1514), calculated first for the meridian of Vienna, and in another version for the town of Oradea in Hungary (*Tabulae Waradienses*).

With the encouragement of Bessarion, who had by this time been made papal legate to the Venetian Republic, Regiomontanus left Rome with him in July 1463, and settled for a time in the neighbourhood of Venice and Padua, where he lectured. At least as early as 1467 we find him working in Hungary – in fact as a professor at the new university of Pressburg he was responsible for selecting an astrologically propitious moment for its actual foundation. (This was a not uncommon procedure. The wonder is that it has not been revived.) In Hungary he collaborated with the royal astronomer Martin Bylica, and dedicated a work on trigonometry to the king, Matthias I Corvinus, a noted patron of humanist scholarship.

In 1471 he decided to settle in Nuremberg, a place that he regarded with reason as the commercial centre of Europe. Its position made communication with other scholars all the easier. He could there obtain fine instruments, and he set up a printing press, notable for its Latinate typeface (as opposed to the German Gothic forms) and for the admirable judgement of its owner. His first publication was the *New Theory of the Planets* of the lamented Peurbach. He followed it with his own planetary ephemeris (almanac) for the period 1474–1506, not of course the first almanac, but the first *printed* work to exploit a vast potential astrological market for pre-computed planetary positions. It is said to have been taken by Columbus on his fourth transatlantic voyage, and to have been used by him to astonish the Jamaican Indians through his foreknowledge of the lunar eclipse of 29 February 1504.

This is perhaps a suitable point at which to digress on a much misunderstood situation. As yet, astronomical methods of ocean navigation were rudimentary, the main principle being that of sailing east–west at a constant latitude, as determined by the altitude of the Sun or the pole star, or sailing north–south with the aid of the Sun or a magnetic compass, and keeping a tally of position, again through measuring latitude. The great explorers of the four-teenth and fifteenth centuries were for the most part no more than pilots, that is, they set known courses and found their way when within sight of land by recognising objects on land. Sacrobosco's simple astronomy, with its teaching of the sphere of the stars – not to mention that of the Earth – was finding its way into books for navigators, how-ever, notably in Portugal in the late fifteenth century. There in 1484 King John II set up a commission to improve tech-niques, especially urgent in the southern hemisphere when the pole star was no longer visible. Simplified solar tables were consequently drawn up on the basis of more extensive tables by Zacuto of Salamanca, a Jewish astronomer of the period. They were tested in 1485 off Guinea, and gave rise to a type of literature known as a 'Regiment of the Sun', a set of simple rules for finding latitude through the Sun's meridian altitude. The mariner's astrolabe by which alti-tudes were usually found was crude, essentially no more than a heavy scale fitted with a rule on which were sighting vanes. On a pitching ship it was virtually useless, and the fortunate sailor made a landfall before using it. Its accuracy was rarely better than half a degree. Techniques and instru-ments generally improved steadily, especially towards the end of the sixteenth century, but the problem of finding the ship's true position at sea – that is, supplementing its latitude by its longitude – was not solved to an acceptable standard until the eighteenth century.

The history of the astronomical part of navigation is one in which the most advanced of all the exact sciences offers help in solving a practical problem. There is a myth preval-ent in some quarters that the debt was owed by astronomy, which was supposedly driven and refined in response to the practical needs of navigation. The main problem of navigation before the middle of the sixteenth century was no more than the problem of educating seamen in the most elementary parts of spherical astronomy, and devising instruments for use at sea. There is a whole family of such instruments – various sorts of cross-staff, quadrant, back-staff and sextant, for instance – for measuring angles, but they were of no great importance to land-based astronomy. Astronomers were rarely driven by mariners' needs, except when they had hope of gain, moral or financial, from solving points of theory (say in regard to Jupiter's satellites or the Moon's longitude) that would help in the problem of finding terrestrial longitude.

To return to Regiomontanus: he died on a visit to Rome in 1476. One hollow story has it that he was poisoned by the sons of George of Trebizond, from whom he had learned much Greek, but whose translation of, and commentary on, the *Almagest* of Ptolemy he had openly criticized in abusive terms. The programme of astronomical observation he had begun was continued by an able colleague, Bernhard Walther, over the period 1475 to 1504 – an early example of a fairly continuous set of systematic observations. Walther's observations were published (1544), but before that, some of them were used by Copernicus for the orbit of Mercury.

Regiomontanus' reputation by the time of his death was considerable, and justifiably so, not only for his powerful connections and his enterprise as a scholar-printer, but for his tireless and usually clear way of setting out existing astronomical knowledge, especially its mathematical foundations. In the 1460s he spent much effort on improving the presentation of spherical trigonometry. His work *On Triangles*, first published in 1533, made use of the so-called 'cosine law' as well as the 'sine law' for spherical triangles. Their originality has been much exaggerated by those unaware of the works of Jābir ibn Afflah, Richard of Wallingford and others. Even Jerome Cardan, in the same century, noted with some justice Regiomontanus' many debts to earlier medieval writers. His work in trigonometry has a somewhat more modern look, and he supplemented it with something of even greater practical value, namely tables of trigonometrical functions. Here he was following in the steps of Peurbach, who wrote a work on the techniques required. Regiomontanus broke with traditional practice, and no longer used a sexagesimal division of the standard radius (that is, one which made the sine of 90° equal to 60 parts). He gradually changed his own practice, first taking 60 000 parts to the radius (as Peurbach had done), then 100 000, and later 10 000,000. The excellence of his tables of sines and tangents helped to bring about, for good or ill, the supremacy of the decimal system.

ASTROLOGY AS MOTIVE

Regiomontanus' *Tables of Directions*, produced in 1467 during his time in Hungary, were not purely astronomical. They included tables for astrological purposes, in particular for calculating the end-points of the twelve 'houses'. This term is often used for the thirty-degree signs of the zodiac – Aries, Taurus, Gemini, and so on – in which the planets are supposed by the astrologer to have their homes, their domiciles. It was also used for another kind of division of the zodiac, dependent on the time of day and the place from which the sky is observed. The division in question is something about which Ptolemy was strangely silent, although at least five different methods of performing the division have been (wrongly) ascribed to him at one time or another. The division usually – but not always – started from the ascendant, the point of the zodiac (ecliptic) where it crossed the eastern horizon. The houses were then numbered in the direction of increasing longitude, that is, with the first six under the horizon. The details do not concern us here, but we must emphasize that many of the eight or so quite different methods for effecting the division were mathematically difficult to apply.

The astrolabe could make the astronomer's life easier, but was not precise enough for the fastidious. Rather as in the case of calculating planetary positions for astrological purposes, the sheer time and energy involved prove the seriousness of purpose of the astrologer. Astrology was not primarily a cynical way of making money or attaining power, but once having presented an intellectual challenge, it took on a life of its own.

Certain names were often attached to the various methods of division, and again the printed word suppressed the admittedly hazy awareness of which astronomer had invented which method. The name of Regiomontanus therefore became attached to a method that is at least as old as al-Jayyānī and al-Ghāfiqī. German astronomers called this the 'rational' method, and complained about the way the French astronomer Oronce Fine treated 'our Regiomontanus' uncivilly when discussing it. Cardan, an Italian, accused him of plagiarizing Abraham ibn Ezra. The Italians tended to favour 'the method of Campanus of Novara', but this too involved a historical misunderstanding. The history of these difficult techniques is a tangle of mistaken ascriptions. What is notable is how much it mattered to astronomers of the sixteenth century, into whose scholarship a kind of nationalism had crept that was quite alien to the centuries before it. Once again, however, the power of the printed book was such that 'the method of Regiomontanus' gradually acquired a large international following.

ASTROLOGY

Astrology, which helped to motivate so many of the best astronomers, had entered a strange limbo in the Latin west. On the one hand, its inherited doctrines, many of them mutually inconsistent, were often accorded the sort of uncritical reverence usually reserved for religious works. On the other hand, the training in Aristotelian philosophy in the western universities of the later middle ages encouraged scholars to rationalize astrological writings. The old texts survived – Manilius, Vettius Valens (who despite his

Latin name came from Antioch) and Ptolemy. They were written before the subject had begun to harden its boundaries, and Ptolemy in particular took astrology to be part of the total rational account of the physical world. Mixed in with the old half-rationalized magic, there are signs that it could still be viewed as an empirical science. The Hippocratic writings in medicine, for instance, leave many hints of astrological medicine (*iatromathematica*), where the human body is placed under the influence and protection of the different parts of the zodiac, and of the planets. Here the logic of the argument, where it can be detected at all, usually rested on analogies. Ptolemy describes Saturn as having a cooling and drying quality because he is furthest from the warmth of the Sun and the moisture exhaled from the Earth. Here is one of many such examples from his *Tetrabiblos*, and a few of the scholastics tried to follow in this 'philosophical' tradition.

The Arabs of the eighth and ninth centuries had begun the process of collecting Greek, Persian, Syrian and Indian astrological materials, but not without opposition from many quarters. The great Islamic philosopher-theologians, al-Fārābī, Avicenna, Averroes, and Ibn Khaldun, all opposed astrology in some measure, often reproducing arguments (for instance on fate, determinism, and responsibility for one's actions) reminiscent of their opposite numbers in the early Christian Church. On the other hand, Māshā'allāh, the encyclopedic writer al-Kindī, his pupil Abū Ma'shar and al-Qabiṣi were all astrologers who wrote in an authoritative way very much to the scholastic taste.

The subject became spiritually dangerous chiefly when it spilled over into magic and demonology. Here too, however, al-Kindī had tried to introduce a rational physical basis, and his book *On Rays* was on the very fringe of Christian respectability. And yet it was much copied and read in the west. The west even learned something of Aristotelian physics from Abū Ma'shar's astrology in the twelfth century, before the arrival of the full body of the philosopher's writings. An influential astrologer in that same period was Abraham ibn Ezra, whom we have already encountered, and whose works quickly became available in Hebrew, Latin, Catalan and French. The treatises of Henry Bate, written around the 1280s, continued and reinforced this popular Hebrew–Latin tradition.

The twelfth century saw a flood of Arabic astrological texts reach Latin readers, and this was soon turned into more familiar literary forms. John of Seville not only translated the Arabic, but produced a much esteemed summary, an *Epitome of the Whole of Astrology* (1142). This type of summary continued to be written, often at enormous length, for instance by Guido Bonatti (after 1261) and John Ashenden (1347–8). An English translation of Bonatti of 1676 – ten years before Newton's *Principia* – and frequent references made to Ashenden at the end of the same century, testify to the esteem in which such works might be held. A medical astrology by William the Englishman and a general digest by Leopold of Austria belong to the best-known works of the thirteenth century. In the fourteenth we find the subject spawning a vast new literature in which the old ideas were given more local colour. Italians such as Pietro of Abano, Cecco of Ascoli and Andaló di Negro were much admired, but the Arabic authors remained 'classical' for most scholars in the universities. Cultural distance clearly lent enchantment to the view.

Astrology developed a strong meteorological association. There were many texts on the astronomical 'prognostication of times' (compare the French word *temps*, 'weather'), and we have already mentioned Richard of Wallingford's. When the Oxford scholar William Merle added to the list, he incorporated many other sorts of consideration, and his empirical outlook is well and truly proved by his journal of weather observations covering the period from January 1337 to January 1344. Of course not all such records of observations were made for the reasons for which we now tend to value them. Here the weather was being correlated with astronomical events, but it was the weather itself that was the thing of final concern.

Many careful records that survive to us from the middle ages are of observations of comets, with classifications of position and colour, and so forth. Again, the focus of interest was not on the comets as such, but on the *disasters* they portended. To see how deeply entrenched were the presuppositions underlying such a question as 'Why comets signify the death of potentates and wars', consider the answer given by so rational a thinker as Albertus Magnus. The connection was not a strict one. It was merely that the comet was associated with Mars, and that Mars was the cause of war and the destruction of peoples.

The fifteenth century is often seen as the period when comets were first systematically observed – for instance by Toscanelli (from 1433), Regiomontanus and Walther (in the 1470s). This is a period in which an interest in the general forms of comets was supplemented by a concern for their precise co-ordinates, and ultimately their orbits in space – which until the late sixteenth century meant sublunar space. (Comets, following Aristotle, were 'meteorological' affairs, in the strictest sense.) There are plenty of earlier examples, however. Peter of Limoges measured the co-ordinates of the head of the comet of 1299, using a torquetum, an instrument with circular scales for right ascension and declination. Geoffrey of Meaux took the co-ordinates of the comet of 1315 with reference to neighbouring stars. Jacobus Angelus found the longitude of the comet of 1402 from that of the Moon. All three men were highly placed physicians with training in astrology, as indeed was Toscanelli. This particular social group, of medical astrologers, is very well represented when we look down the lists of scholars with a demonstrable interest in the making of astronomical instruments.

The Great Plague (or Black Death) of the late 1340s, the long wars between England and France, the ever-present fears of the arrival of the Antichrist, the Hussite heresies, and later the Protestant splits in the Church, all helped to turn the thoughts of astronomers to the kind of astrology that predicts the rise and fall of kingdoms and religious sects on the basis of 'great conjunctions' involving the planets Saturn, Jupiter and Mars. Geoffrey Chaucer worked this theme into his greatest poem, *Troilus and Criseyde*. Here indeed is an instance of a poet who, as we have seen, spent considerable time and energy in adding a hidden dimension of meaning to his poetry, by very carefully calculated astronomical and astrological allusion. Many other poets tried to copy Chaucer's style, for more than two centuries, but none was ever able to equal his arcane skill. On a more obvious level, however, the whole of European literature was gradually becoming coloured by cosmic metaphor and allusion. This is a subject as large as literature itself, but the products of sixteenth-century France – especially Jacques Peletier and Pierre de Ronsard – are especially noteworthy, while in the seventeenth century, John Milton's *Paradise Lost* actually introduces a modification to the standard astronomical system of his time.

Shakespeare's allusions are widely known, even if their astrological antecedents are not always understood. Romeo and Juliet, that 'pair of star-cross'd lovers', suffered the misfortune of incompatible nativities. There are many other comparable allusions, but they seem to lack deep structural importance. Perhaps Cassius was speaking for Shakespeare when he exclaimed that 'The fault, dear Brutus, is not in our stars, But in ourselves . . . ', perhaps not. At all events, these echoes of astrology are growing fainter by Shakespeare's time. In due course the astrologer in literature became a figure of fun, and even the California of the 1960s was unable to redress the situation. The art of turning astronomy to poetic advantage is now virtually lost, although as readers of Algernon Charles Swinburne will know, 'astrolabe' is one of the few words in English that rhymes with 'babe'.

The three great questions facing those who thought deeply about astrology in the fourteenth century were, first, whether the influences it claimed to describe were real; secondly, whether human beings could ever reduce it to a workable science; and thirdly, whether such a science would be licit. It has to be said that few scholars would have hesitated to answer 'Yes' to the first two questions, and most would have passed by the third as quickly as possible.

Of those who spoke against astrology from a position of scientific strength, Nicolas Oresme (1320–82) is one of the most interesting. Norman by birth, and trained in Paris, he became a confidant of the future King Charles V, while Charles was still Dauphin of France. Nicolas died as bishop of Lisieux. His view of the cosmos was in a sense mechanistic – for instance he often used a metaphor likening it to a mechanical clock – and yet he was not ready to break with the Aristotelian division of the cosmos into two regions, one above and one below the Moon. He continued to speak of the spheres as moved by intelligences, in the Aristotelian style. (His master in the university, John Buridan, had said that the spheres could have been given impetuses by God at the time of Creation, so providing them with indefinite motion.) When Oresme wrote against astrology, it was as a subject incapable of explaining events here on earth. They arise, he said, from immediate and natural causes, and not from celestial influence. But he was too deeply immersed in the culture of his age to be ruthless in his criticism. The qualities of the stars, signs, degrees, and so forth, can in principle be known, he thought. Predictions from great conjunctions can be made, but only in general terms, not in detail. As for the weather, the same is true – but farmers and sailors he thought likely to be more reliable. Medical prognostication is relatively safer from the Sun and Moon than from the planets. Here we are dealing with what for him was a question of *Nature*. The three final divisions of astrology, however, dealt with *fortune*: the casting of birth horoscopes, asking whether a thing will happen, or deciding on propitious times for action, these concern the freedom of human will, and are to be avoided.

Oresme's view of the situation was shared by many of the more thoughtful intellectuals of his time. It was not particularly new, but it made clear a distinction that is frequently misrepresented. We often read that astrology and astronomy were one and the same thing in the middle ages and earlier. For some, however, there were *three* astrologies, one mathematical, 'which we call astronomy', one natural, akin to physics, and one spiritual.

Oresme's most original work was in terrestrial physics, and we may leave him with a glance at his suggestion that the Earth may not be fixed at the centre of the Universe, although its centre of gravity strives towards that place. He considered carefully the whole question of the movement of the Earth. It has often been seen as a revolutionary step to discuss this point at all, but since Aristotle had done so, in view of the medieval method of writing commentaries on Aristotle, scores of scholastic authors did so as a matter of course.

In his *Questions on the Heavens* and *Questions on the Sphere*, Oresme emphasized the relativity of motion: the phenomena we observe daily could just as well be explained by the daily rotation of the Earth as by that of the heavens, he argued. In the end he opted for the traditional view that the heavens rotate – a matter for persuasion rather than demonstration. More than a century would pass before the birth of Copernicus, who was to be persuaded otherwise.

ASTROLOGERS AT COURT

Throughout its long history there has been a political dimension to astrology that has had little to do with its scientific content. Babylonian kings, medieval Christian kings, bishops and popes, Renaissance generals, Wallenstein, Hitler, even it seems Nancy Reagan, regularly took astrological advice. The courtly concern for learning in Muslim Spain became known in other parts of Europe very quickly, from at least the tenth century. Petrus Alfonsi, the Jewish convert who served both Alfonso I of Aragon and Henry I of England, was one of those who helped to spread eastern astrological styles of divination. His contemporary, Adelard of Bath – another royal servant – who had travelled as far as the Norman kingdom of Sicily, was no less eager to practise the subject.

Adelard was very probably the author of a series of politically interpreted horoscopes that still survive. He was a great admirer of Arabic learning, and despite his work in the more intellectually challenging realms of astronomical tables he shows his eagerness for astrology, even to the point of spiritual recklessness. He translated from the Arabic not only two astronomical works but three astrological, including one that contained passages on gemstones engraved with magical images. (Such images were to be carved under carefully selected constellations of the heavens.) He tells us that he himself wore an engraved emerald ring. When the astrological form of divination arrived on the scene, it simply stepped into the place of existing magical practice. Both magic and astrology had the same general function of *explaining* things that could not be explained otherwise, and both were involved in practices for *bringing about* things that one wishes to happen – although less often in astrology, and then it was usually coupled with magic.

The Holy Roman Emperor Frederick II (1194–1250) furthered the cause of eastern learning. He had been brought up in Sicily, was an excellent linguist – he had spoken Arabic with his childhood friends – and made a number of contributions to the sciences. Nevertheless, the greatest reversal in his fortunes began with his sheer complacency, induced by astrologers who had assured him victory over the city of Parma. Frederick is also said to have had in his employ as court astrologer Michael Scot, a man whose name passed into medieval currency as a magician of a most dangerous sort. Michael was more at home with astrology when allying it with incantation and the conjuring of spirits. In this he was not at all unusual – a far worse example of a high ecclesiastic with such views was William of Auvergne, bishop of Paris from 1228 to 1249.

From this time onwards, most European courts seem to have had advisers with a knowledge of the new sciences, although not always astrologers as such. Vincent of Beauvais, for example, a Dominican who served Louis IX of France as royal chaplain, librarian and tutor to the royal children, was no expert. He took a critical view of both demonology and astrology, in both of which he obviously believed, condemning them with arguments that had been circulating since the time of the Church Fathers. He practised astrology side by side with medicine, and it seems that here, as so often elsewhere, astrology crept into court life as much on the coat-tails of medicine as from its potency in predicting other personal and state affairs.

Academic astrology was a comparatively sober activity in the middle ages. The number of horoscopes surviving from before the end of the thirteenth century is small, possibly suggesting that the personal element was as yet little developed. All the same, many theologians were beginning to worry about developments, and the most famous moves of censure were those by Etienne Tempier, bishop of Paris, and Robert Kilwardby, archbishop of Canterbury, in 1277. We must not put too much emphasis on the effectiveness of their moves against Paris and Oxford respectively, and we should remember that there was far more than astrology at issue in their edicts, which were hastily prepared and directed against Aristotelian 'scientific' tendencies generally. Nevertheless, what they took to be particularly dangerous was the idea that if the movements of the stars are predetermined then the same holds true of human fortune – at least as long as the one is believed to be caused by the other.

This theme of determinism became an important issue in two works with an anti-astrological flavour written for royal consumption, one by Thomas Bradwardine for Edward III of England and one by Nicole Oresme for Charles V of France. The library of the latter prince in the palace of the Louvre contained numerous astrological works. One especially interesting manuscript from the collection, now in Oxford, contains among other items a collection of five horoscopes, very carefully drafted, one for Charles himself and the others for four of his children. Such natal horoscopes could be consulted by the physician-astrologer at later stages in a person's life to pronounce on the probable course of an illness. It is a sad thought that those of Charles' children must have been studied very closely, since most of them died of disease prematurely.

It became a habit in the late middle ages and afterwards for astrologers to collect together horoscopes of the powerful and famous. (In fact, this type of literature is still being produced.) Among surviving collections are horoscopes for most of Europe's medieval rulers, many accompanied by attempts to find astrological explanations for political events. Some attempt to evaluate the risks of pertinent royal betrothals. Much effort was clearly being put into dynastic astrology, but astrologers were often primarily medical men, and when they were not, it has to be said that their salaries were much lower than those of their medical colleagues.

The life of an astrologer was not without its dangers, especially when magic was involved. This is well illustrated by a case involving the English royal house in 1441. Eleanor Cobham, duchess of Gloucester, was then accused with two clerks, Roger Bolingbroke and Thomas Southwell, and a woman named Margery Jourdemayne, known as the Witch of Eye, of conspiring to bring about the death of the king,

Henry VI. The duchess fled to sanctuary at Westminster, and was later imprisoned for life; Southwell died in gaol; Bolingbroke was hanged, drawn and quartered, after being exhibited 'with the vestments of his magic and with waxen images, and with many other magical instruments'; and the witch was burned as a relapsed heretic.

The crime was the use of necromancy and the black arts, but the men were practised astrologers – indeed they were senior Oxford scholars, while the king's advisers were Cambridge scholars. The duchess at her trial admitted that she had consulted Bolingbroke, but only to get help in conceiving a child by her estranged husband – Duke Humphrey, brother of Henry V. The real crime, however, was not so much the practice of astrology as advising the king's enemies, and for this they paid the price. An astronomer of great talent, Lewys of Carleon, almost met the same end. He was physician to many of the nobility in the Lancastrian faction in the English civil wars, the Wars of the Roses. Lewys escaped with a period of imprisonment in the Tower of London, a time he used to great advantage to produce numerous astronomical tables in peace and tranquillity.

Having first permeated the courts from centres of learning, astrology moved slowly down the social scale. It did not leave the courts, however, until the seventeenth century – in some cases still later. The great astronomers, Brahe, Kepler and even Galileo, received court patronage with a view to their astrological competence. Roughly speaking, with the seventeenth century university astrologers became less speculative, less concerned with the sillier excesses of prognosticating human affairs, and more concerned with empirical matters such as medicine and meteorology, and with new systems of their own making. They made no progress, and eventually, at this scientific level, interest in the subject did die away almost totally, say by the end of the seventeenth century.

As the common people moved in and the academic breed became more circumspect, old-style astrologers flourished, as they do to this day. Old literary habits die hard, however, and a good example of the fact is to be seen in the astrological metaphor of Sun, gold, heart and king, which lived on to affect common attitudes to kingship itself. This very example, with a history going back at least to Babylon, was exploited by the publicity managers of *Le roi soleil*. Although the imagery was by this time more or less empty of scientific pretensions, it was still possible to use it in a very general and non-technical sense with effect in writing treatises on kingship – as did Jean Bédé and Jean Bodin in France, William Pemberton and many others in England. They were involved in a power game that had little to do with astrology, but they exploited language that was by this time so familiar and acceptable to their readers that it was rarely questioned. The political power of the astrologer was not that of a Napoleon or a Rasputin. It worked at a deeper level, in people's minds, and it was a force for tradition.

When beggars die there are no comets seen,
The heavens themselves blaze forth the death of princes,

as Shakespeare has it in *Julius Caesar*. The link between princely powers and astrological conceits was one that was simply taken for granted by most people until long after Shakespeare's death.

A RETURN TO THE GREEKS?

Renaissance humanism, the intellectual movement in which Regiomontanus found himself being swept along, had from its beginnings in fourteenth-century Italy been generally unfavourable to the natural sciences. The Italian scholar Petrarch had at the outset poured scorn on the Oxford logicians who were then becoming fashionable throughout Europe, and in whose writings there are so many of the seeds of the Scientific Revolution to come. There was a pedantic literary spirit abroad that refused to accept medieval versions of classical treatises. Humanism, however, was concerned with the place of humankind in history and in Nature. Many of the humanists had flirted with astrology, even if they did not remain constant in their loyalty to her. This explains in part why the rebirth of classical studies could easily accommodate astronomy, highly esteemed by many who had no understanding of it. That this could be so should be obvious from the relations between Bessarion and Regiomontanus, but there are many other examples to be found. Many humanists, however, grew openly hostile to astrology, notably Marsilio Ficino (1433–99), the first of the great Renaissance Platonists.

Ficino was well read in astrology, not to mention its standard opponents, from Augustine onwards, and had nothing particularly new to add on the vexed question. He accepted that the planets might influence human beings at their birth, but gave to the individual the power to decide freely. 'The stars incline, but do not compel' was the standard adage. Giovanni Pico della Mirandola (1463–94) expressed himself on the subject at much greater length.

Pico was scion of a noble family who was put to learn Hebrew and Arabic under Jewish teachers. Insisting on defending in Rome theses that were judged heretical by a papal commission, he fled to France. After arrest, and return to Italy under the protection of powerful supporters, he settled in Florence under the protection of Lorenzo de' Medici. There, among other things, he wrote his *Disputations Against Astrology* in twelve books, his most extensive work. Central to all his writing was the theme of the dignity and freedom of humans, and this theme he carried over into his *Disputations*. He acknowledged the influence of the stars through such physical effects as light and heat, but dismissed occult influences. We cannot, he thought, as free spirits, be influenced by the stars, which are bodies of a lower nature.

Pico was no proto-modern astronomer: his chief motive for writing was his wish to defend the Church, and his work was much heeded for many decades, indeed centuries. Gian Domenico Cassini (1625–1712), the first of the great family of astronomers, tells us that it was Pico's work that turned him against astrology and to astronomy.

When Tommaso Campanella (1568–1639), a Dominican friar, wrote what many regarded as a classic work of astrology, he was careful to point out that it concerned only what Pico had called 'physical astrology', and that it had been purged of 'the superstitious astrology of the Arabs and Jews'. The same sort of distinction was then being drawn between 'natural magic', under which were included topics that we would acknowledge as physical science, and its occult, demonic and spiritually unacceptable alternatives, to which we now generally apply the name 'magic' exclusively.

The debate on magic and astrology taking place at this period occasionally had recourse to some interesting notions of a psychological sort stemming in some measure from the writings of Avicenna, the most renowned philosopher of medieval Islam. Natural magic was often explained as obtaining its very real but seemingly miraculous effects through the medium of faith and imagination. This is where the stars came into play. As Pico said, 'To work magic is to do nothing other than to marry the universe', by which he meant a psychic union. Many Arabic texts put forward the same idea, but not always ending in the same conclusion. Some, such as Ibn Khaldūn, thought that our innermost thoughts are beyond the reach of the stars. Which stance one took depended on the conclusion one wished to draw. In about 1490 Galeotto Marzio da Narni defended divinatory astrology against attacks by Averroes, and used Avicenna in his defence. The human soul, said Avicenna, has the power to change things through violent longing. Strong personal desires, he thought, could be an agency operating on the lives of people who were on Avicenna's authority affected by the stars at the times of their births. Faith can move mountains, and the actions of such a faith are more effective, Galeotto insisted, than all instruments and medicines. Others went so far as to treat speech itself as an even more innocent tool of natural magic. Although a measure of scepticism was abroad, it is fair to say that most scholars were happiest when they could find such new ways of defending old orthodoxies.

Renaissance humanism began to spread abroad rapidly from Italy at about this time. From an astronomical point of view, the importance of the new translations from the Greek can be easily exaggerated. Some, such as Thomas Linacre's translation of excerpts from Geminus on the celestial sphere (printed 1499; he thought the author was Proclus), could not have been of more than historical interest. Fashions, on the other hand, can be more powerful than reason. When Corpus Christi College was founded specifically to favour such studies in Oxford, provision was made to have astronomy taught there. There was no clearly qualified humanist for the job. The choice fell on Nicolaus Kratzer (1487–1550), who like his friend and compatriot the artist Hans Holbein had served at the court of Henry VIII, and who had for long rubbed shoulders with such scholars at Henry's court.

Courtly fashion has very often favoured the foreign and exotic, and when the king's father, Henry VII, had employed an astrologer, he had opted for an Italian, William Parron. A notable English humanist, Kratzer's friend and one time patron, Sir Thomas More could safely express his distaste for astrology, but there is no doubt that by engaging scholars from abroad – and from Italy in particular – the tone of any subject at this period was somehow thought to have been elevated.

The Greek connection mattered no less. Thus in a letter to Erasmus from a pupil, Kratzer was in 1517 described as a skilled mathematician, bringing with him astrolabes and armillary spheres and *a Greek book*. Such was his passport. He does not seem to have known any Greek himself. When he built sundials in Oxford, they were duly honoured by a colleague's verses in fine neoclassical Latin. Medieval astronomy was being dressed up, but only superficially changed. A new generation of instrument-makers began to produce ever finer examples, beautifully engraved in the new Italic scripts. One of the most skilful of all was Thomas Gemini, who came to Henry VIII's court in or before 1524, from the Netherlands. Many comparable instances of the fashion for classical learning could be cited from other parts of Europe, but few changed the course of astronomical learning significantly. For all his humanist friends, Kratzer's outlook on astronomy was one that involved no break in continuity with the middle ages.

One very strange effect of the new literary fashions was a revival of interest in the Greek doctrine of homocentric spheres. Girolamo Fracastoro (1478–1553) was a physician and philosopher who taught at the university of Padua. In 1501 he made the acquaintance of Copernicus there, who was at that time enrolled in medicine at Padua. Fracastoro's fame stems from a very long narrative poem, written in classical Latin of great elegance on an unlikely subject, *Syphilis, or the Gallic Disease* (the first draft was finished in 1521). The disease was supposedly visited on Sifilo, a young shepherd, by the sun-god, to whom Sifilo had been unfaithful. Fracastoro believed in the standard doctrine of the dangers attendant upon a triple conjunction of Saturn, Jupiter and Mars, and he argued that the spread of the disease was due to the corruption of the air under the malign influence of that event. He later wrote works on medical astrology of a more prosaic sort, but it was in his *Homocentric [Spheres], or Concerning the Stars* (1538) that he set a new astronomical fashion, albeit one of short duration. At Padua he had been a friend of three brothers, of whom one, his teacher Giovan Battista della Torre, had tried to revive the homocentric idea. On his death-bed in 1534, della Torre asked Fracastoro to complete his work. The resulting book was dedicated to Pope Paul III – as was Copernicus' great work of 1543. The obscurity of Fracastoro's might explain its lack of success, although there were instrument-makers who thought its schemes worthy of being modelled in metal.

It is hard to believe that Fracastoro's ideas on homocentrics were entirely independent of similar ideas in a slender work by Giovanni Battista Amico (1512–38). The fact that this was printed three times in four years (in Venice in 1536 and 1537, and in Paris in 1540) says something about the ambitions of the Aristotelians, who were anxious to find a system in keeping with their ideas. Amico was murdered in Padua in the year of publication of Fracastoro's book ('by the hand of an unknown assassin, it is thought, out of envy of his learning and virtue').

Where the older man took only the special case of 'Eudoxan' spheres pivoted with axes at right angles, Amico started with that assumption, and then moved on to the general case of arbitrary inclination. His models are not very coherent, but they make extensive use of the theoretical devices we found Copernicus using, devices that we know had been used earlier by Naṣir al-Dīn al-Ṭūsī. Amico applied the Ṭūsī couple – without any ascription, of course – repeatedly in the theory of planetary longitude and latitude. He also tried his hand at celestial physics, and like Fracastoro was to do, considered the passage of planets through some medium in space of varying density ('vapours') as explaining their changing brightness – a problem for anyone who proposed a model making the distances of the planets constant, of course. He introduced precession into his model in a way that makes clear his debt to the Alfonsine tables.

Fracastoro's system, to which he assigned seventy-seven or seventy-nine spheres in all, was astronomically much inferior to Ptolemy's, from which a number of ideas were in any case borrowed. He seems to have thought that he was using Eudoxus' own ideas precisely, and one cannot help wondering whether the Paduans of the day had been trying to reconstruct Eudoxus' system, and had hit on the Arabs' constructions in the course of doing so.

At one place in Fracastoro's account he likened the effect of varying planetary brightness to the increasing size and clarity with which we can see something through a double lens, in comparison with our view of it through a single lens, or the changes when we look at something through different depths of water. (This is not, as has sometimes been claimed, the invention of the telescope.) In many ways his remarks on the nature of the celestial spheres are as interesting as anything on the subject before him. He observed comets carefully enough to realize that their tails always point away from the Sun. He thought it obvious that comets are nearer than the Moon, for how else could they move freely through the spheres? Tycho Brahe later reversed the argument, and said that since comets were demonstrably more distant than the Moon – by his own measurements – it was the solidity of the spheres that was to be doubted.

Fracastoro's discovery of the direction of cometary tails was published in 1538, two years before Peter Apian's publication of the sam. discovery. In his case it was made while he was observing the comet of 1532. The two discoveries are usually presumed to have been independent, but since Apian's explanation was that comets are spherical lenses, one cannot rule out borrowing. Apian thought that the comet's tail is simply a fan-shaped pencil of rays through the lens. Gemma Frisius and Jerome Cardano later gave publicity to this ingenious idea. Tycho Brahe was one of those who adopted the lens principle, in a work of 1588 concerning the comet of 1577: he decided that in this case the source of the light was not the Sun, but Venus. Kepler for long accepted the idea of the solar lens. Despite the fact that from 1618 he offered several criticisms of it, Descartes took over a version of the idea. Newton criticized it once again, later in the century, but it died a slow death. At all events, it served a very useful purpose, to the extent that it led astronomers to forsake the Aristotelian view of comets as burning fires. From this point onwards, their location – sublunar or celestial – became an open question. Tycho was to solve that question.

All told, while Amico's early death robbed astronomy of an imaginative talent, and Fracastoro's writings had a certain modish quality that made them appealing to contemporaries, both men were attempting to turn the clock back nineteen centuries. Tycho Brahe was later scathing in his criticism of Fracastoro's planetary absurdities. The future lay with other planetary schemes, mathematical and physical. There was still much life left in the fourteen-centuries-old Ptolemy, but the time was near when astronomers would have to contend with a new system, that of Copernicus. To this we now turn.

Copernicus' Planetary Theory

COPERNICUS

With the advantage of hindsight, many recent historians have been tempted to represent the changes in planetary astronomy wrought by Copernicus as self-evident. Even if this were true – which it is not – there would remain his undoubted impact on later history. The so-called Copernican revolution has so often been taken as a paradigm, a model for intellectual change. Is this an illusion? Those who from childhood have been taught that the Earth moves are perhaps not the best judges of the difficulty with which the idea was made a part of the realm of platitude. When the question of the Earth's motion was raised in antiquity by Aristarchus, the charge of impiety was brought against him. When Copernicus came to that same conclusion, the situation was if anything more dangerous, for he was confronting three weighty authorities: the Church, the Aristotelian orthodoxy of the universities, and the astronomers, all of whom were working in a broadly Ptolemaic tradition. His courage, in daring to criticize two of these three pillars of western ideology, was certainly greater than that of many of his latter-day critics.

Nicholas Copernicus was part of the 'establishment', in all these three senses. He was born in Torun, in Poland, in 1473. His father died when he was ten, but an uncle made it possible for him to attend the University of Cracow in 1491. The uncle became bishop of Varmia (Ermland), and so Nicholas was found a salaried position there as canon of the cathedral chapter of Frombork (Frauenberg). In 1496 he enrolled in Bologna as a student of canon law, but astronomy seems to have become his main pursuit, and in a later period of leave to study in Padua – medicine, this time – he came back with a qualification in canon law, from Ferrara. If a hint of wilfulness was already showing itself, the remainder of Copernicus' personal life did not conform to it, for he lived out his last forty years in the service of the Varmia cathedral chapter. His comfortable situation allowed him to build an observing tower, from which he used three principal instruments, an armillary, a quadrant and Ptolemy's parallactic rules.

Copernicus' books at Cracow show that he there learned the use of the Alfonsine and derivative tables, and also some astrology. He read Peurbach on eclipse calculation. In Italy we know that he assisted Domenico Maria de Novara in Bologna and gave lectures (on mathematics or astronomy) in Rome. It is only a matter for speculation, but it was probably at this time that he met with the planetary theories developed in Marāgha, which he somehow acquired, as we have already seen in an earlier chapter. A page of notes in his copy of the Alfonsine tables shows him at work on his own new astronomical system, but the first extensive draft we have of it is to be found in his *Commentariolus* (Brief Commentary). This seems to have circulated anonymously in some quarters, although some must have known its source.

Copernicus' greatest work, *De revolutionibus orbium caeles-tium* (On the Revolutions of the Celestial Spheres), was his definitive statement. It was completed at the very end of his life (1543) and differed in several technical respects from the brief sketch, which was probably written around 1510 (and certainly not later than 1514). The sketch contains a statement that the calculations in it are reduced to the meridian of Cracow, which he took to be the same as that of Frombork. In arguing that the Earth revolves around the Sun like any other planet, he introduced the names of authorities that he knew would carry weight in his own time, namely the Pythagoreans. In his last work he named them specifically as Philolaus and Ecphantus, but we do not have to take this reference as more than a canvassing of historical support. In his last manuscript he actually mentioned Aristarchus in the same connection, but then deleted the passage before publication, perhaps concerned lest he be associated with someone with a reputation for impiety. (He mentioned the name elsewhere, in connection with a figure for the obliquity of the ecliptic, but did so then in error for Eratosthenes.)

In his *Commentariolus*, Copernicus sets out his basic assumptions, and it is clear that he considers the strongest arguments for his system to be that it conforms to appearances while at the same time being more pleasing to the mind than Ptolemy's. By this he means, among other things, that he has adhered religiously to the principle of uniform motion on a circle, *avoiding the use of an equant*. His models, broadly speaking, follow the pattern of Ibn al-Shāṭir's — so avoiding the equant — but geometrically transformed so as to bring the centre of each planet's model to a common point. This common centre was not quite at the Sun, but at the centre of the Earth's orbit. A theory of planetary latitude was added. It was possible to make such a (roughly) heliocentric theory much simpler than Ptolemy's, for reasons explained earlier (that is, because the planes of the planetary orbits do in fact pass through the Sun and not the earth). The parameters on which the models were based, not surprisingly, are more or less Alfonsine.

Such parameters gave reasonable results for longitudes, but were not of the precision needed for one of the easiest observations of all to make, namely the timing of conjunctions of the Sun, Moon and planets. (The uncertainties in the positions of *both* objects, Sun and Moon, enter the calculation; and since their relative speed may be very small, the uncertainty in the time of their meeting will be correspondingly great.) Copernicus therefore set to work to amass the sort of observations he would need to make his schemes more accurate, more acceptable than those they were meant to supersede. His observations were made in the period 1512 to 1529, a period when he was also busy with his ecclesiastical and medical duties.

Over and above his ordinary duties were various defensive actions on behalf of his beleaguered cathedral and city during the wars between East Prussia and Poland. After hostilities ceased, and the Duchy of East Prussia became a fief of the Polish kingdom (1525), he played a useful part in improving a greatly debased coinage (he wrote an economic tract on the subject) and in regulating the price of bread. And finally, from bread to housekeeper: it is customary in biographies of Copernicus to allude to the fact that he had living with him in this capacity a younger divorced woman, Anna Schillings. In 1539 she, and women in the service of two other canons, were ordered by the bishop to leave. One of those canons was later harassed and imprisoned for the Lutheran heresy, a fact that should not be forgotten when we are assessing Copernicus' supposed timidity. It was in May of 1539 that Copernicus acquired his most vociferous young disciple, Georg Joachim Rheticus (1514–74), and as another reminder that these were anything but peaceful times, we note that when Rheticus was fourteen, his father had been executed for witchcraft.

Between his two chief astronomical works, apart from an almanac, Copernicus wrote only one other of importance, an attack on a treatise by a competent Nuremberg mathematician, Johannes Werner (1468–1522), 'On the Motion of the Eighth Sphere' (1522). Copernicus was not above using some of Werner's ideas in his *De revolutionibus*, even so. In 1533, we know that the papal secretary Johann Albrecht Widmanstadt was presented with a valuable Greek manuscript for explaining Copernicus' ideas on the motion of the Earth to Pope Clement VII. From the same source in 1536 Cardinal Nicholas Schoenberg heard of Copernicus' system, and the Cardinal in turn asked the Polish astronomer to make his ideas public. Copernicus printed the letter in his

De revolutionibus, a work he had in any case dedicated to the new pope – a doubly useful defensive tactic.

The young Rheticus had set about designing his own fame by travelling from one distinguished scholar to another, beginning with Johann Schöner, who was then in possession of most of the manuscripts of Regiomontanus, Walther and Werner. When he arrived at Copernicus' door, it was with a gift of books, many printed by Schöner's friend, the great Nuremberg printer Johann Petreius. Copernicus was not reluctant to share his latest ideas with his young visitor, who passed them on to the learned world in the form of a long communication addressed to Schöner. This elegant summary, *Narratio prima* (First Account), was quickly printed twice, in Dantzig in 1540 and Basel in 1541. By this time it was clear that Copernicus would soon be releasing his own work, that he was working hard on the numerical revision it entailed, and that he was worried about the reception it would have from the natural philosophers. In July 1540 he wrote to Andreas Osiander, a well connected Lutheran theologian, on this point, and in April 1541 received a reply to the effect that astronomical hypotheses, theories, were not articles of Christian faith, but simply 'the basis for computation', devices for representing observed phenomena. Their truth or falsity was of no importance, he suggested. Osiander alluded to the well known equivalence of the eccentric and epicyclic models for the motion of the Sun – different models, in his estimation, that produce the same observed results. He advised Copernicus to touch on these matters in his book, in order to placate the Aristotelians and theologians, 'whose opposition you fear'. On the same day he addressed a similar letter to Rheticus.

Osiander's views have a long philosophical history, and one that continues to provoke discussion, but whether or not they were acceptable, they introduced a psychological ingredient into the discussion. There can be no doubt that Copernicus and Rheticus thought the new system *true*, physically true, even if unprovable, and had no wish to water down their claims for it. In May 1542, Rheticus brought the fair copy of Copernicus' *De revolutionibus* manuscript to Petreius in Nuremberg. Printing began soon after, and Rheticus corrected the proofs, until in October he left to take up a professorship in Leipzig. Osiander now took over the supervision of the printing, and added his own anonymous preface to the work. In this, while praising Copernicus, he expressed himself even more strongly on the questions raised in their earlier correspondence. 'These hypotheses need not be true, nor even probable', as long as they fit with the observations, he explained, and he pointed out some apparent absurdities concerning the changing apparent sizes of the planets, saying in effect that they were of no importance to a theory of longitude. The new hypotheses were to be taken alongside the ancient ones, 'which are no more probable'. 'Let no one expect anything certain from astronomy, which cannot supply it, lest he accept ideas as true that were conceived for another purpose, and so leave this study a greater fool than when he entered it.'

When the book appeared in March 1543, Rheticus and Tiedemann Giese were angry at Petreius and Osiander for this betrayal. An unsuccessful action was brought before the Nuremberg city council, to have the printer replace the original with a corrected edition. It is unclear whether Copernicus himself ever realized what had happened, for in December 1542 he suffered a stroke that left him paralysed. Even without a severe loss of his mental powers, he would probably not have seen the offending preface before he saw a copy of the work in its entirety – since prefaces are usually printed last. This event, according to Giese, was on the very day of his death, 24 May 1543.

That Osiander, and not Copernicus, was the author of the anonymous preface was not generally appreciated by most early readers. Kepler discovered the fact from a friend in Nuremberg, and gave the discovery prominence on the back of a title page of his book on Mars, but this was only in 1609, so that for three generations many astronomers could regard Copernicus as having spent a lifetime in developing a theory that he did not regard as physically true.

THE COPERNICAN SYSTEM

By introducing the Sun (or, strictly speaking, a point near the Sun) into the theory of motion of every planet, Copernicus made it possible to represent all in a single system. When Eudoxus, Callippus and Aristotle established their system, it was on the basis of an Earth that provided a centre for all motions. The sizes of the spheres were left as an arbitrary question. When Ptolemy achieved a single system, the sizes of the shells accommodating maximum and minimum planetary distances were settled on the principle that there must be no void, no wasted space, between them. The Copernican system followed from the fact that in each of the separate Ptolemaic models for the planetary motions there was a certain line that represented the same real thing, (roughly speaking) the earth–Sun line. To those not irredeemably steeped in Aristotelian philosophy, this in itself must have seemed a more plausible principle. At least it took the mystery out of something that had been half-appreciated since Ptolemy, namely why the mean Sun plays an important role in the motions of the Moon and planets. Unfortunately, this is not how the changes effected by Copernicus were generally viewed. For most people he was simply the man who set the Earth in motion. The Earth's motion was *not* an inevitable consequence of the single system. Astronomers were quick to find ways of avoiding it, while keeping the system intact, as we shall see.

With these two innovations, Copernicus introduced into astronomy the most far-reaching changes since ancient times, and yet he was in almost every respect a product of the Ptolemaic tradition – purged of the equant principle, that is. His *De revolutionibus* is divided into six books. The first gives a general survey of his system, and ends with two chapters on plane and spherical triangles. The second is a useful textbook on spherical astronomy, but is not in itself revolutionary. The third concerns precession and the motion of the Earth. The fourth deals with the Moon, the fifth with the planets in longitude, and the last with their latitudes. Apart from the actual points of division, the pattern of Copernicus' book is very closely modelled on the *Almagest* of Ptolemy.

Copernicus was a skilful propagandist for his theory. Many of the old arguments for the superiority of circular motion are here, for instance that it makes endless repetition possible. To Ptolemy's fear that a rotating Earth would imply motions so violent that the Earth would fragment and be dispersed throughout the heavens, Copernicus replies that we should fear more for the stability of the *heavenly* sphere. As for the supposedly obvious character of the Earth's centrality, this has been misconceived. It is universally conceded that the distances of the planets from the Earth vary, as do their motions with respect to the Earth. What sort of basis is this for a theory with the Earth at the centre of all motions? To the Aristotelians, who might have felt the lack of the idea of an Earth that is surrounded with a spherical region of terrestrial force (up to the lunar sphere), he hints at the idea that there might be *many* such centres, and not just ours. Gravity might be a natural tendency inherent in all particles to unite themselves into a whole in the shape of a sphere, but not necessarily at the centre of the Earth. This is one of the first hints of dissatisfaction with the old idea that gravitation was always towards the centre of the Universe, but Copernicus does not tackle the problem of why matter should congregate in a relatively small number of places (the Sun, Moon and planets); and not until Newton did so was universal gravitation to be turned into a truly coherent theory.

Copernicus gives much attention to arguments for the general arrangement of the planets, knowing, no doubt, that most readers would not get much further than this part of the book. Even the order of the planets had never been conclusively established. There was general agreement that the Moon, having the shortest period, was nearest the Earth, and that Saturn, with the longest (29½ years), was most distant. Jupiter and Mars, being explicable by the same model as Saturn, were presumed to come next below it, again basing the order on their periods (nearly 12 and nearly 2 years). Mercury and Venus were problematical, however. Plato put them above the Sun, Ptolemy put them below, and some Arab astronomers put Venus above and Mercury below. Copernicus simply presented his heliocentric system in a very general way, with the order of planets around the Sun being Mercury – Venus – Earth – Mars – Jupiter – Saturn, and showed how easily and naturally it explains the relative sizes of the retrograde arcs: Venus' being greater than Mercury's, Mars' greater than Jupiter's, which is greater than Saturn's. His system also explains why the outer planets are brightest in opposition. But not the least of its merits, in his eyes, was the position of the Sun at the centre, the lamp at the centre of the beautiful temple of the world. Some call the Sun the light of the world, others call it the guide or soul. Hermes Trismegistus called it the visible god . . . . And so Copernicus goes on, not averse to employing rhetoric of a pagan tint, if he can thereby persuade the reader of the system's inherent plausibility.

These were only qualitative opening shots. The fine detail was still to come, and this required precise parameters. Some were derived ultimately from the Alfonsine tables. Only about thirty of Copernicus' own observations are mentioned in his final work, but he makes it clear that they were the distillation of a much larger number. They were not of the highest quality, compared with others made in the late middle ages, but they were well chosen for his own special purposes. They include conjunctions and oppositions of planets, positions of the Sun (including equinoxes) and Moon, lunar and solar eclipses (which amount to much the same thing) and zenith distances (or altitudes) of various bodies. The point that quickly emerges when one sifts through all this material is that here we have one of the very few astronomers after Ptolemy who appreciated how to build up a planetary model from first principles, rather than by patching up the work of his predecessors.

Of course he does not begin absolutely from nothing, any more than did Ptolemy. He assumes certain general principles – under which heading we must place even the assumption of uniform circular component motions. He accepts some inherited prejudices with a tenacity that is hard to justify even by the lights of his own time – as in the case of his cyclical theory of precession, alluded to here on a previous occasion.

Copernicus settled on a theory with a double circular motion. These motions amount to perpendicular oscillations of the Earth's equator, and can be regarded as equivalent to two oscillations of the Earth's axis of rotation, one tangent to the Earth's precessional motion (this produced a variability in the precession) and a perpendicular motion varying the obliquity. He assumed that the second period was exactly twice the first, which he set at 1717 years of 365 days; and he decided that the obliquity varied between 23;52° (at some time before Ptolemy) and 23;28° (not then reached). As for the steady precessional term, he somehow derived a figure of 360° in 25 816 years of 365 days, that is, one degree in 71.66 Julian years, an excellent result. Copernicus measures longitudes from a star (gamma Arietis) rather than from the true (fluctuating) equinox.

For the motion of the Earth round the Sun, Copernicus did not strictly need to add much to the simple Ptolemaic model, the simple eccentric (or concentric with epicycle). His observations indicated that for AD 1515 the eccentricity was 0.0323 times the orbit's radius, and that the longitude of the apogee was 96;40°. However, he chose to complicate the final model by making the centre of the Earth's orbit, the 'mean sun' (E in figure 11.1), move in relation to the true Sun (S in the figure). T is the Earth. E in fact was taken

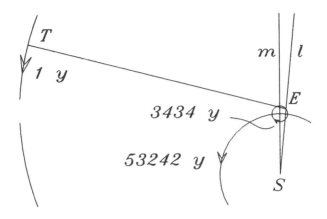

to move round a small circle in the same period as that of the obliquity, with both maxima occurring at the same time (65 BC).

The figure, as drawn here, is to scale in all but the orbit of the Earth, which is about a sixth of its correct size. The lines *m* and *l* are apse lines, *l* giving the direction of aphelion at the given time, and *m* its direction for maximum and minimum eccentricity. The long periods quoted are those for motions in the two central circles.

Why the peculiar complication of a variable eccentricity? Here is a second example – the first was his theory of precession – of Copernicus wishing to preserve as far as possible the work of his predecessors, notably Ptolemy, whose figure for the eccentricity was badly flawed. Copernicus made the maximum eccentricity on his model 0.0417 (417 parts of the standard radius of the Earth's orbit, namely 10 000), only slightly larger than Ptolemy's. Taken together with his own eccentricity, this implies a radius of 48 for the small circle.

Copernicus' generosity towards his predecessors resulted in a curious astrological doctrine, put forward by Rheticus in the *Narratio prima*, possibly with Copernicus' cognizance. It was traditional in astrology to suppose that a planet's strength was increased at its apogee and diminished at its perigee. This happened often – even in the case of Saturn it happened once every thirty years or so. There was also the standard doctrine relating to the rise and fall of sects and religions, based on great conjunctions, which occurred less often. Rheticus saw that the various very long periods inherent in Copernicus' theories lent themselves to a combination of these ideas: he pronounced dogmatically that when the eccentricity was a maximum, the Roman republic was tending towards a monarchy; that as the eccentricity declined, so did the Roman empire; as it reached a mean value there came the rise of Islam; but that the collapse of this empire could be expected in the seventeenth century, with the minimum. The second coming of Christ could be expected with the next following mean value. And all this was essentially as a consequence of the fact that Copernicus wished to preserve the truth of Ptolemy.

In the case of the Moon, the new system could be much simpler. Ptolemy had broken the rules of uniform circular motion, Copernicus thought, and in any case Ptolemy's model had produced much too great a variation in lunar distance. Copernicus' lunar model (figure 11.2) is identical to Ibn al-Shāṭir's, and he had already used it in the *Commentariolus*. (The radius in the figure as drawn here is broken so that the epicycles can be drawn to scale.) He selected parameters that fitted with the Alfonsine tables, and that were indeed of much earlier Indian origins, had he but known it, but of course at the root of all this there was the brilliant work of Ptolemy, in his derivation of the second lunar inequality. Copernicus' discussion of the distance, parallax and apparent diameter of the Sun and Moon is somewhat blemished, although inevitably much better than Ptolemy's. When showing how to compute eclipses, too, he improves greatly on all his predecessors.

FIG: 11.2 *Copernicus' lunar model*

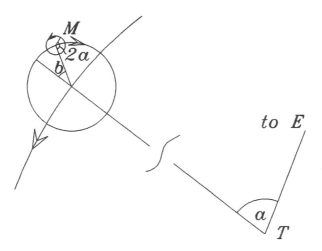

The fifth book of *De revolutionibus*, on the longitudes of the superior planets, includes some of his best work. It is still essentially a rewriting of the Ptolemaic circles, but we should not underestimate the painstaking work of calculating and recalculating the elements of the orbits (by an iterative process involving many hundreds of calculations). In his planetary theories he had the advantage over Ptolemy that he had only the first inequality to consider, accounting for the revolution of the planet with respect to the stars. As we have learned to expect, each model is related not to the true Sun but to the centre of the Earth's orbit (*E* in figure 11.3, which is not to scale for any of the three superior planets).

In fact in all three cases the radius of the epicycle *OP* is more or less equal to a third of the eccentricity *CE*. The equivalent to Ptolemy's eccentricity of the equant can be shown to be now *OP+CE*, and whereas Ptolemy made the eccentricity of the deferent just half of that total, here it is

FIG: 11.3   *The Copernican model for a superior planet*

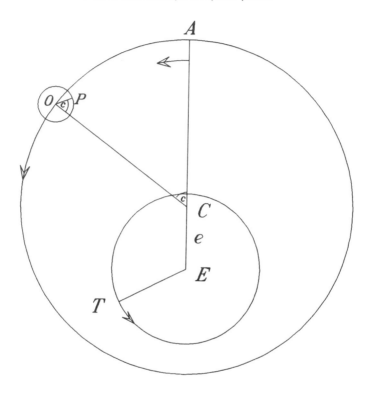

three-quarters. Note the equal angles marked *c*. It is instructive to analyse the planet's path with respect to *E*. It is not a circle, as Copernicus realized.

Copernicus' handling of the inferior planets is less praiseworthy, partly because he could not obtain the observations he needed. The models he now adopted were different from those in the *Commentariolus*, inasmuch as he shifted the two epicycles from the periphery of the model to the centre, where they form a dual eccentricity. An idea of the arrangement may be had from figure 11.4, where *Q* marks the planet and the apparatus for fixing the (anticlockwise)

FIG: 11.4 *The Copernican model for an inferior planet*

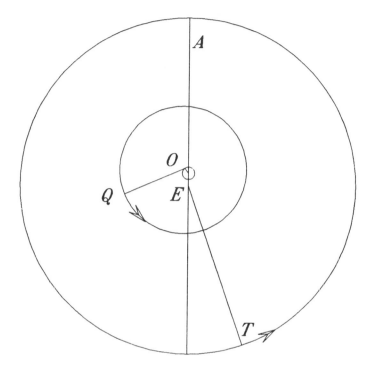

moving centre *O* of its deferent circle in relation to the
centre of the Earth's orbit *E* is shown. The angle at *E* is half
the angle at the centre of the small circle. The models now
correspond to Ibn al-Shāṭir's, duly inverted for a sun-
centred theory. The actual values of the new parameters
are obtained by a process that seems to have relied partly
on Ptolemy and partly on unexplained intuition.

In this account we have said nothing of the motions in
the apse lines (lines to apogee and perigee) that Copernicus
derived. These motions were of course very slow, and
although they left much room for improvement – he was
unfortunate in deciding to measure them from the mean

Sun *E*, rather than the true Sun – we have here again a rare instance of an astronomer capable of deriving the elements of a largely *new* theory from fundamental observations.

We saw in chapter 3 how each planet's model contains a common element, namely the mean radius of the Earth's orbit (*TE*), and how that fact implies that the scale of the entire planetary system may be expressed in terms of the common unit, the 'astronomical unit'. The mean distances (from the Sun) derivable from the Copernican parameters of scale are as follows, with the modern figure in parentheses: Mercury 0.3763 (0.3871); Venus 0.7193 (0.7233); Mars 1.5198 (1.5237); Jupiter 5.2192 (5.2028); Saturn 9.1743 (9.5388). In a sense, this was a free dividend of the model, and almost equally good results can be found from Ptolemy's models (those of the *Almagest*, not the nested spheres of the *Planetary Hypotheses*) by anyone who recognizes the place tacitly occupied by the Sun in those models. The agreement is perhaps surprising, when we consider how it depends on geometrical proportions locked away for fourteen centuries, unappreciated, in Ptolemaic astronomy.

As for *absolute* distances, Copernicus had only minor adjustments to make to the inadequate solar parallax found by Hipparchus. Copernicus gave the mean parallax as 0;03,31° and the mean solar distance as 1142 Earth radii.

Having alluded to this new way of obtaining distances within the solar system, it should be said that the older way of scaling the geocentric system, with a nesting of spheres that allowed for no spaces between them, continued to appeal, *even to Copernicans*, so immersed were they in the Aristotelian dislike of the notion of a vacuum. Copernicus had, so to say, created large empty spaces – for instance between Venus and Mars, and between Jupiter and Saturn – into which astronomers felt themselves encouraged to try to fit comets (Michael Mästlin, Tycho Brahe) or even hitherto unrecognized planets (a speculation of Johannes Kepler).

In the sixth book, Copernicus considered the latitudes of the planet, and here he made little advance over the theory in the *Commentariolus*. Again he penalized himself by trying too hard to reproduce the faulty records of planetary latitude to be found in Ptolemy's *Almagest*. He gave his planetary planes variable inclinations, using Ptolemy's parameters, and seems to have been almost oblivious to the inherent superiority of the heliocentric hypothesis on this admittedly difficult point. Part of his difficulty stems from the fact that his theory is not heliocentric enough. He made his planetary orbits lie in planes passing through the centre of the Earth's orbit (our point E) and not the physical Sun, as Newton's dynamics showed to be the case. Copernicus' was a unified system, in the sense that all the subsidiary planetary models were superimposable, and in the sense that the Earth's motion explained away the second anomalies in the older models. It was still a geometrical system, rather than a physical system that explains appearances in terms of physical laws. Kepler later said of Copernicus that he did not know how rich he was, and that he tried more to interpret Ptolemy than Nature. This much-quoted statement is misleading. Copernicus was trying first and foremost to represent Nature. He often fell back on previous theories, but only when he felt he had no alternative.

A PERIOD OF TRANSFORMATION

In the estimation of leading European astronomers, Copernicus quickly took his place alongside Ptolemy, and yet the ordinary practitioners of astronomy continued to work with the system – and the tables – they already knew. In 1551 Erasmus Reinhold (1511–53) published new 'Copernican' tables in place of the Alfonsine. They were known as the *Prutenic Tables* (Prussian Tables) in honour of Duke Albrecht. They make use of some newly evaluated parameters as well as those of Copernicus, whom Reinhold praises without referring to the heliocentric character of the underlying hypothesis. There is good reason for thinking that he wished to replace it with an earth-centred alternative along the lines later adopted by Tycho Brahe.

Reinhold's tables were widely diffused. In England, John Feild (1520–87) used them to prepare an ephemeris for 1557, in the preface to which the scholar and magus John Dee (1527–1608) expressed qualified admiration for the Copernican system. Feild, however, seems to have accepted the new system without hesitation when calculating his ephemeris. The writer of mathematics texts, Robert Recorde (c. 1510–58), who graduated at Oxford and Cambridge, had previously introduced Copernican ideas briefly into his elementary textbook *The Castle of Knowledge* (1556), but not until Thomas Digges (c. 1546–95), a pupil of Dee, did anyone in England give a simple – albeit flawed – exposition of the system. In 1576, responding to a criticism of him by Tycho, he published a book left unpublished by his father Leonard Digges, and appended an English translation of part of book I of the *De revolutionibus*. This work contains a diagram that advocates the infinite extent of the Universe of stars – not an entirely new idea, but new to the context of the new astronomy.

Such fragments signifying interest in Copernicanism can be roughly paralleled in most parts of Europe, and would perhaps pass unnoticed did we not value the historical phenomenon itself. An English writer whose influence on later astronomy was more substantial was William Gilbert (1540–1603). His major work *On the Magnet* (1600) is first and foremost a work on the physics of magnets. It aims to prove that the Earth is itself a spherical magnet (lodestone), but it also contains much of cosmological concern – it propagates the idea of an infinite Universe, for instance. This work of Gilbert's compares Ptolemy and Copernicus in a general way, and clearly inclines to the latter, but its greatest influence was through the writings of Galileo, who discussed it at length in his *Two World Systems*, and Kepler, who used it as a basis for some of his own cosmological ideas, as we shall see when we come to discuss his *New Astronomy*.

Michael Mästlin was Kepler's teacher at Tübingen. He showed Kepler – as Kepler tells us – that the comet of 1577 moved constantly with respect to the motion of Venus as set down by Copernicus. From his failure to measure its distance, which he realized must be much greater than the Moon's, he decided that it was moved by the same sphere as Venus. (A roughly similar conclusion had been reached by the ninth-century Abū Maʿshar.) Here, and even earlier when he formed the same opinion as to the position of the new star (supernova) of 1572, Mästlin was creating an atmosphere that would eventually help to destroy Aristotelian physics, even though he was perhaps not yet fully committed to Copernicanism. As it turned out, his analysis of the comet was quickly overshadowed by the far more thorough work of Tycho Brahe, but it was Mästlin who most influenced Kepler to follow the Copernican lead.

Tycho placed the comet outside the sphere of Venus 'as if it were a fortuitous and extraordinary planet'. He praised Mästlin, but emphasized the point that there are no real spheres in the heavens, a point on which he said Mästlin would disagree – note his belief that both the comet and Venus were on the same sphere, an odd thing to say in view of the fact that the comet came and went.

Tycho's astronomy developed under the influence of the Copernican system. Throughout his published writings, his letters and his observational records, we find him making comparisons between what he himself recorded and Copernican (or Prutenic) predictions. As a young student at Leipzig University he found the Saturn–Jupiter conjunction of 1563 to be better accounted for by the Prutenic tables than by the Alfonsine, but before long he began to grow dissatisfied with both. Tycho's extraordinary thoroughness becomes evident when we see him revising even the latitude of Frombork as determined by Copernicus: he sent an assistant there in 1584, Elias Olsen Morsing (*d.* 1584). Morsing took with him a fine sextant, believed to be accurate to a quarter of a minute of arc. He then went on to Nuremberg, where Walther and Rheinhold had made their observations, to revise their figures for latitude. From Frombork Morsing brought back one of Copernicus' instruments (parallactic rules), which Tycho found to be very primitive. Astronomy was on the verge of a sea-change in observational technique, and a heroic poem by Tycho to Copernicus could as well be regarded as an epitaph to the age that was ending. He wrote of a man who 'by means of these puny cudgels' had 'climbed the lofty Olympus'. It was by his own example, however, that Tycho flattered Copernicus most effectively, for from Copernicus he had taken the geometrical apparatus of planetary astronomy, before reconverting it to an earth-centred scheme.

The New Empiricism

TYCHO BRAHE AND EMPIRICAL ASTRONOMY

Tycho Brahe was born in 1546 in Skåne, in Denmark (it is now in Sweden). He was of the nobility, and for most of his life was in a strong political and social position, although his fortunes did not endure. His upbringing followed more or less the expected pattern: he was brought up at the castle of an uncle, but unlike his brothers, who were likewise sent as squires away from home, he was academically fortunate. At thirteen he attended the Lutheran university at Copenhagen, where he made the acquaintance of a medical professor, Johannes Pratensis, who encouraged his interest in astronomy. He was not yet fourteen when in 1560 a solar eclipse turned his interests to practical questions of observing, and he obtained a copy of the ephemerides of Stadius, which were based on the Copernican *Prutenic Tables*.

Tycho moved from one university to another – Leipzig, Rostock, Basel, Augsburg – ostensibly at first to study law, and he did not settle in Denmark until 1570. He was unfortunate enough to have the end of his nose sliced off in a duel with another Danish noble in 1566, so that he was obliged to go through life with a metal substitute, about which he was not unnaturally very sensitive. His interest in alchemical experiments might have had some connection with his wish to find a suitable alloy – he seemingly settled on one of gold, silver and copper.

Tycho's concern with astronomy was growing steadily, and his achievements began to attract wide attention when, after sunset on 11 November 1572, returning from his alchemical laboratory, he discovered a new star in the constellation of Cassiopeia. It was brighter than the rest, and he knew the constellations well enough to be able to say that it had not been previously noticeable. His experience with the instruments he had commissioned over the years allowed him to measure its position immediately, relative to neighbouring stars. He kept up these measurements, and recorded its changing brightness and colour, until in the following March this new star, 'stella nova', finally faded from view. (This 'nova', which at its brightest was visible in daylight, would now be classified as a *supernova*.)

This careful work was eventually seen to be of great significance, for the star was not moving as a comet would have done, and Tycho Brahe was in a position to say that its parallax ruled out its being closer to the Earth than the Moon. It twinkled like a star, it did not have a tail like a comet, and its stability seemed to rule out the idea that it was an exhalation from the Earth's atmosphere. The eventual implications for standard Aristotelian cosmological doctrine – even he did not appreciate them at first – were serious: the spheres beyond the Moon were supposed by the Aristotelians to be unchanging, and yet here was evidence to the contrary. By the beginning of 1573 Tycho was convinced of the reliability of his observations, and he wrote a short tract that included an astrological interpretation of appearances. He hinted that perhaps comets might also turn out to be above the Moon, and so might contradict the Aristotelian view of them.

In 1577 he had a chance to put his speculations to the test: a comet appeared then, as bright as Venus and with a tail 22° long. His observations of this, and the use he made of them, were exemplary. He was the first to derive a comet's trajectory in both equatorial and ecliptic coordinates while making a large number of trials in place of a single determination of position. In other words, he deliberately planned to work with an average path, based on data in which there was much redundancy. The modernity of his approach is as striking to us as were the forebodings of danger to the more perceptive Aristotelians.

Between nova and comet, Tycho had continued to move around freely. He had given lectures in Copenhagen, and in 1575 had visited the Landgrave of Hesse, William IV, in Kassel. The Landgrave was an astronomer himself, with a fine collection of instruments, and the two men made systematic observations for over a week, Tycho using semi-portable instruments that were carried around with him. Tycho went on to Frankfurt and then Regensburg, making contacts with astronomers and collecting ideas for instruments wherever he went, and in February 1576, a few months after his return to Denmark, King Frederick II – perhaps as a result of the Landgrave's recommendation – offered him the island of Hven in the Danish Sound, and funds, and asked him to set up an observatory there.

Tycho accepted, and for more than twenty years worked on Hven in what was to become the finest astronomical observatory up to that time. Uraniborg, as it was called, or 'castle of the heavens', was equipped with a full complement of instruments, not to mention a windmill and paper mill, a printing office, farms and fishponds, and the domestic staff needed for its support. The buildings were carefully planned, plumbed for water, and equipped with kitchens, a library, laboratory and eight rooms for assistants. Around 1584 an adjacent additional observatory was built: Stjerneborg ('castle of the stars') had additional instruments on secure foundations in subterranean rooms. On its ceiling was depicted Tycho's own astronomical system, and on its walls were portraits of six great astronomers of the past, from Timocharis to Copernicus, together with two others, one of Tycho himself, and one of Tychoides, his as yet unborn descendant.

This was a research institution in the best traditions of astronomy, but in its excellent instrumentation it outstripped all before it. The instruments included Ptolemy's rulers, armillaries, sextants, octants and azimuthal quadrants, some of wood and some of brass. He had celestial globes, one of them a metre and a half across. On a wall in the plane of the meridian was his finest instrument, a quadrant of radius about 1.8 m, the scale marked with transversal points (reproduced in plate 9) to permit easier measurement of its subdivisions of angle. This mural quadrant too was decorated with Tycho's portrait. Tycho used assistance in taking observations. One observer viewed the object through pinnules on the sighting rule, another entered results in a ledger, and a third noted the time on two clocks beating seconds, clocks that were unreliable but that were checked repeatedly against the heavens. Tycho introduced checks for instrumental error, and cross-checked by comparing results with different instruments.

When using altitudes to obtain the latitude of Hven, he noticed that the pole star gave different results from the Sun. He realized that the cause was atmospheric refraction. As we have seen, he was not the first to appreciate the point, but he studied the phenomenon more intensively than anyone before him had ever done, and he saw that seasonal and temperature effects were important. He tabulated his results for use with the Sun, and included (faulty) solar parallax in the table. His resulting observations were, overall, of unprecedented accuracy. Much of his early work is reliable to three or four minutes of arc, and his later accuracy is often better than a minute of arc.

Tycho Brahe surrounded himself with excellent assistants, some of whom – Willem Blaeu the cartographer, Longomontanus, Paul Wittich and Johannes Kepler – became renowned in their own right. Wittich had a fertile intellect. He had already sketched a scheme resembling a later Tychonic system in his copy of *De revolutionibus*. He made use of a computational technique in trigonometry (replacing multiplications and divisions with additions and subtractions) that Tycho seems to have regarded as something to which he, Tycho, had an exclusive right, and he was angered by its inclusion in a book (1588) by his rival

Nicholas Reymers Baer (in Latin 'Ursus'). Baer, the Imperial Mathematician, had visited Uraniborg in 1584, and according to Tycho then stole other intellectual property. We shall return to this charge.

Among the many socially distinguished visitors to Uraniborg was James VI of Scotland, later to become James I of England.

Tycho's full reasoning about the comet of 1577 was reserved for a Latin work of 1588, published from his own press. Under the title 'Concerning Recent Phenomena of the Aetherial World', this was the second part of a trilogy he planned but never completed. A short German tract earlier announced its implications to a wider public. His observations of the nova and comet had led him to discard the idea that the Aristotelian spheres were real in any strong sense: at least they did not seem to impede the motion of a comet above the Moon, or the generation and decay of a star.

Although this conclusion might have seemed destined to clear the way for Copernicanism, this was not his intention. As early as 1574 Tycho had been lecturing on the mathematical absurdity of Ptolemy's equant and the physical absurdity of Copernicus' moving Earth. Even before this time, Erasmus Reinhold (1511–53) and Gemma Frisius (1508–55) had drawn attention to the ease with which Earth in the Copernican system could be made fixed, allowing all else to turn around it while preserving the geometrical relationships of the system. By 1578 Tycho had been led to the idea that the inferior planets move round the Sun; and by 1584 that the superior planets move likewise. The main objection was that since Mars' orbit seems to cross that of the Sun, the necessary spheres would require matter to interpenetrate matter. The message of the comet – a message that did not strike him forcibly for some years – was that this is no problem, for the spheres are not solid.

The Tychonic system was first published in a chapter hastily added to his work of 1588 (figure 12.1 is redrawn from Tycho's illustration). In the eyes of his contemporar-

FIG: 12.1 *The Tychonic system*

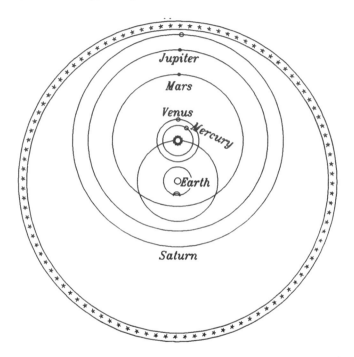

ies, this was his greatest achievement: humankind was once more at the centre of the world. As new evidence accumulated throughout this and the following century that Ptolemy's system was unacceptable, many took refuge in Tycho's, or in related schemes. There were variant schemes by Nicholas Baer, and by Tycho's pupil Longomontanus (1562–1647), for instance. In the following century some Italian astronomers even turned back to Martianus Capella (taking Mercury and Venus to orbit the Sun, the Sun and the other planets orbiting the earth). The Jesuit Giambattista Riccioli in 1651 published a variant of this, in which Mars too went round the Sun. Once the ease of geometrical transformation had been learned, finding such alternatives became an intellectual pastime.

Since the stars, even to Tycho's instruments, revealed no parallax, there is in principle little to choose between alternatives except on grounds of simplicity – for which there were in any case different criteria. Tycho regarded the stars as lying in a shell, centred on the Earth, but not all at quite the same distance. They were taken to lie only a little beyond Saturn. Like so many of his predecessors, Tycho did not believe that God would have created and wasted empty space. For all his innovations, he was a traditionalist in many other ways. His physical description of the Universe owed much to Aristotle – he got rid of the sphere of fire, but merely to extend the sphere of air – and even to traditional astrology. He was convinced of the plausibility of astro-meteorology, and recorded the weather daily for fifteen years to prove his point. His study of the horoscopes of the famous led him nowhere, unless it was to ask himself whether *astronomy* might not be the weak link in the chain of argument.

In astronomy he made some unfortunate decisions, for instance in accepting the ancient value of three minutes for the solar parallax – twenty times too great. This in turn introduced errors into the obliquity of the ecliptic, and detracted from what even then remained the finest theory of solar motion then devised. From 1578 to the mid 1590s he recorded noon (meridian) altitudes of the Sun for approximately one day in three, with different instruments. This habit of his was one he never dropped. His night-time observing grew ever more intensive, and multiple sightings with multiple instruments became the order of the day. With lunar eclipses, for instance, he worked with three sep-arate teams of observers. He would gladly spend a decade on a single task, as when he prepared his star catalogue. This, the finest ever done with non-telescopic instru-ments, was of course largely the work of assistants, in both observation and calculation, and it has many blemishes in the second respect. Its conception was sound enough. By 1588 Tycho had the position of his base star alpha Arietis to within 15″ of its true longitude. He knew twenty additional reference stars almost as accurately, and the mass of the catalogue was done by reference to them.

If the accuracy of his observational work is the thing for which Tycho is most easily remembered, his most lasting single achievement was undoubtedly an aspect of his lunar theory. This was left until a late stage in his overall publishing project, ostensibly devoted to the new star, but in reality taking in most of astronomy as then practised. He made lunar observations systematically from 1581 onwards. He first suspected that something was wrong with traditional theories when he tried to estimate the times of eclipses. He would observe a couple of days before, and so estimate the precise time. In fact he found that in 1590 he missed the first hour of the eclipse, and in 1594 was again late in his estimate: the Moon seemed to have accelerated as it went into opposition. It must, he said, be slowing down on average elsewhere. But where? He settled on the octants, mid-way between syzygies and quadratures, and observed the Moon carefully through these points.

In this way he discovered what is now known as 'variation', the first wholly new astronomical inequality to have been discovered since the time of Ptolemy. With this discovery, Tycho made possible a drastic reduction in the residual errors in lunar longitude.

In 1595 he was led to look more deeply into the Moon's motion, in latitude as well as longitude. Before long he had discovered a slow change in the inclination of the Moon's orbit. He was quick to realize that the old figure of 5° holds only for syzygy, and that in quadrature 5;15° is more accurate. This led him to a model in which the pole of the Moon's orbit moves round a small circle twice every (synodic) month, and before long he saw that there were similarities with Copernicus' model for the ecliptic, and that the nodes of the Moon's orbit must oscillate around their mean positions. This produced another 'correction' in his theory of latitudes. Tycho was no great mathematician, but here we see very clearly how a strong intuitive feeling for such matters can compensate.

Before 1598, Tycho added yet another correction to the Moon's motion, when he introduced the year, a solar quantity of course, into the reckoning. The maximum size of this term is only eleven minutes of arc, and yet he somehow found an accurate figure for it. (He was less fortunate with the timing of its maxima.)

All these new effects he wove into the existing Copernican lunar model, and he was in the middle of printing the result, as the capstone to his monumental book, when — helped by Longomontanus — he decided to correct one of that model's shortcomings. Although much better than Ptolemy's, in regard to the variation in the Moon's distance (and hence apparent size), it was still far from satisfactory. Tycho had learned the Copernican lesson of double epicycles: where Copernicus had one for the first inequality, he now took two. Choosing the parameters of size and velocity carefully, he was able to get a marginally improved, but still far from perfect, model of lunar distance. That he saw the desirability of correctly representing the lunar path in space is as noteworthy as his success, however.

The second lunar inequality (known today as 'evection') had been explained by Ptolemy in terms of the small circle at the *centre* of the lunar model. Copernicus had taken the mechanism away from the centre, but Tycho had two epicycles already, so he moved it back there. The details of the resulting model were more complicated than in Ptolemy's case, and in fact Tycho never managed to fit all of his discoveries into a satisfying lunar model. (He omitted most of the annual equation, for instance.) What he did for lunar theory was nevertheless of immense importance, and it was seen to be so as soon as Kepler succeeded in integrating it into his own, more penetrating theory.

Tycho worked on a new theory for the planets at much the same time as he struggled with lunar theory, but it was left to Longomontanus to carry out this part of his programme, and Longomontanus' *Astronomica Danica* (Danish Astronomy) appeared in 1622, replete with double epicycles in a system that combined the Tychonic and Copernican traditions. That Tycho did not complete his task had much to do with external circumstances. After the death of his patron, the king, and the ending of a regency in which Tycho's brothers played a part, a young new king, Christian IV, ascended the throne. Tycho quarrelled with all and sundry – including a pupil who was engaged to his own daughter, his tenants and even the king himself. Tycho lost his favoured status and looked elsewhere for patronage. He finally left Hven in 1597, first for Hamburg, where he published his *Mechanica* – a highly esteemed description of his instruments that he dedicated to the Emperor Rudolph II, in Prague. After considering settling in various other parts of Europe, Tycho eventually accepted Rudolph's patronage in June 1599. The move meant that he had to ship his instruments, reinstall them, and reorganize his group of assistants. Ensconced in the castle of Benatky, where he began anew to set up an observatory, he had only a year's use of his best instruments. He died in Prague in October 1601. Benatky of course survives, but as for Uraniborg, virtually all that remains is an outline of its foundations, best seen from the air.

In Prague, Tycho took into his service Kepler, his most famous assistant of all. Kepler was assigned to work on the planet Mars, and it was he who finally saw Tycho's great work into print, under the title *Astronomiae instauratae progymnasmata* (First Exercises in a Restored Astronomy). This was in 1602. Another book he was commissioned to write, a defence of Tycho against Baer – who in any case died in 1600 – Kepler was only too happy to drop. It finally appeared in print in the nineteenth century. On his death-bed, Tycho asked Kepler to complete his astronomical tables. The 'Rudolphine tables', dedicated to the emperor of course, were meant to be computed following Tychonic principles. They finally appeared only in 1627, and while the observational basis for them was essentially Tycho's, the underlying theory was Kepler's own.

HYPOTHESIS OR TRUTH, ASTRONOMY OR PHYSICS?

The controversy between Tycho and Baer, to which we have already alluded, well illustrates a philosophical theme that has owed more to astronomy than any other science. Do scientific theories carry with them any implications for the reality of the things they describe? The theme had been given wide publicity through the scurrilous Osiander preface to Copernicus. Baer was a sceptic in much the same mould as Osiander. Copernicus, Tycho and Kepler inclined to the view that there was a true and in some sense real system waiting to be found, and that a sound astronomical theory did more than merely permit the computation of phenomena in advance. Baer knew full well that accurate prediction does not guarantee the validity of a theory, since true conclusions may follow from false premises – a simple logical point that had often been made in the middle ages. This nettled the astronomers, but to understand Tycho's irritation we must remember the colourful stories of Baer's plagiarism of his planetary system on a visit to Uraniborg.

Nicholas Baer was by origin a farmer's boy from Ditmarsch (in the lower part of the Jutland peninsula), and for that reason alone, Tycho thought he should know his place. At all events, while Baer was on a visit to Uraniborg in 1584, as he was sleeping, his pockets were searched by a certain Andreas, a student of Tycho, and incriminating papers were supposedly found there. The evidence later adduced against him was that whereas in Tycho's system

Mars' orbit intersected the Sun's, in Baer's Mars' enclosed the Sun's, as in a diagram by Tycho that had been wrongly drawn.

True or not, Baer's system differed from Tycho's in one important respect: he gave the Earth a daily rotation on its own axis, half-loosening it from its old bonds, so to speak. Baer's is sometimes called the 'semi-Tychonic system', although Baer would certainly have disapproved, and the name could just as well be applied to alternatives mentioned earlier. It has to be said that if we disregard this distinction, there were at least half a dozen writers who claimed to have invented the system independently. Tycho was not pleased by Kepler's remark that it was an obvious step from Copernicus.

The loss of the papers by Baer is said to have begun a mental disturbance that led to his loss of imperial patronage. But what was the nature of the stolen ideas? There were complaints of the appropriation of constructions, inventions, formulae and tables, but not of a fully-articulated world-system. Kepler, later given the task of defending Tycho, saw what was at stake more deeply than either man, namely an integrated viewpoint, a complete system of hypotheses answerable to observations.

Before leaving this controversy it is as well to remind ourselves of the dangers of giving it too modern a colour. Sixteenth-century writers did not usually insist that astronomical theories were 'mere fictions', as is so often maintained. They might, as did Philipp Melanchthon and various astronomers from the Lutheran university of Wittenberg, accept certain of Copernicus' mathematical techniques without embracing the theory as a whole. Like so many others, they were not truly Copernicans, and yet this did not make them 'fictionalists' in a strong modern sense. It was simply, for most of those who considered the philosophical question at all seriously, a case of refusing to allow astronomy to pretend to purvey real physical knowledge.

This is where Kepler came into his own. In a sense, he was on weaker philosophical ground than his opponents, to the extent that he argued for the final truth of certain astronomical hypotheses, or points of view. Although strong in rhetoric, his arguments for preferring Copernicus to Ptolemy, for instance, depend in the last resort on aesthetic arguments concerning simplicity, harmony, elegance, and the like. He had one very potent argument, however, in regard to his elliptical paths. Unlike the systems with multiple circles, his ellipses showed *the final paths in space,* that is, what would be seen by an observer viewing the system over a long period from far out in space.

This is what Kepler often had in mind when he said that his system was the first to avoid hypotheses. (Others before him, such as Toscanelli and Apian, had charted the paths of comets, but there was then no involvement in a deep theory of the orbit.) Baer had said in effect that if two hypotheses do the job equally well, it did not matter which one took. Kepler insisted that the two may be equivalent, but only in certain limited respects, and not in all; and that *physical* respects should not be brushed aside. He was not the first to insist on this point, but he saw more clearly than anyone before him something of great importance to the actual *growth* of scientific theory: he saw how urgent was the need to integrate mathematical astronomy into physics, natural philosophy. He sought for the *causes* of planetary motion, and not only for the geometry of that motion. It could be argued that this traditional distinction is an erroneous one, but the historical fact remains that established physical modes of thinking were capable of supplementing very potently the old geometrical modes.

We might well call Kepler 'the first modern astronomer' on the strength of his perception of the need to combine physics with astronomy, although Ptolemy, and others before and after him, plainly had similar ambitions. Kepler, however, lived at a time when conventional (Aristotelian) physics was under attack – for instance from the writers of several new treatises on comets. This went along with a theological questioning of authority, especially in protestant countries. Michael Mästlin, for instance, Kepler's teacher, had failed to find any parallax for the comet of 1580, and so openly attacked the old physics. Aristotle was well ensconced in the schools of the time, and remained so for well over a century in some places, but by slow degrees astronomy removed what some considered his cosmology's most crucial foundations.

The irony of Kepler's part in all this is that he was largely inspired by types of reasoning that belong as much to astrology as to astronomy, as we now perceive it.

KEPLER AND PLANETARY ASPECTS

Johannes Kepler was born in Weil der Stadt (near Stuttgart) in 1571, the grandson of a mayor of the town, but son of a 'criminally inclined and quarrelsome' mercenary soldier who eventually abandoned his 'garrulous and bad-tempered' wife. The descriptions were Kepler's own, written when he was comparing their characters with their horoscopes. Between 1617 and 1620 Kepler was destined to defend his mother when she was tried for witchcraft.

He eventually attended the university of Tübingen, where he came under the influence of the Copernican expert, Michael Mästlin. After taking his master's degree, Kepler embarked on a study of theology, but then Georg Stadius died, and Tübingen was asked to recommend a replacement for him as mathematician at the Lutheran school in Graz. Kepler was recommended, and in 1594 he left Tübingen to take up his new post. Just over a year afterwards, he happened by chance upon what he thought to be the secret of the structure of the universe.

We recall the curious astrological doctrine, put forward by Rheticus in the *Narratio prima*. Following an established tradition in astrology, he supposed that a planet's strength is increased at its apogee and diminished at its perigee. Kepler was adept at astrology. Not only did he in 1596 compile a collection of horoscopes for members of his family, by which they are compared scientifically in character, but before this he had published a calendar and prognostication for 1595. The peasant uprisings and invasions by the Turks that he predicted were perhaps to be expected, but the excessively cold winter he led his readers to expect was not. His star was in the ascendant. With a gap of only three years, he published annual prognostications until 1606. He took up the work again between 1618 and 1624, this time more cynically, to compensate for non-payment of salary. Kepler made some often-quoted early remarks to the effect that astrologers only get it right by luck, or that astrology is the foolish daughter of astronomy, but they do not really say more about his early belief than that he had been disappointed in his experience.

We have 800 horoscopes from Kepler's pen that could be put down to a need for ready cash, but this does not explain why he drew up so many horoscopes for himself. He thought long and deeply about the question, which he took seriously for most, perhaps all, of his life. He tells us that it was after he met Tycho in Prague that he abandoned *a part* of astrology, but later he said quite explicitly that there was much good in the subject. To Mästlin he wrote that he was 'a Lutheran astrologer, throwing out the chaff and keeping the grain' (1598). He made a standard protest against a forbidden celestial *magic*, not so much against astrology but against the use of spiritual or demonic magic. In a little work on the 'more certain foundations of astrology' he lists three sorts of reasoning underlying astrological prediction: the theories of (1) physical causes, (2) metaphysical or psychological causes, and (3) signs. The first two are valid, the third is not. This is the man of science talking, the physicist-cum-psychologist, the man who no longer believes in the sharp Aristotelian distinction between the regions above and below the sphere of the Moon.

He toys with the idea that *light* might be the physical cause he seeks. Until the telescope had been available for a couple of years he seems to have believed that the planets shine by their own light, and whereas older astrologers had associated the planets with the ancient deities, and had deduced their properties accordingly, Kepler thought that the colours of the lights they sent to Earth were what determined their astrological characters. Here is an example of Kepler the *creative* astrologer, who added new doctrine as Rheticus had done. The irony here is that the two leading Copernicans of the century, Rheticus and Kepler, both added to existing astrological doctrine.

One theme runs right through his life, and that is the idea of 'imprints' on the human soul, received from celestial configurations when a man begins a life independent of his mother. In other words, he approved of the doctrine of *aspects* (synonymous with configurations) and regarded it as a question for *experience* to decide. He showed his allegiance to planetary aspects by his example, again and again, and they are in at the very birth of *Mysterium Cosmographicum* (Cosmographic Mystery), his first important book.

His great experience at Graz was one he had while struggling with the reasons why the numbers, sizes and motions of the orbs were what they were. He was using a standard diagram to explain to his pupils a commonplace of astrological theory concerning the great conjunctions of Saturn and Jupiter. What no one else had previously noticed was that the lines joining the points of the zodiac circle where the conjunctions occur enclose a second circle, and the ratio of the diameters of the two circles was very close to that between the orbits of Saturn and Jupiter, according to the system of Copernican distances. (See the idealized figure 12.2, where the outer circle represents the zodiac, and the places of successive conjunctions, numbered here only as far as seven, are joined by straight lines.)

Kepler could not find similar relationships in the orbits of the other planets until he started looking for analogies in three dimensions, and when he did that, he found an extraordinary geometrical scheme, with nested cube, tetra-

FIG: 12.2   *The pattern of great conjunctions of Saturn and Jupiter*

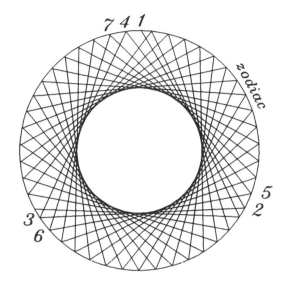

hedron, dodecahedron, icosahedron and octahedron, all fitting the Copernican Universe to within about one part in twenty. (For his own drawing, see plate 11.) That this fits the measured distances so well has even now the power to astonish.

It would be easy to represent all this as a triumph for an astrological starting-point, but in his own psyche the influence must have worked in *both* directions. Having found himself on the track of the correct geometrical scheme of the heavens, as he then thought, he must have been persuaded that the doctrine of aspects was justified by the astronomy that had provided the distances long before. It is not surprising, therefore, that he scrutinized and eventually added to the standard astrological doctrine of aspects. These aspects, he thought, act on the soul, but only when it has a harmonious instinct. The soul reacts by means of this instinct, to certain harmonious proportions, subdivisions of the zodiac. It is not necessary to consider how he added to the collection of traditional aspects, angular separations of the planets, such as quartile (90°, hostile) and trine (120°, friendly). The important point is that this was not an aberration of his youth. He discussed these matters in the *Harmonice Mundi* (Harmonies of the World), and there reminded his readers of another theme that was for long in his mind, the theory of musical consonances. Until about 1610 he thought he could link this with his astrological harmonies, but he gradually separated the two. Another link he tried to forge was between astrology and alchemy, which might have endeared him to Tycho, but in the end led nowhere.

The *Harmonice Mundi* was a long time gestating. It was Kepler's favourite of all his writings. He started it around 1599, worked at it intermittently at the same time as he developed his astronomical theories and proceeded with the Rudolphine tables, and it only saw the light in 1618.

The strange theory of nested regular solids has been of great influence down the centuries, and in due course was replaced by other simple laws that seemed to fit the planetary distances, but for no obvious reason. The Titius–Bode law of planetary distances, for instance, is scarcely better than Kepler's, but it is still often quoted as a mystery worth explaining. Presented by J. E. Bode in 1772 as having been found by J. D. Titius, this gives the orbits of the planets from Venus outwards as $0.4 + 0.3 \times 2^n$ astronomical units (0.7, 1, 1.6, and so on, with the minor planets 2.8 and eventually the newly discovered Uranus 19.6). For Mercury (0.4) we simply drop the second term in the expression – not the most elegant of mathematical procedures. Later, in a dissertation of 1801, the 'dialectical' philosopher G. W. F. Hegel (1770–1831) criticized the arguments that were currently being used to suggest that the gap between Mars and Jupiter must be occupied by another planet. He gave another formula involving simple numbers that fitted roughly the distances as far as Uranus and that did not fill the gap in the same way. Unknown to Hegel, he had backed a loser, and did so after it had lost, for on the first day of 1801 Giuseppe Piazzi had located an asteroid in the gap. Hegel's point was not very clear, but it was not as foolish as it is often presented. He did not 'try to prove that there can only be seven planets', as is often said. He said in effect, 'I can find an alternative law which casts doubt on your way of adding more planets'. The point he missed is that we need more than a few casual fits with observation before accepting a law in isolation from other parts of a science, whether that law is the Titius–Bode law or his alternative.

There were to be many similarly mystical and 'Platonic' studies and interpretations of Kepler in the nineteenth century, especially in Tübingen, by scholars looking for quick returns, laws without affinities with other parts of science. Kepler would have disabused them of the possibility of quick returns.

Kepler had a clear *physical* vision of the astrological workings of the planets: he seems to have thought of them as somehow disturbing a finely balanced, 'equilibrium', situation – for example, with a spring atmosphere of saturated water vapours, the right planetary aspect may be strong enough to trigger off showers of rain. He thought such things confirmable by experience, and made very many observations of the weather to confirm his ideas, for instance the idea that a conjunction of Saturn and the Sun causes cold weather. He repeatedly claimed that the efficacy of his new set of astrological aspects was supported by observation of the weather.

It is usual to describe such matters as instances of Pythagoreanism, a searching for harmonies of a sort beloved in the Renaissance, but Kepler was doing something more: he was matching his geometry against measured quantities. If his harmonies do not quite please us – perhaps we are dissatisfied with that maximum error of one part in twenty – that does not affect their *character*. This point is easily lost if we allow ourselves to be shocked by his strange remarks about the Earth's soul, but here too he is thoroughly empirical in his outlook. The argument goes like this: an aspect (the angle separating two planets, viewed from the Earth) is a purely geometrical relationship, even when light travels in along the directions that define it, say two rays 120° apart. The aspect affects human beings because they may perceive it, or their soul may somehow receive it, but how does it affect the *weather*? The answer is that the Earth has a soul too, and it stretches up to the Moon.

Kepler in this way gives a new meaning to the sublunary realm. Of course for him the more important world-soul is situated in the Sun. The harmonies and aspects we perceive are appropriate to our earthly position; those relative to the Sun are what decide the velocities of the planets and their orbits (*Harmonice Mundi*).

Regarded in this way, Kepler might seem no longer to be quite the 'first modern astronomer' of traditional history, but here a word of warning is needed. In the end it was an abstract doctrine marrying geometry and physics that he accepted, and not the full panoply of traditional astrology, about which he was highly sceptical. He had a strong aversion to *astrology as a theory of signs* and helped to diminish its popularity in scientific circles. The fact that a great conjunction occurs in what was traditionally 'the *fiery* trigon' meant for most astrologers that it would be accompanied by *fire* of some sort, for instance the fires of war, and droughts. He saw that the characters of the signs had been dictated by a series of feeble and random human analogies, and this in the end he entirely rejected. He spoke against the signs and the mundane houses, as residences of the planets, and yet he somehow left himself enough doctrine to *practice* as an astrologer. In 1608 he had drawn up a horoscope which, unbeknown to him – or so the story goes – was for the great military leader Wallenstein. This he did according to the method of computation that was then ascribed to Regiomontanus. He later 'purified' his astrological principles, but the fact remains, had he not been an astrologer he would very probably have failed to produce his planetary astronomy in the form we have it.

KEPLER AND THE LAWS OF PLANETARY MOTION

Kepler's *Mysterium*, printed in 1597, did much in the following decade to aid the Copernican cause. Galileo acknowledged receipt of a copy, saying that he had read, as yet, only the preface. Kepler's reply urged him to proclaim his belief openly. Tycho expressed admiration for it, but this he could do with impunity, for although Copernican in character it was written in a very different style from his own empirical astronomy. Kepler's search for harmonies, however, went further than merely wanting to know the harmonies of *scale* in the Universe. His was a search for *causes*, and he decided that the central position of the Sun must be the key to understanding the causes of the planetary motions. He knew that the planetary periods lengthened with distance from the Sun. He tried to find the relationship between the two quantities, and in the *Mysterium* he settled on a law that made the period (*T*) proportional to the square of the radius of the orbit (*a*). The correct power is 3/2, as he was later to discover.

On 28 September 1598, without clear warning, the Catholic authorities – a commission of the Counter-Reformation – ordered all Lutheran teachers to leave Graz at once. Kepler was treated more generously than most, and was allowed to re-enter the city, but he needed employment elsewhere, and in February 1600 he arrived in Prague, hoping to work for Tycho Brahe. Overcoming initial irritation at Tycho's paternalism in astronomical matters, Kepler settled down in Tycho's employ, and was assigned to the theory of Mars.

He soon discovered that the position of Mars was better referred to the *true* Sun, and not the centre of the Earth's orbit, as Copernicus had taught; and he reintroduced the idea of the equant, now no longer one that was tied to the eccentricity, but an equant that he could vary in position until he found the best fit with observations. The Tychonic data allowed him to obtain a rather accurate orbit for Mars, and his equant model (which he called his 'vicarious hypothesis') matched the observed longitudes better than any before it. It was still unsatisfactory for latitudes, however, and this suggested to him that his *distances* were wrong. Longitudes were predicted to better than two minutes of arc.

Tycho's death left Kepler with obligations – the completion of the Rudolphine tables, for instance – but it meant that he had access to observational records that Tycho had guarded jealously, and it meant that his loyalty to the Tychonic system could not be enforced. He now struggled with the orbit of Mars. Getting the right distances meant adjusting the relative placement of the equant and the deferent centre, and upset the accuracy of his longitudes, now six or eight minutes in the octants. Where others would have let matters rest there, Kepler's confidence in Tycho's observations made him persevere.

His interest in books on the magnet by Jean Taisnier (1562) and William Gilbert (1600) had given him the idea that magnetic forces emanating from the Sun might explain the planetary motions. Kepler was by no means alone in pursuing this idea, for there was a widespread belief that Gilbert had proved the motion of the Earth magnetically. In 1608 the Dutch scientist Simon Stevin published a work written some years earlier, *De Hemelloop* (Movement of the Heavens), in which a complex cosmology was based on a theory of cosmic magnetism deriving from the fixed stars. Kepler's work was independent of this. He argued that if magnetic forces spread like light in three dimensions, they would diminish *in proportion to the square of the distance*, and this idea he could not reconcile with the velocities. If the forces were confined to the plane of the orbit, he thought, like thin spokes of a wheel, their effect should diminish *in proportion to the distance.*

On the slender basis of such reasoning he reached a 'distance law', according to which orbital velocity is inversely proportional to the distance from the Sun. After struggling with problems that we should now solve easily enough with the help of the differential calculus, but for which he was obliged to invent his own methods, he came up with the law of areas, that we now usually call 'Kepler's second law of planetary motion': *the radius vector from the Sun to the planet sweeps out equal areas in equal times.* (This is illustrated in figure 12.3, where three specimen areas marked as *a* are all sup-

FIG: 12.3   *Kepler's second law*

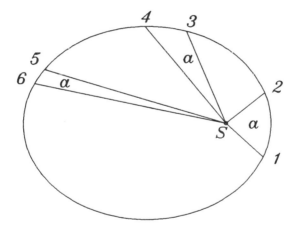

posed to be equal. The planet on the ellipse moves from 1 to 2, from 3 to 4, and from 5 to 6 in equal times.) Strangely enough he does not appear to have felt very pleased with his demonstration, and gave the law no prominent place in his writings until his *Epitome astronomiae Copernicanae* (Epitome of Copernican Astronomy) of 1618–21.

By itself, the areas law was not enough to eliminate that eight minutes of error. The orbit of Mars seemed to be non-circular. He tried an epicycle, which with the deferent was equivalent to an oval, and he tried different sorts of oval. Everything depended on the precise form of the orbit, the extent of its deviation from a circle. He knew that if it were an ellipse, an entire branch of geometry (the geometry of conic sections of Archimedes and Apollonius) would be available to him, but he seems to have thought that fate was unlikely to be so kind, for by 1605, with hundreds of trial calculations behind him and with fifty-one chapters of a book on Mars completed (*Astronomia nova*, 'The New Astronomy', was 'a commentary on the planet Mars'), he still gave the ellipse no place. He could not reconcile the ellipse to the magnetic hypothesis; but then, seven chapters later – for his book was almost a diary of his thoughts rather than a heavily revised treatise – he decided that he could do so, and 'Kepler's first law' was born, according to which *a planet travels in an elliptical orbit with the Sun at one focus of the ellipse.*

In 1606 Kepler had published a work of more popular appeal, *De stella nova* (On the new star), with musings over a wide range of physics and cosmology, but it was his *New Astronomy* that marked him out as a new intellectual force to be reckoned with. This eventually appeared in 1609, after he managed to settle disagreements with Tycho's heirs about his right to use the observations in a non-Tychonic manner. Throughout these years he was harassed by circumstances, public and private: his wife took less kindly to his moves than he, and had little regard for astronomy, no doubt with excellent reason. She had wealth tied up in estates near Tübingen. Obtaining his due salary was one of Kepler's constant headaches. By 1611, spurned by the Tübingen professoriate for overtures he had made to Calvinists, he found promise of employment in Linz to please his wife, but she died of typhus before leaving Prague. Prague was in a state of war and chaos reigned, leading to Rudolph's abdication. Even in Linz, Kepler was involved in endless religious turmoil. This, which at one stage resulted in his being denied access to his own books, and the illness and death of children by his first and a second marriage, weighed heavily on him.

In 1604 and 1611 he had published two of his most important works, both on optics, regarded as a necessary part of astronomy. However, in neither his 'Optical part of Astronomy' nor his 'Dioptrics' did he have the sine law of refraction. (This, 'Snell's law' as it is usually called in the English-speaking world, was first found by Thomas Harriot no later than 1601. In 1606 Harriot sent Kepler a table of angles of refraction for many different substances, but did not give him the sine formula.) In the later work, however, Kepler gave a thorough mathematical treatment of the formation of images by lenses, and the arrangement of two converging lenses into what we now call the 'Keplerian' or 'astronomical' telescope, which produces an inverted image. The 'Dutch' or 'Galilean' telescope was then a couple of years or so old.

Kepler wrote a number of minor works, but for a time his prodigious energies were diverted from astronomy. The *Harmonice mundi*, in the tradition of his *Mysterium cosmographicum*, appeared in 1619. In 1625 he wrote a defence of Tycho against Scipione Chiaramonti, who was trying to uphold Aristotle in the matter of the interpreting of cometary appearances. And then, between 1618 and 1621, came what was to be for many decades the most widely used treatise of theoretical astronomy, Kepler's *Epitome*.

This was a spirited defence of the Copernican world, but of course as modified by his own ideas. It introduced analogies between the arrangement of the world and the Holy Trinity, and no doubt this, as well as the doctrine of the Earth's motion, explains why it was almost immediately placed on the Catholic index of prohibited books. It contained, even so, ideas that could not be confined. It has a derivation from magnetic principles of 'Kepler's third law', which makes *the planetary periods proportional to the $\frac{3}{2}$ power of the semi-diameter of their orbits.*

Here the driving force of the motion was at last reckoned to be inversely proportional to the *square* of the distance; but the reasoning, which introduced the notion of the emission of magnetic emanation and its absorption by the planets, and the related densities of the planets, was only the scaffolding by which the theory was erected, and was not destined to survive.

With the Moon, satellite of the Earth, Kepler got into still deeper waters with his magnetic theory, for it should be under a double influence, from the Earth and the Sun. In the last analysis, the physical explanations he was offering were what he needed to stimulate his geometrical imagination, but it was the imagination that counted for most.

In book V of the *Epitome* Kepler introduced an equation that is fundamental in the solution of planetary orbits. It relates two angles, known as the eccentric anomaly (E) and the mean anomaly (M). The former is shown in figure 12.4, on which P is the planet, S is the Sun, and A is aphelion, the planet's point of nearest approach to the Sun. The angle M is not shown. It can be thought of as the angle between SA and a line through S that rotates at the same average rate as the planet, coinciding with it every time it passes through A. If e is the eccentricity of the ellipse, Kepler's equation is simply

$$E - e \sin E = M$$

where angles are in radian measure.

It should be obvious that M is easily found if E is known, but that the problem facing a calculator of planetary positions will begin with a solution of the inverse problem: to find E given the time (or given the angle M, steadily increasing with time). There is no simple and exact explicit solution.

The solution of the equation by different approximation techniques has a long history. Before modern computing methods took away much of the importance of finding a mathematically elegant solution to the equation, tables for

FIG: 12.4 *The basis of Kepler's equation*

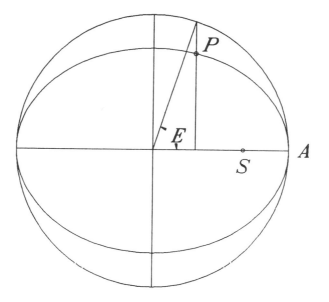

such were refined from time to time (J. J. Astrand's of 1890 and J. Bauschinger's of 1901 have been much used), but Kepler himself produced some very creditable procedures for use in his *Rudolphine Tables*.

Before these tables were ready, he had the good fortune to discover John Napier's invention of logarithms (1614), from a work by another author. Appreciating the underlying principle, but not having formulae for their construction, he prepared logarithms of his own to a somewhat different convention. The logarithms much lightened the labour of calculation, as they were to do in astronomy thenceforward, indeed until long after mechanical calculators began to displace them in the nineteenth century. The practical subject of computation is one that may lack the glamour of theoretical astronomy, but it is of course of great practical importance, even at the stage of refining a theory, as Kepler would have been the first to maintain. Logarithms, be it noted, were usually for *decimally* expressed numbers, while astronomers traditionally expressed angles sexagesimally. A little-known fact is that in due course logarithms of *sexagesimal* numbers were produced, especially for navigation at sea using astronomical methods.

The *Rudolphine Tables* were far more reliable than their predecessors. In the case of Mars, for instance, where errors might approach 5°, they were now less than a thirtieth of that figure. He was able to predict, for the first time in history, transits of Mercury and Venus across the Sun's disc. He died in 1630, a year before this unusual pair was to be seen, but the Mercury transit was indeed seen by Pierre Gassendi in Paris, Johann Baptist Cysat in Ingolstadt, and Remus Quietanus in Ruffach. (The Venus transit was invisible from Europe.)

Kepler hoped to use the tables to recoup unpaid salary, but printing was interrupted when Counter-Reformation forces occupied Linz. Only in 1627, with Kepler now settled in Ulm, was the printing finally completed there. His life was almost to its end interrupted by the wars of religion, forcing him to move repeatedly, and it is easy to appreciate the flights of fantasy that took him above it all when revising his *Somnium seu astronomia lunari*, his 'Dream, or lunar astronomy', of 1609. This science-fiction story of a voyage to the Moon was a literary device used to argue for a Copernican arrangement of the planets, which an observer on the Moon is well fitted to appreciate. It was finally seen through the press by his son-in-law Jacob Bartsch. When Kepler died, he was still owed more than 12 000 guldens by the state in whose service he had spent most of his life.

THE FIRST TELESCOPES

Of all astronomical instruments, none has had a more dramatic immediate effect on the course of astronomy than the telescope. Kepler's design was eventually to occupy the centre of the astronomical stage for more than a century, but it was not the first, and it did not come into its own immediately. The idea of the telescope certainly long preceded its realization. Simple lenses, as we might now call them, had been known from very early times, not merely as drops of water and ice, but created inadvertently in the process of polishing transparent gems, rock crystal, and in due course glass. It requires no great genius to observe that in certain circumstances lenses may magnify, reduce, or invert an image, and by the end of the thirteenth century converging (convex) lenses were in use for reading glasses, spectacles. The Latin word *spectaculum* was used for a single lens at the beginning of the same century. There are numerous imprecise references in medieval literature to the possibility of seeing distant objects clearly, as though they were near at hand. By the seventeenth century there was a well established trade in spectacle lenses, and in some ways it is surprising that the discovery of a method of combining them into a telescope – and later into a compound microscope – was so long in coming. No doubt this has something to do with the relative scarcity of weak converging lenses, lenses of long focal length, and of strong *diverging* lenses.

It is even more surprising to find that the first well attested telescopes were 'Galilean', that is, that they used a diverging eye lens and a converging object lens, as in the modern opera glass. The Galilean combination gives an upright image, unlike the 'Keplerian' telescope, with its converging eyepiece and objective; but a magnified image is a magnified image, and the astronomer can work easily enough with an inverted view of the heavens.

Claims for prior invention have been made on behalf of various sixteenth-century scholars, such as John Dee, Leonard and Thomas Digges, and Giambattista della Porta, but they are without foundation, and rest on an excessively generous reading of ambiguous texts. Some confusion has been created by the existence of medieval illustrations of philosophers looking at the heavens through tubes.

Aristotle himself referred to the power of the tube to improve vision – it may improve contrast by cutting down extraneous light – but the tubes in question were always without lenses. The sixteenth century gave much attention to the theory of 'perspective' – the science of direct vision, reflection and refraction – and 'perspective glasses' were recommended to astronomers and military men alike. Kepler, for instance, observed the comet of 1607 (Halley's as it happens) *per perspicilla*, probably an eye-glass, and certainly not a telescope. There was much experimenting with lenses at the time, and one suspects that published accounts were even then misconstrued, so that an inventor might have believed that he was reproducing the work of another.

Whatever the truth of the case, the first unambiguous evidence that an effective telescope had been made appears in the form of a letter dated 25 September 1608. This letter is from a Committee of Councillors in the Province of Zeeland in The Netherlands to their delegation at the States-General in The Hague. In the letter it is said that the bearer 'claims to have a certain device by means of which all things at a very great distance can be seen as if they were nearby, by looking through glasses, and this he claims to be a new invention'.

Matters then moved very fast indeed. A week later Hans Lippershey, a native of Wesel but spectacle-maker in Middelburg in Zeeland, applied for a patent on his invention. The States-General entered into negotiations with Lippershey at once, and although he was not granted his patent he was given a lucrative commission for several *binocular* telescopes. He was denied the patent on the advice of unknown parties, who held that the instrument would be all too easy to duplicate. We know that at least two other men were in fact making the instrument within three weeks of the original letter – Sacharias Janssen of Middelburg and Jacob Adriaenszoon of Alkmaar (in the province of Holland). There is a possible fourth reference from the same period, but now the stream of history is muddied by the fact that the writer, Simon Mayr, had earlier quarrelled disastrously with Galileo, and was to do so again, on questions of scientific priority relating to the telescope.

Simon Mayr (Marius) was a competent German astronomer who in his *Mundus Jovialis* of 1614 tells us that his patron had been offered one of the Dutch instruments at the Frankfurt Book Fair in 1608, but that he would not pay the asking price, especially as one of the lenses was cracked.

The Spaniards, the old occupying force, were in The Hague negotiating a peace settlement, and various other foreign delegations were also there in force, thus ensuring that the spread of the telescope's fame throughout Europe was extraordinarily fast. The Spanish Commander-in-Chief, Ambrogio Spinola, was evidently alarmed when he was shown the instrument before the end of September, but is said to have been assured by the gentlemanly Prince Frederick Hendrick that although the Spanish forces would now be visible from afar, the Dutch would not take advantage of the fact. Lippershey refused to make telescopes for the French, but the secret was out, and the instrument was on sale in Paris by April 1609, in Milan by May, and in Venice and Naples by July. Galileo learned of it by report in the middle of July 1609, made it his own, and greatly improved its performance. Its value to a maritime power like Venice was very great, and his improvements earned him lifetime tenure of his chair in mathematics at Padua, and an unusually high salary.

These simple telescopes had at first magnifications of two or three, but by improving his grinding and polishing techniques Galileo made objectives with ever longer focal lengths, and by 1610 his magnifications were 20 and 30. He made other improvements, in particular in his use of aperture stops between the lenses. Above all, Galileo recognized the great scientific potential of his new possession, with which he quickly observed the mountains of the

Moon, four satellites of Jupiter, and the starry constitution of parts of the Milky Way that had not previously been resolved.

He was not the first to turn the instrument to the sky. A report of the visit of an embassy from the King of Siam to Prince Maurits in The Hague – the source of the Spinola story – relating to 10 September 1608 also noted that the new invention revealed stars that were invisible to the naked eye. In England, Thomas Harriot was mapping the Moon as early as 5 August 1609, before Galileo's serious studies had begun. What characterized Galileo, however, was the sheer energy he threw into the whole enterprise. He rushed into print with his first telescopic discoveries in his *Sidereus Nuncius* (The Sidereal Messenger) in March 1610, and the excitement it aroused led to a second edition from Frankfurt, before the year was out. Kepler was quick to publish two short books on the new discoveries, the first before he had even observed these things for himself.

How much Galileo knew of similar activities in other parts of Europe we do not know, but in England Thomas Harriot had been covering much the same ground, and in some matters pre-empted Galileo, but without publishing his findings. As a young man, Harriot had been tutor to Sir Walter Raleigh and in 1585 had been on Raleigh's second expedition to Virginia, but now he was a member of the household of Henry Percy, duke of Northumberland, and as such had an awareness of current international affairs. Both he and Nicolas Claude Fabri de Peiresc (1580–1637) observed Jupiter's satellites systematically as soon as they heard of Galileo's observation. Others, such as Clavius, had to await better instruments than were available to them. Harriot's first (1609) map of the Moon was done with a 6× telescope, as he records on the drawing.

One of Harriot's finest pieces of work is a systematic study of sunspots, first observed telescopically at the end of 1610. (Ancient Chinese and medieval European annals record spots seen on the Sun with the naked eye; and Kepler in

1607 saw with a camera obscura what he took to be a passage of Mercury across the face of the Sun. In reality that too was a sunspot.) After making numerous observations, and carefully documenting his findings, Harriot gave sunspot totals, and the first reasonably accurate figure for the rotation of the body of the Sun itself. He observed their growth, decay and change of relative position. Before long, of course, he had heard of Galileo's findings, and must have been inspired by them to study these matters yet more intensively. There was a new excitement in the air, and a letter from a friend, William Lower, explains how they are 'so on fire with thes things', and asks for as many 'cylinders' as would be needed for such observations.

These 'things' were given many names in the early years, and it would be tedious to list them all, but commonplace were 'glass', 'instrument', 'trunk', 'perspective cylinder', 'organum', 'instrumentum', 'perspicillum' (Latin) and 'occhiale' (Galileo's favourite Italian word).

The typical reaction to news of Galileo's discoveries was to send for one of the new telescopes and to attempt to reproduce his experiences. Kepler, no less excited than the common run of astronomers, chose instead to make an intensive new study of the theory of optical systems. He had become expert in the subject in 1600, beginning with the erection of a pinhole camera in the market square at Graz for the observation – after the manner of Tycho Brahe – of a partial solar eclipse. This had led him to study the eye as an optical instrument, and so on to compose the first of the two works in optics we mentioned earlier.

Kepler, by this time a recognized authority and mathematician to the Emperor Rudolph, was sent a copy of the *Sidereus nuncius* at Galileo's request, through the Tuscan ambassador. Galileo asked for his opinion. (We recall an earlier exchange when Kepler was less important.) Kepler replied at length, and published his views as *Dissertatio cum Nuncio sidereo* (1610), a 'Conversation with the Starry Messenger', which was both a generous affirmation of faith in

Galileo and a summary advertisement for his own views on matters as diverse as the finiteness of the Universe, possible life on the Moon, the planetary orbs in relation to the Platonic solids, and his writings on optics. Galileo was content, and even more so when the Imperial Mathematician, having in the meantime borrowed a telescope from a noble acquaintance (the Elector Ernst of Cologne), followed with a short tract on the satellites of Jupiter (*Narratio de Jovis satellitibus*, 1611).

As soon as he had the *Sidereus nuncius*, in the space of two months – August and September 1610 – Kepler composed his second optical tract. The counter-intuitive, inverted, image produced by his telescope design, developed as an alternative to Galileo's, worked against it at first. However, it has the advantage of producing a real image (one that can be focused on a screen beyond the eyepiece), and within a decade or so it was in regular use for projecting images of the Sun on paper, for instance for the charting of sunspots.

GALILEO, THE TELESCOPE AND COSMOLOGY

To suppose that the excitement caused by the invention of the telescope was entirely the excitement of a peep-show would be to underestimate the seriousness of certain questions that the instrument seemed to answer. Aristotelian cosmology was being seriously questioned. The cometary observations worked against Aristotle. As it happened historically they worked in favour of Copernicus, but that was primarily because he and Aristotle were two leading authorities who conflicted on so many points that a reverse for one was seen as an advance for the other. The rational astronomer could in principle *reject* Copernicus – indeed Tycho did so – and at the same time *accept* the cometary evidence. Kepler added to the list of troublesome phenomena. He mentioned a haze in the sky seen in 1547, and the halo round the Sun – the solar corona – seen during the eclipse of 12 October 1605. He mentioned too the new star (*stella nova*) of 1604, which reinforced in no uncertain manner the doubts induced by the new star of 1572. That of 1604, in the constellation of Ophiuchus, had given rise to innumerable university and public debates on the incorruptibility question, and in them Galileo had played an active part. (The new star of 1604 was unusually bright, and – like Tycho's – in modern terminology was a *supernova*.)

The name of Galileo is certainly better known than that of Kepler, and it was so in his own time. No doubt the fact that Galileo more or less avoided attempting to improve on the intricacies of traditional mathematical planetary astronomy during his entire career did him no harm in this respect: his talents were of a different sort, and more easily appreciated by scholars and the world at large than were Kepler's.

Born in Pisa in 1564, seven years before Kepler, Galileo Galilei outlived him by eleven years, dying in Arcetri in 1642 – the year of Newton's birth. Galileo is often regarded as the figurehead of the marked change in attitude towards the empirical sciences that characterizes the seventeenth century. He was sent to the university of Pisa by his father Vincenzio – a knowledgeable writer on musical theory from whom Galileo learned much about the art of matching theory to experiment – to study medicine. His interests turned to mathematics, and in 1585 he left Pisa without a degree. He began to study Euclid and Archimedes privately in Florence, while teaching mathematics there and in Siena. In 1588 he was passed over for a chair in mathematics in Bologna, this going to the more experienced astronomer Giovanni Antonio Magini, but a year later he was given the vacant chair at Pisa.

Galileo now began to criticize many parts of Aristotelian natural philosophy as it was then taught, and of course his laws of falling bodies – usually associated with the mythical demonstration at the Leaning Tower of Pisa – are the best-known symbol of his dissatisfaction. He soon moved to the chair at Padua, however, this time defeating Magini in the contest.

Padua was the leading Italian university of its day, and a focal point for European scholarship. He taught and wrote much on mechanics at this time, and his support for the Copernican theory, expressed in correspondence with Kepler, for example, shows his appreciation of its physical implications – it fitted well with his own ideas on the tides, for instance. The new star of 1604 gave him an opportunity to lecture to large Padua audiences on the difficulties inherent in Aristotelian ideas about the incorruptibility of the heavens. Galileo was nothing if not a great self-publicist.

Those lectures spawned much acrimony when they led to the publication of a pseudonymous attack on the Paduan professoriate in 1605. Written in a rustic dialect, this is not now thought to have been from Galileo's pen, but it was from a sympathizer, and did not endear him to his colleagues. In 1606 his talent for controversy was exercised yet again, when he brought a charge of plagiarism against Simon Mayr, then at Padua, and one of Mayr's pupils, Baldassar Capra. This case concerned the so-called proportional compass, a calculating instrument Galileo had devised in 1597, and from the sale of which he made a small income. (As it happens, the instrument was merely an improvement on an earlier one by Guidobaldo del Monte.) Mayr returned to Germany, and Galileo succeeded in having Capra expelled from the university. In this way was the ground prepared for future intellectual disputes. Mayr would soon be claiming to have seen three of Jupiter's satellites before Galileo. The first printed reference to Mayr's claim, however, is in the introduction to Kepler's *Dioptrice* (1611), and so appeared long after Galileo's *Sidereus nuncius*.

The many discoveries made by Galileo, described earlier in the present chapter, affected his own outlook on the world very deeply. The material advantages of a new post as mathematician and philosopher to the grand duke of

Tuscany were no doubt the chief reason for his return to Florence in the summer of 1610, but there is no doubt that he found the idea of continuing to teach the old Aristotelian philosophy uncongenial. It is true that his first telescopic discoveries did not undermine the foundations of the old ways of thinking: they were all explicable, more or less, in Aristotelian terms. That there are mountains on the Moon, and that the Milky Way is an association of separate stars, have no strong bearing on the fundamental principle of Aristotelian cosmology that had been brought into question by the new stars of 1572 and 1604, and by the absence of parallax in comets. That Jupiter has satellites was more problematical, for it suggested at least that there were centres of rotation in the Universe other than the Earth. Before the end of 1610, however, he was able to announce further spectacular discoveries. He found that Saturn appeared non-circular, as if it were a globe with two handles. He could not resolve what were later found to be Saturn's rings, and took them to be satellites very close to the planet. At first he refrained from publishing his findings, except as an anagram that many tried unsuccessfully to resolve. It spelled out the Latin sentence 'I observed the furthest planet to be triple'. Soon the appendages disappeared for a time – the rings, as we know, were at the time turned edge-ways on to the observer – and this greatly puzzled all concerned. 'Does Saturn devour his children?', Galileo asked. It was to be another forty years before astronomers had instruments good enough to be able to identify Saturn's rings clearly.

Of more immediate importance to the wider cosmological debate were Galileo's discoveries of the seemingly endless aggregate of point-like fixed stars and the phases of Venus. The former did not offer any grave difficulties, for as Clavius had noted, the Bible speaks of innumerable stars, and in any case, what was seen through a telescope in this respect did not offer any great problems of interpretation. The phases of Venus present much greater problems. The various possible arrangements of the Sun (S), Venus (V) and the Earth (E), when they are more or less in line, are limited. According to *all* of the three principal planetary systems in play in the early seventeenth century – the Ptolemaic, the Copernican and the Tychonic – they may lie in the order *EVS*, that is, with Venus lying between us and the Sun. (Of course the orbital planes of Venus and the Earth rarely coincide, or are even near enough to produce a transit of Venus across the Sun's face, as seen from the earth.) Venus will in these circumstances not be visible to the naked eye, since it will be lost in the Sun's rays, but the case will be closely analogous to that of the new moon. The case of the fully illuminated Venus, the analogue of the full Moon, is more interesting. This is *ESV*, and *the Ptolemaic model as traditionally interpreted does not produce it.*

One might imagine that there is another possibility of a 'full' Venus on the ancient model, namely *SEV*, but the Ptolemaic motions are so designed that the centre of the Venus epicycle is more or less in line with the Sun, so that the planet cannot be in opposition to the Sun – and everyone knew that this was the case in reality.

The upshot of this is that *all* models produce a set of phases of Venus, but the Ptolemaic model *does not have the full set.* For traditional astronomers, Venus should at best show a crescent shape. The opponents of Copernicus had pointed out that the variation in Venus' appearance was not enough to support the idea of a full set of phases. Galileo, however, saw a full set with the help of his telescope, and regarded what he saw as a proof of Copernican astronomy. Kepler later, in an appendix to his *Hyperaspistes* (1625), his defence of Tycho's work on comets against the views of the Aristotelian Scipione Chiaramonti, pointed out quite correctly that the Tychonic system could explain the phases of Venus just as well as could the Copernican. By then, however, the battle lines had long been drawn, and Copernicanism was slowly but visibly threatening to topple tradition. And Kepler, of course, was an ally who made his remarks – in the context of criticisms of Galileo's somewhat ill-advised views on comets – more out of loyalty to his old master than in support of the Tychonic world view.

In 1611 Galileo demonstrated his telescopic discoveries to the Jesuits in the Collegio Romano in Rome. The views of the Jesuits were ambivalent, but at last he won most of them over. A key figure in this episode was Roberto Bellarmine, theological adviser to the Pope. Bellarmine, who was then nearly seventy, asked the astronomers in the college – of which he was the former rector – to verify Galileo's findings. They did so, and Bellarmine responded warmly to Galileo, although without accepting Copernicanism. He later argued that until it could be rigorously demonstrated, scriptural texts should continue to be accepted at their face value. Perhaps the greatest of all compliments from the Jesuits at this time, however, was the remark by the great scholar Christoph Clavius in the last edition of his commentary on Sacrobosco (1611; Clavius died in the following year). Astronomers, he said, would need to find a new system in agreement with the new discoveries, for the old would serve them no longer. And this after a life of opposition to Copernicanism.

The Jesuits were certainly not converted as a whole to Galileo's way of thinking. One idea many of them seized upon at first was that the sunspots were of the nature of groups of unchangeable stars or planets. When, in a public debate in the college in 1612, a Dominican noted that stars and planets were round, or at least regular in shape, and sunspots were not, the Jesuits countered with the observation that a group of fifty or so might create an irregular appearance. Galileo later said that he found it hard to accept that fifty stars, like fifty boats joining together at various speeds, would stay together. Besides, he had observed these things with great care, their changes in shape and size, their growth and disappearance, their evident opacity and shading. He had one trump card in such debate. What was the value of endless citations of traditional authors when –

however great their intellects had been – they could not have had access to the new sort of observations?

Back in Florence Galileo became embroiled in yet more of those running controversies for which he had such a talent. The first concerned floating bodies, another question where the authority of Aristotle was at stake, but now the alternative doctrine was Archimedes'. While working on a book on the subject – for which he was attacked by at least four Aristotelian professors in Pisa and Florence – Galileo received a copy of a pseudonymous work on sunspots, sent to him for his opinion by the author, Marcus Welser of Augsburg. Welser had been prompted to write by the Jesuit Christoph Scheiner, one of the principal authors of the idea that the spots were small planets. Galileo replied to Welser in three long letters, claiming priority of discovery and attacking Scheiner's ideas. In this way he added another to a growing list of enemies. Scheiner (1573–1650) was in many ways conservative, but he was an intelligent scholar and a good practical astronomer. He taught Hebrew and mathematics at the university of Ingolstadt. His technique for drawing the Sun by projecting its image through a Keplerian telescope was carefully developed – which is hardly surprising, when we remember that he was the inventor of the pantograph, a set of hinged rods of a sort still occasionally used for scaling drawings.

Galileo's most influential telescopic work was done. He was becoming slowly entangled in a theological net. On 26 February 1616 he was ordered by a papal commission in Rome to abandon the view of the Earth's motion, and not to defend it. A week later, Copernicus' *De revolutionibus* was put on the Catholic Church's *Index of Prohibited Books*, where it stayed until 1835.

Galileo's response was to turn to more practical matters, in fact to the determination of longitude at sea through the use of Jupiter's satellites as a sort of universal clock. For this he recommended an instrument he had designed (the Jovilabium, a sort of satellite equatorium) and tables of eclipses of the satellites, and their motions generally. Galileo applied to the States-General in The Netherlands for a prize they had offered for a solution to the navigational problem, but all he received was the offer of a gold necklace, and this he declined.

He wrote on mechanics and on comets, and in his classic polemic *Il Saggiatore* (The Assayer, 1623) he even threw a sop to the Aristotelians in the form of a possible defence against the anti-Aristotelian argument from the negligible parallax of comets. One cannot discuss their parallax, he said, unless one is sure that comets are not purely optical in nature, such as they would be if they were formed by refractions in clouds of vapour. (At least he held to Tycho's belief that the new stars were celestial.) That comets are far beyond the Moon was to be well and truly settled only after more accurate measurements by Hevelius, while the observational evidence for the idea that they follow parabolic paths with the Sun at one focus was first provided by Georg Dörffel in 1680.

A somewhat chastened Galileo now turned to the writing of perhaps his finest literary creation, the *Dialogue Concerning the Two Chief World Systems, the Copernican and the Ptolemaic.* The book is presented as a series of discussions between three men, by name Salviati, Sagredo and Simplicio. Salviati represents the author, Sagredo an intelligent listener, and Simplicio an obtuse Aristotelian. (While Galileo could always say that he had the great sixth-century Aristotle commentator in mind, there was a clear double meaning in the name.) Galileo spent six years in writing the book, but he sailed too close to the wind, and after its publication in Florence in 1632 he was finally ordered to come to Rome, to account for his actions to the Inquisition. After much prevarication, he finally did so in 1633.

There followed one of the most notorious trials in history. Lest it be thought that this was no more than an intellectual exercise, it is as well to remind ourselves that at one stage the examination was held under threat of torture, and that although there was clearly no intention of carrying out the threat, Galileo is unlikely to have taken much consolation in that thought. He continued to maintain that after its condemnation by the Congregation of the Index, he had never held the Copernican theory. He was condemned, however, to life imprisonment and certain penances. The sentence was signed by seven cardinals, but was not ratified by Pope Urban, who seems to have thought Galileo more rash than heretical. There is a legend according to which Galileo, rising from his knees after repeating the formula renouncing his supposed offence, stamped on the ground and declared 'Eppur si muove!' (And yet it [the Earth] moves!). The legend is truer to Galileo's character than to history.

The sentence was immediately reduced to one of permanent house arrest. He was nearly seventy, and lived for another eleven years – blind for the last four of them. Henceforth in Italy Copernicanism reached the printed page only as the object of criticism. An excellent example of the ease with which a highly talented individual could fall into line with orthodoxy is Giambattista Riccioli: in his exceptionally scholarly survey of the entire history of astronomy – the finest then made – he came down in favour of a modified Tychonic system.

Galileo was not one of the great mathematical astronomers. He remained loyal to the circles used by Copernicus in planetary theory, and even in his *Dialogue Concerning the Two Chief World Systems* he made no reference to the elliptical orbits introduced by Kepler, although he certainly knew of Kepler's work. His mathematical talents were not insignificant, and his decision was not a question of incomprehension. He also argued long and effectively for the importance of mathematics in natural philosophy, a thesis that was by no means universally acknowledged in his time – although several contemporary Jesuits were of the same mind. One of his greatest strengths, however, had little to do with the mathematization of Nature: it was his

Above A fair approximation to the overall appearance of Stonehenge early in the second millennium .

Left The Egyptian goddess Nut, supported by the god Shu and separated from her lover Sibu (the Earth). Her feet are in the east, and the stars move along her body in the course of the night. (After Papyrus Nisti-ta-Nebet-Taui, eighteenth dynasty, 14th century BC)

Part of a Babylonian tablet from Sippar, *c.*870 BC, now in the British Museum. An adjacent text records the restoration of the ancient image (seen here on the altar) of the Sun-god Shamash.

Left From a tenth-century Chinese star-map (Tunhuang manuscript,British Museum). On the left the Purple Palace and Great Bear (in polar projection), on the right a 12-degree segment with parts of our Sagittarius and Capricorn (in'Mercator's' projection).

Left Part of the Dresden Codex, a Maya astronomical text,that deals with the revolutions of Venus. This is perhaps a thirteenth-century copy of a much older original. The middle scene represents the planet's heliacal rising. Spears (also in the lower scene) represent the planet's death-dealing shafts of light.

Above and opposite Selected Arabic constellation figures in the Greek tradition from *The Book of the Fixed Stars* of the tenth-century Muslim astronomer al-Sufi *(upper figures)*, together with their later representation in a fourteenth-century western manuscript *(lower figures)*.

Right Halley's comet as depicted on a section of the Bayeux Tapestry. The tapestry celebrated the Norman victory (1066) over the English king Harold, which the comet supposedly foretold.

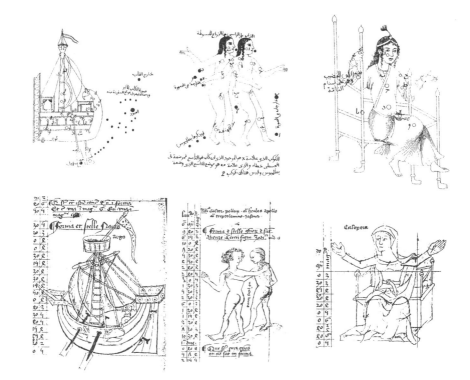

A signed and dated Persian astrolabe made for Shah Abbas II (seventeenth-century) showing *(left)* the front, with the moveable rete and the fixed plate beneath, and *(right)* the back, with alidade (observing rule) and various trigonometrical scales.

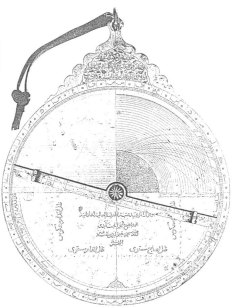

Copernicus

Tycho Brahe

Sir Isaac Newton by G.B. Black after William Gandy Jnr, 1706

Galileo

Johannes Kepler

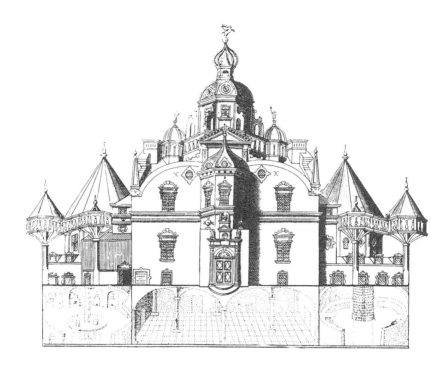

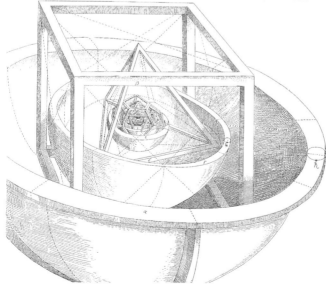

Above Elevation of the central observatory building at Tycho's Uraniborg (1580). Note how the instrument-bearing pillars pass through the cellars.

Left A detail from a drawing of Kepler's nested regular solids, by which in his *Mysterium cosmographicum* he explained the relative sizes of the planetary orbits.

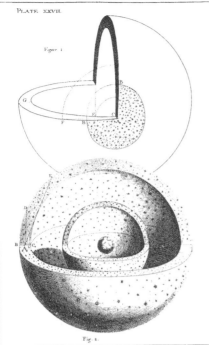

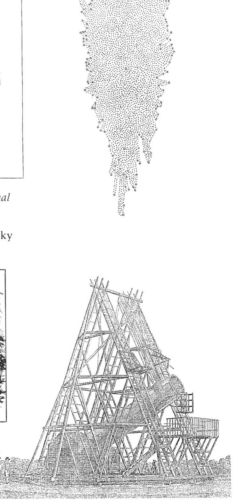

Above One of the several conjectural models in Thomas Wrights's *An Original Theory of the Universe* (1750).

Above right Herschel's Chart of the Milky Way, arrived at from his star-gauges.

William Herschel's 40-inch telexcope at Slough *(right)*, with John Herschel's telescope of similar design at the Cape of Good Hope.

A nineteenth-century view of the assemblage of observatory buildings at Greenwich. The main building is an extension of Flamsteed's.

Below left Fraunhofer's finest reactor, built for Struve at Dorpat (1826, aperture 24cm). This type of equatorial mounting became known as the 'German' mounting.

Below right The 100-inch (2.5m) Hooker telescope at Mt Wilson, completed in 1918. Weighing over 100 tonnes, it is carried between mercury floats, in drums at the end of the polar axis.

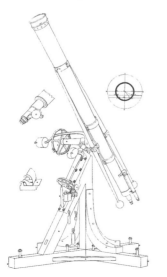

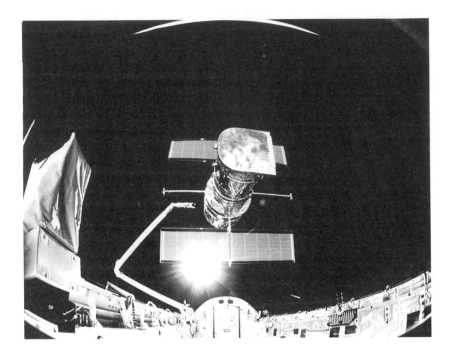

Above The Hubble Space Telescope, moments after its release from the space shuttle *Discovery* on 25 April 1990.

Left The 250-foot (76 m) diameter dish of the Mark 1A telescope at Jodrell Bank Observatory, the world's first giant dish radio telescope.

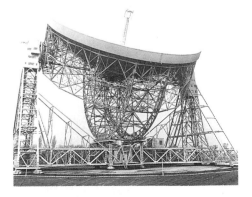

ability to demolish the insupportable nonsense put forward by so many of his opponents, and of course not only in the name of Aristotle.

GALILEAN INHERITANCE AND TELESCOPE DESIGN

Since the telescope was so quickly and widely diffused, it is not surprising to find that numerous priority disputes arose. We have already met with Simon Mayr, whom Galileo had long before forced to leave Padua. He claimed priority on the sighting of three of the satellites of Jupiter (December 1609; Galileo saw four in the same month), and also in the matter of the production of accurate tables of the satellite motions. His measurements do not seem to have been plagiarized, since they are more accurate than any Galileo had then published, and he eventually tried to take movement in latitude into the reckoning; but Mayr's work as a whole is certainly inferior to Galileo's. Not until his *Il Saggiatore* (1623) did Galileo launch an attack on Mayr in print on these questions. One suspects that what irritated Galileo above all else was Mayr's having named the satellites the 'Brandenburg stars', after his own patrons, thus jeopardizing Galileo's standing with the Medici, after whom he had named them.

Since in some human beings the urge to prove priority of discovery seems to take second place to no other, it is amusing to find an eighth-century Chinese work in which the seventeenth-century observations of Jupiter's satellites were pre-empted. *The Kaiyuan Treatise on Astrology,* compiled by Qutan Xida between 718 and 726, refers back to some observations made in the fourth century BC by the astronomer Gan De. He is quoted as having said that, in the year of chan yan, Jupiter was seen in the zodiacal division Zi, rising and setting with certain named lunar mansions (Xunu, Xu and Wei). 'It was very large and bright', he added, and 'apparently there was a small reddish star appended to its side'. This observation, evidently of either Ganymede or Callisto, has been dated to 364 BC, two thousand years before those of Galileo and his contemporaries.

Thomas Harriot had a copy of *Sidereus Nuncius* by June of 1610, and began systematic observation of the satellites in October. At first he could only see one, but by December he was charting the motions of four, and for more than a year he worked at analysing their motions, combining his own observations with Galileo's, and wisely preferring to establish sidereal rather than synodic periods, as the other had done. (He derived a very good figure for the first satellite, 42.4353 days, comparable with the modern figure of 42.4582 days.)

Other astronomers in plenty made sporadic observations, but this was the best work before Gioanbattista Odierna's highly professional *Medicaeorum ephemerides* of 1656 – a rare book, published in Palermo in his native Sicily. Odierna injected a certain theoretical stiffening into the analysis, using three sorts of periodic inequality by analogy with existing planetary theory. Nine years later the Florentine Giovanni Alfonso Borelli tried to draw similar analogies with Keplerian theory, when the Medicis acquired a very fine telescope, and asked Borelli to improve on Galileo's tables. The analysis was beyond his powers, and a book he published in 1666 is disappointing.

One thing that is abundantly clear is that instruments alone were not enough. Even the satellite motions derived by the great Dantzig observer Johannes Hevelius (1611–87) and published in his *Selenographia* of 1647 were decidedly inferior to those derived by Galileo, Mayr and Harriot more than thirty years before.

The phases of Mercury were perhaps first recorded by the Jesuit Ionnes Zupo in 1639. There were several unusually good Italian telescope-makers in mid century, so that for example the Neapolitan Francisco Fontana had an instrument capable of revealing the phases of Mars which others simply could not make out. The finest work of the century on the Jovian satellites was done with the help of what were surely the finest telescopes then available. Gian Domenico Cassini, around 1664, obtained excellent telescopes from the great telescope-makers Giuseppe Campani and Eustachio Divini. Within weeks of obtaining them, and while using one with a length of about 1.5 m, he was observing the shadows of satellites II and III on Jupiter's disc when he noticed an uncharted spot. Some days later he saw two or three movable dark spots, which he took to be clouds, and some bright marks, which he thought were volcanoes. This was the beginning of an extended analogy with the solar system, and the theory of planetary motions. The possibilities excited Cassini, as they did others when he visited the newly created Académie Royale des Sciences in Paris, two years later. Cassini had just published the most accurate ephemerides of the satellites to date, and he was offered such generous terms of employment there that he never returned permanently to Italy. (He was indeed the founder of a Cassini dynasty in French astronomy).

The accuracy of Cassini's work was so great by the time he was issuing tables in 1693 that we can understand Edmond Halley's excitement: the problem of finding the geographical longitude of distant places, on land at least – where the telescope could be held rigid – was solved. The 'universal clock', as we may term the satellite system, was accurate at its best to better than a minute of time, so that longitudes could be had to better than fifteen minutes of arc. The astronomer was meant to observe the first satellite (preferably) as it entered or left the shadow of the planet, and a whole series of such observations could improve the results greatly. Well into the eighteenth century the method remained a standard one for the positions of land bases. This practical problem explains in part, no doubt, the fact that so many attempts were made at the time to improve upon Galileo's Jovilabium: Nicolas de Peiresc, Odierna, John Flamsteed, Cassini, William Whiston and Jerome Lalande were among those who devised variants on the basically simple device, to take into account improvements in the understanding of satellite motions. In due course this meant taking into account the velocity of light, which in 1676 Ole Rømer proved finite, and measured, on the basis of satellite observations.

As an aside on this general question of the acceptance of Rømer's ideas, which were of course to be of great consequence to future large-scale astronomy, it has to be said that this was rapid and widespread. To quote Francis Roberts, an early writer who had taken the message to heart, 'Light takes up more time in Travelling from the Stars to us, than we in making a *West-India* Voyage (which is ordinarily performed in six Weeks)'.

After the first wave of telescopic discovery the first spectacular new material came with Christian Huygens' observations of the planet Saturn. In 1655 he found a satellite to Saturn (now called Titan), using a 50× telescope with a length of 3.5 m. Johannes Hevelius and Christopher Wren had seen it earlier, but had thought it to be an ordinary star. There had been a gradual improvement in the resolution of Saturn's 'two-handled teacup' appearance from the 1610s, and yet drawings by Gassendi, Odierna and Hevelius show that the ring structure was not understood even by the late 1640s. Christopher Wren (1657) proposed an elaborate theory of an elliptical corona, but in fact a year and more earlier, using telescopes first of 7 m and later of 37 m, Huygens had a broadly correct interpretation of what was to be seen, namely 'a thin flat ring surrounding the planet and not touching it'. This led to the discovery of shadows cast by the ring on the planet's surface that helped to satisfy astronomers of the truth of the theory in the 1660s and 1670s (Giuseppe Campani, Gian Domenico Cassini). In 1675 Cassini discovered a gap in the ring, another instance of the rapid advance in observing technique and instrumental power. (In 1980, photographs taken from the spacecraft Voyager 1 revealed hundreds of subdivisions, not to mention a rope-like braiding of the outer rings.) Cassini was fortunate to have instruments by Giuseppe Campani of Rome (1635–1715), with which he found surface-markings on Mars and Jupiter, allowing him to derive rotation periods for both. After moving to Paris he discovered Jupiter's oblate form, the belts on Saturn's surface, and four more Saturnian satellites (Rhea and Iapetus in 1671–2, Tethys and Dione in 1684).

Almost within the lifetime of a single human being, the telescope had completely transformed the nature of planetary astronomy. A new style of publication was now to be found, representing celestial appearances, not by schematic diagrams but pictorially. The Moon lent itself to this treatment above all other visible bodies, and the magnificent copper engravings of Hevelius' *Selenographia* (1647) provide the finest example of this common new art. At the same time, however, refracting telescopes had more or less reached the limits of their potential, without achromatic lenses.

It was at first assumed that Newton had shown the impossibility of combining lenses in such a way as to avoid spurious colours in the image, and yet it seemed to several people that the human eye – which was wrongly supposed to be perfect in this respect – was proof to the contrary. An English land-owner, lawyer and amateur optician, Chester Moor Hall (1703–71), by a series of experiments that he did not publish, found that different sorts of glass (crown and flint) could be combined to produce an achromatic combination. He worked out details in the period 1729–33, and contracted out the work of making the first achromat to professionals – George Bass made the first in 1733, and Hall had at least two achromatic telescopes made at about the same time. He told the English instrument-maker John Dollond of his success in the mid-1750s. Euler too saw the possibility, and wrote to Dollond to ask him to experiment.

The Swedish physicist Samuel Klingenstierna (1698–1765) published his own theoretical account in 1754, and sent it to Dollond, but was not mentioned by Dollond in an account of his own investigations published in 1758. Klingenstierna, nettled by this, in 1760 published by far the most thorough account then available of lens systems avoiding chromatic and spherical distortions, a work of great value to later makers of large astronomical telescopes. The commercial production of achromatic lenses was begun, however, by Dollond. Denmark was a distant land, and a patent was a patent. Dollond later suffered litigation from rival tradesmen, who argued that Hall had been first, but Hall stayed well outside the controversy and the judge ruled that those who first brought inventions forth for the public good should profit from them. The contrast with the experience of Hans Lippershey is enlightening.

To achieve greater magnification without spherical distortion or spurious colouring of the image, in the refractors of the seventeenth century, apertures had been kept small and focal lengths had been increased, making for extremely long, often tubeless, telescopes. Hevelius built a 46-m example in Dantzig. Huygens had a 37-m instrument. Huygens replaced the tube with a length of wire that connected the objective to the eyepiece, and that when taut pulled the objective – hinged on a universal joint at the top of a high pole – in the right direction. He soon discovered an unpleasant fact about open telescopes: the quality of the image was much distorted by air currents. The work of Hall, Klingenstierna and Dollond made these 'aerial telescopes' a thing of the past.

In France, Alexis Clairault simplified the achromat, reducing Dollond's three-component lens to a two-component lens with the two components in perfect contact. This was made up by several Paris opticians in 1763, and became the commonest type. Dollond's son Peter lent their first achromatic telescope to Maskelyne, the Astronomer Royal, in 1765, and results were so impressive that the Dollonds' fortune was assured.

The improved quality of object lenses guaranteed the refractor a long history, and compound eyepieces helped, by greatly improving magnification, but light-gathering power depends on aperture, and in the long term, descriptive astronomy needed something new. It came in the form of the reflecting telescope.

The Scottish mathematician and astronomer James Gregory had designed such a telescope, and he published details of it as early as 1663. It made use of a parabolic mirror that reflected rays from a distant object back to a small secondary concave (elliptical) mirror outside its focal point. From there the rays were again reflected in a narrow pencil that passed through a hole in the centre of the main mirror. And from there they passed to a plano-convex eye lens and into the eye. In the same year, Gregory commissioned a London optician, Richard Reive, to make a six-foot telescope to this design, but the workmanship fell below what was needed, and when Isaac Newton, around five years later, looked into the problem, he settled on a simpler design, using a plane mirror just within the first focus to reflect the converging rays out of the side of the main tube, and so into the eyepiece. (A small reflector to Newton's design, and held to be his own, made for the Royal Society, is still in the possession of the Society. It is about 30-cm long.) A very similar design to Gregory's, but using a convex mirror just inside the prime focus, was announced in 1672 by a Frenchman called Cassegrain – about whom almost nothing is known.

The Royal Society took an especial interest in the perfection of the telescope and other scientific instruments, and long continued to do so. London soon became the centre of a large trade in such material. One notable maker was James Short (1710–68), an Edinburgh man settled in London, an astronomer of minor repute as well as a specialist in making large and very accurately figured mirrors. First these were of glass, but later of speculum metal – a highly reflective alloy devised by Newton. Short made no fewer than 1370 reflectors, and well over a hundred of them still survive. He himself observed with an instrument of about 1.5 m focal length. Most of his reflectors were appreciably smaller, but the largest was of about 3.6 m focal length.

The fourth classical type of reflector was simplicity itself. It was an arrangement used by William Herschel almost a century later with a very large mirror, and involves the observer looking somewhat obliquely at the main mirror itself, through an eyepiece at the edge of the open end of the tube. This eliminates the need for an intermediate reflection.

After Dollond's manufacture of achromatic lenses, the reflector tended to pass out of favour with astronomers for a time. The speculum mirrors were fragile, they tarnished easily, and accuracy in grinding and polishing them was found to be a more critical matter than in the case of lens surfaces. The spectacular discoveries made by Herschel with his reflectors, however, brought such instruments back into favour with those astronomers whose main concern was magnification and light-gathering power, as opposed to precision measurement.

TELESCOPE, MICROMETER AND SOLAR DISTANCE

When Gassendi observed the transit of Mercury across the face of the Sun in 1631, he was astonished at the small angular diameter of the planet, which he estimated at twenty arc seconds. Kepler, who had predicted the transit, but who had died in the meantime, had made a much larger estimate. A young and talented English astronomer, Jeremiah Horrocks (1618–41), who had made some slight changes to Kepler's elements of the orbit of Venus, and who had predicted a transit of Venus for 1639, observed it with his friend William Crabtree and wrote a short treatise around the event. (This, because of his untimely death, remained unpublished until 1662. Many of Horrocks' papers survive, including some remarkable materials on lunar theory, but many are lost.)

An ardent supporter of Kepler, Horrocks was trying to modify Kepler's magnetic theories in the light of Galileo's mechanics, but he did not live long enough to make much progress in this direction. He absorbed the spirit of Kepler's doctrine of harmonies, and having proposed that the diameters of the planets were proportional to their distances from the Sun he was naturally delighted to be able to measure the angular size of Venus. This he did by projecting the Sun's image upon a white screen in a camera obscura, and he found 76"±4", close to the correct value. Once more, therefore, he was able to improve Kepler's parameters for the model.

Here are two important examples of reasonably accurate measurement of small angles by telescopic means, but they were not the first. The field of view to be seen through the earliest Dutch (Galilean) instruments was very small – less than the diameter of the Moon, in most important cases. The usual way to improve magnification was to increase the focal length of the objective, and so (for lenses of a given diameter) to restrict the field even further. Angles were at first then simply gauged in relation to the field of view. Galileo explained how to make various object-glass diaphragms, and calibrate them. Harriot gave much attention to such angular measurement. Alas, the method is not reliable, for reasons that no one understood. With this type of telescope the result depends on the size of the pupil of the eye, which may change appreciably according to the overall illumination.

In the late 1630s William Gascoigne hit on the idea of measuring apparent planetary diameters with an astronomical (Keplerian) telescope fitted with a micrometer, with its cross-wires moved, by a screw, in the focal plane of the eyepiece. Gascoigne died in the Civil War, at Marston Moor in his native Yorkshire, but Richard Towneley preserved his micrometer, and it was used by the Royal Society in a priority dispute with the French, when Adrien Auzout and Jean Picard developed the micrometer into a truly invaluable instrument for fine angular measurement. They had developed their screw micrometer from a simpler design by Huygens. In the meantime Eustachio Divini had used a gridiron micrometer for lunar mappping, but this was a relatively trivial affair.

By the late 1660s, the measurement of apparent planetary diameters was reasonably reliable. Kepler, building on Copernicus' ideas, had provided the means of specifying the *relative* distances of the planets at all times with high precision, so the *relative* (actual) sizes of the planets were known, with the exception of the Earth. Some tried to make use of a Horrocks-type principle, setting the Earth somewhere in size between Mars and Venus, and when they did so they found an astonishingly large distance for the Sun — say twenty or thirty thousand Earth radii. Huygens suggested 25 086, implying a solar parallax of 8.2″, which is very accurate but based on quite impressionistic reasoning. We recall that the traditional value was 180″, while Kepler's was 60″.

One of the most pressing of tasks for positional astronomy was now clear, namely to discover the solar distance (or equivalently its parallax), and so the linear scale of the solar system.

The key figure in this great enterprise was Gian Domenico Cassini. Even before he left Bologna for Paris, he had taken a strong interest in solar theory, when he had been asked to restore the 'gnomon' in the church of San Petronio. Egnatio Danti (1536–86) had there arranged for a small hole to be placed, so that when the Sun crossed the meridian, an image of it was thrown on a calibrated strip, itself in the meridian plane, bedded in the church floor. This arrangement is capable of giving very accurate solar altitudes, and so the obliquity of the ecliptic and the latitude of the place. Cassini found that the latter did not square with what was found from systematic observations of the pole star. He knew that parallax and refraction effects were significant, but even then he could only explain his observations on the assumption that the solar parallax was less than 12″, and this went so strongly against received opinion that he put the matter aside, only to take it up again after 1669, when he had moved to Paris.

Cassini had earned much fame from his tables of the motions of the satellites of Jupiter. To add lustre to the newly founded Academy of Sciences in Paris (1665), Colbert invited a number of renowned foreigners there – Huygens and Cassini among them. Almost from the moment of his arrival, Cassini began to participate actively in the affairs of the Academy – sometimes to the displeasure of other members. Under his eventual leadership, the Observatory, which was associated with the Academy, for a time took the lead in European astronomy. With its lavishly funded instruments, on a large scale, and soon to be equipped with micrometers, it was the first to outstrip unequivocally the observatories of Tycho Brahe.

It was clear that accurate measurements required a detailed knowledge of atmospheric refraction and solar parallax, and an expedition was called for to a place on the Earth's surface from which the Sun could be observed at high altitudes. An expedition was therefore sent to the French colony in Cayenne (South America) under Jean Richer (1630–96).

A movement had already begun to find accurate figures for the spacing of leading observatories, notably now Uraniborg and Paris, and the size of the Earth itself. Jean Picard had attempted the first measurement, and he had also measured the length of a degree on the Earth's surface, in northern France. One way of attempting the solar distance (or parallax) was to measure the related parallax of Mars – we have already seen the pattern of the interrelations between these quantities. A long base-line is needed for any parallax measurement. A 'diurnal parallax' is a change in position as seen from a *single* place on the Earth's surface, as between one time of day and another. (That is to say, it has a value less than the maximum parallax. See the right-hand side of figure 4.10, which can of course be regarded as applying to any planet.) Of course there are many calculations, adjustments and corrections needed, but essentially the idea is that the observer is carried from one end of the base-line to the other by the rotating Earth. Tycho had found the diurnal parallax of Mars, when the planet was at its nearest approach to the Sun (perihelion) in 1582. John Flamsteed, the future English Astronomer Royal, did the same in 1672, deriving 10″ for the solar parallax. Cassini found a similar figure: for long he favoured 9.5″.

When Richer's expedition returned, in 1673, it was with an enormous stock of new observational data. It had become clear, for instance, that the period of swing of a pendulum was different as between Cayenne and Paris – a fact explained in due course by the oblate shape of the Earth, which bulges at the equator. The expedition found that the solar altitudes measured could be reconciled with parallax and refraction corrections only if the solar parallax was less than 12″. The implications of this were that the obliquity too needed revision (23;29° was now favoured). The figure derived from Mars was therefore roughly confirmed.

It was to be improved yet further as a result of work by Edmond Halley (1656–1743) – a well connected scholar and astronomer of wide experience who by the age of thirty had visited Hevelius in Dantzig, assisted Flamsteed, catalogued stars on his own initiative off the coast of West Africa, seen Newton's *Principia* into print, and written a paper that is a classic in the history of geophysics, on solar heating as a cause of trade winds and monsoons. In 1663, James Gregory had drawn attention to a method of finding the solar parallax by noting the timing and manner of Venus' crossing the Sun's disc, during any of the rare occurrences of that event. The details are then to be compared with results obtained by another observer at a different latitude. Halley had tried out the scheme with Mercury in 1677, but he saw that Venus, being much nearer to us at its closest, would give more reliable results. In three separate publications (1691–1716) he calculated the details of what would be seen after his death if the transit of Venus were to be observed.

In the event, this influenced the French astronomer and geographer Joseph-Nicolas Delisle (1688–1768) to lead an attack on the problem, using the transit of Venus. (He also made several attempts using Mercury's transits.) Delisle co-ordinated a network of world-wide observation on an unprecedented scale – sixty-two stations were in use altogether, many equipped with the new achromatic telescopes. A year after Delisle's death, the 1769 transit yielded further results from sixty-three stations – many of them using reflectors made by James Short. (Short died in 1768, but he had helped to organize the Royal Society's participation in this venture.)

The human history of this magnificent venture is rich in incident. There were several cases in which astronomers concerned were accused of fabricating evidence to prove a point. The reputation of the Viennese Jesuit astronomer Maximilian Hell (1720–92), who had observed the 1769 transit in northern Norway, suffered greatly when Lalande hinted that he had manipulated his observations to fit those of other observers. A successor, Karl von Littrow, claimed that he had found proof, in the form of ink of different tints. Not until 1883 did Newcomb disprove these allegations – afterwards discovering too that Littrow had been colour blind. Less easily disproved was one account of the Rev. Nevil Maskelyne, sent by the Royal Society to St Helena for the 1761 transit: his personal account for liquors topped £141, out of a total expenditure of less than £292.

Britain and France could collaborate in matters scientific and yet be engaged in the Seven Years War. In that context, Charles Mason's and George Dixon's expedition got off to a bad start when their ship was attacked in the English Channel by a French frigate, which killed eleven of their crew. Threatened by the Royal Society with disgrace, not to say legal action, if they called off their expedition, they eventually made their observations from the Cape.

Guillaume le Gentil arrived in Pondicherry (near Madras) for the 1761 transit, to find that the town had been taken by the British. He waited in the area eight years, trading for a time, before setting up his instruments in Pondicherry for the second transit, only to find the Sun obscured by cloud – neither before nor after, but only at the time of the transit.

Jean d'Auteroche, to take the example of the leader of another French party, was more successful in 1761 (in Russia), and again in 1769 after trekking across Mexico to a base in southern California. Alas, he and two other members of his party of four astronomers died of disease almost immediately afterwards, and the fourth set out for home alone, on another hazardous journey, with the precious records.

Many comparable illustrations could be given of the difficulties encountered by the numerous groups engaged in obtaining this essentially rare sort of information – rare in the sense that not until 1874 would there be a similar opportunity to obtain it. One of the saddest discoveries of all was that the image of Venus, at the onset and end of transit, was indistinct. (This is due to the combined effect of the atmosphere of the planet and the corona of the sun.) By the last quarter of the eighteenth century, however, the Venus transits had settled the solar parallax, at a figure within about 1.4 parts in a thousand of its presently accepted value of 8.80″. Seldom has an astronomical parameter been so hard won.

The Rise of Physical Astronomy

PLANETARY THEORY FROM KEPLER TO NEWTON

The sound physical basis Kepler had tried in vain to find for his planetary astronomy was eventually provided by Newton, with the theoretical mechanics he built on foundations laid by his predecessors, and a theory of gravitation that he had every right to claim as his own. So overwhelming have been the implications of Newton's work for astronomy, that the feverish nature of activity in the intervening period has often been overlooked. To understand its character we must first appreciate how inconspicuous Kepler's laws then were, and likewise the tables that rested on them. Except in the cases of Mercury and Mars, the known planetary orbits did not diverge greatly from circular paths. The law of areas could be assessed only very indirectly, and was not much heeded by astronomers at first. The third law (relating periods to orbit dimensions) was more easily seen, and Jeremiah Horrocks and Thomas Streete made use of it to derive the (relative) dimensions from the easily measured periods. The real test of Kepler, for most of those who were not so prejudiced that they could not even consider accepting his ideas, was the unprecedented accuracy of the *Rudolphine Tables*, and yet even a Copernican like the respected Belgian astronomer Philip van Lansberge (1561–1632), with his follower Martinus Hortensius (1605–39), could fail to appreciate Kepler's merits. Van Lansberge prepared tables that were as inaccurate as they would have been had Kepler never lived, and yet they were republished twice, were modified slightly by others, and remained in use across Europe for thirty years and more. They were torn apart mercilessly in a brilliantly devastating little book by the Frisian adherent to Keplerian principles, Johan Phocylides Holwarda (Jan Fokkens Holwarda) of Franeker, published in 1640, and again in another book by him in 1642. Few seem to have taken much notice of the attack at first, although it was renewed by Jeremiah Horrocks, whose posthumous works (published in 1673) included even more devastating criticism of Lansberge. One might have felt sorrier for him had he not been so scathing in his criticism of his predecessors.

(Holwarda (1618–51) deserves mention for his measurement of the period of variation in the brightness of the star Mira Ceti – a period he set at about eleven months. Boulliau later set it at 333 days, and proposed a model: he thought that the star was rotating, and that only half of it was luminous. Many others observed the fluctuating brightness with care during the later seventeenth century and after, but the study of variable stars made little real theoretical progress before the alliance of astronomy with spectroscopy, in the nineteenth century.)

Horrocks' criticisms, not to mention his generally positive support for Kepler in matters of planetary theory, seem to have had rather more repercussions. One of the outstanding difficulties left by Kepler was in regard to his equation (for the eccentric anomaly), to which as already explained could be found no exact solution of a sort that interested astronomers. Horrocks, like Bonaventura Cavalieri at about the same time, tried repeatedly, and seemingly independently they found similar formulae – but these were not exact, from Kepler's standpoint. Horrocks argued against Kepler's magnetic theories, and proposed the conical pendulum (a plumb-bob swinging in an oval orbit) as a model. In the 1660s Robert Hooke, Secretary of the Royal Society in London, revived the analogy. Had Horrocks not died in

1641 at such an early age, it might well have yielded something more valuable.

Among the leading astronomers of Europe, many were dissatisfied with Kepler's work for reasons often having little to do with the prediction of planetary positions. Ismael Boulliau (1605–94) was a notable example. A French Calvinist from Loudun, who converted to Catholicism and was ordained a priest, he entered Parisian astronomical circles around 1633, at the very time of the Galilean crisis in the Catholic Church. This did not prevent him from joining forces with his friend Gassendi in speaking up in Galileo's favour. He accepted Kepler's elliptical orbits, and in 1645 published tables based on them, but with a different law of motion from Kepler's.

Kepler had referred to the inverse square law of illumination, and Boulliau took this over for light in a work of 1638, before modifying it. His first attempts in print were in a work called *Philolaus* (1639), named after Philolaus of Tarentum, the supposed author of the Pythagorean astronomy that had displaced the Earth from the centre of the Universe. (In the ancient world even Plato had been accused of plagiarizing Philolaus, when writing his *Timaeus*.)

If there were a planetary moving force, Boulliau said in his more polished *Astronomia philolaïca* of 1645, then it too should be an inverse square law. He did not make real use of this rather vague thesis, which was later mentioned politely by Newton – who remarked, surprisingly, on the accuracy of Boulliau's tables. Boulliau was a member of the Royal Society, but he was never elected to the Paris Academy, and this is consistent with the expressed dissatisfaction of Huygens and Picard with his tables, which they regarded as still inferior to the *Rudolphine Tables*. They also knew from the work of Horrocks that his figure for the solar parallax (141″) was far too large.

Boulliau's own planetary theory was kinematic, that is, force-free and descriptive in the way that all before Kepler's had been. It is extremely complex, and it contains many mathematical errors. Some of these were pointed out by Paul Neile and by Seth Ward, the Savilian professor of astronomy at Oxford. In his *Astronomia philolaïca* Boulliau had used certain principles that were at best arrived at intuitively. As an example: in moving 90° in mean motion from aphelion, the planet reaches the average of its speeds at aphelion and perihelion. He decided that only a path that was a section of a cone – he took an oblique cone – satisfied this rule. Of course an ellipse is a conic section, and this is what he accepted as the path. His laws of motion, that is, the laws for calculating the eccentric anomaly, were very different from Kepler's. As Ward showed, however, they simply did not follow from Boulliau's own principles.

Ward showed that from these principles the planet should move uniformly around the empty focus of the ellipse, the focus not occupied by the Sun. In a strange act of deference, Ward now adopted this principle, which makes the empty focus an *equant* point, as it were, for the motion. The different way of calculating the planetary motion was published in Ward's *Astronomia geometrica* of 1656, which is actually dedicated to Boulliau, among others. Boulliau counter-attacked with another book in 1657. This acknowledged some errors, and scored off Ward chiefly by noting how impractical were his proposals for deriving planetary parameters. The whole episode proved only one thing: astronomy was a theoretical science working at the very limits of the observational techniques by which it was justified or rejected.

Many others, lesser minds, failed to see that Boulliau's results were inferior to Kepler's, and it is hard to avoid the conclusion that this was because Boulliau was a more blatant adherent of astrology, whose works were therefore respected uncritically by others of the same persuasion. Jeremy Shakerley (1626–55?), John Newton (1622–78) and Vincent Wing (1619–68) all produced works and astronomical tables for London that were greatly influenced by Boulliau's.

One of the more regrettable of the patterns in seventeenth-century astronomy was that – common enough in the middle ages – whereby in the drafting of planetary tables parameters were drawn from inconsistent theories. In drawing up his *Astronomia Carolina* (1661), the London Irishman Thomas Streete (1622–89) made use of parameters from Kepler, Boulliau, Horrocks and others, in a careful blend, but one that could only be justified in terms of the fact that Streete's planetary tables made more accurate predictions possible than most rivals. No truly systematic comparison with the heavens was ever made, and it is doubtful whether Flamsteed, for instance, who praised them greatly, realized how best to go about doing this. They were reissued in 1689 with slight amendments by Nicholas Greenwood, put into Latin by Johann Gabriel Doppelmayer in 1705, and between 1710 and 1728 reissued five times in all by Edmond Halley and William Whiston, in Whiston's case with his own book. It was from Streete's *Astronomia Carolina* that Newton learned Kepler's first and third laws of planetary motion. The second law he probably learned from Nicholas Mercator's *Institutiones astronomicae* (1676).

It was Mercator – born in Denmark as Niklaus Kauffman, but settled in England – who effectively put an end to the Boulliau–Ward hypothesis of an ellipse with equant at the empty focus. He showed that Cassini's method of determining the line of apsides of a planetary orbit was dependent on it, and so that too must go. Mercator was more important as a mathematician than as an astronomer, but in astronomy he helped to clear away some of the last traces of mist that drifted between Kepler and later theoretical astronomers.

Lunar theory was long in yielding to substitutes for the theories of Tycho, perhaps because his ideas were so complex as to seem impenetrable, despite the inaccuracies of his predictions. An important breakthrough came in 1672 with Flamsteed's publication of Horrocks' lunar theory. Incomplete though this theory was, it revealed – with much intermediate reliance on Kepler's account – both a variation in the eccentricity of the orbit, and an oscillatory motion in the line of apsides. Horrocks' manuscripts had been scattered, and some were lost, but Hevelius had printed those on the Venus transit, and the Royal Society published most of the remainder in 1672–3. Flamsteed added tables to this edition, in doing so making some improvements, and a mistake that Halley later corrected. It is to Flamsteed's great credit that he recognized the talents of Horrocks, who had died thirty years before.

DESCARTES, VORTICES AND THE PLANETS

When Tycho Brahe and those of like mind dismissed the solid spheres of the Aristotelian heavens, they left an intellectual vacuum. Most scholars were unhappy with the idea of banishing material from the heavens entirely, and few were truly satisfied with the magnetic cosmology that Kepler had introduced. The theoretical movements discussed in the last section took little account of physical arguments: Kepler was a rare example of a scholar who could marry the two streams of thought. When the French philosopher René Descartes came on the scene with an elaborate substitute for the Aristotelian theory of matter, he gave cosmologists exactly what most of them wanted, a Universe in which movements took place under the action of matter on matter. Alas for Descartes, some of the best mathematical minds in Europe were unable to turn it into a theoretical scheme to rival Newton's. Endless attempts to do so were made, and in some quarters the struggle persisted well into the eighteenth century. As with Aristotle's physics before it, Descartes' was psychologically satisfying, but in the end it proved to be a superfluous psychological luxury.

René Descartes was born in La Haye in Touraine, France, in 1596. A member of a family of minor nobility, and of independent means, he was well educated, even in the latest scientific matters, at the Jesuit college at La Flèche. Having graduated in law from the university of Poitiers, he became a volunteer in the army of Prince Maurits of Nassau, and in 1618, while stationed in the town of Breda, he had the good fortune to meet Isaac Beeckman, assistant headmaster of a school on the island of Walcheren.

Beeckman had a lively interest in the natural sciences, and was at that time making astronomical observations with Philip van Lansberge. He introduced Descartes to a number of recent problems in mechanics, and some of Descartes most important insights into algebraic geometry stem from this time. The philosophical career for which he is more often remembered did not really begin in earnest for another ten years, during which time he travelled widely. From 1628 to 1649 he lived for the most part in the Netherlands. Persuaded then to serve as philosopher to Queen Christina of Sweden, he died in Sweden in 1650, victim of a cold climate and a Utrecht physician.

Descartes' best-known work, his *Discourse on Method*, was published in 1637 in a single volume together with the treatises *Meteors, Dioptrics,* and *Geometry.* In the *Dioptrics* he extended the theory of lenses set out by Kepler, but now with the sine law of refraction, and this treatise was of great importance for the later development of a subject on which astronomy was becoming increasingly dependent. In cosmological matters, his influence was of a different sort. He argued in ways not unlike Aristotle's against the existence of a vacuum, and he insisted on the idea that mechanical effects be explained by the action of matter on matter.

Descartes treated motion in a straight line as a *state*, just as *rest* is a state; and since a cause was needed to change a state of rest, so it followed that a cause was needed to change a state of motion in a straight line. This, a form of the 'law of inertia', was to assume great importance when adapted by Newton.

The same is true of Descartes' law of the conservation of 'quantity of motion' (a product of the separate magnitudes and velocities of bodies in a closed system). The steps by which this was transformed into – as we should say – a law of the conservation of momentum are not our main concern. In this connection, Descartes developed a theory of colliding bodies, but it was highly unsatisfactory, and Huygens greatly improved on it in the 1650s.

Of more immediate importance for any theory of planetary motion was Huygens' theory of centrifugal force, developed by him in the late 1650s, but not then connected with any idea of a gravitational force in the Newtonian sense.

In the years 1629–33, Descartes developed a system of the world that rested on a theory of celestial vortices, whirlpools of subtle matter. He explained terrestrial gravitation as an effect of these vortices. A treatise he wrote on the subject, *Le Monde, ou traité de la lumière* (The World, or, Treatise on Light) was ready for publication, but hearing of the condemnation of Galileo he decided not to publish, and it was first printed only posthumously, in 1664. As he gradually realized, a Universe of vortices is one in which every natural body may be at rest with respect to the local matter and yet moving with respect to distant bodies. This seemed to him to be an answer to the problem of contenting both the Copernicans and those who took the Earth to be stationary. One could say that in a sense they were both right. Emboldened by this insight, he made public his vortex theory in his *Principia philosophiae* (Principles of Philosophy) in 1644, a book that was soon translated into French, and became highly influential.

His cosmology made use of the idea that there are three different forms or elements, luminous, transparent and opaque. The first was the finest, with particles moving at great speeds, and it made up the Sun and stars. The Earth and planets were of the coarse third element, and the second element, filling the spaces between these different sorts of bodies, was of globules in rapid motion. Celestial matter was supposedly capable of penetrating the pores of terrestrial matter. It was no easy task to rationalize the vortex motions in three dimensions. Each vortex has an equator and poles, and it is no easy matter to explain how they are fitted together. He developed a theory of the movement of matter, from the equator of one vortex to a pole of another, for instance, with collisions modifying the shapes of particles. The shapes were supposedly so designed as to facilitate the passage of particles through gaps between others. Magnetism was seen as evidence for the vortices, and so magnetism was worked into the cosmic scheme, as were sunspots – elements of the third type, floating on the Sun for a while in the course of the vortical process. The comets too were found a place, as were the satellites round planets, and the Moon of course, and the Earth's diurnal motion. This was a theory truly meant to solve all physicists' problems. Gravitation towards the Earth's centre was seen as analogous to the tendency of floating bodies to move to the centres of whirlpools on water. It was all very ingenious, but almost entirely qualitative, and in several places inconsistent. It is not known whether Descartes even knew of Kepler's laws of planetary motion. If he did, he seems to have made no attempt to explain them.

The Cartesian ideas were welcomed at first, with other parts of his philosophy, in the universities of the Netherlands, and later in informal discussions held among scientists and scholars in Paris. At the beginning of the century atomism, of the type taught by the Greeks, was widely regarded as spiritually dangerous, and linked essentially with atheism. Descartes was anxious to avoid being associated with the atomists, but eventually the link was made, and it was made all the more easily since Pierre Gassendi – in fact an opponent of Descartes – had sweetened the pill by making the causes of atomic movements spirits or minds internal to them. The popularity of Cartesianism snowballed, supported by such disciples as Henricus Regius (1598–1679) in Utrecht, Jacques Rohault (1620–75), Pierre Sylvain Régis (1632–1707) and Nicolas Malebranche (1638–1715) in Paris. They added new phenomena to the list of those treated by Descartes, but their treatment was still qualitative, and like Descartes, none of them ever made a serious attempt to explain Kepler's laws – which indeed received only passing mention by Cartesians even in the eighteenth century.

Huygens was that rare phenomenon, an early Cartesian capable of introducing quantitative argument. It is significant that by the time he and others – for example Gottfried Wilhelm Leibniz – were beginning to achieve a measure of success in explaining gravitation in the solar system on Cartesian principles, they were making use of certain principles developed by *Newton*, and so were working within a very different intellectual system. They were the exceptions to the general rule. Most of those who wrote in the Cartesian tradition seem to have been overwhelmed by the idea that cosmological argument was within the reach of all who had a clear and distinct idea of change coming about by the action of matter pushing matter.

ISAAC NEWTON AND UNIVERSAL GRAVITATION

Another largely qualitative style of cosmological thinking that paralleled Cartesianism, and occasionally overlapped with it, was an extension of the magnetic philosophy of Gilbert and Kepler. This survived, and was long actively pursued, in England, where it was kept alive especially in discussion groups centred at first on Gresham College (London), and later the Royal Society. John Wilkins (1614–72) and Christopher Wren (1632–1723) were both spokesmen for Gilbert. In 1640 Wilkins published two easily comprehended books concerning the possibility of a voyage to the Moon, both affording some publicity to Gilbert's and Kepler's ideas, namely *Discourse Concerning a New Planet* and *Discovery of a World in the Moone*. In 1654 Walter Charleton, a follower of Wilkins, published a blend of these ideas with Gassendi's atomism. Even so, it is hard to believe that, without Wren, Edmond Halley and Robert Hooke, the English discussion would have remained focused, as it did, on the laws governing the *precise* forms of the orbits, and their relation with laws of mechanics – such as a law of inertia and some sort of law of a central (sun-directed) force of attraction.

The discussion was stimulated by the appearance of a comet in 1664. Wren was at this time Savilian professor of astronomy at Oxford, and John Wallis was Savilian professor of geometry. Wallis was advancing the cometary theory of Horrocks, who had analysed the motion of the comet of 1577 into a straight-line motion, modified by the Sun's magnetic action. Wren now attempted to derive the new comet's path on the basis of four observations, assuming with Kepler that it followed a straight line at constant speed.

Within months of the first, a second comet appeared, which served to stir flagging interests in a problem that was evidently beyond the powers of all concerned. Hooke tried the hypothesis of circular motion, but tended to favour the rectilinear alternative, with some sort of solar attractive power. He suggested that there might be an all-pervading ether that was vibrating, with the vibrations diminishing as the distance from the Sun increased. Here was a mechanism that might have owed something to Descartes' idea of a circulatory ether, but that was much more in tune with the law of a central attractive force that was in the thoughts of Hooke and his friends at the Royal Society.

Hooke, Curator of Experiments there, was beginning in fact at this time to conduct a long series of experiments into gravitation. Ten years later, he had made no appreciable theoretical progress, if we are to judge by the theories of Newton that were soon to emerge, but we may use him as a measure of changing attitudes towards the old Aristotelian Universe. For Hooke, as he claimed in a lecture of 1674, *all* celestial bodies have an attraction or a gravitating power towards their own centres, which binds their parts together and also acts on other heavenly bodies that fall 'within the sphere of their activity'. This last qualification suggests that he thought the force to fall off to zero at some finite distance. He admitted that he had not yet verified the law of force – but he did not say what he thought it to be.

In or around 1677 Newton discussed these matters with Wren, and took it that Wren was assuming a law that made the force fall off as the inverse square of the distance between the attracting bodies. His own announcement of the inverse-square law was first published in his *Principia mathematica philosophiae naturalis* (Mathematical Principles of Natural Philosophy) in 1687. It seems likely that he had become convinced of its truth only about three years earlier, but the law was only one piece in a complex jigsaw puzzle, and for the rest we must look to earlier stages in his career.

Isaac Newton was born on Christmas Day 1642 in Woolsthorpe, Lincolnshire. His father died before he was born, and his early years were spent with his grandmother, his mother having married a clergyman for whom Isaac seems to have had no great affection. After school in Grantham he entered Trinity College, Cambridge, in 1661, and four years later he returned home when the university was closed because of the plague. He was already developing an interest in current mathematical and scientific affairs. He read widely – he learned much mathematics from the works of Descartes and Wallis – and the plague interlude led him to a number of original discoveries of his own. By 1669 his qualities were such that he succeeded Isaac Barrow as Lucasian professor of mathematics at Cambridge. He went often to meetings of the Royal Society in London, but left Cambridge only in 1696, when he was made Warden of the Mint. He died in 1727, a national figure with an unrivalled international scientific reputation. He wrote much on religion, and on a variety of other subjects peripheral to those for which he is best known, namely mathematics, physics (especially optics) and theoretical mechanics. And it is through these that he influenced the course of astronomy most materially.

One of Newton's student notebooks, begun in 1661, shows that he was then aware of Kepler's third law, and of some of Horrocks' observational records, and that he had studied the methods for finding planetary position given in Thomas Streete's *Astronomia Carolina*. By 1664 he had improved on Descartes' work on the 'conservation of motion' – he realized that the directions of the 'motions' had to be taken into account – and he had developed a theory of centrifugal force – which Huygens announced independently only in 1673. Combining this with Kepler's third law, he applied it to the case of the Moon, which is about sixty Earth radii distant, and to an object at the Earth's surface (the famous falling apple, for instance). Doing this he found – or perhaps we should say that he confirmed a previous conjecture to this effect – that the force acting on such bodies diminished inversely as the square of the distance.

His data were poor, and did not confirm the idea as well as he could have wished, and it is for this reason that he is usually supposed to have set the question aside – for twenty years, in fact, until 1685, when he was writing the *Principia*. In the meantime we know that he read Borelli, who in 1666 had written of the planets that their curved orbits implied a centrifugal force that could be regarded as equal and opposite to a force of attraction by the central body. This passage would certainly have meshed very well with his own ideas, but it is not clear that it was new to him.

It was at the very end of the 1670s, or even after 1680, that Newton, having gained a mastery of his newly developed dynamical principles, encountered Kepler's law of areas. This, combined with an exchange of correspondence with Hooke, set Newton on what turned out to be a very fruitful course. Hooke wanted to know the law of central force that would turn the straight-line motion of a planet into an ellipse of the sort Kepler said the planets followed. Newton was in possession of precisely the tools needed – namely the methods of his own infinitesimal calculus and his dynamical principles – to show that the law of areas implies that the force is indeed directed to a single centre, and that it is as the inverse square of the distance. An essential step in the proof was that a homogeneous solid sphere exerts a gravitational force exactly as if all its mass is concentrated at its centre.

In December 1684 Newton asked Flamsteed for data – distances and periods – on Jupiter's satellites, and Flamsteed replied, in effect, that they are in agreement with Kepler's third law (the law relating period to orbital size). Newton asked him too whether there was anything to suggest that he was right in a hunch that Jupiter might perturb the orbit of Saturn. Flamsteed explained some errors he had detected in Kepler's parameters for these planets. Both answers pleased Newton, for in the first case the implication was that the effect of the Sun on the satellites could be ignored. The second answer meant that Kepler's data were not above reproach, and freed him, as he thought, from an obligation to take other types of force than gravitational into account. They could not, at least, be proven on the strength of Kepler's data.

It was in 1684 that Halley visited Newton in Cambridge to ask what path a planet would follow under the action of an inverse-square force. He explained that Wren, Hooke and he had failed to solve the problem. Newton's answer was that it was an ellipse, and that although he could not find the proof, he would send it on to Halley. (After correspondence with Flamsteed, he had decided that cometary orbits were parabolic, a consequence of the inverse-square law. This was no less important a conclusion.) This led Halley, after seeing some of Newton's remarkable writings on mechanics, to press him to publish. The *Principia* was the result, written in a remarkably short time.

No sooner was the work published than a number of minor disputes as to priority arose. Hooke, for instance, claimed priority in the inverse-square law. In a letter to Halley (1686), Newton was not boorish enough to draw attention to Hooke's mathematical inadequacies, but gently pointed out that what he himself had proved was with Hooke no more than a hypothesis. He went further, and said that even Kepler had only conjectured that his ovals were ellipses. Newton, however, could now settle the matter at issue, using the 'correct' law of force.

Newton did not mention Kepler's name in his *Principia* until the third book, but there was no thought of concealing a debt, and as Halley noted, his first eleven propositions were in full conformity with the '*Phenomena* of the Celestial Motions as found by the great Sagacity and Diligence of *Kepler*'. Book III of the *Principia* is called 'The System of the World', and there is a strong sense in which it is the very first complete explanation of material movement in all parts of the Universe under the action of a single set of physical laws. The motions of the planets and their satellites, of the comets, of the Earth and the tides in its seas – all are explained in terms of a *universal* gravitation. The planet attracts the Sun as the Sun attracts the planet. All matter attracts all other matter, and the force is independent of the type of matter. Only the 'quantity of matter' and the separation were significant. (He performed experiments with pendulums made of different materials, and found no consequent difference in their behaviour.)

As we saw from his correspondence with Flamsteed, he realized that while the gravitational forces between the central body and the planets are considerably greater than those between the planets themselves, the latter cannot be ignored, especially where planets make close approaches to one another. The perturbing action of the Sun on the Moon is another important non-central force that cannot be ignored. The important theory of planetary perturbations thus entered Newtonian celestial mechanics at its very birth.

With his powerful dynamics and theory of gravitation at his disposal, Newton was in a position to explain the flattened shape of the Earth, and how the Sun's force of attraction for the near side of the bulge is very slightly greater than that for the far side. This discrepancy produces a turning effect (couple) on the axis of the Earth. This couple, as he could then show, leads to a precessional, conical, motion of the Earth's axis, equivalent to a precession of the equinoxes. For the first time in history, this phenomenon had been explained in terms of physical laws.

There was much in the *Principia* concerning comets, and their virtually parabolic and elliptical orbits. Newton realized that comets shine by the reflected light of the Sun, and that the space through which they move could not offer significant resistance to their motions. In the first edition he gave much attention to the comet of 1680–1; in the second edition (1713) he gave Halley's recalculation of it; and in the third (1726) he gave more calculations by Halley, based on the idea that the comet was periodic, and identical with the comets of 44 BC, AD 531 and 1106. The hypothesis was disproved in the nineteenth century, but by then the periodicity of comets had been proved brilliantly by Halley in connection with another comet entirely.

Among the most compelling parts of Newton's work, however, at least for the few experts who were capable of reading it with understanding when it first appeared, were those dealing with the Moon's motion. Newton explained in general terms the gravitational causes of the known inequalities in the Moon's motion, the motion of the nodes of the orbit, and the reason for the Moon presenting always the same face to us. In the second and third editions he supplemented his lunar theory, having been asked by Halley to continue with his work on it. By the time he had finished he had as many as seven 'equations' of lunar motion, some of them possibly found from Flamsteed's observations rather than from fundamental gravitational arguments. His data were such that lunar tables based on his work, by such as Flamsteed, Charles Leadbetter and Halley, were hardly better than those based on Horrocks' methods. It was in its potential that the real merits of New-ton's theory lay.

In the 1690s, Newton desperately needed Flamsteed's observational data, but the two men, disagreeing strongly as to whether theory should lead observation, or follow its lead, quarrelled violently. The fact that there was friction between Halley and Flamsteed did not help, but as Newton had aged he had become increasingly autocratic. As Warden of the Mint he had distributed largesse to Halley, in the form of comptrollership of the Chester Mint (1696). In 1699 matters grew more serious when he told Flamsteed that he was interested in his observations, not his calculations. Flamsteed had always felt that his observations, made with instruments largely financed out of his own pocket, were his own. Newton and Halley took the view that the Astron-omer Royal's work was public property, and in 1712 they published a considerable part of his work without his approval. He had the sombre pleasure of burning most of the edition before publishing his own work, *Historia coelestis Britannica* (British History of the Heavens). Here 'history' has the sense of 'data', in modern parlance.

A curious footnote to Newton's theory of the Moon con-cerns his estimate of its average density in relation to the Earth's, based on the relative tidal effects of the Sun and Moon. In the first edition of the *Principia* this was over-estimated by a factor of three. The fact that the Earth seemed so relatively light led Halley to the conclusion that four-ninths of it must be hollow. He was not the first with the idea, which will be found for instance in Burnet's *Sacred Theory of the Earth* (1681, first in Latin), but there one finds little more than an old tradition of grottoes and caves, as found in ancient myth. Halley was making a valiant attempt to find a theory of the Earth's magnetism. From the time of his voyage to St Helena in 1676, he had studied magnetic declination, and by 1683 had concluded that the Earth had four magnetic poles. He now thought that they might best be understood on the hypothesis that the Earth is a system of *a sphere within a sphere* – perhaps even more spheres were involved – *in relative rotation, each carrying magnetic poles.* There are slight similarities between this and an earlier model of the Earth put forward by Hooke, but in Halley's case the model had, as it seemed, a double justification, and like his paper on monsoons it gives him an honourable place in the history of geophysics, a science that has always maintained strong links with astronomy. A portrait of him as Astronomer Royal in 1736 shows Halley holding a hollow Earth. William Whiston had by then propagated the idea widely in a book of 1717, and had even provided biblical evidence for the notion that the cavity is inhabited.

Newton's *Principia* is often described as the most important work ever published in the physical sciences. The criteria for such judgements are hard to define and easy to vary, but the work can certainly be seen as marking the end of one historical era and the beginning of another. It gave physical reasons for Kepler's descriptive laws of planetary motion, and in that sense legitimized them, or as Newton would have said, changed them from speculation to fact. And it presented a programme for future astronomical research that in some respects still continues. Its demonstrations were not always complete – indeed, in many cases Newton had not yet developed the necessary mathematical techniques to give compelling proofs. As later astronomers were to discover to their surprise, however, he had a remarkable instinct for correct conclusions, even when he had to paper over the cracks in his arguments.

When first published, Newton's work aroused much hostility on philosophical grounds. Leibniz, for instance, objected to Newton's ideas about absolute space and time, and 'action at a distance', which he deemed an occult quality. There were many in England who felt uneasy at the idea that gravitation could act through empty space, and Cartesian vortices had a wide currency there. Leibniz and Newton had of course quarrelled over priority in the invention of the calculus, and when an embittered Leibniz, through the mediation of the princess Caroline, began a philosophical exchange with Samuel Clarke, a supporter of Newton, he must have suspected that Newton was colluding with Clarke.

Such scruples as Leibniz's have never quite disappeared from philosophical discussion, but astronomers decided that they could ignore them with impunity. Could Leibniz or his followers quantify a seventh inequality of the moon? When relativistic arguments surfaced in astronomy at the beginning of the twentieth century, they were only very indirectly inspired by the tradition in which Leibniz was arguing his case.

New Astronomical Problems

COSMOLOGY IN BENTLEY, NEWTON AND OTHERS

By 1685, Newton had written a little work with the title *De mundi systemate* (On the System of the World) intended as a tail-piece to his *Principia*, but destined to be published only in 1728, a year after his death. In this he used a technique due to James Gregory (1668) to show that the stars lie at much greater distances from the Sun than had previously been supposed. The method was 'photometric'. It depended on a comparison of the brightness of the Sun with that of a star, and on the inverse-square law of photometry. The comparison could not of course be made directly, but was done by considering the sunlight reflected off Saturn. Certain assumptions had to be made, for instance about the nature of the reflection and the absence of light loss in space, and that the star considered is equal in brightness to the Sun, but these seemed plausible enough. When Newton used the method on Sirius, for instance, he found its distance to be a million times that of the mean distance of the Sun from the Earth (the astronomical unit). The figure is too great, but there is a case for counting this as the first acceptable determination of a star's distance.

If the stars were at enormous distances, Newton felt he could assume that their gravitational attractions on one another were minimal. This was a vague conclusion, but it was important to him, for he was perplexed by the fact that the world did not collapse on itself, under gravity. By the time his *Principia* was finished, however, he had developed a test for external forces acting on the solar system: large forces would produce detectable rotations of the apse lines of the planets. Such were not observed at a significant level, so external forces must be negligible, and this fact fitted with the idea that the stars were very distant indeed.

Late in 1692, Richard Bentley, a brilliant young classical scholar who was then chaplain to the Bishop of Worcester, was giving the first series of Boyle lectures. One of his themes was that 'the observed structure of the Universe could only have arisen under God's guiding hand'. Before going into print, he asked Newton for advice. What would happen if matter were spread uniformly throughout space and allowed to move under gravity? If space is limited, said Newton, it would fall into one large spherical mass. If infinite, into infinitely many masses. But surely, said Bentley, if matter is *evenly* spread, there is no sufficient cause for a particle to move one way rather than another. Newton's reply was that this evenness in relation to even a single particle is unlikely, as unlikely as that one could make a needle stand on its point on a looking glass. How much more improbable to find *all* particles so placed. God could have made them so, however, and then they would have stayed in place. But then, said Bentley, consider the Universe to be divided by a plane into two parts. A particle in the plane will be pulled by an infinite gravitational force to one side of the plane, and this will be balanced by an infinite force pulling in the opposite direction. Why should the presence of the Sun in the particle's neighbourhood have any effect on its behaviour? Its attraction would simply be incorporated in one of the infinite forces. Newton's reply was that not all infinites are equal. A particle in equilibrium will, he said, be moved by an extra force. The two men were getting into deep waters, waters that had proved themselves too deep for most philosophers for over two millennia. Bentley sent a summary of his seventh sermon. The Universe was not homogeneous, and the conclusion was more or less that if the Universe is in equilibrium, it is God who keeps it so.

Newton was engaged around this time in revising his *Principia* for a second edition. It must have seemed that he was glad to drop the subject, but in reality it continued to occupy his thoughts, as his unpublished papers show. He tried to find a geometrical model of the Universe in which the stars are distributed in an *exactly* regular way, so as to be in equilibrium. Of course even the simplest of observations of our uneven world throw doubt on that idea, but such a cosmological model is meant only as an approximate representation. He tried out the idea that the stars are all on spherical surfaces, at one unit distance from neighbouring stars on their sphere, the spheres being centred on the Sun. He took their radii to be one unit, two units, three units, and so forth. The advantage of the scheme is that for large radii the distribution will correspond to a thin uniform shell of matter, and in *Principia* he had shown that the net gravitational attraction on any star within such a shell is zero – a very comforting result, bearing in mind the perplexing discussion with Bentley.

Newton investigated the properties of his model. How many stars can be put at unit distance on a sphere of unit radius? Kepler had examined the problem, and thought that the answer was at most twelve. Newton thought possibly thirteen. There will be four times as many on the sphere of radius two units, then nine times as many for radius three, and so on. Newton at first assumed that those on the innermost sphere were stars of apparent magnitude 1, that the next were of magnitude 2, the third of magnitude 3, and so on. (Herschel, a century later, took more or less the same position.) An observational test is therefore not only possible but simple. One merely counts the stars of successive magnitudes in the best catalogues available. In a rough-and-ready sort of way, the stars of the six visual magnitudes seemed to fit the scheme, although there was a tendency to accumulate faster than in the model, so Newton crossed out magnitudes 5 and 6.

There was another problem remaining: was he not making the Sun the focus of the universe? Newton tried more adjustments to his model. The details are less interesting than a paradox that might seem to have escaped him — except that he was presiding over the Royal Society at a meeting in 1721 when it was raised by Halley. On Newton's model, stars accumulate as the areas of the surfaces of the spheres, with a certain number on the first, four times as many on the second, nine times as many on the third, and so on, as explained. At a distance of two units, however, each star is a quarter as bright as a star at one unit; and at three units a ninth as bright, and so on. In other words, the total light from the stars at any particular distance is constant, so that in an infinite Universe the sky should be ablaze with light, the sum of an endless series of constant totals. (We are here assuming point sources, and that stars do not stand in the way of light from other stars. If they do, we shall still have an entirely bright sky.)

It is not known who first appreciated this paradox, but Halley said that he had 'heard it urged' by someone he did not name. David Gregory has been suggested as a possible candidate, for we know that in 1694 he was engaged in discussions with Newton of the cosmological problems Bentley had raised, and later wrote of them in a book of his own. Perhaps a stronger candidate is William Stukeley. Halley's paradox was for long ascribed to W. H. M. Olbers, as we shall see, but Olbers' formulation of it came more than a century later.

THE EIGHTEENTH CENTURY

Looking back over the history of eighteenth-century astronomy, from the perspective of one living at the end of the nineteenth, Agnes Clerke wrote that it 'ran in general an even and logical course'. She saw the age of Newton as one lasting almost exactly a hundred years, as having ended, in fact, in 1787, when Laplace explained to the French Academy the cause of an acceleration in the Moon's motion. The only anomaly in her description, as she believed, was the rise of William Herschel, whose work did so much to influence the course of later events, but whose starting point was not that of Newtonian dynamics.

There is much to be said for this simple account, but it is too narrowly focused. There were other forces at work, other motives to study astronomy than a wish merely to extend the monumental system of the *Principia*. Many important discoveries were made in the course of practising astronomy in a perfectly traditional way, but with new instruments – and new intellects. There was an ever-present desire to make new telescopic discoveries, and although here Herschel was pre-eminent, he was not alone. Astronomy no longer occupied the compulsory place that it had formerly held in the university arts curriculum, but it was a subject in which there was a strong cult interest, creating a demand for travelling lecturers, for example. Without astronomy, no gentleman regarded his children as properly educated, a fact that produced new types of popular literature and educational instruments. There were simple telescopes, of course, and globes – terrestrial and celestial were usually paired – and simple orreries eventually became commonplace. (These moving models of the solar system were named after Charles Boyle, Earl of Orrery, who merely happened to commission a particularly fine example.) The orrery tradition merely continued at a popular level one that stretched back through centuries of astronomical clock-making to the planetary models of antiquity.

One sort of literature that achieved a new popularity, and that required no special expertise, was natural theology, the attempt to argue from Nature, and especially from the harmony of the cosmos, to the existence and attributes of God. One of the most influential works in this genre was the *Astro-Theology* (1714) of William Derham (1657–1735).

William Paley (1743–1805) was much indebted to it in his even more influential writings at the end of the century. Paley's best work, his *Natural Theology* (1802), provides an important measure of a change that took place in the intellectual atmosphere in the course of the century. For Newton's contemporaries, the ordered celestial Universe gave proof of God's existence. For Paley, it was necessary to introduce biological considerations, although the Universe was a benevolent one still. Some of Paley's writings were required reading at Cambridge when Charles Darwin was an undergraduate, and he took much pleasure from them, but in the middle of the nineteenth century they created a climate that worked against his theory of evolution, in which a benevolent God was conspicuous by his absence. This conflict of intellectual and religious interests was one of the less obvious legacies of centuries of discussion of the cosmic harmonies, a discussion to which Plato, Kepler, Newton, Leibniz and scores of lesser scholars had contributed.

THE INSTRUMENT-MAKERS

On a more pragmatic note, the eighteenth century in European astronomy is characterized by a rapid growth in the number of official observatories, that is, observatories maintained by states, universities and scientific communities, and religious groups. Medicine apart, no other science could boast such large numbers of people professionally engaged on research, even though the research was of a routine nature. Ever greater precision in the recording of celestial co-ordinates paid great dividends in the long run. It was an expensive business, but it could be justified in terms of its practical utility for navigation, the surveying and mapping of country and empire, and still even for theology, the mapping of humankind's place within creation. National observatories became symbols of power and principle. They had to be well provided, however, and it was no longer enough to employ local artisans on what had become highly specialized work.

We have already seen how the face of astronomy was drastically changed by the Paris Observatory and the employment of Cassini there. The founding of the Greenwich Observatory in England was due to French influence, but of an unusual sort. In late 1674, the master of the king's ordinance was raising promises of money for an observatory when one of Charles II's mistresses, Louise de Kéroualle – a Bretonne who had recently been made Duchess of Portsmouth – recommended to the king a certain Sieur de St Pierre. He was claiming to be able to find terrestrial longitudes 'from easy celestial observations'.

The longitude problem reduces to that of finding a universal clock that will allow a comparison of local celestial phenomena with what would be seen at a standard meridian, such as Greenwich. (For example, if the Sun is on the local meridian, what is its position as seen at Greenwich? To know the answer is one way of finding one's relative longitude.)

One such universal clock is a transportable time-keeper. This was not available in a reliable enough form until after 1763, when John Harrison was awarded a first Board of Longitude prize for one of his chronometers. Another 'clock' is comprised in the satellites of Jupiter, as we have already mentioned, for from their relative positions around the central planet the time can be found from tables prepared at one of the great observatories.

St Pierre kept his method secret at first, but Flamsteed and others guessed correctly that it was to use the rapidly moving *moon* as time-keeper. (It was not original, and he probably took it from Jean Morin.) The Royal Society was instructed by the king to collect the necessary lunar data, and the services of John Flamsteed (1646–1719) were obtained. His verdict was that neither lunar nor stellar positions were well enough known to make the method reliable. At all events, the king was thus moved to found the observatory, and he appointed Flamsteed his 'astronomical observator', with the task of improving tables of motions and star positions for the benefit of both navigation and astronomy. Flamsteed moved into the new building, designed by Sir Christopher Wren, in July 1676. He was more fortunate in his architect than Cassini had been in Paris: Claude Perrault's building there was much more splendid, but less functional by far.

We have seen something of Flamsteed's relations with Newton and Halley. He would not have been pleased to know that on his death in 1719 he would be succeeded at Greenwich by Halley, although he might have taken consolation in the fact that his heirs kept his instruments, on the grounds that he had paid for them. Flamsteed gave very much attention to the Moon throughout his career, but he was a perfectionist in such fundamental matters as the parameters of the Sun's motion, matters that almost all other astronomers were prepared to consider settled. His measurements of angle were of unprecedented accuracy, his errors being typically a tenth or even a twentieth those of Tycho's measurements. His 'British Catalogue' of 3000 stars (in volume 3 of his *Historia*) was for long the best available.

Not only did Flamsteed have to provide most of his own instruments, but he was often more than a year in arrears in obtaining his modest salary from the king – his history is at times reminiscent of Kepler's. In the long term, however, the relatively large investment of the British government in the fabric of the observatory at Greenwich provided a stimulus for the growth of a profession of instrument-making there that for the best part of the eighteenth century became the supplier to all of Europe. The time was past when professional astronomers could be their own instrument-makers, although they continued to have a strong interest and involvement in the art; and the makers themselves were often passable astronomers, capable of designing new instruments from first-hand knowledge.

Of course new designs came from other quarters too. To take one valuable new idea: Roger Cotes (1682–1716), first Plumian professor of astronomy at Cambridge and editor of the second edition of Newton's *Principia*, sent Newton a design for a heliostat. This allowed a solar image to be reflected from a moving mirror, driven by clockwork, into a static telescope. The principle is still in use. The first satisfactory heliostats on a large scale followed the design of Jean Foucault. The first great heliostat was built at the Mount Wilson Solar Observatory (from 1903). A notable recent example is that at Sacramento Peak, New Mexico, and another is that at Kitt Peak National Observatory, Arizona, where by reflecting the image down the polar axis it is possible to use only a *single* upper mirror.

George Graham (*c.* 1674–1751) was one of the first great specialist makers to produce a series of designs covering almost every aspect of observatory installation – mural quadrants, transit instruments, zenith sectors, astronomical regulators (precision clocks) and many more. Graham was a generation later than Flamsteed, and his first great instrument was in fact a large quadrant made for Halley in 1725. Graham was famous as a watch- and clock-maker who achieved much publicity from the Halley quadrant, the fame of which was spread when it was described and praised by Robert Smith in his widely read textbook of optics. The idea of a quadrant mounted on a solid wall (usually) in the meridian was of course not new, but Graham added a central telescope and an axis in the form of a double cone that made his work far superior to anything before it, and the design soon became standard. It was Graham who introduced the two main technical improvements to the long-case (pendulum) clock that made it acceptable as an astronomical regulator – these were the mercury compensation pendulum and the dead-beat escapement.

Graham's fame benefited greatly from one of the most important astronomical discoveries of the time, made by

James Bradley (1693–1762). (Bradley succeeded Halley as Astronomer Royal, but much later, on Halley's death in 1742.) It had long been realized that the stars should show a parallactic displacement due to the Earth's shift in position over a period of six months, that is, a shift of the diameter of the orbit. In 1669 Hooke tried to measure the shift, but failed. In 1725 a wealthy amateur, Samuel Molyneux, made careful measurements on the star gamma Draconis, using a very long (24 foot) zenith sector, a telescope arranged to measure angles over a small range in the neighbourhood of the zenith, where refraction can be effectively ignored. With Bradley's help he did indeed observe displacements, but they were too large and *in the wrong direction*, even though it became clear eventually that they followed an *annual cycle*.

Bradley took over the observations, and in 1727 extended his work to other stars, using a smaller zenith sector by Graham. He tried various hypotheses, but the story goes that he hit on the right one only when travelling on the Thames in a pleasure boat, watching the shifts in direction of the pennant at the masthead. The shifts in a star's position were due, he decided, to the changes in the combined effect of the Earth's orbital velocity and the large but finite velocity of the incoming light from the star. The velocity of light was known approximately, following Ole Rømer, and so was the Earth's. The explanation fitted excellently with his observations, and so the discovery of the 'aberration of light' was announced to the Royal Society in 1729.

The announcement was doubly important, for Bradley was able to make a very powerful negative statement about the parallaxes of the stars. Had they been as great as a second of arc, he would have been able to detect them. The stars were clearly very much more distant than had been generally supposed. (The parallax of gamma Draconis is in fact less than a fiftieth of a second.) A third implication of his work was that the Earth is in motion relative to the frame fixed by the distant stars, or round the Sun, as the usual interpretation would have it. This 'Copernican' implication might have been heeded more had it not already been so widely accepted.

In 1727 Bradley noticed that the declinations of certain stars seemed to be erratic. Five years later he had found an explanation: the Earth's axis is 'nodding' as a result of the Moon's attraction on its equatorial bulge. From this nodding there results an apparent displacement of the stars such that each seems to describe a tiny ellipse about its true (mean) position, in a period of about 18.6 years – the period of the lunar nodes. Nutation, as he called the effect, was discovered with the same instrument as aberration.

Bradley expressed his gratitude to Graham in extravagant terms, and Graham's flourishing trade in instruments in Europe began, as a result, with the zenith sector. There had previously been something of a fashion for the French portable quadrant, but now zenith sectors were all the rage, although totally unsuitable for most types of routine observation.

When Pierre Louis Moreau de Maupertuis (1698–1759), sponsored by the French Academy of Sciences, led the famous expedition to Lapland in 1736 that settled the controversy over the shape of the Earth, the principal instrument was one of Graham's zenith sectors. This was an important episode, inasmuch as it made many French converts to Newtonian principles. Two generations of excellent astronomical observers trained in the Paris school of the Cassini family had formed a nucleus of anti-Newtonian natural philosophy, and the general opinion there was that the Earth was a prolate rather than an oblate spheroid – a rugby ball rather than a flattened orange, as we might say. Expeditions were sent to Peru and Lapland, to measure a degree of terrestrial longitude in both places, for comparison with that in France. After a difficult expedition that involved shipwreck on the return journey, and a long period during which the observations had to be evaluated, Maupertuis pronounced in favour of Newton. Voltaire, one of Newton's few supporters in France, congratulated him on having flattened both the poles and the Cassini. The precise form of the Earth continued to call forth both practical and theoretical energy of large numbers of scientists, and some of the best mathematicians of the following century – Clairaut, d'Alembert, Legendre, Laplace, Gauss and Poisson, for instance – gave the problem a central place in the theory of gravitation.

In England the tradition of fine instrument-making was extended by John Bird (1709–76), who fitted out the second most important English observatory of the time, the Radcliffe Observatory of Thomas Hornsby (1733–1810) at Oxford. Bird wrote an influential account of the method of dividing the scales of instruments. By the mid-1760s, he had made large instruments for Greenwich, Paris, St Petersburg, Göttingen, and Cadiz. Jonathan Sisson was another maker, who had worked under Graham's direction, and who made instruments for European observatories, one of them indeed loaned for a time by Le Monnier to the Berlin Academy, to allow them to supplement lunar parallax observations made by Lacaille at the Cape of Good Hope.

Later in the century the pre-eminent maker was Jesse Ramsden (1735–1800) – who supplied Piazzi and Zach, for example. Ramsden supplied many European observatories with achromatic telescopes equipped with finely graduated scales, read by micrometer microscopes that he had developed. Giuseppe Piazzi's 'Palermo circle' (a 1.5 m refractor in an alt-azimuth mounting) was his most famous instrument. It soon proved to the world that proper motions of the stars (see p. 395) were the rule rather than the exception, and it gave evidence of the extraordinary stability and accuracy that might be achieved in a fine instrument.

It would be foolish to pretend that this trade was economically as important as, for example, the London trade in clocks and watches of the same period, nor should we forget the rapid expansion in the trade in sextants used for astronomical navigation. Ramsden, for instance, with a staff of sixty artisans, had produced a thousand sextants by 1789, quite apart from his other work. Astronomical methods of navigation have a history of their own, but it is worth noting here that in 1756 Bradley had reported to the Board of Admiralty that the new lunar tables by the Göttingen astronomer Tobias Mayer (1723–62) should give terrestrial longitude to an accuracy of half a degree. (Unlike longitude, geographical latitude is easily found. Mayer, incidentally, used a Bird quadrant.) After sea-trials, Bradley decided that this was over-ambitious, but after correcting Mayer's tables he decided that an accuracy of better than one degree was attainable.

Every country at every period of history had its instrument-makers, in one sense or another, but hitherto their influence had usually been strictly local. The London makers were noteworthy for the way in which they set an international standard and style of practice, and this with the encouragement of the Royal Society, to which the best were elected as members. The Royal Observatory at Greenwich provided the third vertex of this fortunate triangle. Delambre was exaggerating, but mildly, when he wrote that if all other materials of the kind were to be destroyed, the Greenwich records alone would suffice for the restoration of astronomy. There was scarcely an astronomical textbook of note at this time without either a description or an illustration of the work of the English makers. Lalande's books are notable examples. He had an observatory at the École Militaire equipped with a Bird quadrant superior to anything then to be found at the Paris Observatory, despite the far more lavish funding of the latter.

Bird's friend George Dixon, together with Charles Mason, took some of his instruments on the voyage already mentioned, that ended with observations of the Venus transit of 1761 at the Cape of Good Hope. They made observations that turned the Cape into one of the best surveyed places in the world, until, that is, two years later they surveyed the boundary between Pennsylvania and Maryland – the

'Mason–Dixon line' – and in the process provided an extremely accurate figure for the size of a degree of terrestrial latitude. Here are a few of the consequences of the new instrument industry that had grown up under the patronage of astronomy; and in return for patronage, the London trade raised astronomical standards, and so supplemented the new wave of excellence in astronomical writing that had followed in the wake of Newton's work. Together they created a strong international sense of astronomical purpose, and it is one that has lasted, more or less, to the present day.

As a postscript it might be added that Prime Minister William Pitt and the British government were responsible for almost destroying the London optical industry with the introduction of a punitive tax, at first on windows and then on glass itself. Another factor was the invention by Pierre Guinand, a Swiss bell-caster and glass-maker, of a new technique for stirring molten optical glass to achieve a homogeneous mix. In 1805 he moved to Munich, and was shortly afterwards joined by an assistant, Joseph Fraunhofer. Fraunhofer (1787–1826), the son of a glazier, and trained as such himself, was a man of outstanding technical ability, and although he died at a relatively early age, was to become a dominant influence on nineteenth-century astronomical practice, as we shall see.

MATHEMATICS AND THE SOLAR SYSTEM

At a theoretical level, astronomy after Newton profited greatly from the very rapid advances being made in mathematics. These advances were again largely due to the foundations Newton had laid. One of the worthiest of his early followers was the Scottish mathematician Colin Maclaurin (1698–1746), who investigated the equilibrium of ellipsoids (such as the Earth) and the tides. British mathematicians were unfortunately often too loyal to Newton, when sticking to certain of his techniques on which continental mathematicians had improved, and the lead soon passed to the continent. The Basle mathematician Leonhard Euler (1707–83) was one who advanced virtually every branch of the mathematics of his time, pure and applied. Time and again he won prizes from the Paris Academy of Sciences for his work. Without any title to practical experience, he gave astronomy some of its most useful mathematical procedures, for instance in the theory of instrumental errors, the determination of orbits – whether of planets or comets – from a few observations, and ways of finding the solar parallax. The problem of the lunar perigee was a worthy test of his abilities.

Following Newton's principles, Clairaut and d'Alembert had derived a value of about eighteen years for the period of revolution of the Moon's perigee, the nearest point on its orbit to the Earth. (This is not to be confused with the period of 18.6 years for the Moon's *nodes*.) From observation, the figure was known to be only about half as great, and for long Euler and others believed that the only remedy was to make adjustments to Newton's law of gravitation. In 1749, Clairaut found a mistake in the method of approximations all had been adopting. Euler did not at first agree, with the result that he composed a treatise on lunar theory that outshone everything before it. This *Theoria motus lunae exhibens omnes eius inequalitates* (Theory of the Motion of the Moon, Showing all its Inequalities) was published in 1753, and included a method for an approximate solution to the three-body problem (in this case the problem of the sun–earth–Moon system). The work makes use of a new technique that was destined to prove of enormous value to future mathematical astronomy and physics, the 'method of variation of the elements'. More immediately comforting was the fact that Clairaut and he had shown that Newtonian gravitation and dynamics passed this stringent test.

Euler devoted much labour to the three-body problem, and the problem of the perturbation of planetary orbits, which is essentially similar. His greatest work on lunar theory appeared in 1772, a work that was not properly appreciated for more than a century, when it was rescued from neglect, and further developed, by the New York mathematical astronomer G. W. Hill (1838–1914). (Hill was the leading person in his field. He seems to have had difficulties precisely the reverse of Kepler's and Flamsteed's, for he insisted on *returning* his salary to Columbia university.)

One of the most difficult problems confronting mathematical astronomers of the eighteenth century, and one that acted as a constant spur to progress, again concerned an inequality in the Moon's motion. The Moon's mean motion, averaged over a reasonably long period (say a century rather than a millennium) is not constant over very much longer periods of time, but *accelerates*. This was first suspected by Edmond Halley, around 1693, on the basis of a comparison of ancient eclipse records with what the best modern tables gave for the same eclipses. In 1749, Dunthorne revived the subject, and added further ancient data to confirm Halley's suspicions. The acceleration was extremely small, and indeed a measure of the progressive refinement in astronomical accuracy: Dunthorne fixed it at only 10″ per century, and others later in the century (such as Mayer and Lalande) settled on figures between 7″ and 10″. But what was its physical *cause*? The Paris Academy offered a prize for a solution in 1770, which Euler took, jointly with his son Johann Albrecht, thinking, however, that they had proved that the steady ('secular') acceleration could not be explained by Newtonian gravitational forces.

Here once again was something of a crisis in Newtonian science, and the subject was proposed for the Academy prize of 1772. This was awarded to Euler jointly with Lagrange.

Joseph Louis Lagrange (1736–1813) was born of an Italian family of French ancestry in Turin, Italy. (His French name represents only the last version of a continuously varying quantity.) Before he was twenty he drew attention to his mathematical talents through correspondence with

Euler. Following Euler's lead in the 1760s, he introduced some brilliantly original methods of his own into a study of the motion of the Moon and another of the perturbations of Jupiter and Saturn, which won him the Paris Academy prize for 1766, and much fame. He was found a position in Berlin, thanks to d'Alembert's friendship with Frederick II of Prussia. Euler, who was on the point of leaving a post in Berlin for one in St Petersburg, did not manage to persuade Lagrange to join him there, but in Berlin he had several stimulating colleagues, including Johannes Lambert, whose cosmological ideas we shall mention shortly. His prodigious mathematical talents soon became obvious to all. In 1772 he shared the Academy prize with Euler for an essay on the three-body problem, that is, on lunar motion. Euler, in his essay, now maintained that gravitation could offer no explanation for the secular acceleration of the Moon, but that there must be some sort of etherial fluid in space, offering resistance to the motion of the Moon and Earth. Lagrange offered a new solution to the three-body problem, but did not explain the secular acceleration.

Again in 1774 the Academy prize was offered for a solution, and again Lagrange was successful, with an account of how the *shape* of the Moon has an effect on its motion – and similarly for the Earth. Still he could find no explanation for the secular acceleration, and in a study of the historical evidence for it, he pronounced it a dubious idea that should be abandoned.

The series of Academy prizes continued to attract work of the highest quality, but Lagrange was tiring of the constraints they placed on his work, and he preferred to write independent memoirs. His last entry was for the 1780 prize, which he won with an important study of the perturbation of cometary orbits through the action of planets. He contributed several additional memoirs of great value on Newtonian planetary theory. Brought to Paris in 1787, he managed to survive through the turbulent years of the Revolution. He became a member of the Bureau of

Longitudes, and was able to assist in the practical needs of astronomy, such as the production of ephemerides – of which he had experience in Berlin. He was honoured by Napoleon, and when he died in 1813 his funeral oration was spoken in the Pantheon by Laplace – who had by this time solved the problem that eluded him and others for so long.

Pierre Simon Laplace (1749–1827) was born in Normandy, where he studied at the university of Caen before leaving for Paris in 1768, with a recommendation to d'Alembert. Within five years, a brilliant series of mathematical papers had resulted in his election to the Paris Academy of Sciences. He wrote on the integral calculus, celestial mechanics and the theory of probability. The great works for which he is best remembered are on those two last themes. Successive volumes of his *Mécanique céleste* (Celestial Mechanics) appeared between 1799 and 1825, and like his important writings on physics, made much use of a number of mathematical techniques that he had himself developed, and that are still widely used attached to his name.

Laplace was not above pointing out his own incomparable genius, and lost many friends as a result, but he was conscious too of the need to make the mathematical sciences available to a wider audience, and one of his most popular works was his eminently readable *Exposition du système du monde* (Exposition of the System of the World, 1796 and later editions), which dealt with cosmological matters over a very wide range. His work in mathematical astronomy reached its peak during the Revolutionary period in France, and he was able to exert very great influence over the structuring of intellectual life in France at every level. During the empire he was honoured in various ways by Napoleon, with whom he discussed astronomical matters – in one case on the field of battle, or so it is said. It was the returning Bourbons, however, who eventually made him a marquis.

When Laplace came to study the possible acceleration in the Moon's motion, he began by dismissing the claims of sceptics, that the historical evidence for it was unreliable. He dismissed too a solution that was being offered, that the effect was no more than an illusion, caused by the slowing down of the Earth's rotation due to friction, for instance in the winds. Why, he asked, in this case, did the planets' mean motions not also increase? Euler's notion of an etherial fluid he rejected for want of independent evidence. In short, he faced up to the problem as it had stood three generations before.

But still he could not solve it, and so it was that he tried to modify Newton's law of gravitation. It had been generally assumed that the gravitational force exerted by one body on another acted instantaneously, but what if it required a finite time to act? Laplace showed that this could result in a secular acceleration of the Moon, but only if the speed of transmission of gravity were more than eight million times as great as the speed of light. (And he showed that if the secular acceleration *could* be explained in other ways, then the speed of gravity must be fifty million times that of light for it not to be otherwise in evidence.)

He was not happy with this solution, which like the ether was not exactly obvious elsewhere; but then in 1787 he gave a much more acceptable alternative. He had found that the shape of the Earth's orbit was changing, in fact the eccentricity of the ellipse was decreasing, and he was able to connect this with the gradual shortening of the length of the month. The analysis was supplemented by his study of the motions of Jupiter's satellites. (Jupiter actually enters the calculation of our own Moon's behaviour.) He calculated a theoretical expression for the secular acceleration of the Moon that for his own day gave a figure of approximately 10.1816", close to the best historical evidence; and he showed that after about 24 000 years the secular change would reverse, and the month would lengthen.

When Lagrange read the paper announcing these findings, he looked over his own earlier work (1783) and found an omission which, when put right, yielded almost exactly Laplace's results. Long afterwards, John Couch Adams (1819–92), 'discoverer of Neptune', demonstrated that Laplace's theory could not account for all of the known effect, but for long Laplace's achievement was regarded as the acme of dynamical astronomy.

Laplace had many other results to his credit in gravitational theory. For instance, he found a relationship between the shape of the Earth and certain irregularities in the Moon's motion, and he introduced the rotation of the Earth into tidal theory. One of his greatest achievements was to explain certain fluctuations in the orbital speeds of Jupiter and Saturn: he found that they resulted from a curious relationship between the periodic times of the planets, five times Jupiter's period being very nearly twice Saturn's. One of his greatest achievements, however, seemed to touch on a subject of great concern to those who were trying to relate astronomy to natural religion. This was his work on the stability of the solar system. Will it continue forever, without intervention by a divine clock-maker? Leibniz had taunted Samuel Clarke with the imperfection of a Newtonian Universe, which he said would need winding up by God from time to time, so implying that God was an inferior artisan.

Laplace made much use of Lagrange's method of introducing variations into the six elements of a planet's orbit (the eccentricity, direction of aphelion, and other parameters that define it), and in 1773 he was able to prove that, even if one planet's elements are perturbed by another planet, its mean distance from the Sun will not change appreciably, even over millennia. Over the next few years he followed this with more complex theorems relating the distances, eccentricities and angles of the orbital planes, and again these seemed to point in the same direction: the solar system is highly stable. He showed that there is a certain plane in the solar system about which the whole system oscillates. In more recent studies, the frictional effects of the tides have been introduced into the account, and again it has been found necessary to qualify Laplace's claims, but the skeleton of his analysis remains, a remarkable testimony to the achievements of Newton's successors in the century following his death.

MOVEMENTS OF THE STARS

With rare exceptions before the eighteenth century, the stars were regarded as fixed – at least relatively to one another. Hipparchus' discovery of precession merely introduced the need for catalogue-makers to add to *all* ecliptic longitudes a constant appropriate to the date. Ptolemy's catalogue of over a thousand stars was repeatedly revised in this simple way, and even when alternative catalogues were drafted – by Ulugh Begh, Tycho Brahe, Hevelius, Flamsteed, and the rest – there was always the assumption of their internal constancy. The 'new star' phenomenon, which helped to make Tycho's reputation, did nothing to change this view. The situation did change, however, with Halley's discovery that some at least of the stars were in relative motion.

The career of Edmond Halley (1656–1743) was so intimately bound up with those of Newton, Flamsteed and other leading figures of his time that it is easy to treat him only as a satellite to them, but he was a man of great originality and his contributions to astronomy were substantial. We have already encountered his probing question as to why, if the heavens are infinite, and are filled with a uniform distribution of stars, the sky is not ablaze with light. He produced editions and Latin translations of Apollonius and Menelaus. In the course of analysing cometary orbits he decided that the comets of 1531, 1607 and 1682 were one and the same object, and that this would return – he allowed for perturbation by Jupiter – in December 1758. It did, but of course he did not see it.

It almost goes without saying that Halley's is the best known of all comets, not only because it served to underwrite Newtonian science. It is associated with the best-remembered date in English history, 1066, for it appeared to William of Normandy, who took it to be a favourable omen for his conquest of England. It is depicted on the Bayeux Tapestry. Comets had from ancient times been associated with the downfall of princes. It made a spectacular appearance in 1910. In 1985–6 it returned, but in relation to expectations was no more than a damp squib.

A practical man, Halley was as much at home writing on gunnery and annuity tables as on the properties of thick lenses. In 1676 – at the age of twenty – he made a voyage to St Helena, off the African coast, to catalogue southern stars. Between 1698 and 1700 he captained a mutinous ship across the Atlantic, and as we have already seen, charted magnetic variation, with the result that he developed an ingenious theory of the Earth's structure. He was sixty-four when he finally succeeded Flamsteed as Astronomer Royal, but he promptly set in motion a programme for observing the Sun and Moon over an eighteen-year cycle, and he lived to see the cycle through.

Halley's scholarly talent assisted him in one of his finest achievements – one that in retrospect proved his instincts to have been as important as his statistics. In a paper published in 1718, he explained how he compared modern observations with those of the Greeks. He had been studying star catalogues, especially Ptolemy's, since about 1710, and he came to the conclusion that precession and observational error were not enough to explain the discrepancies. He was convinced that a southerly motion for the bright stars Aldebaran, Arcturus and Sirius was proven, that there were 'proper motions' of the fainter stars, and that these would have been more obvious had the stars not been so distant.

Jacques Cassini (1677–1756) confirmed Halley's claim in 1738. In his case he could detect a shift in the position of Arcturus even from as recent a measurement as one made by Jean Richer (1632–96) at Cayenne in 1672 – of course a much more precise and certain measurement than Ptolemy's. Cassini argued that this was indeed a true proper motion of Arcturus, rather than a consequence of some shift in the ecliptic, since it was not shared by a faint star nearby. (Many years earlier Cassini had expressed irritation with Halley, who found a mistake in his claim to have measured the parallax of Sirius.)

The discovery of proper motions in the stars opened up for serious discussion a completely new vista in astronomy. Stellar parallaxes were as yet undetected in anything approaching a direct way, but the fact that the stars have different motions, as seen from the Earth, seemed to confirm the conjecture of several writers of the late-sixteenth and seventeenth centuries that the stars were *scattered* thoughout space – whether finite or infinite space was another question. This would seem to follow whether the observed motions are due to the Earth's motion or to motions intrinsic to the stars themselves. As for the idea that only the solar system moves, that is, through a system of stars at relative rest, Bradley, at the same time as he announced nutation (1748), suggested that it would be many ages before there was evidence to decide between this and the alternatives. In this he was mistaken.

Astronomers soon began a programme of systematic measurement of proper motions. Tobias Mayer, in Göttingen, published the proper motions of eighty stars in 1760, based on a comparison of his own and Lacaille's measurements with Ole Rømer's of 1706. Mayer stated clearly an important consequence of a motion of the solar system through the stars: those in the general direction towards which we are heading (the direction of the 'solar apex') will seem to become more widely spaced, and in fact will seem to radiate from the apex. Those in the direction away from which we move (the 'solar antapex') will seem to close up on one another, and to draw towards the antapex.

Mayer could not see any such pattern in his proper motions, but in 1783 William Herschel found exactly such a pattern from a limited number of stars, observed by the fifth Astronomer Royal, Nevil Maskelyne (1732–1811). In that same year the French astronomer Prévost showed that Meyer's own data yielded a similar result. Herschel placed the solar apex at a point in the constellation of Hercules (a little to the north of the star lambda Herculis). Prévost's apex was about thirty degrees away, while Georg Simon Klügel (1739–1812) in Berlin found an apex only about four degrees from Herschel's. Together it seemed that these convergent results provided a remarkable new perspective on the large-scale structure of the universe.

It seemed so at the time, as it does today, and yet in the first two decades of the nineteenth century a number of leading astronomers – notably Biot and Bessel – claimed that the evidence justified no such conclusion. Later in the century, however, as more and more data were collected, not only was the result confirmed qualitatively by the best observers in both northern and southern hemispheres, but Herschel's original position for the apex remained respectably close to the newly derived positions. And when eventually data were obtained for the *distances* of the stars whose proper motions were analysed, it became possible to quote a figure for the speed of the solar system, as well as its direction. (Otto Struve (1819–1905) gave the speed as about 60 million miles per annum, that is, about 96 250,000 km per annum.)

Of course it went without saying that this entire exercise was only of *statistical* value, since there may be – and indeed were eventually proved to be – motions among the stars themselves, regardless of our motion through them. At first this began as pure conjecture, even before Herschel's work. Thomas Wright, as we shall see, believed the solar system to be in motion around a central body, and various alternative schemes involving the stars and system of the Milky Way were devised, before Herschel at length tempered the more ardent imaginings with the chill of observation.

WILLIAM HERSCHEL

If any astronomer of the eighteeenth century turned the subject in the direction it would follow in the nineteenth, that person was William Herschel. In an age when astronomy was becoming on the one hand allied with the most advanced mathematics then practised, and on the other increasingly institutionalized, it seems paradoxical that an amateur standing entirely outside both traditions should have been able to achieve this. Friedrich Wilhelm (William, as he later became known) Herschel (1738–1822) was born in Hanover, and first visited England at the age of eighteen as an oboist in the Hanoverian Guards, his father's regiment, in British service. A year later, fleeing the French army, who had defeated the Guards at Hastenbeck and captured his father, he moved permanently to England. He began at first to earn his living by copying and teaching music. He made himself acquainted with astronomy and telescope-making through Robert Smith's renowned textbook of optics – which we mentioned in connection with Halley and Graham – and over the years he became so adept at lens and mirror grinding that by the 1770s he was in possession of telescopes on a par with the best available in the country.

What drove him on was not a clear vision of what he would discover, but a desire to see what others saw, and more, and to excel in an art that was highly valued by the society of his day. His ambitions were more than fulfilled. In 1782, after a comparison done at Greenwich, Maskelyne acknowledged that Herschel's was a better instrument than any in use there.

In 1772 Herschel brought his younger sister Caroline over from Hanover, and she helped him greatly with the work of grinding and polishing large mirrors. She searched for comets, found three new nebulae in the course of doing so, and in 1787 was granted a salary by the king. A year later, William – who had been similarly rewarded six years earlier – married a young widow. (Their son John Herschel was destined to be almost as famous as his father.) Between 1786 and 1797, equipped with their excellent telescopes, Caroline found no fewer than eight new comets. In 1798 she revised Flamsteed's star catalogue for publication, and after her brother's death she edited much of his work. She herself lived on until 1848, when she was nearly ninety-eight.

William Herschel set his sights on the study of distant stars and nebulae, one of his objectives being nothing less than a plan of their distribution throughout the entire Universe. By 1779 he had surveyed the sky down to stars of the fourth magnitude, and he was undertaking a second review – of which more in due course – when on 13 March 1781 he found an object that he knew was not a star. At first he supposed that it must be a comet.

It is noteworthy that after Herschel had announced his discovery, Maskelyne could not measure its position relative to the reference stars Herschel had named, for the stars were so faint that the light from the cross-wires of his micrometer made them invisible, while Hornsby in Oxford could not even locate the object. Herschel's quoted diameter (five seconds of arc) was likewise unverifiable, so inferior were other telescopes even to his modest reflector of 6.2-inch aperture. French astronomers spent much time calculating the orbit, but the data were difficult to handle, and it became clear that observations would be needed over a very long period before the matter was settled. The scales were more or less finally tipped when in the summer of 1781 Anders Johann Lexell (1740–84), the Imperial Astronomer at St Petersburg who was then visiting London, calculated the first broadly acceptable elements of the orbit.

Most professional astronomers henceforth accepted that it was not the orbit of a comet, and that in fact Herschel was the first to have discovered a planet in historic times.

Herschel called it Georgium Sidus, in honour of the Hanoverian king of England, George III. Other names were suggested. Erik Prosperin (1739–1803) of Uppsala even suggested 'Neptune', and Lexell liked the idea of 'Neptune de George III' or 'Neptune de Grand-Bretagne'. Following a proposal by the Berlin astronomer Johann Bode, who no doubt thought a Hanoverian name too parochial, astronomers eventually settled on Uranus, father of Saturn and grandfather of Jupiter. In France, largely under Lalande's influence, 'Herschelium' was long favoured. In fact three different names for the planet were used simultaneously for at least sixty years, and as a fossil of the controversy, two different symbols for it are in use to this day.

Herschel was now famous. He was granted a royal pension – but appreciably less than his salary as organist in Bath – in return for occasional instruction and display to the royal family. He moved from his old abode to the neighbourhood of Windsor Castle, and finally to Slough nearby. He went into the business of manufacturing telescopes for sale, and at the same time worked at ever-larger mirrors. The king financed the largest, a 40-foot telescope with a 48-inch mirror that cost much trouble and energy. The mirror was four times the area of that in his previous telescope, until then the largest in the world. The first two attempts resulted in failure, in one case with molten speculum metal spilling on to the floor and exploding the stones like shrapnel. The final mirror weighed over half a tonne. A team of twenty-four workmen hired for the grinding and polishing produced poor results, and eventually Herschel built a large machine to finish the work. Completed in August 1789, on the second night of its use it revealed a sixth satellite round Saturn (Enceladus). Herschel had a paper in press on nebulae. He wrote to Sir Joseph Banks, President of the Royal Society, modestly asking that he add at the end of it: 'P.S. Saturn has six satellites. 40-feet reflector.'

The telescope was vast, but difficult to manoeuvre, and on the whole less useful to Herschel than his 20-foot reflector. It was used until 1815, and when it was finally laid to rest in the garden at Slough, the family held a requiem mass inside its 12-metre tube.

In the early 1780s, Herschel's interest in the white patches of light in the sky known as nebulae (Latin for 'mist' or 'clouds') was aroused when he was given a copy of Messier's catalogue of a hundred such objects. In many respects this marks the beginning of a completely new phase in cosmological thought. To map the Universe in three dimensions, a knowledge of stellar distances was needed, and virtually the only thing known of such distances was that they were too great to yield measurable parallaxes. Under these circumstances, it was necessary to make conjectures, judging the plausibility of these by considerations having little to do with astronomy as such – considerations of symmetry and analogy, for example. Herschel was not the first to enter this territory, which was opened up from two different directions. On the question of the possibly infinite extent of the Universe, considered in a very general sense, Newton and Bentley had considered difficulties with gravitational equilibrium, while Halley and the Swiss Jean-Philippe Loys de Cheseaux had noted the paradoxical consequence for the illumination of the sky at night, which on too simple a picture leaves no dark spaces at all. Complementing such abstract approaches were others that began from the actual structure of the system of stars we see as the Milky Way.

The Greek word for the Milky Way is *galaxias*, hence our word 'galaxy'. Distant star systems were only called galaxies at a relatively recent date, that is, when there was evidence that they somehow resemble whatever it is within which we are situated. The path to this knowledge was long and arduous. Before the telescope, most who touched on the subject at all spoke of the Milky Way as a cloud, but some – even as early as Democritus – treated it as a conglomeration of tiny stars that were so close as to be indistinguishable. Even after Galileo's telescope had resolved much of it into

stars, the structure of the *system* was not at first a subject
of great concern. But after Wright, Kant and Lambert had
written on the subject in the mid eighteenth century, it
became an open question as to whether there might not
even be systems of higher order, that is, systems of milky
ways.

NEBULAE AND CLUSTERS BEFORE HERSCHEL

The early history of attitudes to the nebulae has no very
profound chapters before the advent of the telescope. Ptolemy used the descriptions 'cloud-like' or 'misty' of six or
seven items in his catalogue, four of them being now classified as true (galactic) clusters and the rest as asterisms. Since
most later catalogues of any importance before Tycho
Brahe's followed Ptolemy's to a greater or lesser extent, the
list remained virtually unaltered.

Tycho's half-dozen had only one in common with most
of his predecessors, but he missed the Andromeda nebula,
which had been included in al-Sūfi's tenth-century list.
Simon Mayr recorded it in 1612, and of course could see it
more clearly through his telescope. He described it as like
the flame of a candle shining at night through transparent
horn. When in 1656 Huygens found another nebula in the
sword of Orion – which was to become co-equal in fame
to that in Andromeda, although as we now know of a completely different character – the area of sky around it was
so black that he thought he was looking through a hole in
the heavens into the luminous region beyond.

The number of nebulae recorded grew only slowly. Hevelius' and Flamsteed's lists, for instance, included only fourteen and fifteen respectively. Few astronomers seem to
have taken any great interest in individual nebulae for their
own sake.

Halley was one of those who did so. He spoke of patches
of light seen in his telescope that look like stars to the naked
eye, but that are in reality light coming from 'an extraordin-

ary great Space in the Ether; through which a lucid *Medium* is diffused, that shines with its own proper lustre' – that is, without the aid of an embedded star. He knew of others that seemed to shine because there was a star within them. In these vast and very distant places, he noted, there would be perpetual day. This thought caused him to defend Moses against those who had criticized the biblical account of the creation of the world, saying that Moses was wrong to speak of light before God's creation of the Sun.

Not all speculation was so restrained, and when the seventy-five-year-old clergyman William Derham (1657–1735) culled material from various sources and added a few ideas of his own, he brought to general notice the old idea that stars may be openings in the heavens to a brighter region beyond. Might the same idea not apply rather to the nebulae, as Huygens had suggested of that in Orion's sword? His supposition that these were plainly much more distant than the fixed stars was without foundation, as Maupertuis pointed out in a justly critical reply, which also emphasized our helplessness in discussing luminosities in the absence of firm knowledge as to distance.

As telescopes improved, astronomers began to search for and to list nebulae as objects of interest in themselves. Cheseaux listed twenty objects in 1745–6, at least eight of which were true nebulae or clusters not recorded previously. Guillaume le Gentil added others, including the elliptical companion to the (elliptical) Andromeda Nebula. Lacaille, Messier and Bode added more, and yet by 1780, a hundred and seventy years of telescopic observation had increased the number of 'genuine' clusters and nebulae from nine to only around ninety. By the end of Herschel's programme, he had charted 2500.

Gradually, nebula hunting was becoming a pursuit in its own right, although for long it was held in less esteem than comet-hunting. Charles Messier (1730–1817), whose catalogue led Herschel in 1783 to launch his twenty-year programme of searching for nebulae, was first and foremost a comet-hunter, indeed the leading European adept at this sport in the 1760s. He was an observing assistant at the Marine Observatory in the Hôtel de Cluny in Paris. He had more than a dozen new comets to his name, and it is said that the need to be at his wife's death-bed cost him another, to his sorrow.

Messier began his catalogue of nebulae simply to make comet searching more reliable, for nebulae and comets were all too easily confused. When, at the end of the century, he referred to two thousand nebulae charted by Herschel, he added that the knowledge gained would not simplify the comet-hunter's task. So much for the priorities of the age. He resurveyed the old lists and eliminated much spurious material from them. His last published catalogue had 101 distinct items, and for all his lack of interest in them on their own account, these, the most conspicuous nebulae, are still often known by their 'Messier numbers'. (The nebula in Orion's sword is M42, for example, that in Andromeda is M31, and its companion is M32.)

THE MILKY WAY: FROM WRIGHT TO HERSCHEL

Newton may have started from very general cosmological principles, but they were certainly more securely founded than most of those that followed in the eighteenth century. But almost by accident, an intellectual lead was given by a man with very different credentials, Thomas Wright (1711–86) of Durham. Wright was at first apprenticed to a clock-maker, but taught himself practical astronomy effectively enough to teach navigation and to work as a land surveyor. This last occupation led him to write successful works on English and Irish antiquities, and in the second half of his life he followed the career of architect, and made no contribution of influence on astronomy. That he is remembered in this connection today is chiefly due to the fact that his ideas were seized upon and embroidered by others.

In 1742 Wright prepared a 'key to the heavens', a volume explaining a large (2.2 m²) plan of the Universe, as he thought it might be. In 1750 he published his most influential work, *An Original Theory or New Hypothesis of the Universe.* Wright was anxious to include a religious dimension in his schemes. As early as 1734, in a lecture-sermon illustrated by one of his large paper plans, he had identified the divine centre of the Universe with the gravitational centre around which he believed the Sun and stars all move in orbit. This movement, *for which Halley's proper motions seemed to provide slight evidence*, gave an ingenious explanation of why the Universe does not collapse into a single body under gravitation, the problem that had worried Bentley.

The Milky Way in Wright's early model was the cross-section of the Universe we see when looking in the direction of the grand centre. This was not well thought out, and in 1750 he changed the model, placing the stars (including our Sun) in a thin spherical shell. Looking inwards and outwards from the shell we see relatively few stars, but looking along any direction parallel to the tangent plane to the shell, we see stars at a high density. Here was a simple hypothesis that explained appearances more or less. (The Milky Way is of course irregular, and of uneven density, but that could simply suggest irregularities in the shell of the universe.) As he saw, however, there are other possible models. One he considered was a flat ring, a mere slice of the spherical shell, as it were. Again the Milky Way is the effect of looking along tangents to the ring, and again looking inwards or outwards produces a thin spread of stars, but now the variations in the actual Milky Way are slightly better represented. Finally, Wright considered that our Universe might contain many such star systems, each with its supernatural centre.

A manuscript ('Second Thoughts') survives showing that he was not content with this scheme. He now proposed an infinite set of concentric shells around the divine centre.

He supposed that each looks like a fiery Sun from outside, but that from inside it is pierced with volcanoes, which we see as stars, and as the Milky Way. He supposed that divine punishment was achieved by God moving the soul from one shell to a more confined shell.

With these schemes Wright had shown, if nothing else, how numerous are the hypotheses open to a vivid imagination, and how slender was the evidence. But neither consideration seems to have worried the great philosopher Immanuel Kant unduly. Kant (1724–1804) learned of Wright's book only through a review of it in a Hamburg newspaper (1751), and the reviewer had unfortunately misunderstood some essential points. In his *Universal Natural History and Theory of the Heavens* (1755), Kant shows that he did not recognize the supernatural nature of the centre in Wright's theory, and therefore took Wright's ring model to be a *disc* model. In view of the shaky foundations of Wright's ideas this was of no great consequence, and Kant did add a number of new speculations to the existing stock.

He thought that the Sun and planets of the solar system might originate by condensation from some thin primordial matter – we are reminded of the Cartesian theory. He gave a rough qualitative 'Newtonian' explanation of how under gravity such diffuse matter could form a disc, before condensation occurred, and he considered that this process is going on throughout the Universe. The Universe was thus, for Kant, a non-static affair. Bodies evolve, suns condense, then heat up to a point when they explode into fine matter within which the process can repeat itself. This process, he thought, goes on throughout an infinite space and over an infinite time – concepts that were destined to cause him a great deal of trouble in his later so-called 'critical' philosophical writings.

When Johann Heinrich Lambert (1728–77) first seriously turned his thoughts to the structure of the Milky Way, around 1749 if we are to accept his own account, he had not heard of the work of Wright and Kant. Rather as Kant was to do, Lambert took the Milky Way to be a (convex) lens-shaped structure, but he took the Sun and stars in its neighbourhood to be a subsystem of the whole, and one of many such. In similar fashion he proposed that the Milky Way is a member of a higher-order system of milky ways. Unlike Wright and Kant, Lambert was a good mathematician. He was aware of the difficulties that Euler (at a later time his colleague) and others were having with the theory of perturbations of Jupiter and Saturn, and he decided that there must be forces at work within the solar system coming from outside it. (He himself tried later to represent the motions of both planets by empirical equations, and he even anticipated some of the results that Lagrange later obtained on theoretical grounds.) This encouraged his vision of a hierarchical Universe, and although purely speculative, this is an early example of an astronomer trying to introduce the large-scale distribution of masses in the Universe into an analysis of specific *local* effects.

Lambert's ideas were published in 1761 as *Cosmologische Briefe* (Cosmological Letters), and became popular in Germany and abroad, for they were translated into French, Russian and English. When Herschel first came across this work is unclear, but there was in any case a widespread discussion of the problem of the nebulae: were they, too, resolvable into stars? His telescopes were better fitted to answering that question than any in the world, when he studied the nebulae in the early 1780s, and he soon found to his delight that many (in his enthusiasm he said 'most') of them *were* resolvable. (In 1790, however, he confirmed a suspicion that there was another class of nebulae, for then he found one that was plainly a cloud of luminous gas with a single central star.)

Clusters of stars fitted well with the idea of a process of a drawing together through gravitation, at places where the stars had originally been somewhat closer together than average. He found clusters too within the Milky Way. In a series of papers (1784–9) he presented the evidence, and he announced the preliminary findings of his study of the Milky Way as a whole.

How was he to chart its outline, in the absence of star distances? He made two assumptions: first, that his telescope could reach to the farthest limits of the Milky Way, and secondly, that the stars are distributed regularly within its limits. On these assumptions, his programme was one of counting stars within identical segments of the sky of the same (solid) angular size. These star-counts he called 'gages'. The 20-foot reflector telescope he used had a field of view of about fifteen minutes of arc. Occasionally he registered no stars at all, or just two or three, while in one segment he registered as many as 588. He took specimen gauges from well over three thousand places in the sky. Of course he knew that the assumptions behind his star gauges were not impeccable, but he hoped that they would give statistically acceptable results, and indeed, were we ignorant of the method by which it was obtained, we should today judge his first chart (1785) a quite remarkable success.

It was in a sense unfortunate that as his techniques improved, and he made use of the more powerful 40-foot telescope, his confidence in his early work was shaken. Uniform distribution was unacceptable, and the new telescope introduced so many more stars into the gauges that he could not guarantee that a bigger telescope would not add yet others. He still clung to a belief in condensation and clustering through gravitation, however, for the more he observed, the more examples he found of both – that is, of nebulae having the brightest patch in the middle and stars in resolvable clusters getting closer to the middle. He found regions at opposite sides of the sky that seemed to be unusually densely packed, and this and other items of evidence he produced would have fitted well into what we now know to be the spiral form of our Galaxy. In a paper of 1814 he drew attention to the way clusters of stars seem to favour the plane of the Milky Way – the galactic clusters, as we now realize, are a part of our own galactic system. He saw them as threatening to close in eventually on our galactic centre (by gravitation) and so bring about an end to the Milky Way. This conclusion did not, however, take possible dynamical (rotatory) effects into account. At all events, observational cosmology was at last under way.

Herschel's gauges were continued by others. Otto Struve combined Herschel's observations with others by Bessel and Argelander to find a formula expressing the clear tendency of brighter stars to favour the plane of our Galaxy. John Herschel continued his father's work in the southern hemisphere – at the Cape of Good Hope – with a similar telescope and using the same methods. (John, an excellent mathematician who would probably have earned distinction there had he not turned to astronomy out of duty to his father, had learned the art of mirror-making from William, and made several large telescopes of his own.) By 1838 he had catalogued 1707 nebulae and clusters, 2102 binary stars, and for his gauges had counted 68 948 stars in 2299 fields of view. He noted that the southern hemisphere is richer in stars than the northern. Struve and he formed similar conclusions, in particular that the Sun is not in the galactic plane but a little to the north of it.

MICHELL, HERSCHEL AND STELLAR DISTANCE

The concern of both William and John Herschel with double stars was connected with the hope that star distances might somehow be found. In 1767, John Michell (1724–93) had published a paper in which he had pointed out that close pairs of stars occur much more frequently than one would expect on the assumption that stars are uniformly distributed in space. The conclusion – an interesting early use of statistical reasoning in astronomy – was that there is a high probability that stars that seem to be close together are in reality so. Michell, who was the rector of a church near Leeds, decided to leave the reader with two alternatives: there was a general law at work, perhaps the law of gravitation, or perhaps a law of the Creator. Astronomers have tended to accept the first alternative and to speak of Michell as the man who 'proved the existence of physical binary stars'.

Michell, in whose house William Herschel was often a guest before he moved to Bath, has a second claim to distinction, for in 1784 he offered an early argument leading to a plausible distance for a star. Saturn, he said, when at opposition (to the Sun), is about as bright as the star Vega. Saturn is known to be at such a time about 9.5 times as distant as the Sun (9.5 astronomical units), and the angular size of Saturn is about twenty seconds of arc. Assuming that Vega is of the same intrinsic brightness as the Sun, and that Saturn returns to the Earth all (or a certain fraction) of the light that falls on it, it is a simple matter to calculate its distance. It is necessary to use the inverse-square law of brightness, but this had been established by the excellent work of Pierre Bouguer (1698–1758) half a century before. In fact Kepler had stated the law, but Bouguer, royal professor of hydrography in Paris, was the first to turn astronomical photometry into an exact experimental subject.

The derived distance of Vega was approximately 460 000 astronomical units. In fact Vega is now known to be much more luminous than the Sun, and in 1837 F. G. W. Struve (1793–1864) showed from trigonometric measurements that its distance was about four times Michell's estimate, but no remotely accurate estimate of any star's distance had been made earlier, unless we count Newton's figure for Sirius.

Michell's treatment of double stars was in stark contrast to another, which indeed set Herschel off on his study of doubles. Galileo and others had hoped to use 'optical doubles' – to use our term for stars that are only accidentally in line as seen from the Earth – in the hope that the faint distant component might be used as a standard of reference. It would be a token stationary point, with reference to which the fluctuating position of the much nearer star would be measurable. The hoped-for fluctuation would be that due to the Earth's orbiting round the Sun, and would, it was thought, provide the star's parallax.

Herschel published three catalogues of double stars (1782, 1785, 1821), with 848 examples. After the first catalogue, Michell drew attention to his own earlier reasoning, and in 1802 Herschel remeasured his doubles, finding to his surprise that many of them showed relative movements that had nothing to do with parallactic shifts (that is, they were not due to the Earth's annual movement; see figure 4.10). This was the first step on the road to a proof that the stars are moving around one another in gravitational orbits, and that gravitational attraction is at work beyond the solar system.

It was Herschel's hope that relative brightnesses of stars would show their relative distances, applying the inverse-square law. For long he followed a common assumption that stellar *magnitudes* were a guide to relative distance, with third-magnitude stars, for instance, three times as distant as first-magnitude stars. In 1817 he devised an ingenious procedure for comparing brightnesses: he pointed near-identical telescopes at two stars, and cut down the aperture of that pointing to the brighter until the two stars appeared equally bright. His observations of double stars produced some unsettling results. If we are certain that two stars are orbiting around one another, and one seems much brighter than the other, then it is surely intrinsically so. Alas, the data Herschel was now obtaining from his photometric comparisons led him to a distribution of the stars in space that he saw to be unacceptable. He had himself discredited one of the chief assumptions that led to his old plan of the structure of the Milky Way.

We have already seen how the motion of the Sun through the stars was related to their proper motions, how the solar apex is the point away from which those tiny motions seem to radiate, and how the antapex is the point to which stars in the other half of the sky seem to converge. There is a simple relationship between the proper motions due to this effect (say in seconds of arc per century), the speed of the Sun and the distance of the star. The proper motions can be measured; the Sun's speed can be found if we know the star's distance, but at all events the speed and direction have unique values with respect to the star system as a whole. Ignoring local motions, one would imagine that *all stars must yield the same answer for the velocity*. When Herschel derived relative distances from brightness alone, however, he found very discrepant results, and the best he could do was find a statistically acceptable answer for the solar velocity.

Turning the argument round, from a given value for the solar velocity one might expect to be able to derive stellar distances from the proper motions of the stars, if these are produced by our motion (with the Sun) through the star system. Here Herschel was at first disappointed. There were bright stars – which one would expect to be close – that seemed to show no proper motion, and that should therefore be at a very great distance. At length he saw the reason: his argument was valid only in a generally static Universe of stars. If some stars *shared* the Sun's motion, their distances would not show up on the argument previously presented. Herschel had in this way received the first intimation of the phenomenon of 'star streaming', but it has to be said that his discovery made little impact at the time, and that not until Kapteyn's announcement of 1904 was the phenomenon properly appreciated.

After Herschel had been using his largest reflector for some years, he learned to distinguish between nebulae that appeared nebulous simply because the stars constituting them had not yet been resolved, and nebulae made out of continuous luminous matter, each with one or more stars within its boundaries. The more examples he accumulated, the more certain he was that he possessed visual evidence for the way stars condense out of diffuse matter, and then cluster under gravitational forces. In other words, by the mid 1810s he had introduced yet another argument for change in the remotest parts of the known universe.

Herschel never lost his interest in the solar system – for instance he destroyed J. H. Schröter's claims to have found colossal mountains on Venus, he studied Mars and found its rotation period and axis, and he made measurements of the newly discovered asteroids Ceres and Pallas. In observing the Sun in 1800, he noticed that the sensation of heat in the image projected by his lens did not correspond with the light at the same place; and in this way he was led to experiment with thermometer and prism, and so to discover infrared radiation. (He found no heating effect at the violet end of the spectrum. Ultraviolet light was detected soon afterwards by R. W. Ritter (1776–1810), from its blackening effect on silver chloride – and this long before photography proper.) In short, Herschel was a man of an irrepressible experimental talent that counterbalanced the brilliance of contemporary theoretical astronomy, especially that in France. His interests took in the whole of astronomy, but his most lasting contributions concerned the Universe of the stars and the nebulae. In that last respect, after his death, many decades passed before his work was substantially extended by others.

The Rise of Astrophysics

BESSEL AND STELLAR PARALLAX

Astronomy in the first part of the nineteenth century benefited greatly from the industrial advances that had diffused from England and France to Germany and Europe generally in the latter part of the previous century. There was now much worthy competition for the London workshops from Germany especially – for example from the firms of J. G. Repsold (Hamburg, 1802) and G. Reichenbach (Munich, 1804). These firms modified very successfully the English designs produced by such firms as Dollond, Ramsden and Cary. A strong market for instruments for surveying and navigation continued to provide a technological basis for the grander requirements of new, and often rich, observatories, and Germany was fortunate inasmuch as the leading mathematician of the age, Carl Friedrich Gauss, took an active interest in the practicalities of instrument-making.

For exact measurement, the transit instrument replaced the mural quadrant in astronomers' affections. With its refracting telescope mounted on an east–west axis so as to align on the meridian, it was fitted with *complete*, finely graduated, circles that were read by micrometer microscopes. The fact that the circle was complete made centring and other pivoting errors easier to assess. Accurate time-keepers ('regulators') were used for the measurement of certain angles by the timing of stars' crossing reticles of illuminated wire (or better, spider's web) in the focal plane of the transit telescope. New theories of instrumental errors were developed, and, as a consequence, angular measurement became very much more precise.

The timing was done electrically after the chronograph was introduced from America in 1844, but the first electric chronographs were not particularly accurate. As Simon Newcomb said of the first, at the Naval Observatory in Washington, 'its only drawbacks were that it would not keep time and had never, so far as I am aware, served any purpose but that of an ornament'.

If one name stands out above all others in the question of precision, it is that of Friedrich Wilhelm Bessel (1784–1846). Bessel was apprenticed at the age of fourteen to a large Bremen merchant company, but privately applied himself to astronomical calculation so thoroughly and with such success that he came to Olbers' notice. It is curious to find that he had been greatly stimulated by reading of Harriot's 1607 observations of what became known as Halley's comet, and with the help of works by Lalande and Olbers, Bessel – at the age of twenty – wrote a masterly account of how he had improved the elements of the comet's orbit. With Olbers' support (Olbers modestly called it his 'greatest service to astronomy') Bessel found employment next in a private observatory near Bremen. There he developed his techniques further, before turning to a reduction of Bradley's observations of over 3000 stars, at Olbers' suggestion. So successful was he in the first stages of this that when Friedrich Wilhelm III of Prussia founded a new observatory at Königsberg, Bessel was appointed director (1810). He remained there for the rest of his life, constantly regretting the climate, but refusing invitations to move elsewhere.

Bradley's published observations were valuable very largely because Bradley himself had determined the errors of his instruments, or observations from which they could be deduced. For his reduction, Bessel needed accurate values for fundamental astronomical constants such as aberration and refraction. He began with a study of astronomical refraction, and this to such effect that in 1811 he was presented with a prize by the Institut de France for his tables. (Even a modest correction for refraction amounts to two or three minutes of arc, which might be several thousand times as great as the probable error claimed for the declination co-ordinate. The greatest difficulty is in allowing for atmospheric temperature and pressure.) His first task completed, he published the results of his reductions in his 'Fundamentals of Astronomy' in 1818.

This brought Bessel's name to the attention of an astronomical world that was grateful for proper motions far better than had ever been available before. Soon after, the Dollond transit instrument and Cary circle with which the observatory was equipped were replaced by a fine new Reichenbach–Ertel meridian circle (1819). This was followed by a Repsold circle in 1841, by which time Bessel's *Tabulae Regiomontae* (Königsberg Tables, 1830) were in use throughout the world's observatories. Bessel's influence was immense twice over, however, for by his example he gave to the meridian circle almost the status of a cult instrument. Every new observatory, large or small, had to be equipped with one. Greenwich fell into line, and there G. B. Airy acquired one on the German model in 1835. From this period onwards, astronomers regularly quoted star declinations to a hundredth of a second of arc, and right ascensions to a thousandth of a second of time. The measurement of star co-ordinates became almost an end in itself – rather like cleanliness and godliness – regardless of potential utility, although of course this was often forthcoming.

In 1844 Bessel himself made an important discovery from his measurements of the places of the stars Procyon and Sirius, which are on Maskelyne's list of fundamental stars: he found that their proper motions are variable. He drew the conclusion that each had an invisible companion, massive enough to make the motion of the brighter component around their centre of mass visible. He was not the first to make such a claim, and in fact a controversy over the same sort of claim had been simmering since John Pond at Greenwich had made it (in 1825 and 1833) for a large number of stars. Bessel successfully dismissed Pond's argument – his stars were far too distant for such an effect to manifest itself – but as for Sirius, the companion was seen telescopically by Alvan Clark in 1862 when testing a new telescope. It was an eighth-magnitude star, and yet with a mass – derived later from its orbit – of about half the mass of Sirius. (Its orbit had been analysed by C. A. F. Peters in 1850, before it was seen.) The dark companion to Procyon was found by Schaeberle at the Lick telescope in 1895. It was even fainter – of the thirteenth magnitude. When eventually the distance of Sirius was found, it was possible to deduce the size of its orbit and hence the masses of the components. They were of the order of one and two solar masses; and when similar calculations were applied to other binaries, it became clear, to the great surprise of those engaged in the work, that while the luminosities of the stars may differ by a factor of millions, the masses were rarely as much as ten times greater or smaller than the Sun's mass.

As a curious postscript, underlining the great value of the raw materials bequeathed by Bradley, we note that more unpublished work of his was used later in the century by G. B. Airy, who supplemented it with his own observations at Greenwich. They were in turn taken over in the 1860s by A. J. G. F. Auwers (1838–1915), who had also worked at Königsberg but was by now in Berlin, and Auwers made yet another reduction of the material, improving on Bessel's. Auwers' three-volume catalogue (1882–1903) set new standards of accuracy, and, with Bessel's catalogue of 75 000 stars of brighter than the ninth magnitude, occupies an important place in the trunk of the family tree of later material.

Bessel's work was continued by a man whose work he long encouraged, F. W. A. Argelander (1799–1875), who similarly followed the ninth-magnitude limit. The combined work was the basis of the *Bonner Durchmusterung*

(Bonn Review), which from its inception in 1859 has continued to serve as a standard reference work, not least in providing a comprehensive system for identifying stars. (The old Greek–Arabic names, even Bayer's Greek-letter labels, apply only to a thousand stars or so.) The positions of over a hundred thousand stars were recorded with reference to certain fundamental stars whose co-ordinates had been measured to a very high degree of accuracy – whether the thirty-six stars Maskelyne had first selected, which Bessel increased to thirty-eight, or over 400 so-called 'Nautical Almanac stars', or Bradley's much longer list.

Although far too extensive even to list here, later fundamental tables were assembled in ways that mark out astronomy as the science of international co-operation *par excellence*. In 1871 the German Astronomical Society organized co-operation between thirteen observatories (later sixteen) in different parts of the world, each of them assigned a zone of declination north or south of the equator covering all the sky between both poles (say 20° to 25°, or −35° to −40°). This was the basis of a continuing programme of proper-motion measurements that improves in quality as time passes. It has to be said, however, that when the results of the collaboration began to come in, cross-checks between co-ordinates assigned to the same stars by different observatories showed up numerous discrepancies and unsuspected errors. Clocks contributed some of the most serious errors, and – true to history – astronomers were closely associated with improvements in methods of time-keeping. As for personal errors in observation, Bessel was perhaps the first to take them seriously as something unavoidable but possibly systematic. Airy might not have dismissed an assistant whose observational records differed from his own had he recognized the inevitability of the 'personal equation', that is, of differing physiological delays in registering any reading.

Bessel's most memorable achievement was to do what astronomers had tried and failed to do for centuries, namely measure the distance of a star by relatively straightforward trigonometrical means. We recall how Bradley had hoped to detect the shift in star positions that results from the movement of the Earth, in particular when it moves from one extreme of its orbit to the diametrically opposite extreme (figure 4.10). As we have seen, he found an annual variation, but one that was due to the aberration of light rather than to simple parallax. Bradley's work made astronomers aware of how very small were the parallactic shifts they should expect – smaller than half a second of arc, say. John Brinkley (1763–1835), the first Astronomer Royal for Ireland, announced the parallaxes of a handful of bright stars in the period 1808–14, all in the region of 2″ of arc, but Pond, at Greenwich, disputed his findings over a period of many years. Attention had already turned to measuring star positions relative to much fainter (so presumably much more distant) stars. Doing this, as we saw, Herschel had also discovered something other than what he had set out to find, that is, pairs of stars physically related. He and others gradually began to realize that there were many faint stars with large proper motions.

Bessel now assumed that a large proper motion was a surer sign of proximity to the Earth than faintness, so he focused attention on the star 61 Cygni, which had the largest proper motion then known, 5.2″ per annum. (Actually this is a double star, with 16″ between the components, so the line joining the pair gives a useful indication of direction in the sky.) He observed 61 Cygni for eighteen months, in relation to two much fainter stars nearby, and by the end of 1838 had found the parallactic shift he sought. He found it to be less than a third of a second of arc (0.314″ ± 0.020″; it is still quoted as at the lower end of this range). In astronomical units (semi-diameters of the Earth's orbit) this is about 657 000.

Within a year or two, other parallaxes were found by F. G. W. Struve at Dorpat (Russia; now Tartu, Estonia) and Thomas Henderson at the Cape Observatory. Henderson's figure for the star alpha Centauri showed that the parallax of that star was well over double the figure for 61 Cygni, that is, its distance seemed less than half as great. His measurements had actually been made in 1832–3, for another purpose, but were not used for a derivation of parallax until after Bessel's announcement.

The instrument used by Bessel to measure the small angular separations of 61 Cygni and the reference stars was the *heliometer*, designed by John Dollond but in this case made by Joseph Fraunhofer. The name of the instrument comes from its use for measuring the Sun's diameter. The principle of the heliometer is simple: the telescope objective is sliced into two semicircular halves, which can slide sideways (by amounts measured on a vernier screw) relatively to one another. An image of the first star is seen through one half-lens, and an image of the second through the other. The components of the split lens are then adjusted until the two images coincide. The movement needed can be translated into an angle.

UNIVERSAL TIME

Throughout history, astronomers have made important contributions to methods of time-keeping, for scientific and civil purposes alike, although influences have always worked in both directions. A great improvement in time standards came in 1879 with the design of the Repsold pendulum. Its designer, J. A. Repsold – one of a dynasty of Hamburg instrument-makers who as it happens also wrote much on the history of astronomy – was chiefly concerned with the needs of astronomers. Attempts to use electricity in horology were first made in the 1830s, and electricity as an aid to automatic recording of observations eventually became universal, but not until 1921, with the invention of the Shortt pendulum, did electricity make for another marked improvement in clock accuracy. This was the outcome of an eleven-year collaboration between the Edinburgh civil engineer W. H. Shortt and the Synchronome Company. Atomic and quartz clocks are instances of later advances made outside astronomy that, like the Repsold and Shortt pendulums, were found immediate application in observatories.

Greenwich and other national observatories were the principal keepers of time standards during the nineteenth century, but here, unlike most other parts of astronomy, it was society at large that called the tune. Local time is decided of course by the Sun's place in relation to the local meridian, with noon as the moment when the Sun (more strictly the mean Sun) crosses the meridian. It follows that since the Earth turns once in twenty-four hours, noon will be an hour later for every fifteen degrees of longitude one moves westwards. Before it became necessary to regulate travel and communication accurately over long distances, the discrepancies between time at distant places was of no great importance in the daily lives of ordinary people. Astronomers were well accustomed to converting between local meridians, especially those of the more famous astronomical tables, and there was no great pressure for a universal standard from that quarter.

This came rather with telegraphy and the railways. Most of the British railway companies agreed to keep Greenwich time from 1847. Pressure was strongest where distances east–west were great, especially in North America. Railways there at first standardized on the main cities along a line, but in 1878–9 Sandford Fleming, of Toronto, gave much publicity to the use of time zones, fifteen degrees wide, starting from some prime meridian. (The scheme seems to have been worked out by Charles F. Dowd of Saratoga Springs.) Even before a Washington conference was called in 1884, many railway managers had adopted the system, with Greenwich as the prime meridian, although Detroit stood out against change until 1900. State laws were gradually tabled giving the system legal standing in banks and in the courts.

The first strong call from the sciences for such a standard system came from a Geodetic Conference in Rome in 1883. The astronomers agreed on the Greenwich system at a specially convened Washington conference in the following year, but there was much wrangling among them as to the choice of base meridian, with strong pressure being exerted in favour of Berlin and Paris. In Greenwich's favour was the world-wide use of the *Nautical Almanac*, produced there for ocean navigation. Sweden, the United States and Canada used the new scheme at once, and most of Europe soon followed suit.

In all this, pressure from the civil authorities was paramount. In Germany, for example, the civil change became general in 1893, with German (Central European) time exactly an hour ahead of Greenwich. The famous Prussian military strategist the Count von Moltke had thrown his weight behind the change when, shortly before his death in 1891, he had observed that the system would make for greater efficiency in the mobilization of the army. What more convincing argument did one need? In France, Paris time continued to be used for another twenty years, although even in Paris itself matters were complicated by a five-minute difference between station time and city time. In due course, however, France became in a sense the guardian of international time, with the establishment of the International Congress on Time (1912) and of a Bureau International de l'Heure in Paris (1919), the chief promoters here being the astronomers Camille Bigourdan and Gustave Ferrié.

OPTICS AND THE NEW ASTRONOMY

As a result of the serious reverses suffered by the English optical industry in the late eighteenth century, here again Germany seized the commercial initiative. Joseph Fraunhofer's name became synonymous with the new and vital scientific activity that resulted. In 1806 he joined the

Munich firm of Utzschneider, Reichenbach and Liebherr. At first he worked with Guinand, between 1809 and 1813, improving the quality of the different mixes of optical glass. Most previous makers had used trial and error with whatever discs had been cast, when grinding and polishing the components of achromats, but in 1814 Fraunhofer proceeded more scientifically, determining the optical properties of glasses in advance, using the bright yellow lines in a flame spectrum as his source of light. He made accurate measurements of the angles of light entering and leaving a prism by adapting an ordinary theodolite to create what is now a standard piece of laboratory apparatus, the spectrometer.

In the course of comparing the flame spectrum with that produced in a prism by the Sun's light, Fraunhofer noticed that the latter was crossed by innumerable dark lines. (Some of these had been recorded by William Hyde Wollaston in 1802.) He later realized that some of the lines had counterparts in the spectra obtained from flames in the laboratory, and that while there were similarities between the spectra of the Sun and bright stars, there were many subtle differences between them. These observations were the basis of research by a number of physicists over the following decades, until in 1859 a revolutionary interpretation of spectra was offered by Kirchhoff and Bunsen. This interpretation marks nothing less than the beginning of a new type of astrophysics, astronomical spectroscopy.

Robert Wilhelm Bunsen (1811–99) was one of the century's leading experimental chemists, while Gustav Robert Kirchhoff (1824–87) was a distinguished theoretical physicist who followed Bunsen to Heidelberg in 1854. John Herschel and W. H. Fox Talbot, in 1823 and 1826 respectively, had advocated chemical analysis from spectral observation, and by the 1850s it was widely known, although rarely used. Bunsen was experimenting with such methods for analysing salts, that is, by the coloration of flames in which they are burned – just as when sodium salts were used to produce the bright yellow flame that Fraunhofer had associated with the 'D-lines' in the Sun's spectrum. Kirchhoff introduced an element of precision into Bunsen's work, by measuring the colours on a spectrometer. By 1860 they had shown that *each metal has a characteristic line spectrum*. This led Bunsen to analyse alkaline compounds, and then to the discovery of two new elements (caesium and rubidium).

A year earlier, Kirchhoff had been surprised to discover that if weak enough sunlight was passed through a sodium flame before entering the spectrometer, the dark spectral D-lines ('Fraunhofer lines') were replaced by bright lines from the flame, but that with bright sunlight the flame could make the dark lines relatively darker. His interpretation of the situation was that a substance capable of emitting a spectral line – that is, light of a particular wavelength – is also able to *absorb* light of the same wavelength. Ten years earlier, William Stokes had come to a similar conclusion concerning the D-lines in the laboratory flame and in the Fraunhofer spectrum, and had even given a theoretical interpretation in terms of atomic resonance, but his ideas remained dormant, and it is odd that Kirchhoff later wished to have nothing to do with this fundamental explanation of the phenomenon.

Kirchhoff's findings led to the dramatic conclusion that not only was sodium present in the Sun, but many other identifiable elements too, as judged by the appropriate Fraunhofer lines. What many had held to be a perfect example of unattainable knowledge, that is, the chemistry of celestial bodies, was at last seen to be within reach.

Within a few weeks, Kirchhoff had developed a quantitative theory of the emission and absorption of light, not as fundamental as Stokes', but in many ways more useful for the development of physics. His conclusion was that the ratio of the power of a surface to absorb radiation to its power to emit radiation is the same for all bodies at a given temperature. (Within a given interval of wavelength, the same is true.) This law of radiation became one of the principal pillars of the thermodynamcs of radiation. What it gave to astrophysics was an opportunity to add temperature to the list of measurable celestial parameters. In yet another respect, therefore, astronomy was revolutionized by a discovery that had little to do with astronomy as such.

THE ASTEROIDS, NEPTUNE AND PLUTO

There was still room for astronomical discovery of the old familiar sort, of course. It seems that the whole world is excited by the idea of a new planetary neighbour. The two that have been found since Herschel's discovery of Uranus are Neptune, found in 1846, and Pluto, found in 1930. There are so many points in common in the manner of their discovery that it is instructive to take them together. They were found only after predictions had been made on the basis of Newtonian planetary dynamics. Before they were found, however, another sort of planetary neighbour had been found, and a very different line of reasoning had been applied to it.

The Titius–Bode law of planetary distances (see chapter 12), which to most astronomers seemed to have been pulled out of thin air, without any real justification, had received support of a sort by Herschel's discovery of Uranus. The radius of the orbit predicted by the law agreed with the observed radius, within two or three parts in a hundred. This reinforced a belief in the existence of a planet between Mars and Jupiter – suspected even by Kepler on different grounds – which corresponded to a value for n of 3 in the law. Led by von Zach, a group of German astronomers even went as far as to found a club in 1800, 'the Lilienthal detectives', to search for the missing planet. The gap is in fact occupied by the asteroids, the 'minor planets', and the first of these was found in 1801 by Giuseppe Piazzi (1746–1826) of Palermo, Sicily. Piazzi was not searching for a missing planet when he found what he at first took to be a faint

star, but one that proved to be moving, first with retrograde
and then with direct motion. He confided in his friend Bar-
naba Oriani in Milan, saying that he had found a new
planet, but when he wrote to Lalande in Paris and Bode in
Berlin he was more cautious in saying that he believed he
had found a *comet*. The Lilienthal detectives were nevertheless convinced that here was the missing planet. Whatever
it was, Piazzi's object was for a time lost in the glare of the
Sun. Soon the young Gauss was able to calculate a new
orbit, and on the last night of 1801 the object was found
again, where Gauss had predicted. Piazzi chose the name
'Ceres Ferdinandea', after the goddess Ceres and the ruler
of Naples and Sicily, Ferdinand IV. When Herschel measured its size, however, he and the others were in for a
surprise: he estimated a diameter equivalent to only 259
kilometres.

Another surprise came in March 1802, when Olbers, one
of the Lilienthal group, found another object. This he
named Pallas, and again Gauss found for it an orbit placing
it between Mars and Jupiter. Herschel estimated its size as
about two-thirds that of Ceres, and noted that the Titius–
Bode law would be overturned if both were to be regarded
as planets. Olbers, on the other hand, pointed out that one
could save the law if the two asteroids (Herschel's word)
were remnants of a single object that had disintegrated in
the past. This suggestion led to a minor industry, later in
the century, when yet more asteroids had been found (Juno
1804, Vesta 1807, etc.), that of calculating the time when
all of them would last have been in one place. Before this
phase was entered, however, Gauss, prompted by his researches into the asteroids' path, had published his 'Theory
of the Motion of Celestial Bodies Orbiting the Sun in Conic
Sections' (1809). This was the most exquisite mathematical
analysis ever written on the general problem of determining
an orbit from any number of observations greater than
three. For this purpose Gauss developed the so-called
'method of least squares', which is of use today in almost

any science that requires a matching of theory and measured observation. In a sense, Gauss' theoretical work leaves the discovery of the asteroids quite in the shade.

Even before the planetary character of Herschel's new discovery (Uranus) was recognized, attempts had been made to give the parameters of its orbit. Old sightings of it helped here: Bode found that Tobias Mayer had seen it – without appreciating that it was a planet – in 1756, and Flamsteed in 1690. Almost a score of similar old 'star' sightings that were really of Uranus were found over the next forty years. These sightings supplemented modern observations, and allowed half a dozen astronomers to specify its orbit, and to draw up tables of its motion (from 1788 onwards), but as the years went by it became clear that Uranus was a most intractable planet, for even the best of tables – Delambre's of 1790 long held that position – ran quickly into difficulties. After a time, even so, interest in the planet's orbit seems to have waned. The Napoleonic wars did not entirely close down scientific communication, as we have seen, but on the continent of Europe, at least, they were not conducive to tranquil research. Piazzi's patron was deposed, and Gauss' died in the war soon after the battle of Jena.

After Napoleon's defeat, Uranus came once more under attack. New tables were drawn up in 1821 by Alexis Bouvard (1767–1843), a farmer's boy from the Alps by origin, but one with a genius for computation, who had become an invaluable assistant to Laplace. Within eleven years we find the future British Astronomer Royal George Biddell Airy (1801–92) complaining that they are nearly half a minute of arc in error. Various possible explanations presented themselves. Had Uranus been struck by a comet since its discovery? Did it have an invisible but massive satellite? Was there an interplanetary fluid, impeding planetary motions? Did Newton's law of gravitation cease to operate at great distances? Or was there, perhaps, an invisible planet that was exerting a perturbing effect on Uranus?

Clairaut had long before put forward this last proposal to explain oddities in the movement of Halley's comet, and now an English amateur, T. J. Hussey, made the same suggestion for Uranus. Niccolo Cacciatore (1780–1841) in Palermo thought he had seen a planet beyond Uranus in 1835; Louis François Wartmann (1793–1864) in Geneva later said he had seen one in 1831. Both claims were shown to be misguided, but gradually the belief spread, that some perturbing planet existed. In 1842 the Göttingen Academy of Sciences offered a prize for the solution of the Uranus problem. Under these circumstances, the fact that two men came up with a solution quite independently of one another – often presented as an example of some sort of mystical 'synchronicity' – is no real cause for surprise.

Urbain Jean Joseph Leverrier (1811–77), the son of a local government official in Normandy, had distinguished himself in mathematics as a student at the *Ecole Polytechnique*, and had further studied chemistry under Gay-Lussac, when he turned to celestial mechanics. He studied the perturbation of cometary orbits, extended Lagrange's general theory of perturbation, and in 1845 revised Bouvard's theory of Uranus. His successes were such that he was seen in some quarters as a successor to Lagrange and Laplace. On 18 September 1846 his calculations were complete: he wrote to J. G. Galle at the Berlin Observatory asking him to search at a specified place for a new planet. On 23 September the planet was found, within a degree of the calculated position. (The name eventually given to it, Neptune, had already been proposed for Uranus, as we saw earlier.) There was much jubilation, not least in France, but this was quickly soured when it was learned that Leverrier's calculation had been anticipated.

John Couch Adams (1819–92) was the son of a Cornish tenant farmer, whose struggle to send him to Cambridge was justified when he headed the class lists in mathematics in 1843. On graduating, he was made a fellow of his college, St Johns, and at the beginning of the long vacation he set to work on the Uranus problem. By October he had an approximate solution, and in February 1844 he applied through James Challis to the Astronomer Royal Airy for more exact data on Uranus. Using Airy's figures he calculated values for mass, heliocentric longitude and elements of the elliptical orbit of the presumed planet. His findings were in close agreement with those of Leverrier later. These results he gave to Challis in September 1845. After twice trying to see Airy in person he left a copy at Greenwich on 21 October. Airy wrote to Challis with a misconceived criticism some weeks later, getting Adams' name wrong, and clearly misconstruing his status – he supposed that Adams was an older clergyman. He pressed Challis to begin a search for the planet in July 1846, but irritated Challis by his language. Leverrier's later investigation led to the discovery of Neptune, as we have seen, in September of that year.

It was John Herschel who first drew the public's attention to Adams' achievement, in a letter to the London *Athenaeum* in October 1843, and this gave rise to a bitter controversy, partly over the facts of priority, and partly concerning the behaviour of Airy and Challis in the affair. Challis, who had followed Airy as Plumian professor of astronomy at Cambridge when Airy became Astronomer Royal in 1836, announced that he had been searching for the planet since the end of July, and had recorded the necessary star positions, one of them corresponding to the planet, but had not gone through the necessary process of elimination – finding which star was moving – since he was busy with comets and did not wholeheartedly believe in Adams' prediction. Some took this to amount to a claim for priority, and a claim to have 'found' the planet on 12 August, rather than as a confession of negligence. At all events, Leverrier was showered with honours and Adams received very few. In 1848 the Royal Society gave him its highest honour, the Copley medal. It had given this same honour immediately after the discovery to Leverrier alone.

The two great peacemakers in this unhappy episode were John Herschel in England and Jean Baptiste Biot (1774–1862) in France. Chauvinistic controversy centred on Adams did not end here, however. In the 1850s he discovered a substantial error in Laplace's treatment of the secular acceleration of the Moon's motion. This was decided in Adams' favour, but not until 1861, after yet more French recriminations. As a result, the figure for the acceleration was reduced by about half, from 10.58″ to 5.70″ . The year 1861 saw Adams succeed Challis as director of the Cambridge Observatory.

For many years the story of Neptune was an object lesson in armchair discovery: one might be too poor to afford a great telescope, but not to learn mathematics. The discovery of yet another planet, beyond Neptune, taught a somewhat different moral. Although it had to wait for the year 1930, the search was certainly encouraged by the earlier efforts of a rich New England amateur, Percival Lowell (1855–1916). Lowell was convinced that he had a key to the mystery of a missing planet just as surely as the Lilienthal group had: he believed he knew of certain 'resonances' in the motions of the planets. Lowell had set up an observatory in Flagstaff, Arizona, where the air – or so he had been told – was the most stable in North America. He had earned much publicity for his observations of the surface of Mars, and for a book entitled *Mars as the Abode of Life* (1908). His idea that Mars experienced seasons, and that he could detect changes in its vegetation, held sway in some quarters until the Mariner space missions of the 1950s. His idea about resonances fell out of favour much sooner, but not before it had achieved remarkable effect.

Lowell accepted the hypothesis of Chamberlin and Moulton, that the planets were formed out of material pulled from the Sun after the near-collision of a star with the Sun. (We return to this theme in chapter 16.) His idea was that after the formation of one planet, the next would tend to be formed at a place where its period of revolution was in a simple ratio to that of the former – say five to two, in the case of Saturn and Jupiter. Uranus follows Saturn with a period three times as long; and Neptune follows Uranus with twice its period. (Lowell juggled the numbers slightly by talking of perturbations in a rather vague way.) Should there not be a 'planet X' beyond Neptune, with a period of revolution fitting such a scheme of simple ratios?

He tried to follow the example set by Adams and Leverrier, now adding Neptune to the list of perturbing planets, but he was not their mathematical equal. With the help of C. O. Lampland, who took photographs of the appropriate regions, he searched unsuccessfully for planet X from 1905 to the end of his life. The search continued as an act of piety at his observatory after his death, with the advantage of increasingly refined calculations, and it was eventually found by Clyde William Tombaugh in 1930, using a telescope especially built for the purpose, with a wide field of view. It was through its *movement*, not its appearance, that the planet, called Pluto, was detected. Pluto's symbol is formed from the letters PL, which are of course Lowell's initials.

For all the theoretical approach to the subject, there is good reason for regarding this discovery as mathematically accidental, and simply a result of systematic searching. Pluto's mass is now known to be simply too small to produce the supposed perturbations of Neptune and Uranus. Neptune's mass is in fact ten thousand times as great as Pluto's, which underlines the futility of Lowell's attempt to emulate the procedures of Adams and Leverrier; but he was not alone in this. William Henry Pickering (1858–1938), younger brother of the more famous Harvard astronomer Edward Charles Pickering (1846–1919), was Lowell's friend and adviser, and he too had been searching with photographic help for a planet beyond Neptune from 1907 onwards, using calculations hardly more meaningful than those from Flagstaff. After Tombaugh's discovery, however, the planet was found on plates taken for Pickering in 1919.

The charge that the discovery was in some sense accidental provides a strange echo of a similar charge brought against Adams and Leverrier by Benjamin Peirce and S. C. Walker in America. After Neptune had been under observation for some months it became clear that the orbit was very different from what had been predicted, starting from the assumption of a distance fitting the Titius–Bode law. It was pointed out that *several different solutions* fitted the known data, and that in the period when Neptune was found, the 'real' planet by chance just happened to coincide roughly in position with the calculated planet. Many European astronomers, not least Leverrier, were at pains to rebut the charge, while others thought that it simply proved the unremitting honesty and openness of American science. It has much to recommend it, but was presented in a misleading way. The Americans seem to have implied that they somehow 'knew the real planet' because they knew better values for Neptune's parameters – such as distance – than their predecessors. At the foot of this slippery slope is the conclusion that astronomers can never know the real Neptune or, for that matter, specify any other 'real' celestial body.

Whether there are planets beyond Pluto remains to be seen, but the difficulties of locating even Pluto, however 'accidental' we may judge the discovery, should not be underestimated. Its angular size as seen from the Earth is less than a third of an arc second, well below the limit at which any surface detail would be visible. And there are twenty million stars in the sky that appear as bright as Pluto.

REFRACTORS AND REFLECTORS

Bessel's parallax measurements confirmed professional astronomers in their preference for the achromatic refracting telescope over the large and unwieldy reflector, as a tool for precision measurement. A professional whose astronomical reputation came to equal Bessel's was

Friedrich Struve, and Struve, astronomer to the Tsar of Russia, was able to patronize the best makers of refractors. In 1833 Struve left Dorpat to found a new and glittering imperial observatory at Pulkovo, near St Petersburg.

Since there were destined to be no fewer than six astronomers in the Struve family, several of them distinguished, it will be as well to note their relationship here, for they are occasionally confused. Friedrich Georg Wilhelm (1793–1864) was the father of Otto Wilhelm (1819–1905) – not to mention seventeen other children, by two wives. Otto's sons included Karl Hermann (1854–1920) and Gustav Wilhelm Ludwig (1858–1920). Each of the last had an astronomer son, Georg Otto Hermann (1886–1933) and Otto (1897–1963) respectively. Friedrich Struve had been sent from Germany to Russia to avoid conscription. His son followed him at the Pulkovo Observatory, but on retirement moved to Germany. In the third generation, Gustav took various astronomical posts in Russia, while Karl Hermann remained there until 1895, when he accepted the directorship of the Königsberg Observatory. Karl's son's career was consequently in Germany – and of lesser distinction, although he made important studies of Saturn. Otto, youngest of the six, suffered great hardships when his studies were interrupted by the civil war in Russia, for he joined General Denikin's army, only to be driven out of his homeland by the Red Army. In 1921, by way of Turkey, he managed to reach America and the Yerkes Observatory (Wisconsin), of which he was eventually to become director.

When Friedrich Struve, founder of this dynasty of astronomers, left Dorpat for St Petersburg, he took with him the splendid Dorpat refractor made by Fraunhofer, and the Tsar bought for him numerous other fine instruments, by such makers as Ertel, Repsold, Merz, Troughton, Dent, Plössl and Pistor. Within a decade or so, he had at Pulkovo what was for a time the best-equipped observatory in the world. He obtained the biggest refractor then built (a 38-cm achromat by Merz and Mahler, in a mounting by Repsold). Later, when his son learned of the excellence of the workmanship of the American firm of Alvan Clark and Sons, a 30-inch object lens was commissioned from them, and was for a brief time the largest in the world. It was outstripped by the Clark glasses at Lick (36 inch; 91 cm) and Yerkes (40 inch; 102 cm); but then, this last instrument was destined to fall eventually within the domain of a Struve.

Apart from fulfilling the duties of a state astronomer whose assistance was sorely needed for the charting of a vast empire, Friedrich Struve set himself the task of continuing William Herschel's study of double stars, and by the time he published his 1827 catalogue, he had recorded the positions of 122 000 stars, of which 3112 were double. By 1847, he and his staff had covered the northern sky, and a catalogue of 1852, making comparisons of the positions of 2874 stars with those recorded by his predecessors – from Bradley to Groombridge – is a monument to nineteenth-century thoroughness, comparable with Bessel's. Struve looked farther afield, however. He wanted to solve the problems Herschel had set. Are the stars distributed in a recognizable pattern? Is there a meaningful relationship between the distances of the stars and their magnitudes? He came to the conclusion that the loss of brightness of the stars examined by Herschel, and now by the staff at Pulkovo, was only partly due to the fact that light falls off following an inverse-square law. It was partly due, he thought, to the absorption of light in interstellar space. The actual figures he quoted are reasonably close to modern figures in the neighbourhood of the Milky Way, the object of his study.

Despite the preference shown by professional observatories for the refracting telescope, especially in precision measurement, the reflector had all the advantages where light-gathering power was needed. Advances here came slowly, however, and would never have come at all had everyone accepted the consensus among most telescope-makers. According to William Parsons, they seemed to think that 'since Fraunhofer's discoveries, the refractor has entirely superseded the reflector, and that all attempts to improve the latter instrument are useless'. When in 1833 John Herschel took to the Cape his father's favourite reflector, that of 47-cm aperture (quoting its length, this was usually known as the '20 foot' telescope), there was no better reflector in the world, and yet it was half a century old.

The revival in the fortunes of the reflector came about in the 1840s as the result of the work of three amateurs: William Lassell (1799–1880), an English brewer, who mounted his reflectors (of 23-cm and 61-cm aperture) in equatorial mountings of the Fraunhofer type; the Scot James Nasmyth (1808–90), one of the greatest engineers of the century, who built three fine instruments (25 cm, 33-cm and 51 cm) with excellent mechanical properties and a new type of viewing arrangement; and the Irish land-owner William Parsons (1800–67), third Earl of Rosse, who achieved his ambition to build a mirror larger than any of Herschel's.

Lassell designed and used machines for grinding and polishing his mirrors of speculum. With his larger instrument (1846) he discovered a satellite (Triton) to the newly discovered planet Neptune. Two years later, simultaneously with W. C. Bond of Harvard, he found the eighth of Saturn's satellites. He took the instrument to Malta, where he charted over 600 new nebulae, hoping to make further satellite discoveries, but he made none.

Nasmyth chose to concentrate his attentions on the Moon and Sun. His drawings of the Moon were outstanding, and he was the first to appreciate – and to cause some controversy by announcing – that the surface of the Sun is patterned, as he said 'like willow leaves'. Nasmyth's main contribution to astronomy, however, was a mechanical one. After reflection at the secondary mirror of a telescope of Cassegrain type, rather than pass the light through the middle of the main mirror he reflected it at right angles to

the tube, through the middle of the axle bearing the telescope. This ingenious arrangement meant some slight loss
of light, with bright objects not particularly important, but
made it possible for the observer to sit or stand at the same
height, merely moving round with the telescope. The 'Nasmyth focus' is still found on most large telescopes, and if
we are to judge by earlier accounts of mishaps with large
reflectors, must have saved many broken limbs.

After graduating at Dublin and Oxford, William Parsons
(Lord Oxmantown, until after his father's death he took
the title Earl of Rosse) became a member of the House of
Commons. Renowned as that rare thing, an honest man
in Irish politics, his obsessions were rather with making
telescope mirrors of the finest quality. He worked with
speculum metal, constantly cooled while being ground and
polished on a steam-driven machine. The metal is brittle,
and Herschel had included more copper to make it less
likely to shatter, but in doing so had lowered its reflectance.
Rosse tried mirrors cast in segments, these being held on a
brass back-plate with the same coefficient of expansion as
the speculum metal. The fine lines of division in the
resulting mosaic were very troublesome, and the image was
at first of low quality, but after repeated experiments he
found another solution. He saw that the problem in casting
was that of controlling loss of heat. He experimented with
different sorts of mould, partly of sand and partly of metal,
and allowed his 90-cm mirror no less than a fortnight to
cool. The mirror was of excellent quality. The weather at
his home, Birr Castle, was not, and yet in rare intervals of
clear sky he saw much new detail in clusters and nebulae.
In the nebula M57 in the constellation Lyra, a disc-like
object, he found at its centre a very faint blue star. This
was another 'planetary nebula' of the sort that in 1790 had
persuaded Herschel of the existence of true nebulosity.

Rosse was so encouraged by his discoveries that with his
estate workers and others he now set to work on a mirror
of 183-cm aperture and nearly four tonnes weight. After

five attempts and four disasters in five years, the mirror of 'the Leviathan of Parsonstown' was successfully cast, ground and polished. The suspension of the massive telescope tube was extremly difficult: it was elevated in the gap between two bearing walls about eighteen metres high in the meridian, so that stars could only be followed for an hour or so at most. To prevent the vast mirror from flexing under its own weight, this was supported on felt-covered cast-iron platforms.

The Leviathan, now by far the biggest telescope in the world, was brought into use in February 1845. By April, Rosse had made his most momentous discovery, that of the spiral structure of the nebula M51, of which he prepared an excellent drawing. Ireland was in the throes of a potato famine, however, and the telescope had to be neglected for some months. (Rosse gave the major part of his rents to alleviate the poverty of his tenants at this time.) When observing resumed, he had a clear programme in front of him, namely to investigate the shapes of the nebulae. He was assisted in this by his eldest son, who continued the programme after his death, by the Rev. Thomas Romney Robinson, and by various other friends and hired observers.

He was convinced that the spiral structure of nebulae must hold important dynamical information – he had graduated in mathematics, but it would be many decades before the necessary information could be found, and applied to the problem. The Birr Castle astronomers thus engaged on a systematic search for spiral nebulae, and they found many. Several were of a type since known as Seyfert galaxies. (These have highly condensed centres, and it has recently been thought that they are powered by black holes.) The detail in the drawings produced at Birr often continued to be of very great value, even after the rise of astronomical photography. The great reflector continued in use in the same place regularly until 1878, when the Danish-born J. L. E. Dreyer resigned as Birr astronomer to accept a post at the National Observatory at Dunsink.

Dreyer, well known as a historian of astronomy, was also later to produce the famous *New General Catalogue of Nebulae and Clusters of Stars.* Its 'NGC numbers' are still often used to identify such objects. (Herschel's planetary nebula, for example, is our NGC 1514.)

Among other things, Rosse's work helped to change fashions in telescope-making, in favour of large reflectors. The great refractors of course continued to produce their discoveries – as when Asaph Hall in 1877 used the Washington instrument with a Clark lens to discover satellites around Mars. In addition to examples of refractors mentioned earlier, in the 1880s the Dublin firm of Grubb made a 63-cm lens for Cambridge, while the brothers Paul and Prosper Henry in Paris made refractors of 76 cm for Nice and 83 cm for Meudon. Ever better lenses were produced – for example in France at the workshop of Mantois, using glass from the factory of St Gobain, and in Germany at the firm of Carl Zeiss at Jena, founded by a famous theoretician in optics, Ernst Abbe. The future was with reflectors, however, and this simply because with lenses over about a metre in diameter, the glass at the centre is so thick that the absorption of light is intolerable. Physical deformation under the glass's weight is also a problem.

A new medium was needed to replace the unworkable speculum metal. Glass mirrors were not new, and glass grinding and polishing was of course a highly developed art, but methods of silvering the glass were crude. In the 1850s, however, C. A. Steinheil of Munich and J. B. L. Foucault of Paris made use of a technique of depositing a thin and uniform layer of silver on fairly small glass telescope mirrors. (The technique had been shown at the Great Exhibition in London in 1851, but was seemingly reinvented by the chemist Justus Liebig.) Foucault introduced another innovation, the 'Foucault test' for the image quality produced by the mirror. Lassell's second (Malta) telescope followed their example. The last large mirror to be cast in speculum was a 120-cm example for Melbourne (1870), built by Grubb with Royal Society advice, and this turned out to be a failure, for reasons both optical and mechanical. The mirror was shipped with a protective coating of shellac. In removing it, the surface was ruined. The director of the observatory had to try to learn the technique of refiguring one of the biggest telescopes in the world in a few months. He failed, of course; but then, Grubb had been doing the work for over thirty years.

The age of the amateur builder of the largest instruments was coming to an end. The firm of Grubb eventually redeemed its reputation with a 51-cm glass reflector made for Isaac Roberts in 1885, and with it Roberts scored a notable hit, with a photograph of the Andromeda nebula, clearly showing its spiral structure. By the end of the century, the tide had well and truly turned. George Hale's experience at Yerkes had taught him the limitations of refractors. He was fortunate to have the services of an optical worker of great expertise, George Ritchey. And when a new telescope was begun at Mount Wilson, with the help of a grant from the Carnegie Institution, there was no doubt that it had to be a reflector. Twice in the course of building, Hale had a change of heart, and plans for a 1.5-metre mirror gave place to the reality of a 2.5-metre mirror – the 100-inch Hooker telescope that produced the countless spectacular astronomical photographs so widely used in the first part of the twentieth century. Completed in 1918, and weighing about a hundred tonnes, the open-girder 'tube' is held in a yoke polar mounting of great rigidity, although it does not allow the tube to point to stars near the pole. At last, an instrument was available that was worthy to continue the work on the forms and distribution of the nebulae begun so long before by Herschel and continued by Rosse.

THE PARADOX OF THE DARK SKY

One aspect of the more general problem of the distribution of matter in space takes us back to Friedrich Struve's motives for studying interstellar absorption, and further back still to the paradox noted by Halley and others. When it was believed that the stars were finite in number – whether or not they were situated on a sphere made little difference – there was no mystery about the general darkness of the night sky, but as Kepler wrote in his *Conversation with the Starry Messenger* (1610), 'in an infinite Universe the stars would fill the heavens as they are seen by us'. Intuitively considered, no matter what the direction in which one looks, if the stars are not specially arranged for the particular viewpoint (for instance so as to leave straight avenues of empty space), then in an infinite Universe with uniformly distributed stars, sooner or later the line of sight will encounter the surface of a star. The sky should be filled with light. This very simple way of presenting the case does not seem to have been appreciated before Olbers. We have already discussed Halley's more intricate route to a similar conclusion, and his possible sources. Halley published two papers (1720) on the problem of the dark night sky, and his solution to the paradox was that light from distant stars is not indefinitely divisible, and that at great distances it diminishes faster than the inverse-square law suggests, so that distant stars are simply too faint to be detected by the human eye.

A generation later, Cheseaux again asked why the sky is not filled with light of the average surface brightness of stars (1744). He now explained the dark night sky in terms of interstellar absorption. This effect might seem to have the same consequences, but as John Herschel was to point out in 1848, absorption alone is no answer, for the interstellar absorbing substance will heat up until it re-emits as much energy as it receives.

In 1823 the Bremen physician and astronomer Heinrich Olbers repeated the explanation in terms of interstellar absorption, and since most of those who discussed the problem in the first half of the twentieth century knew of it through him, it became known as the 'Olbers paradox'.

Olbers was an acquaintance of Struve, and this was the context of Struve's study of interstellar absorption. Absorption was not the only way of resolving the paradox, however. In 1861, J. H. Mädler favoured another explanation, in terms of a finite age of the Universe. The idea was that light from distant stars has simply not had time to reach us, assuming that the time of the light's travel is less than the age of the Universe. Since there was no firm and independent astronomical knowledge of many of the distances, or of the processes of creation, this was no more than conjecture, but it foreshadowed future explanations in a curious way, as we shall see. In 1901 William Thomson, Lord Kelvin, took a similar route to the solving of the puzzle. He also considered that the individual stars might have a finite lifetime. His solution was carefully worked out, and can be extended to cover an expanding Universe (of finite age). Not until later in the century did this problem of the dark night sky occupy the centre of the astronomical stage. Before it could do so, astronomers had to learn to take seriously the physics of the Universe as a whole. This they were largely forced to do by scientific developments that were taking place outside astronomy as it was then generally conceived. We shall return to this subject in chapter 17.

THE PHOTOGRAPHIC REVOLUTION

The introduction of photography into astronomy in the nineteenth century is in many ways reminiscent of the introduction of the telescope in the seventeenth, for both brought new phenomena to light. The first photographs of the Sun and Moon showed almost instantly a degree of detail that would have required many hours of careful drafting by hand. As time progressed, and materials became more sensitive, ever-fainter objects were recorded on photographic plates, and by the end of the century numerous discoveries had been made that could not have been made without photographic aid. (Even before workable photographic processes were available, it was known that certain materials were sensitive to light invisible to the eye.)

The history of photography was far more protracted than that of the telescope. It has been common knowledge since ancient times that substances may be bleached, and others darkened, by sunlight, and many chemists of the seventeenth and eighteenth centuries – among them Joseph Priestley (1772), Carl Wilhelm Scheele (1777) and Jean Senebier (1782) – investigated the chemical action of light, including light of different colours. In 1802, Humphry Davy published a report of the experiments of Thomas Wedgwood on methods of copying glass paintings or drawings on paper or leather treated by silver nitrate or silver chloride. He had no means of making them permanent. The first real progress came only after decades of experimenting by Joseph Nicéphore Nièpce. The camera obscura, or darkened chamber with a pinhole or lens through which an image was cast on a screen within it, was in widespread use for display purposes, and for assistance in sketching. For long Nièpce struggled to produce a camera-made image that would serve as a plate that would be of use to a printer. His first moderately satisfactory images were produced in 1816, although of course they did not satisfy his original need.

In 1829, still experimenting, he went into formal partnership with a scenic painter and showman of similar interests, Louis Jacques Mandé Daguerre. There is scope for endless controversy as to the precise nature of their numerous processes, who was responsible for them, and whose was the priority of invention; and matters are complicated by the work of William Henry Fox Talbot, who by 1835 was quite independently producing tiny negatives in England. Talbot's friend John Herschel took a strong interest, and among other things suggested to him the use of hypo as a fixing agent. (Herschel has to his credit the invention of the photographic terms 'positive' and 'negative', 'snapshot', and even, in 1839, the word 'photograph' itself.) Through Daguerre's adoption of it, this became the classic fixing agent. This entire period was one of a steady rise in sensitivity, both human and photographic. Talbot was open with his inventions, but patented them and expected royalties, although often to have his claims contested. It was Daguerre, however – never particularly sensitive to the rights of his rivals – who provided the enterprise that brought photography into the consciousness of the whole world, not least because he published a clear handbook of instruction that went into numerous languages and editions, and that advertised his apparatus.

Just as astronomy needed maximum apertures for light-grasp, so in photography, exposures could be shortened by increasing aperture. (We should here speak strictly of *relative* aperture. All photographers will know that more light is admitted to the film by decreasing 'f/number', the ratio of focal length to aperture.) In 1840, hoping to maximize the light from the subject, and following astronomical practice, Alexander Wolcott opened the world's first portrait studio in New York using a camera with a large concave *mirror*. A more important turning point came in that same year, when the Viennese mathematician Josef Petzval designed a double lens with an unusually large aperture of f/3.6. Earlier exposures had usually been of the order of minutes. To give an idea of what the new lens made possible: by 1841 a Viennese photographer was able to take a military parade in bright sunlight with an exposure of one second.

A photograph of the Sun itself clearly needed a shorter exposure than this, and indeed, Foucault and Fizeau in Paris in 1845 found it very difficult to make exposures *short* enough. A typical refractor might have a focal ratio of about f/8. The faint images of the stars were for long out of the question, but a few crude 'daguerrotypes' of the Moon were obtained by J. W. Draper of New York in 1840. In July 1850, W. C. Bond (1789–1859) at the Harvard College

Observatory used the 38-cm refractor there – co-equal in size with the world's largest – to obtain a very much finer daguerrotype of the Moon, a photograph that obtained much publicity for both sciences.

Astronomical photography was thus launched, but the first photographs of substantial value were those taken by Warren de la Rue (1815–89), a Guernsey-born paper manufacturer in England. He began in 1853 with the Moon, using Archer's collodion process (1851). The need to follow celestial objects during the necessarily long exposure time meant following the object with a telescope accurately driven mechanically, rather than by hand, and this mechanism de la Rue did much to perfect. By 1858, commissioned by the Royal Society, he began a series of solar photographs at Kew Observatory that he continued for fourteen years. Since the mid 1870s, photographs of the solar surface have been taken virtually every day from some part of the world.

De la Rue's photographs were not confined to simple snapshots of the Sun. Stereoscopic photographs of ordinary scenes were then much in vogue: one looked through a viewer at two photographs, each seen by one eye only, and the scene took on a three-dimensional appearance. The photographs had been taken simultaneously by two cameras spaced more or less at the spacing of the eyes. De la Rue applied a method by which he could obtain a three-dimensional image of the Sun's surface, something that the eyes cannot appreciate. Rather than photograph the Sun from two widely separated stations, he simply took one shot, and waited for the Sun to turn on its axis before taking another, so providing the view that would have been seen from a far-distant viewpoint. He found 26 minutes a suitable interval.

By this means de la Rue found that the bright faculae (Latin, 'little torches') are high in the Sun's photosphere (visible layer), and that the dark part of a sunspot seems lower than its surrounding penumbra, above which the faculae seemed to float. This at once disposed of a host of fanciful ideas about the Sun, not only those of the Derham type, which made out sunspots to be volcanic, but also Lalande's notion that spots are rocky islands in a luminous sea, with the penumbrae, so to speak, the sandbanks on the shore.

De la Rue cannot be said to have touched William Herschel's strange ideas about the Sun as a 'lucid planet', probably inhabited by people protected by a thick cloud canopy from its intensely illuminated upper atmosphere. (It was Schröter who gave this the name 'photosphere'.) This strange suggestion died a natural death. It should certainly not be allowed to detract from the many properties of the solar surface that Herschel was the first to describe.

De la Rue's was not the only approach to the problem of assessing three-dimensionality, for the profiles of the spots can be studied as the Sun rotates and they reach the visible edge. Alexander Wilson of Glasgow had studied a very large spot in this way in 1769, and found perspective effects that led him to study the problem intently; and many others did the same. By 1866 a completely new kind of evidence was called upon, when – as we shall see later – Norman Lockyer applied the spectroscope to the study of sunspots.

There were two curiously isolated achievements in solar spectroscopy in the 1840s. First John Herschel devised a way of noting the spectrum in the infrared, by moistening black paper in alcohol, and observing the 'dark' lines by virtue of the fact that they dried last. Others found his published results unrepeatable, but they were seen to be real enough in the 1880s, when the wavelengths deduced were found to be those corresponding to absorption in the water vapour of the Earth's atmosphere. The second unparalleled achievement was in 1842, when Edmond Becquerel (1820–91) used a daguerrotype plate to record the whole Fraunhofer spectrum of the Sun. This he recorded even into the ultraviolet, where the plate is naturally sensitive, and he was able to extend Fraunhofer's system of labelling the dark lines. Becquerel was the son and the father of talented Parisian physicists. His feat seems not to have been repeated for more than thirty years, although in the meantime (1852) the ultraviolet spectrum of the Sun had been observed by George Stokes (1819–1903) using its ability to make certain substances fluoresce.

Photography was soon applied to the problem of the fiery corona ('crown') of the Sun seen at the moment of its total eclipse by the Moon, and the small pink prominences in it. These had been described as early as 1185, in a medieval Russian monastic chronicle, as issuing forth from the ends of the crescent of the eclipsed Sun like red-hot charcoals. But did they *belong* to the Sun? Were they an optical illusion, a mirage, perhaps? B. Wassenius, who had seen them in 1733 from Gothenburg in Sweden, described them as red clouds in the *Moon's* atmosphere. The eclipse of 1851, seen from Sweden, was generally thought to have indicated that the place of the prominences was truly the *Sun*. De la Rue observed the eclipse of 1860 from the upper Ebro valley, while Father Secchi viewed it from Desierto de las Palmas, 400 kilometres to the south-east, and the similarity of what they saw, and of what was now recorded for all to see on a photograph, settled the question of the solar nature of the prominences once and for all.

The camera was used increasingly in the study of the fine detail in the spots, as well as the granulation of the Sun's surface. P. J. C. Janssen (1824–1907) was a pioneer in this respect, but he had earned some fame from a most ingenious device he had devised after observing the total solar eclipse of 18 August 1868 from Guntur, near the Bay of Bengal. Pointing the slit of his spectroscope at two great prominences, when the Sun was in total eclipse, he found intense spectral lines appropriate to hydrogen. It occurred to him that if he only admitted light of this particular wavelength (that is, occupying this position in the spectrum), then by rapid scanning of the Sun with the slit of his spectroscope he should be able to produce a picture of the Sun, and so follow its changes on a regular basis, instead of having to wait for eclipses. In fact he gave his spectroscope a rotary movement, to form the solar image out of its slit-like components. If this reminds us of the mechanics of the cinematograph, another invention of his, the photographic revolver, came even closer. This was designed to take a rapid succession of photographs during the transit of Venus in 1874.

If Janssen's name was known to the French public in his lifetime, it was surely as the man who left the besieged city of Paris by balloon during the Franco-Prussian war to observe the eclipse of 22 December 1870. Lockyer had obtained a permit from the Prussians to allow Janssen to pass through their lines, but honour would not allow him to use it, since he intended to carry war dispatches. Having asked the Academy of Sciences to support his attempt, Janssen was provided with a balloon, the *Volta*. Together with an assistant, he reached an altitude of 2000 metres and was carried westwards by the wind, landing safely with his instruments – and precious dispatches – near the Atlantic coast. He reached Oran (Algiers) to observe the eclipse, but the weather was less co-operative than the Academy of Sciences and the Prussians had been, and his journey had a symbolic rather than a scientific value.

Janssen's studies, which would now be thought perfectly straightforward astrophysics, were not then perceived as falling squarely under traditional astronomy, and it is worth noting that he found great difficulty in obtaining serious government support. It was the minister of education Victor Duruy who strove to equip him with an observatory and who eventually turned the tide in his favour. At the end of seven years, in 1876, Janssen was granted a site, and he chose Meudon. He was eighty years old before the astronomical staff was increased beyond a total of two; but in the meantime he had conducted a highly important research programme, producing an atlas of solar photographs covering the period

1876–1903. And so began one of the world's most distinguished solar research centres.

There is a doubly interesting parallel between Janssen's career and that of Norman Lockyer (1836–1920) in England. Lockyer, a civil servant without university training, had attached a spectroscope to his 16-cm refractor in the mid 1860s, and in 1866 conceived the same idea as Janssen (1868), that is, of viewing the Sun only in light of the colour of the prominences. He obtained a government grant in 1867, but received a suitable high-dispersion spectroscope only a year later, observing the prominence as planned on 20 October 1868. He tried an oscillating slit, with poor results, but helped by William Huggins he came to see that a wide slit was enough, and he waxed eloquent over the strange shapes in the forest-like solar atmosphere. Stranger still was what happened when he communicated his findings to the French Academy of Sciences: his letter and another from Janssen, explaining the same method, arrived within minutes of each other. The sad story of Neptune was only twenty years old, and the French government celebrated this instance of near-simultaneous discovery by striking a medal carrying the portraits of the two astronomers.

For all his many maverick achievements in astrophysics, Lockyer, like Janssen, never really found a niche in establishment astronomy. A man of enormous vitality who worked with missionary zeal to improve public scientific awareness, he was founder, and for half a century editor, of the famous journal *Nature*. He achieved very limited success, even so, in persuading the government to found a national astrophysical observatory. A Solar Physics Observatory was set up in South Kensington, but the site was later needed for the Science Museum. At the age of seventy-six, a disappointed man, he built himself a new observatory at Sidmouth, where he worked until his death eight years later.

It is easy now to be critical of this apparent government lack of concern for the advancement of science, but it is as well to remind ourselves that science was regarded as the province of the universities, which in Britain – in contradistinction to many other European countries – were not primarily a government concern. Astronomy was in a borderline position. Airy, a generation earlier, had been the first Astronomer Royal to be entirely dependent on his official government salary. (Halley had received a naval pension, and all other incumbents were in holy orders, and so had drawn small stipends from the Church.) But Greenwich was not seen as a research institute, any more than was its American counterpart, the Naval Observatory in Washington, where Lockyer's contemporary Simon Newcomb was pursuing a clear vision of the need to improve planetary and lunar theory. They were founded to serve the state, to supply tables and astronomical constants as a public utility for purely practical ends.

Huggins (1824–1910), Lockyer's senior by twelve years, resembled him in some ways. Both came to astronomy as amateurs, without any university education, and both were eventually knighted for their services to science; but Lockyer was the speculator and Huggins the generally cautious observer, who became at length President of the Royal Society (1900–5). His reputation was made by his pioneering work in spectroscopy. Learning of Kirchhoff's findings, he enlisted the help of W. A. Miller (1817–70) of King's College, London, in investigating stellar spectra, and by 1863 was in a position to publish lists of spectral lines in stars. A year later he found two green lines in the spectrum of the Great Nebula in the constellation of Orion, and believed that he had found a new element there, to which he gave the name 'nebulium'. In 1928 I. S. Bowen showed that the lines were the so-called 'forbidden lines' of oxygen and nitrogen. The fact that Huggins was mistaken should not be allowed to obscure the great importance of his observation, which proved that the nebula was *gaseous*, and not, for example, a solid or liquid, as was sometimes supposed.

Huggins' pseudo-discovery of nebulium contrasts with the very real discovery of the element helium in the Sun by Norman Lockyer. (The discovery of terrestrial helium did not follow until 1895, when it was isolated by Sir William Ramsay.) Strangely enough, however, it was Lockyer who made most speculative capital out of nebulium, for he thought it confirmed a theory he had of celestial evolution. There was much friction between the two men over the question of whether the green lines corresponded to part of the spectrum of a magnesium spark in the laboratory. Huggins here was right: they do not. Another question, however, was whether Huggins' 'green nebulae' were the breeding-ground of stars, following the old idea developed by William Herschel, for instance. Again Lockyer seized on the idea, and wove it into his so-called 'dissociation hypothesis' of stellar evolution, but Huggins was more circumspect. He knew that the spectra of stars contained signs of many chemical elements, and those of the gaseous nebulae very few. Not for many decades was a theory forthcoming that could reconcile these differences with an evolutionary theory.

Huggins seems to have been happy to leave the spectra of bright objects to others, and to have concentrated on such faint objects as comets and stars, including the nova of 1866. In 1868 he made one of the most useful of all extensions of spectroscopy to astronomy. The Austrian physicist Christian Doppler had in 1841 given theoretical reasons for a change in the wavelength of a source moving relatively to an observer – a change of pitch in a source of sound, or of colour in a light source. A. H. L. Fizeau – better known to physics than astronomy – had seen the possibilities of using the dark Fraunhofer lines as reference colours, and Huggins had enough knowledge of their overall patterns to make comparisons betweeen laboratory sources of light and the feeble spectra of stars. In 1868 he found for the first time a stellar velocity by this means: he stated the velocity of Sirius as 29.4 miles per second away from the Sun (*away*, since he found a shift of the spectrum towards the *red*, indicating a lowering of frequency or lengthening of wavelength). This visually acquired figure was on the high side, and he later revised it downwards, but the application of the Doppler principle in astronomy was to prove of the utmost importance, especially in cosmology, when it was later applied to entire galaxies.

Huggins had been trying to use photography on the spectrum of Sirius since 1863, but at first he obtained mere streaks of light on his plates. In 1872, Henry Draper in New York got a photograph with four lines crossing the spectrum of Vega; and then from 1875 Huggins began to obtain better and better results, pioneering a new 'dry-plate' photographic process, with sensitized gelatine in place of the old wet collodion. Four years later he was able to record ultraviolet spectra, which, with others obtained earlier by H. W. Vogel in Berlin, and later by M. A. Cornu in Paris, showed that white stars were abundant in hydrogen. This was the beginning of a general awareness of the overwhelming preponderance of hydrogen in the universe.

The year 1875 was an important turning point in Huggins' life, not to say in astrophysics, for then he married Margaret Lindsay Murray, of Dublin. Although only half his age, she quickly became an invaluable intellectual partner, both in astronomical observation and in the joint publication of their findings. Together they worked on the spectra of stars: it was no easy matter, working alone, to observe a spectrum while holding the image of a moving star on the spectroscope's slit less than a tenth of a millimetre wide, for an exposure lasting an hour. Together, around 1889, they obtained a photograph of the spectrum of light from the planet Uranus. Father Secchi had first observed it by eye in 1869, and in the intervening decades others had repeated the observation, some suggesting that the spectrum gave signs that Uranus was shining partly by its own light. The Huggins pair scotched this idea, showing that the spectrum was more or less that of light from the Sun, so that there was no reason to discard the assumption that its light came by simple reflection.

From the eclipse of 1882 onwards the spectrum of the solar corona was regularly photographed, showing, for instance, the spectral lines appropriate to hydrogen that had been found in the prominences by Janssen at the eclipse of 1868. Also now photographed was the spectrum of the chromosphere (the narrow pink layer between the photosphere and the corona), with its strong lines appropriate to calcium. The first photograph of the corona's spectrum was in fact taken by yet another 'outsider' to regular astronomy, Arthur Schuster (1851–1934), a Jewish emigré from Frankfurt who held a post at Owens College in Manchester. Schuster's straightforward photographs of the complete corona during that eclipse were hardly bettered before the twentieth century. They showed its great extent, which had been appreciated visually at the 1878 eclipse (seen from America): parts of it in the neighbourhood of its equator were seen around two full diameters from the solar centre. Schuster was even fortunate enough to catch the image of a comet in the same shot.

Inspired by these records, Huggins experimented on photographing the corona *without* an eclipse. His idea of using a restricted part of the spectrum was probably inspired by that set down in Lockyer's original request for government aid. Huggins noticed that Schuster's negatives showed that there was a great concentration of light from the corona in a certain region of the spectrum (a region in the ultraviolet). Could he not photograph the Sun with photographic plates sensitive to just this region? He chose silver chloride as the sensitive material, and after several trials, at about the time of the eclipse of 6 May 1883 obtained results very similar to those obtained during the eclipse itself by standard methods. Unfortunately it was another three years before the method gave further photographic evidence of the same sort, and there were many who doubted its authenticity. The reason for difficulty in repeating Huggins' results was partly the filtering of light through volcanic matter thrown into the Earth's upper atmosphere by an eruption in the Straits of Sunda in August 1883. Agreement was long in coming. Gradually other techniques for separating out the light of appropriate colour were developed, and Huggins' ideas were well and truly vindicated.

As the century wore on, photography became integrated into advanced astronomical practice, which indeed eventually became unthinkable without it. If the nineteenth was the century of star catalogues, then it was only proper that photography should be applied to their production too. Juan Thomé in Cordoba (Argentina) laboured until his death (1908) to extend the Bonn catalogue southwards, reaching to latitude 62°S. Not until 1930 did others extend the lists of his *Cordoba Durchmusterung* to the south pole. One of the weaknesses of this work was in the estimation of star magnitudes, and here photography, and theories of image density on the plate, came to the rescue. It was found, paradoxically, that poor lenses gave better results than the finest; and later, techniques were developed that required a slight defocusing of the image.

Even more useful was the rapidity with which *star co-ordinates* could be measured from photographic plates. The idea occurred to the brothers Paul and Prosper Henry of Paris, but neither they, nor the members of an international conference convened in Paris in 1887, appreciated all the errors inherent in the technique. Nevertheless, the great *Carte du Ciel* project was born, with the aim of mapping the sky photographically down to the fourteenth magnitude (that is, to mag. 15.0). It was planned that there would be an *Astrographic Catalogue* of stars down to the eleventh magnitude. Only after several decades was the work put on a really reliable footing. Even before it had begun, however, Jacobus Cornelius Kapteyn (1851–1922) was at work, producing one of the century's great monuments of cataloguing, by a most simple and elegant technique, again photographic.

Kapteyn was in some ways fortunate to be at a university – Groningen – reluctant to provide him with a large telescope. (The Netherlands, the United Kingdom and the Irish Republic redressed the injury, however, by establishing a 'Jacobus Kapteyn telescope' of 1.0-metre aperture at La Palma on the Canary Islands in the 1970s.) He used instead a set of photographic plates taken by David Gill (1843–1914) at the Cape Observatory between 1885 and 1890. By the ingenious use of a theodolite in his laboratory, viewing singly the stars on the plate, which he placed at a distance equal to the focal length of Gill's telescope, he could measure each star's co-ordinates (right ascension and declination) *directly*, and even more accurately than had been done for the Bonn catalogue. He also found stellar magnitudes by measurements on the star images, so that in the space of thirteen years (ten were needed for the measurements), in two small rooms of the physiology laboratory in Groningen, the *Cape Photographic Durchmusterung* was produced, with its 454 875 stars between 18°S and the pole, down to the tenth magnitude.

Later, the Harvard Observatory went one stage further, and distributed its 'atlas' in the form of boxes of plates. It is amusing to compare Kapteyn's with the Harvard tradition of reducing measurements, which was to use the services of faculty wives and other ladies with arithmetical abilities and time on their hands. Kapteyn, living in a more conservative society, persuaded the governor of the Groningen state prison to lend him the services of selected male guests of that place.

Hardly a branch of astronomy remained untouched by photography. The camera was used from several stations to observe the transit of Venus in 1874. Planetary photography as such required good atmospheric conditions to obtain a steady image. Mars was photographed by B. A. Gould from Cordoba (Argentina) in 1879; and in 1890, in a succession of photographs, W. H. Pickering, at Wilson's Peak (California), showed the southern polar cap. To everyone's surprise, and to the excitement of the many who were then speculating on the possibility that Mars, with its canals, was populated with intelligent beings, it appeared that the polar cap was changing in area.

Jupiter was systematically photographed with the great Lick telescope in 1890–2, when near opposition, and at a time when the 'great red spot' on its surface, long a subject of study through the telescope, seemed to be threatening to disappear. Even asteroids became subjects of the photographic lens. They were now being detected in ever greater numbers by virtue of the simple fact that they may leave trails on photographs of stars. The first asteroid so found was discovered by Max Wolf of Heidelberg in 1891. In the fifty years from 1890 to 1940, the number of recognized asteroids increased from fewer than 300 to almost 1500. Among the more notable now known are Eros, discovered by Gustav Witt in 1898, and Icarus, which Walter Baade discovered by chance in 1949 as a streak on a photographic plate made with the newly completed Schmidt telescope at Palomar. Icarus can approach close to two-thirds of Mercury's distance from the Sun, and yet its aphelion is far outside the orbit of Mars. Eros was the first asteroid known to come within the Earth's orbit, but asteroid Hermes may come uncomfortably close. In 1937 it was only about twice the Moon's distance.

On a point of definition: asteroids are usually distinguished from meteorites in terms of their visibility. The latter are very loosely defined as those natural planetary objects too small to be seen in space by reflected light with the best telescopes.

In the post-War era there was much speculation over the best course of action should an asteroid seem to be heading for planet Earth, the hawks favouring a nuclear attack on it, the doves a rocket engine that would put it off course. The 1979 film *Meteor*, starring Sean Connery, presented the hawkish alternative, and together with the cult of computer war-games may have much to answer for. In 1991 NASA set up an 'Interceptor Committee' that is said to have proposed a battery of laser guns on the Moon, an orbiting fleet of nuclear warheads, and 'nuking' a few specimen asteroids for target practice. These proposals, and the Meteor Crater in Arizona, remind us that the danger need not come from an asteroid at all. Icarus is of the order of a kilometre in diameter. The object responsible for the Arizona crater was almost certainly less than a hundred metres across. As for the sizes and velocities of the fragments of the unfortunate asteroid, they are anyone's guess. And as for the difficulties of detection, most methods rely on minute changes in the appearance of the sky, that is, movements *across* the field of view of the detector. But the real danger will be from an object that is *not* moving across the field of view.

Accidental discoveries have often been made with the camera's help. For example, working at the Lick Observatory, Edward Emerson Barnard (1857–1923) – a man with considerable reputation for finding comets by honest searching – was photographing stars in the constellation Aquila in 1892 when he found a cometary trail on the plate. This comet was not the first to be photographed, but was the first to be *discovered* by photography. Barnard's systematic photography of regions of the Milky Way, and of comets, were of material help in advancing a knowledge of both. As the speed of photographic plates improved, the task became easier, but that simply meant pushing out to more distant regions, where it was less so. And not all advances came with the fastest of cameras. Barnard, who was then a junior member of staff under the autocratic direction of E. S. Holden, produced his fine Milky Way photographs with poor apparatus requiring exposures of up to six hours each, using a guiding telescope without illuminated reticle. As the nineteenth century ended, it had in Barnard an excellent example of a new type of astronomer, one who not only grasped at the opportunity to introduce photography wholesale into astronomy, but who was prepared even to place it above the methods of observing with which the century had begun. In the twentieth century, the character of astronomy was destined to change in many more respects as a result of the discovery of yet other observational techniques.

Cosmogony, Evolution and the Sun

COSMOGONY AND THE SOLAR SYSTEM

Optics was not the only branch of physics to make a welcome intrusion into astronomy in the nineteenth century. In mid-century, a new and coherent theory of heat, thermodynamics, was assembled from laws that had been derived in relative isolation – the conservation of energy, the law of entropy, and so forth. Once accepted, these laws were accepted as *universal*. No sooner had they been established by a succession of physicists, including S. Carnot, J. R. Mayer, J. P. Joule, H. von Helmholtz, R. Clausius and W. Thomson, than they were applied to the heavens, and in particular to the Sun. Once it was realized that there is a balance-sheet on which all forms of energy must be accounted for – energy of motion, energy of position, heat, electrical and chemical energy, and so on – then it became clear that most of the energy that manifests itself on the Earth is ultimately derived from solar radiation, with some of it lost through tidal friction. But what is the origin of that solar energy? It could in principle be transformed from energy of another type, but from which?

In a privately published work of 1848, Julius Robert Mayer (1814–78) suggested that it came from the mechanical energy released with the continuous bombardment of the Sun by meteors. Mayer's work was not well known, but the same idea was later put forth independently by John James Waterston (1811–83), and for a short time the 'meteoric hypothesis' attracted much attention. It was possible to calculate how much mass would need to fall into the Sun to produce the heat that it is found by measurement to radiate. John Herschel and the French physicist Claude-Servais-Mathias Pouillet had independently measured the heat received from the Sun fairly accurately, and had made estimates of the amount absorbed in the Earth's atmosphere. The annual infall of mass was calculated differently by different authors, but William Thomson's figure is representative: the annual infall was said to be of the order of one seventeen-millionth of the Sun's mass.

Here is an excellent example of the interrelatedness of physical evidence. Celestial mechanics had been brought to such a state of perfection that even this minuscule quantity could immediately be ruled out as much too great: as Thomson showed in 1854, it would imply a shortening of the time of revolution of the Earth, say by a couple of seconds a year, an easily detectable quantity over the interval between the Babylonians and the nineteenth century. Hermann von Helmholtz (1821–94), on the other hand, had a more subtle mechanical theory. Helmholtz suggested that the Sun's heat came from the conversion of gravitational energy to heat energy in the process of condensation of material that began as a vast cloud. (This version of a 'nebular hypothesis' was also sometimes confusingly called the 'meteoric hypothesis'.) The Sun might at present seem to be a well formed object, but the idea was that the same processes of contraction continue, the gravitational energy ('potential energy') of its material continuing to be reduced, and the energy released still being transformed mainly into heat energy.

This hypothesis seemed very much preferable to an alternative theory of solar energy supply often then proposed, namely that *chemical reactions* in the Sun were the source of its heat. As Thomson pointed out, the most energetic chemical reactions then known would not keep the Sun radiating for more than about 3000 years. Even the theologians wanted more time than that.

There are many slight variants of the data quoted, and the following calculation is meant only to show the path to an important new conclusion that was then reached. The contraction needed to provide the enormous radiant energy of the Sun would reduce its diameter by only about seventy-five metres per annum, far too small a quantity to be measured, even over centuries. Helmholtz's contraction hypothesis was thus safe from criticism on this account. It led on, however, to a conclusion as to the vastness of the time-scale of solar activity that seemed theologically so dangerous that astronomers frequently apologized in advance for it. There is no difficulty in accounting for *twenty million years* of heat, by the theory. Ten million years, said Thomson, is a minimum requirement, and fifty or even a hundred million years would be quite arguable.

This was the first coherent physical argument for the age of the Sun, and of course it provided a lower limit to the age of the Universe. Needless to say, those who believed that God created the world around four thousand years before Christ were displeased by such reasoning. Some geologists were admittedly by this time demanding a much longer period, to account for the geological changes that the Earth had seen. Even Buffon, in the mid eighteenth century, had estimated 75 000 years on the basis of the rate of the Earth's cooling, but three million years on the basis of the deposition of sediments, although he did not publish the second figure. By the time of Thomson's calculation, however, geologists such as Charles Lyell (1797–1875) were quoting figures far *in excess of his*, and this paradoxical situation remained a thorn in the flesh of solar astronomers for more than half a century, although they did not find it difficult to expose errors in the geologists' standard methods of calculating cooling times for the Earth. How the geologists came to terms with their problem falls outside the scope of this book, but broadly speaking there was always one easy way of shortening time-scales, that is, by introducing some or other catastrophe into the world's history.

There was another awkward conclusion to be drawn from thermodynamics, however. This predicted a running down of the Universe, 'a state of universal rest and death', as Thomson called it, assuming that the Universe was 'finite and left to obey existing laws'. He avoided this by arguing for 'an overruling creative power' that was responsible for introducing living creatures to the Universe, and that removes the need for 'dispiriting views' as to human destiny. He wrote these words in 1862, two years after Charles Darwin had published his theologically controversial *Origin of Species*, in which a biological theory of the evolution of living forms was presented that had some interesting parallels in the law of entropy in thermodynamics.

This, Clausius' second law of thermodynamics, amounts to saying that natural processes as a whole move in one direction, that the entropy of the world can only increase, that time is an arrow, so to speak, pointing in the direction of the 'heat death', when all mechanical energy will have been turned into heat, and the temperature will be the same throughout the Universe. For generations, counterarguments were offered that seem to have been motivated chiefly by an emotive rejection of the idea that the Universe could possibly have such an ungodly end.

Another important step in the development of a theory of the Sun came when astrophysicists began to consider the detailed structure of its interior, and other possible sources of energy within it. Jonathan Homer Lane (1819–80) was a rather shadowy figure on the edge of American science, for some years an examiner in the US Patents Office in Washington. In 1869, Lane developed Thomson's arguments further, assuming that there are convection currents in the Sun. He investigated the conditions for the Sun to remain in equilibrium, and found that its temperature should change in inverse ratio to the radius. If, when it contracts under gravity, part of the heat generated by contraction is radiated, the rest can be kept to increase the temperature of the sphere, so as to keep it in equilibrium. The Sun may thus lose energy and yet grow hotter.

Although it was soon appreciated that 'Lane's law' breaks down when the contraction eventually produces a gas of very high density, it stimulated others to investigate the Sun's structure, and the structure of stars generally, and it forced Kelvin to reconsider his own argument. A similar argument to Lane's was presented independently by August Ritter in 1872. It is a curious fact that both took *meteorological* models as their starting point. When in 1907 the Swiss physicist Robert Emden (1862–1940) published what was to become a classic textbook of this branch of astrophysics, he applied his theory of spherical distributions of gas to both cosmological and meteorological problems in the same work.

Another example of interplay between astrophysics and other scientific subjects was when George H. Darwin, the son of Charles Darwin, considered friction in the tides as an agent of cosmic as well as biological evolution. In a series of studies, beginning in 1879, he projected the motions in the earth–Moon system back in time, and found, for instance, that there was a time when the daily rotation and the monthly revolution were equal. Going further back still, he was led to a situation when it seemed that the Moon and Earth were a single body, destined to split into two, a conclusion fitting closely with theories of equilibrium in rotating fluid bodies developed around 1885 by the great French mathematician Henri Poincaré. Darwin explained how the split might have taken place, and his work led others to investigate the wider questions of the formation of the entire planetary system. There were some serious difficulties here. The simple Laplacean model of condensation from a rotating cloud did not explain why Jupiter has so much rotational momentum (angular momentum), nearly two-thirds of that tied up in the solar system as a whole, when the mass of Jupiter is only a thousandth of the whole. The Sun, on the other hand, with most of the mass, has only a fiftieth part of the angular momentum.

In 1898 the American astronomer Forest Ray Moulton (1872–1952), still a graduate student at the university of Chicago, and the chairman of the geology department there, Thomas Crowder Chamberlin (1843–1928), had begun a study of the formation of the planets, breaking with the full Laplacean view. After studying photographs of the solar eclipse of 28 May 1900 they were eventually led to their hypothesis of 'planetesimals', lumps of matter that had supposedly solidified out of the original condensing nebula. By 1906, considerations of the oddities of angular momentum led them to the idea that the planetary system had originated with the close approach of another star to the Sun, which had drawn matter out gravitationally and provided the motions in a way they could explain, at least approximately. Planetesimals near the Sun collected and formed small planets, and indeed the energies released in their collisions were said to be responsible for the high temperatures inside the planets, known at least from study of the Earth.

It should be clear that there are at least three quite different explanations possible for the formation of the planetary system. It could have been the result of a rare accident, such as the approach of another star to the Sun; or during the typical evolution of a star from gas and dust; or in the course of an unusual sort of stellar evolution. The rarity we ascribe to our own solar system hangs on the choice made: in the first case planetary systems would be rare, in the second common, and in the third less common but widespread.

In the 1920s James Jeans expressed a preference for the first alternative, and thought the odds against a star being surrounded by planets to be about a hundred thousand to one. At the same period, Eddington went even further, and hazarded the suggestion that our world, with its living beings, might be unique. This has made many commentators uneasy, for no better reason than that it seems to go against the 'Copernican' trend in history. Humankind began by taking first the Earth, then the solar system, and then our Galaxy, as central to the Universe. Uniqueness in character is uncomfortably similar to spatial centrality.

More modern theories have tended to start from the observation that clouds of dust are frequently observed around stars in the process of formation; and that a very small fraction of matter remains in the cloud, as opposed to the central star; but from this point on, two different sorts of theory are still in vogue. One supposes a rapid formation of giant proto-planets. The other supposes a slow building up of planets out of solid chunks of ever-larger size. Current majority opinion inclines perhaps more in favour of the second, for which strong arguments have been marshalled by the Russian astronomer Safronov, but the question is still an open one.

Astronomers gradually became disillusioned with Thomson's age for the Sun, since it was so very much smaller than the age of the oldest rocks in the Earth's crust, not only by the old criteria, but also following estimates made from rates of radioactive decay. It became clear that there was some other, much more abundant, supply of energy available in the Sun. After Einstein's special theory of relativity (1905), astrophysicists gradually came to realize that there might be some process for the conversion of mass into energy. No one appreciated this better than Arthur Eddington, who from 1917 was working on a theory of the internal constitution of stars. He included in it a formula for the relation between the mass of a star and its luminosity. James Jeans challenged the correctness of this, because it ignored *intrinsic* sources of energy in the star.

At this stage Jeans thought that these must be independent of temperature, and the product of some sort of radioactive transformation, involving massive atoms. One by one he dismissed doctrines that had been generally assumed, such as that the material in stars obeyed the gas laws of the laboratory physicist. He was searching for suitable energy sources at about the same time as F. W. Aston (1877–1945) was investigating the properties of isotopes, that is, in the early 1920s. Aston was showing how 'uranium lead' 'thorium lead' and 'ordinary lead', for example, differed in their atomic weights, that is, in one important identity tag, and yet were indistinguishable in their chemical properties. Jeans and others saw that in the development of stars, the transmutation of elements, even between other states than mere isotopes, may take place on a colossal scale, with the corresponding release of very large amounts of energy. Many of his ideas were short-lived, but he and Eddington gave a considerable impetus to this entire branch of astrophysics.

Another physicist to do so was Jean Perrin (1870–1942), who as early as 1919 perceived that what would now be called 'thermonuclear fusion reactions' – that is, reactions in which atomic nuclei join to form heavier elements – might be the source of energy in the stars. This speculative insight, only expressed in qualitative terms, was ultimately to prove correct. It was not until 1939, however, that von Weizsäcker and Bethe described the fusion reaction that converts hydrogen to helium, and so produces the main source of energy in most stars.

SUNSPOTS AND MAGNETISM

Solar astronomy in the nineteenth century was conducted, as astronomy always has been, at two different levels, the one merely observational, the other largely theoretical. Observation, especially in amateur circles, was often no more than a pious recording of curiosities. Even moderately acceptable theories often followed observation at a considerable distance in time, if at all. There are many illustrations of this. In 1826, Samuel Heinrich Schwabe, an apothecary from Dessau in Germany, was hoping to discover a planet below Mercury. For this reason – and notice how closely the story parallels that of Messier and the nebulae—he registered sunspot positions, simply in order to eliminate them from his searches. Reviewing his records after twelve years of observations he suspected that *sunspot totals fluctuated over a period of about ten years.* To make sure, he went on with his work, and in 1843 tentatively published that ten-year periodicity.

Little attention was paid to this until in 1851 Alexander von Humboldt (1769–1858) published Schwabe's table of results, with some supplementary data. In 1852 the Swiss Johann Rudolf Wolf (1816–93), working first in Berne and later in Zurich, assembled all the historical material on sunspots he could discover, and stated the average period as 11.11 years. He continued to publish reports on the numbers of sunspots almost up to his death. In 1851, John Lamont (1805–79), a Scottish-born astronomer who had left his native country for Bavaria in 1817, published a discovery he had made, that the Earth's magnetic field also seems to vary with a period of about ten years, and that the alternate periods are weak and strong – in short, that the full cycle is twice that of the sunspots. Immediately Wolf, and Edward Sabine in England, noted that, broadly speaking, sunspots do indeed follow magnetic changes (including the aurorae) in all their irregularities. That there was some strange connection between solar phenomena and terrestrial effects could not be doubted, but a century would pass before any plausible explanation would be found.

Even more elusive was a theory to explain the periodic changes in sunspot behaviour discovered by the wealthy English amateur astronomer Richard Christopher Carrington (1826–75). Carrington was for a time a salaried observer at the newly founded university of Durham, but he resigned in order to be able to set up a worthy private observatory of his own near Reigate in Surrey, to complete the zone surveys of Bessel and Argelander within 9° of the north celestial pole. For seven and a half years between 1853 and 1861 he systematically and meticulously observed sunspots by a simple and accurate method, and so discovered that the period of rotation increases with their distances from the solar equator. This showed that they could not be regarded as fixed to a solid solar object. He found that when spots were generally most numerous, they tended to approach the equator, and to become extinct around latitude 5°, at the time of sunspot minimum. At the same time, the first spots of the new cycle start to appear.

Prompted by such findings as these, the Greenwich Observatory, under the direction of W. H. M. Christie, at last decided to enter into a programme of astrophysical observations, appointing Edward Walter Maunder (1851–1928) as photographic and spectroscopic assistant in 1873. Maunder was one of a class of 'computers' recruited from the ranks of society who did not have a university training, and some of the consequent social tensions at Greenwich are no doubt in evidence in anecdotes told of Airy by Maunder— for example, that Airy stuck the label 'empty' on boxes that were indeed empty.

As an aside on this social question, it should by now be obvious that throughout history the social position of those who worked at astronomy was as variable as the motives for doing so. There is a clear sense in which there was a moderately well defined astronomical profession in several early societies—for example, those of Babylon, Alexandria and the medieval universities—but the subject was usually practised in combination with other activities, whether of philosophy, a priesthood, teaching, or something more mundane. Even those in the great observatories had other strings to their bows. The Tycho Brahes of previous history were few and far between when the change eventually came with the founding of the great national observatories—Paris, Greenwich and the rest. Personal enterprise could turn Herschel from professional musician to professional astronomer, but again he was an exception, and the upper echelons of the professional hierarchy in his day were for the most part occupied by men trained in mathematics, with a few leading instrument-makers and amateurs having an honorable associate position. The amateurs have remained a force to be reckoned with, and even now make occasional contributions of importance, but by the nineteenth century astronomy was becoming increasingly dependent on the work of salaried professionals. To describe their training would be to write the history of the democratization of education itself, and of women's part in that process. We shall come across many other examples of self-evident social changes, which hardly need to be spelled out on each and every occasion, but to continue with the case of Maunder: from 1891, Annie Russell was recruited as 'lady computer' to assist him. An excellent mathematician, she contributed many new ideas. They married in 1895. Not all social changes come about by Act of Parliament.

Maunder's name is still remembered. A programme of daily photographic sunspot recording, which began in the following year, allowed him to tabulate the various aspects of sunspots, and the butterfly-shaped 'Maunder diagram', showing the numbers of spots to be found at all latitudes at any time, has become a standard means of representing the situation graphically.

The relationship between the Earth's magnetic field and the sunspot cycle was more problematical. It had been under investigation by several astronomers since Wolf's and Lamont's work of the 1850s. In 1896 Olaf Kristian Birkeland of Norway had put forward the idea that peaks of geomagnetic activity, 'magnetic storms', were caused by beams of electrically charged corpuscles from the Sun, and he conducted a number of laboratory experiments to support his belief. His idea was that these charged particles, drawn in to the Earth's magnetic field near the poles, were what gave rise to the aurora borealis and aurora australis (the 'northern and southern lights'). Two discoveries of great significance were now made by Maunder, and published in 1904. He found that the greatest geomagnetic storms were associated with large sunspots near the central meridian of the disc – on average they had passed the central meridian about twenty-six hours earlier. He also found that geomagnetic storms as a whole recur at intervals of about twenty-seven days, which is the Sun's period of rotation with respect to the Earth. His conclusion was that they were due to some streaming effect from localized regions of the Sun, and that this streaming took about a day to reach the Earth. In the case of the great storms, sunspots seemed to be the cause, but lesser storms could occur when the disc was virtually blank. The twenty-seven-day cycle seemed to settle the solar origin of all of them, however. Matters were left here for over two decades before others – notably W. H. M. Greaves and H. W. Newton – re-examined the relationship, and while Maunder's results were confirmed and extended, the question of cause remained a mystery.

THE SUN AND G. E. HALE, TELESCOPIC VISIONARY

A breakthrough came at last as a result of work done with the spectrohelioscope in the 1930s by George Ellery Hale (1868–1938), to whom practical solar astronomy already owed innumerable debts. Hale had invented a photographic form of this instrument for photographing the solar prominences by daylight, in 1889, after a number of others had tried and failed to get good practical results. (All were effectively following the lead of Janssen and Lockyer, with their *visual* methods. Hale could not perform the more difficult task of *photographing* the corona of the non-eclipsed Sun. This was first done by Bernard Lyot in France in 1930.) This first instrument he called a spectroheliograph. The idea is simple. The telescope forms an image of the Sun on the slit of the spectrograph. The key to the design was a series of levers and linkages that scanned the solar image with the slit, keeping the relation of slit to prism unchanged, so building up a picture of the Sun in light of a single wavelength (colour).

Out of his early success with the spectroheliograph came Hale's concern with solar research. Having graduated at the Massachusetts Institute of Technology, he moved back to his home in Chicago where his father financed a 12-inch (30 cm) refractor, and with this in 1892 he achieved excellent results, for instance photographing bright clouds showing calcium lines, and prominences round the entire disc. At the end of the century he designed a spectroheliograph for the great Yerkes refractor, and with it he found the dark hydrogen clouds, as well as investigating the processes of circulation of calcium at various levels.

A man of great drive, and well placed financially, Hale persuaded his father to pay for a 152-cm mirror to further his research. The university of Chicago would not finance its mounting, and the disc remained unused for twelve years, until in 1908 it was set up on Mount Wilson, above Pasadena in California, as part of what was then the largest reflector in the world – but not for long. Even before this date, he had persuaded a wealthy Los Angeles businessman, John D. Hooker, to pay for a 254-cm mirror, and this was eventually made and mounted in a telescope built with funds from the Carnegie Institution of Washington and completed in 1917. These telescopes, the 'sixty inch' and especially the 'hundred inch', became household names throughout the world of popular science in the period following the First World War.

No sooner had the 'hundred inch' been proved successful than Hale was planning something more. In 1928 his plans were more or less outlined, and he was able to raise six million dollars from the Rockefeller Foundation for a 200-inch (5.08 m) telescope. This was donated to the California Institute of Technology – an institution that already owed much of its greatness to Hale – and set up on Palomar Mountain in southern California. Hale died before it was finished, and the Second World War interrupted progress. The mirror blank was cast by 1934 and shipped to Pasadena in 1936, but only finished and installed in 1947.

Dedicated in 1948, the instrument was with some justice designated 'the Hale telescope'. It was an extraordinary engineering project. The tube assembly, for instance, weighs 520 tonnes, and the dome, 41 m high, almost a thousand tonnes. The telescope was the first to allow the observer to ride at the prime focus. It is worth noting that the Corning Glass Works who cast the Pyrex blank for the mirror worked their way up to a 200-inch disc through discs of 30, 60 and 120 inches, and that the last of these, used for testing the 200 inch, eventually became the main mirror of the Shane reflector at Lick.

To return to the beginning of the century: ambitions of other sorts had then occupied Hale's horizon. In 1904 he received $150 000 from the recently founded Carnegie Institution to set up a solar observatory on Mount Wilson. The materials for this new observatory were transported under great difficulties up the mountain, by burro and mule, and the whole enterprise soon became for many a symbol of a type of astronomy done by pioneers, living and working in cabin and bivouac outside the reach of courts and cities, academies and universities. With the Snow telescope brought from Yerkes Observatory, a solar telescope driven by a coelostat, Hale in 1905 obtained the first photograph of a sunspot spectrum. With his colleagues in his mountain laboratory he found that the spectral lines that are strong in sunspots are those which are strong in laboratory sources, at relatively low temperature, so that sunspots are – as many had long suspected, but not proved – cooler than neighbouring regions of the disc.

The Snow telescope suffered from distortions due to solar heating, so in 1908 Hale designed a second, and in 1912 a third, still larger, each a tower topped by a two-mirror guidance system feeding its image into the telescope (18 m/45 m focal length) and thence into an underground spectrograph (9 m/22 m length). With the first he was able to detect vortex motions in the hydrogen clouds (flocculi) near sunspots, and he felt sure that this could be the source of magnetic fields, and that the widening of the lines in sunspot spectra was due to those magnetic fields. Certain double lines had been seen earlier in sunspot spectra, but misunderstood.

Perhaps Hale's most inspired single discovery was that this was an example of the 'Zeeman effect', the splitting of spectral lines that occurs when light is passed through an intense magnetic field. (The effect had been observed in his Leiden laboratory by Pieter Zeeman (1865–1943) in 1897.) In Hale we have a fortunate example of an astrophysicist trained in university physics, who was able to make such associations easily. In this case he followed with a study of the polarity of sunspots (the orientation of their magnetic poles), and so to the discovery that at the end of the eleven-year sunspot cycle the polarity reverses: the true periodicty could therefore now be stated as one of twenty-two or twenty-three years.

Hale's 1912 instrument was aimed at solving another problem, that of the *general* magnetic field that the Sun was thought to have, as judged, for instance, by the shape of its corona seen during an eclipse (F. H. Bigelow, 1889). We have already mentioned Birkeland's notion of an inflow of charged particles from the Sun. Could the Sun's magnetic field drive such particles? Hale, with a number of colleagues, made measurements of the field, but this seemed to vary, and little real progress was made. The reason is simply that the magnetic system is enormously complicated. It is now known that there is a complex three-dimensional whirlpool structure to the solar magnetic field in which all the planetary system is situated. (Kepler would have been delighted.) The field is dragged, as it were, by plasma ejected from the corona. The electrically charged particles themselves move in tight spirals, like springs wrapped round the lines of force.

In the 1930s it was regularly found that brilliant flares in the Sun's chromosphere were connected somehow with fade-outs that had been experienced on long-distance short-wave radio and even telephone communications.

Around 1930 Sydney Chapman and V. C. A. Ferraro calculated the speed of ions thrown out by the Sun, and realized that they would reach the Earth in one or two days, and have the disturbing effects observed. And then, towards the end of the 1940s, an important discovery was made by Scott Ellsworth Forbush, when studying cosmic rays – energetic charged particles that reach the Earth from outer space, inclusive of those from the Sun. (They include electrons, positrons, ions, alpha-particles and protons.) These were found to be of low intensity when the Sun was active, and were much reduced during magnetic storms, as though a magnetic field was blocking the path of cosmic rays from the Galaxy and perhaps beyond. Another piece of evidence was provided by Ludwig F. Biermann. Around 1950 he showed that the usual assumption that comets' tails point away from the Sun because of the pressure of the Sun's light was mistaken. The pressure was not enough. He showed that the force must be provided by a stream of matter moving at some hundreds of kilometres per second. Studies of cometary tails thus made possible a study of the supersonic streaming of material, and in the late 1950s the first satisfactory model was presented by Eugene Newman Parker.

As though to prove that the rocket programmes of the superpowers were drafted primarily with astronomy in view, space probes now came opportunely to the aid of solar physics. The first probes on board Soviet spacecraft Luna 2 and Luna 3 in 1959 provided confirmation of what had been painstakingly pieced together about the outpourings of the Sun over more than seventy years. Parker's model has been well confirmed in outline by many probes since that time. Many properties of the 'solar wind' are the object of current research, for example its relationship with the Galaxy in which it is embedded. The mechanism for heating the corona is of course an aspect of the overall model of the active Sun. How, for instance, are the complex patterns of sunspot behaviour to be explained? There have been many explanations for the 'solar dynamo' that powers the system, but our first concern is with the situation in the late nineteenth century.

STRUCTURE OF THE SUN AND STARS

In no other branch of astronomy have so many disciplines been brought into play as in the study of the Sun. As the solar scheme has been pieced together over the last century or so, it has been found that the sunspot cycle coincides with fluctuations in growth in plants, as evidenced by tree rings, radioactive dating by carbon-14 residues, silt deposits, fish stocks in the oceans, and so on. It is now accepted on the basis of geological evidence that the sunspot cycle has persisted for 700 million years, and this will obviously be of great significance to any theory of the character of the Sun. Astrophysics, which is rarely in a position to experiment with its objects, must always be heavily dependent on laboratory physics, which usually can. On the other hand, as Hale pointed out when canvassing support for the 200-inch reflector, the extremes of mass, density, pressure and temperature available in the heavens so far transcend those of the laboratory that 'many of the most fundamental advances in physics depend upon the utilization of these conditions'. In this he was peering into the future. What both physics and astrophysics sorely needed, at the end of the nineteenth century, was something nearer at hand, namely a sound theory of radiation.

This slowly emerged with a succession of physical laws that are now a standard part of physics, even though they have known limitations: Stefan's law (1879), Wien's law (1893) and—comprehending them all – Planck's formula (1906). Wien's, which relates radiation to temperature, allowed meaningful statements to be made about the temperature of the Sun's surface, which had previously been assigned wildly differing values. Planck's took more detail into account, relating as it did radiation, temperature and wavelength. The new figure for the temperature of the photosphere was about 6000K (kelvins can be roughly thought of as degrees Celsius from an absolute zero of −273°C). The temperatures of the corona and chromosphere were much more difficult to chart, and in estimating them, little of importance was achieved until techniques were developed for receiving radiations outside the visual range – radio waves, ultraviolet rays, X-rays and so on. When this was achieved, in the second half of the twentieth century, the great surprise was that in the transition zone between the chromosphere and corona, only a few hundred kilometres in thickness, the temperature increases dramatically from 10 000 K to about a million kelvins.

In the absence of an acceptable theory of radiation, astronomers fought shy of speculation about the interior of the Sun. Exceptions to the rule were August Ritter and Robert Emden, mentioned above. Emden's model made use of heat transfer in stars (supposed gaseous) by conduction and convection. The fact that the Sun's surface is granulated had long been known, and his interpretation of the granulation as the visible parts of convection elements was essentially correct. The most remarkable studies of the structure of stars made in the early years of the century, however, were those by Karl Schwarzschild, the most talented German astronomer of his generation.

Born in 1873 in Frankfurt-am-Main of a Jewish family, Schwarzschild died in 1916 from a disease contracted at the Russian front. In 1896 he took his doctoral degree *summa cum laude* with a dissertation applying Poincaré's theory of stability in rotating bodies to a variety of pressing problems, among them the origin of the solar system. His interests gradually focused on a subject that had never been given the attention it deserved, namely stellar photometry, the measurement of the radiant energy received from the stars. Apart from the introduction of some rough photographic techniques – which Schwarzschild supplemented and refined – measurement was chiefly done as it had always been done, estimating and comparing brightnesses with the human eye. Working then in Vienna, Schwarzschild applied his photographic methods to the magnitudes of 367 stars, some of them variable, and in following one variable star (eta Aquilae) he found that the range of change in its magnitude was much greater when estimated photographically than when judged by eye. He realized that this was an indication that its surface temperature fluctuates, an important discovery relating to a type of star – cepheid variables – that was soon to assume great importance in the development of astronomy.

In June 1899 he returned to Munich, and in 1901 he moved on to Göttingen, where he was director of the observatory built and fitted out by Gauss. Here he continued his photometric work, and when visiting Algeria to observe the total solar eclipse of 1905 he obtained a remarkable series of ultraviolet photographs of the solar spectrum – sixteen in thirty seconds – that led him into a study of the transfer of energy in the neighbourhood of the Sun's surface. Discarding the old cloud theory of the photosphere, he took the Sun to be layered, and considering the net energy absorbed from below and that emitted upwards, he was led to a series of equations, amounting to the 'Schuster–Schwarzschild model for a grey atmosphere'.

The two men had arrived at the model independently, Schuster in 1905 and Schwarzschild a year later. In this model, the temperature and the density of solar matter increase with depth. Later practical investigations showed that it did not account well for the flow of energy, and not long after Schwarzschild's death it was abandoned in favour of other solar models, first by E. A. Milne (1921) and then by A. S. Eddington (1923).

In 1909 Schwarzschild became director of the Potsdam Astrophysical Observatory. (He married in the same year, and one of his sons, Martin Schwarzschild, was later to become a notable American astronomer.) After volunteering for the army and serving in various scientific capacities in Belgium and France, he moved to the Russian front. From there he wrote two papers on general relativity that were to become his most lasting monument. The first concerned the gravitational effect of a point mass in empty space, on Einstein's theory. (It was the first exact solution of Einstein's 'field equations'.) The second paper concerns the gravitational field of a uniform sphere of material – not a model true to the Sun, of course, but an important beginning. Again he found an exact solution, and this time it was one with a very surprising property. This concerns a certain distance from the centre of the sphere, 'the Schwarzschild radius', which is related to the mass of the sphere in a very simple way. If a star collapses under gravitational forces so that its radius becomes smaller than this critical Schwarzschild radius, it becomes incapable of emitting radiation. It becomes a 'black hole', a concept of which we shall say more in our last chapter. The Sun would constitute a black hole if it shrank to a radius of 2.5 km.

This idea of a star that cannot emit radiation, and so cannot be seen, by virtue of the high concentration of its mass has interesting parallels in the eighteenth century. In 1772, Joseph Priestley discussed in print unpublished ideas by John Michell, ideas that Michell put into print in 1783. The central idea was that just as any ballistic missile can leave the gravitational field of the Earth only if it is projected with a sufficiently high velocity, so light can only leave the Sun if it has a sufficiently high velocity, a velocity that is easily calculated. (In fact Michell computed that the known velocity of light was 497 times greater than was needed for escape.) Then again, in 1791, William Herschel suggested that the nebulous character of the matter he could see through his telescope might be explained by the gravitational opposition to light in its attempt to leave or to pass gravitating matter, that is, making its escape difficult. And then, best known of all, came the statement by Laplace in 1796, in his *Exposition du système du monde*, to the effect that any star with the density of our Sun, but with diameter

250 times as great, will be able to recapture all of the light it radiates. In 1799 he published the calculation, which was new only in its details. Laplace dropped the topic from the third edition of his book, perhaps because he appreciated a difficulty it encounters: if light is like any ordinary projectile, it loses speed as it is shot away from a gravitating body. The speed of light, however, was generally supposed to be *constant*.

The star that is supposedly invisible because light cannot leave it does not have to be *undetectable*. That an invisible object can be seen by its *gravitational* effects was appreciated from Bessel's discovery of the invisible companion to Sirius (1844). In this case, admittedly, the companion was directly observed twenty years later – it is now classified as a white dwarf, Sirius B – but the principle was plain enough.

To round off this theme we must mention a brilliant application of some new ideas in the science of quantum mechanics, made in 1928 by the Indian astrophysicist Subrahmanyan Chandrasekhar (1910–). He was then a young graduate student on a voyage from India to study with Eddington. He saw that as a star contracts, in general there will come about a situation where the *inward gravitational attraction is balanced by a certain repulsion* of a type demanded by what is known as the 'exclusion principle'. (This law of quantum mechanics was due to Wolfgang Pauli, an Austrian physicist, in 1925.) He showed that the mass of the star must be less than 1.44 solar masses. White dwarfs were known to belong to the category of extremely dense objects above this limit. Following work by E. Stoner, extended by Chandrasekhar, it was soon realized that there is a limit to the second type of force, a limit set by relativity theory. It was shown that if the mass of a star is greater than the Chandrasekhar limit of 1.44 solar masses, the load on the upper layers will be so great that it will collapse catastrophically. Such a collapse is a possible fate for any star with mass above the limiting value.

The whole idea of the Chandrasekhar limit seemed so strange that Eddington, for one, could never bring himself to accept it, and Chandrasekhar's work moved in another direction for a time. Eddington, after all, was a leading authority on stellar structures. When Chandrasekhar was awarded the Nobel prize for physics, jointly with William A. Fowler (1911–) in 1983, it was for research into the 'origin, evolution and composition of stars', and his earlier ideas had become more or less standard doctrine.

In 1932 another young researcher, the Russian physicist Lev Davidovich Landau (1908–68), proved similar results to Chandrasekhar's, but more simply. He showed that there is another possible final state for a star, even smaller than that reached by a white dwarf. If the 'exclusion principle' repulsive force were due to protons and neutrons rather than electrons, the stars might be even smaller and denser. Neutron stars were not to be detected until a much later time.

In summary, the important point to be deduced from Schwarzschild's result, since the critical radius is proportional to the mass, is that the horizon around a collapsed star (or black hole) can be *on any scale*, as long as the condition of compactness is satisfied – say the mass of a mountain in the volume of an atom, the mass of the Earth in an aniseed ball, the mass of the Sun in the volume of an asteroid, the mass of a small Galaxy in the space of the solar system, and so on.

Until the 1960s, such ideas failed to occupy the thoughts of many astronomers. Eventually, however, highly energetic phenomena were found, both on the scale of a star and on the scale of a Galaxy, that seemed to call for black holes as their explanation. Using satellite telescopes, systems of binary stars were discovered in which one exceedingly compact component, optically invisible, is emitting a very great flux of X-rays, while the other is visible. It was suggested that the massive flux of X-radiation comes from the conversion into energy of material – for instance from its companion's atmosphere – falling into a massive star, such as a neutron star or a black hole. By settling the mass of the invisible component, a decision is made between the alternatives: a black hole is more than about three solar masses. There are now several candidates for this distinction, with masses well above this limit.

On the galactic scale, there are objects – Seyfert galaxies and quasars, for instance – that emit very much more energy than normal galaxies at all wavelengths. In such cases we may be dealing with black holes of the order of a billion star masses, drawing in gas and dust of many star masses annually. To put this idea to the test will require techniques as yet unproven, the most promising being, perhaps, the detection of gravitational waves, which are theoretically emitted with the infall of material into the black hole.

SPECTROSCOPY AND STELLAR EVOLUTION

The use of the spectroscope to study the Sun slowly yielded knowledge of its structure and composition. In due course the instrument was applied to the stars, allowing the findings of solar physics to be extended, but also providing distances and velocities. In this way it provided a key to the structure of the Universe at large. We shall first explain how the latter techniques are related to other ways of determining distance. We have already encountered Bessel's distances, 'annual parallaxes', trigonometrically obtained. Unfortunately the method is of use only with the nearest stars, say up to 100 parsecs, beyond which the annual displacement is too small to measure. Beyond this limit the method based on proper motions was of use. Herschel and later astronomers knew the Sun's motion in space, and its direction. This gives nearby stars a large proper motion and distant stars a lesser one, judged against very distant stars, just as nearby objects seen from a train seem to move faster than distant objects. Of course the objects in question, like the stars, may have other motions of their own, but if we use suitable averaging procedures, and independent reasoning as to what these other motions might be, we can estimate distances from proper motions, given the Sun's velocity.

This last method, bearing in mind the necessary averaging, is called the method of statistical parallaxes. It was applied by several astronomers before Kapteyn, but at the very beginning of the twentieth century he used it with great effect to ascend one rung higher on the distance ladder. From the proper motions at hand, he analysed the relative frequency of stellar (absolute, intrinsic) magnitudes in the Sun's neighbourhood. (From a star's distance, and its apparent magnitude, its absolute magnitude could be found.) From there, assuming the same proportions elsewhere, he could examine groups of distant stars and give probable values to their brightnesses, and so, statistically, estimate their distances. Even then, however, the method would not take him to great distances. The key to further expansion came unexpectedly from spectroscopy.

Stellar spectroscopy had proceeded by fits and starts after Fraunhofer first described lines he had seen in 1814 in Sirius, Castor, Pollux, Capella, Betelgeuse and Procyon. There were isolated attempts by, for example, J. Lamont in the late 1830s, W. Swan in the 1850s and G. B. Donati in the early 1860s; and then came a flurry of activity in 1862–3, with the work of Lewis M. Rutherfurd (1816–92, an American amateur), Airy, Huggins and Secchi. Out of Airy's work there developed the Greenwich programme for measuring star velocities through the Doppler displacements in their spectra.

Rutherfurd's and Secchi's publications were of importance in a quite different way, for they turned astronomers' thoughts to the *classification* of stars through their spectra. Rutherfurd's was a simple threefold division: stars with lines and bands like the Sun, white stars like Sirius with very different spectra, and finally white stars with apparently no lines – 'perhaps they contain no mineral substance', he wrote, 'or are incandescent without flame'. Secchi's first classification, published marginally later but introducing more specific spectral criteria, distinguished two classes; and then in 1866 he decided on three. Briefly stated, these were: coloured stars whose spectra have wide bands in them; white stars with fine lines; and stars with fine lines but a broad band in the blue. Secchi elaborated further on these divisions until his death in 1878.

In the collection of stellar spectroscopic data, especially photographically, Huggins thereafter was long pre-eminent, although increasing numbers of astronomers were entering the field, for example the many who were excited by his study of the spectrum of the nova of 1866. New spectral types were found. The astronomers Charles Joseph Étienne Wolf (1827–1918) and Georges Antoine Pons Rayet (1839–1906), for example, at the Paris Observatory in 1867 announced that they had found three very faint (eighth-magnitude) stars in the constellation of Cygnus, with several broad emission lines on a continuous background spectrum. Then, and indeed now, this spectrum seemed very peculiar: it is now known that Wolf–Rayet stars are as much as ten times the solar mass and are surrounded by an envelope of matter, matter that they are ejecting at very great velocity, of the order of a thousand kilometres per second and more. They are young stars at a certain evolutionary stage that seems likely to be short-lived, and their rarity is due to the shortness of that phase. Fewer than two hundred are known.

New schemes for classifying stellar spectra multiplied as the century wore on, but what was conspicuous by its absence was any deep, unifying, principle, such as might have been provided had more been known of the physics of stars and their production of light. The man who came nearest to this insight was Hermann Carl Vogel (1841–1907), who in 1870 was appointed director of the privately owned Von Bülow Observatory near Kiel in Germany – which then sported the largest refractor in the country. With W. O. Lohse (1845–1914), Vogel began a massive survey of spectra of visible stars, and tried his hand at a sequence of spectra that would lead to an understanding of an *evolution* in the stars. The spectrographic *'Durchmusterung'* went from strength to strength, and its detail was exemplary. Its fame in the long term has far outstripped that of Vogel's classification, even that of a second version he produced to take into account the helium lines that were recognized after Lockyer's discovery of the element. Vogel's schemes are of interest, however, for their emphasis on subtle variations in the character of the hydrogen lines, and later the helium lines, which would eventually be brought into the evolutionary picture in no uncertain way.

When progress came, it came rapidly as the result of a technique that seemed in retrospect to be astonishingly simple. The widow of the Harvard College physician and astronomer Henry Draper established a fund to support the photography, measurement and classification of star spectra, and the publication of results, as a memorial to her late husband. Work began around 1886 at the Harvard College Observatory. Edward Charles Pickering (1846–1919), who had been trained as a physicist, had been director there since 1869. He devised a new and simple technique for producing spectrograms of many stars at once: he placed a large narrow prism of low dispersion in front of the object lens of his telescope, so that each star created a tiny spectral strip on the photographic plate instead of a point image. Substantial bodies of data for the *Draper Catalogue* were ready before the end of the century, although publication of the main work was in nine volumes that appeared only between 1918 and 1924.

There were many changes of direction in the technique of classifying the resulting spectra. At first they were classified by the strengths of hydrogen absorption lines, and an alphabetical series of labels (A, B, C, . . .) was then used.

This sequence seemed to make no sense of the absorption lines for other elements. Pickering was assisted on his project by Williamina Fleming, Antonia Maury (Draper's niece), Annie Jump Cannon and more than a dozen women who helped with the computing. It was Annie Cannon (1863–1941), who joined the team in 1895, who found a way of rearranging the spectra (subdividing some and reclassifying types C and D) so that the progressive changes in the lines seemed orderly over more or less all types. This resulted in a disorderly jumble of the old letters, but it was too late to change them easily, and the sequence that emerged became the basis of that still used. The old sequence had become: O, B, A, F, G, K, M, R, N, S. Henry Norris Russell provided the useful mnemonic *O*h *B*e *A* *F*ine *G*irl, *K*iss *M*e *R*ight *N*ow, *S*weetheart, but the circumspect male astronomer now only dares to speak it inwardly, and never ever refers to 'Pickering's harem'.

Already by 1901 Annie Cannon was able to publish spectra of well over a thousand brighter stars; in the nine later volumes of the *Draper Catalogue* she listed 225 300 spectra, many of stars as faint as magnitude 10. She developed a rare skill in rapid classification, and in spotting peculiarities, but she did not develop any theory to explain the sequence she detected. Several astrophysicists realized very quickly that surface temperature was somehow related to the sequence, with O-type stars the hottest, and Antonia Maury was the first to appreciate that even within a given type there were various possible differences of line-width in the spectra. Here were the raw materials for an extremely important discovery, made almost simultaneously by two astronomers working elsewhere, Ejnar Hertzsprung (1873–1967) and Henry Norris Russell (1877–1957).

A Danish astronomer who had first been trained in chemical engineering, Hertzsprung was later to work in Germany with Karl Schwarzschild, at both Göttingen and Potsdam. His visit to Mount Wilson in 1912 was to prove particularly important, as we shall see, but it was in two separate papers published in 1905 and 1907 in a journal devoted to photo-graphy and its allied chemistry that he showed how Antonia Maury's line-widths could be related to the bright-nesses of stars. Using proper motions to determine distance, and thence intrinsic brightness (absolute magnitude), Hertzsprung showed that her c-type stars, with sharp and deep absorption lines, were more luminous than the others. He thus developed the idea of 'spectroscopic parallax', the idea that line-width generally could be correlated with absolute magnitude, so that the latter could be found dir-ectly from a simple observation of the former. Since Maury types a, b and c were subdivisions of a given spectral type, all stars of which were taken to be at the same temperature, he drew the conclusion that what made some (c-type) brighter than others was their actual physical size. In effect, he had discovered giant stars, hidden in the data produced and duly analysed at Harvard.

Hertzsprung went further, and decided that stars could be divided into two series, one now known as the 'main sequence' and one a sequence of high-luminosity giant stars. His first diagram of this was done in 1906, for the stars of the Pleiades cluster, and was not known in America, where remarkably similar ideas were being developed by Russell.

After studying astronomy at Princeton, Russell worked for a time in physical laboratories in London and Cambridge (England), and at the Cambridge Observatory. There he worked with Arthur Hinks on determining stellar parallax photographically, and he continued this work after returning to a post at Princeton in 1905. By 1910 he had assembled a mass of data allowing him to correlate spectral type with absolute magnitude, as Hertzsprung had done. The branching graph showing the correlation, now known as the Hertzsprung–Russell diagram (or H–R diagram, see figure 16.1), was not generally known until Russell pre-sented his results to the Royal Astronomical Society in London in 1913.

FIG: 16.1   *The Hertzsprung–Russell diagram*

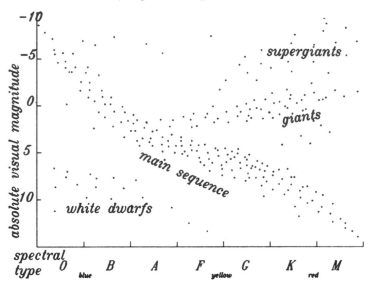

A fuller version than the earliest of the 'H–R diagrams'. Eddington's careful redrawing (1914) of Russell's figure of 1913 is reproduced in plate 14.

With his version of it, Russell gave an interpretation differing, as it was to turn out, from Hertzsprung's. Both took the diagram to show stars at different points in a general pattern of evolution. Russell thought that stars started out as red giants that heated up as they contracted to bright blue stars; and that they then cooled without much further change in size. Hertzsprung at first treated the double graph as indicating two different evolutionary paths. The theoretical study of these questions very soon made great progress, when it passed to a more mathematically inclined class of astrophysicists. The most notable of these was Arthur Stanley Eddington (1882–1944), who had heard Russell deliver his 1912 paper in London.

Eddington, educated at Manchester and Cambridge, was an excellent mathematician with a sound knowledge of observatory practice obtained when he worked for a time at Greenwich. From 1913 until his death in 1944 he was Plumian professor of astronomy at Cambridge, from which place he acted as an incomparable stimulus to world astrophysics. He now took as his theoretical starting point Schwarzschild's theory of the outer atmosphere of a star, which explained how the outward pressure due to radiation could be in equilibrium with the inward pressure due to gravitation. Eddington also took gas pressure into account, and extended the study to all depths within the star. The 'Eddington model' turned out to have some unexpected properties. So rapidly did radiation pressure increase with increasing mass that Eddington decided that stars with masses ten times greater than the Sun's would be relatively rare. In 1924 he was able to publish a theoretical relation between the mass and luminosity of a star. By then it was known that dwarf stars could have very high densities, and many supposed that they could not be gaseous, but they fitted Eddington's model so well that he decided otherwise. With data from G. E. Hale and W. S. Adams at Mount Wilson, his model triumphed over widespread scepticism when he applied it to the companion of Sirius, to which he assigned an extraordinary density, 50 000 g/cm^3. His theory was followed in 1926 by R. H. Fowler's studies of the super-dense gas – or plasma, as it would now be called – using ideas drawn from the new branch of physics known as wave mechanics.

Eddington presented his ideas in a masterly book, *The Internal Constitution of the Stars* (1926). He was able to introduce time into his theory, and so to interpret the H–R diagram as a template of stellar evolution. It seemed, however, that it implied a time-scale for evolution unparalleled in other astronomical theories, in fact one of the order of a million million years. It seemed now that a massive star, say of type O or B, would require as much time as this to bring it down to the mass of a white dwarf. This whole question was one that would later link the theory of stellar evolution with theories of the age of the Universe. Eddington, one of the leading exponents of Einstein's theories of relativity out of which so many new cosmological ideas were springing forth, was no stranger to the equivalence of mass and energy, and had earlier (1917) developed a theory of the sub-atomic origin of stellar energy (electron–proton annihilation). He later developed alternative explanations (especially electron–positron annihilation), and lived to see them incorporated in a solution to the problem later found by Hans Bethe in 1938, that is, the carbon-nitrogen-oxygen-carbon cycle.

CEPHEID VARIABLES AND THE MILKY WAY

At a time when astronomers so desperately needed techniques for determining distances, yet another came to hand unexpectedly from work at Harvard. In 1786, John Goodricke had noticed the variability of the star delta Cephei, and the star became an object of interest, but little more. In 1908, while studying stars of fluctuating brightness in the southern group known as the Small Magellanic Cloud, Henrietta Swan Leavitt (1868–1921) noticed that, for the sixteen stars whose light variations she had measured carefully, the longer the period over which the change took place, the brighter the star. Four years later she found a simple mathematical relationship that fitted observations very well: the apparent magnitudes were nearly proportional to the logarithms of the periods. Since the stars were clearly all at much the same distances, the same sort of relationship should hold for the absolute magnitudes of the stars. She saw the potential value of the stars as distance indicators, but could take the technique no further for the time being.

The first of her announcements had attracted little attention, but after reading her second, Ejnar Hertzsprung realized that the pattern of variation in the luminosity of the stars she was studying resembled those of cepheid variable stars. He saw that since such stars seemed to proclaim their identity by the pattern of their light curves, they should make excellent distance indicators in the Universe at large, granted that the Leavitt period–luminosity relationship could be calibrated, that is, could be made to yield a measure of absolute luminosity from the periodic time of fluctuation.

This was a very promising idea, but unfortunately there were no known cepheid-type variables near enough to the Sun for their distances to be measurable by trigonometrical methods for parallax. Hertzsprung was able to use proper motions for the method of 'statistical parallax', however, and so went on to the necessary calibration of Leavitt's curves. Although there were to be many subsequent revisions of the calibration, this new method opened up a new era in the measurement of distances, for now a single cepheid-type star, for instance in a distant nebula, could yield a plausible distance from a relatively simple measurement of its periodicity. Revision was certainly necessary, for Hertzsprung had made an arithmetical error that made his distances too small by a factor of ten. (The distances were so large by the standards of the time that the error went unnoticed, and persisted in some quarters for twenty years.)

At much the same time, but independently, H. N. Russell was working out the absolute magnitudes of the cepheids in the Milky Way, and he found values close to Hertzsprung's, but without appreciating the technique that rested on Henrietta Leavitt's discovery. Also at this time Harlow Shapley was a young doctoral student working with Russell. They saw that the cepheids could not – as some had argued – be eclipsing binaries. They were very bright and very large. In 1914 – after a tour of Europe, some months in Princeton to complete his work on eclipsing binaries, and marriage to Martha Betz, who was to become an authority on that subject – Harlow Shapley moved to work at Mount Wilson. There he began a study of variable stars in those clusters of tens of thousands of stars whose apparently spherical form has given them the name 'globular clusters'. (Many globular clusters are easily seen with binoculars. The number associated with our Galaxy is of the order of a hundred – but their status was as yet unknown.)

Working with the 60-inch reflector there, he found a few cepheids among the variables, and these seemed to follow the pattern of Leavitt's law. Their apparent magnitudes differed from one cluster to another, but this was only to be expected if the clusters were at different distances. In fact here he had a means of determining the *relative* distances of clusters, even *without calibrating* the Leavitt graph. In 1918 he produced a calibration in the way Hertzsprung had done, and announced that the typical globular cluster was of the order of 50 000 light years distant. He found that they were not symmetrically arranged, however, and decided that the centre of the system of globular clusters was probably identical with the centre of the Milky Way system of stars, some tens of thousands of light years distant from us.

This result, placing the Sun on the edge of the system, was much disliked by many astronomers for a variety of reasons. It was not that many of them expected the Sun to be at the centre of things, but that the distances seemed to fit badly with existing, hard-won, ideas. In a series of papers beginning in 1884 and covering two decades, Hugo von Seeliger (1849–1924), having developed a number of new principles of stellar statistics, made counts of stars of different apparent magnitudes in various regions of the sky, and found from them a flattened-disc model of the Milky Way. In overall form this was not unlike Herschel's model, with the Sun not far from the centre. In 1901, Kapteyn used his statistics of proper motions, as already explained, to provide a distance scale for Seeliger's work, making the system about 10 kpc (kiloparsecs) in diameter and 2 kpc in thickness. Kapteyn saw that one unknown quantity, interstellar absorption, might have strongly affected his data, and he made various attempts to measure absorption, but with limited success. By 1918, working with his assistant Pieter

Johannes van Rhijn, he had concluded from the absence of serious reddening of starlight that the effect was not very significant, and that his basic model was a reasonable representation of the Milky Way.

The model, while circulating widely among his many acquaintances, was not published in detail until 1922, the year of Kapteyn's death. The overall shape of his scheme was an ellipsoid – a squashed sphere rather than a rugby ball – with its axes in the ratio of about 5 to 1, its major diameter being about 16 kpc. Having a statistical idea of stellar distances, he could make estimates of the thinning out of stars with distance. He took the Sun to be about 0.65 kpc from the centre and off the central plane, but for convenience often spoke as though the Sun was at the centre. He thought that at 8 kpc distance from the centre there were only about a hundredth as many stars in a given volume as in the Sun's neighbourhood, while at 4 kpc the density was about a twentieth that near the Sun. Broadly speaking, his and Van Rhijn's estimate of the pattern of thinning of stars with distance was reasonably correct in directions well away from the central plane of the Milky Way, but was in serious error in the central plane, where interstellar matter congregates.

That this was so was first conclusively shown by the Swiss-born Robert Julius Trumpler (1886–1956), studying open clusters (less tightly bound than globular clusters) at the Lick Observatory in California, in 1930. Trumpler's method can be very simply explained. He sorted his clusters into a small number of categories, using criteria that need not concern us. Assuming that clusters within a given category were all of much the same physical size and character, the relative distances of his clusters could be estimated twice over, once on the basis of their faintness, and once on the basis of their apparent overall size. The two methods in practice turned out to give quite different results, and he decided that the brightness criterion was being upset by interstellar absorption. Expressed otherwise: if we accept distances based on the size criterion, then brightness provides a measure of absorption in the direction in question.

In all his work on distances, Kapteyn had shown a preference for proper motions over apparent magnitudes, as distance indicators, since the magnitudes depend on the intrinsic properties of stars, which vary widely. Proper motions are in large part a reflection of the Sun's motion through space, and it was generally assumed that *on average* they depended on nothing else, that is, that the motions of the stars have a random character, like the molecules of a gas. Kapteyn found, however, that this assumption gave various inconsistent results. It seemed to him at an early stage in his work that stars belong to two different groups, populations, that are intermingled.

His discovery of the two 'star streams' was announced at a congress at St Louis, Missouri, in 1904, and caused a great stir in astronomical circles. Karl Schwarzschild, master of stellar statistics, not to be outdone, in 1907 devised a model that explained the measured proper motions on the assumption of a carefully chosen relationship of velocity and position within the Milky Way, avoiding the assumption of intermixed populations. These were the beginnings of a long and continuing study of motions within our Galaxy, which even in Kapteyn's work made it possible to introduce gravitational considerations, and so strengthen the models derived from counts of stars with measured magnitudes and proper motions.

Even before Kapteyn's final model had been made known, a rival was being worked out by Harlow Shapley, after his move to Mount Wilson. As we have seen, he had noted that globular clusters, for which he had tentative distances, are not symmetrically arranged in the sky. He noted an idea put forward by Bohlin in 1909, without much evidence. This idea, which was simply that the Sun's position is not central, while it seemed to explain this asymmetry, did not fit with his own estimates of their distance, at least taken together with conventional wisdom on the size of the Milky Way system. By 1916 he had found from cepheid measurements that the globular cluster Messier 13 was 30 kpc from the Sun, far beyond the limits of Kapteyn's galaxy. A year later, with more evidence from other globular clusters, he returned to Bohlin's idea, and decided that they were indeed all associated with the Galaxy (Milky Way), centred on the invisible nucleus of our Galaxy, somewhere in the direction of Sagittarius. A third of them were in only a twentieth part of the sky in that direction. He concluded that *the Galaxy must be ten times larger than was generally believed.*

Broadly speaking, Shapley had overestimated by a factor of two, since he had ignored interstellar absorption. He had also wrongly assumed that cepheids in globular clusters are like those in the Sun's neighbourhood, but by a stroke of luck he also underestimated the brightness of the latter by a factor of four, so that these factors cancelled out. Even had the extent of absorption been appreciated, he would still, however, have been regarded by most astronomers as wildly wrong.

ACCOMMODATING THE SPIRALS

The ultimate reasons for dismissing Shapley's claims did not rest on Kapteyn's great authority, but had to do with the spiral nebulae, and beliefs about their status. Spectroscopy had enlarged knowledge of them: after Huggins' first excitement at finding that they seemed to have the line spectra of a luminous gas, he found ever more with continuous spectra. Distinguishing between the two sorts as 'green' and 'white' nebulae, astronomers struggled to resolve the white nebulae into stars telescopically. The first success on this score was not until 1924, with the resolution of the nebula in Andromeda, M31 – although its spectrum had been shown to resemble that of a cluster of stars by J. Scheiner, shortly before the end of the previous century.

Now although, with the advantage of hindsight, it is possible to distinguish between two separate fields of study, that of our own Galaxy and that of the 'white', predominantly spiral, nebulae that we now regard as co-equal galaxies, such a clear-cut division was not possible as long as there was a possibility that the second were merely appendages to the first. Kapteyn's model of the first was carefully devised on a statistical basis, but there were earlier alternatives based on impressions received from visual and photographic evidence. John Herschel had noted what appeared to be streams of stars between us and the main body of the Milky Way. Giovanni Celoria of Milan had in 1879 proposed a model making the Milky Way equivalent to two more or less concentric rings of stars. Von Seeliger marshalled evidence against these notions, and yet in 1900 Cornelis Easton, a Dutch amateur astronomer in Dordrecht, reintroduced Celoria's ideas, but varied the position of the Sun and the centres and inclinations of the rings. By 1913, Easton had drawn together much photographic evidence to argue that the Milky Way was like a spiral nebula, in fact resembling M51 and M101. But still those other spirals were not comparable to ours. They were 'small eddies in the convolution of the great one'.

Despite forceful arguments to the contrary, especially by Eddington, there was a strong feeling that the Milky Way was a unique focus of the Universe. The arguments for this view were broadly five in number. The dimensions of the nebulae were thought to be insignificant by comparison with the size of the Milky Way. The star-like spectra of the white nebulae (with Fraunhofer lines) could be explained by starlight reflected from a diffuse nebula – such as V. M. Slipher found in 1912. Thirdly, for those prepared to swallow Shapley's pill, the Milky Way was big enough to include at least some of the white nebulae, according to distances estimated for them at the time. Fourthly, the white nebulae seemed to avoid the plane of the Milky Way, suggesting that they were symmetrically arranged around it. And last, new evidence as to their velocities was coming in that likewise suggested symmetry.

The first type of argument seemed at the time particularly strong, especially in a version that began from evidence as to the internal motions of the nebulae. Adriaan van Maanen (1884–1946) had studied with Kapteyn before taking a post at Mount Wilson in 1912. From its completion in 1914, he used the 60-inch reflector there, and in 1916 was able to publish the results of measurements concerning the rotations of spiral nebulae, deduced from photographs taken over an interval of time. Checked by an experienced colleague, Seth Nicholson, motions outwards along the spiral arms of nine nebulae were surprisingly large. The actual velocities of matter along the arms could in some cases be measured by spectroscopy (the Doppler shift), and taking the two results together for M33, a distance of only 2 kpc was deduced, putting the object well within a Kapteyn universe.

Alas, Van Maanen's results were soon rejected. Knut Lundmark measured his plates again in 1927 and found movements only a tenth as great as he had said. Even he later halved the movement, but stuck to his general conclusion, and his friend Shapley accepted the same, and argued for nearby nebulae. Hubble later showed his findings to have been influenced by some sort of systematic error, and indeed, although an excellent practical astronomer, in claiming to measure movements well below the limits of his apparatus Van Maanen was far too sanguine. He also had a way of brushing aside the findings of other astronomers – C. O. Lampland of the Lowell Observatory, W. J. A. Schouten of Groningen and H. C. Curtis, for example. Knut Lundmark struggled hard to refute his inferences, and to uphold the idea that the nebulae were 'island universes'. (Alexander von Humboldt gave this last expression – *Weltinseln* in German – a wide currency in his book *Kosmos*, in

1850.) Van Maanen's findings could not be easily refuted, however, without coming close to questioning his integrity, and many continued to accept them until around 1933.

This entire episode well illustrates, as it happens, how other criteria than the purely scientific may play an important but largely invisible role in the evolution of a science. Here there were clear personal forces working for and against the island Universe idea. It was no secret that Hubble had a strong dislike of the socially ebullient bachelor Van Maanen, and that although Shapley and he were from Missouri, during his time as a Rhodes scholar in Oxford Hubble had acquired mannerisms that impressed the ladies in the vicinity of Mount Wilson more than they impressed Shapley. In his autobiography, Shapley tells of how he heard that Hubble, refereeing a paper he had written for a general readership, had simply scribbled 'of no consequence' across the top. The editors told Shapley that his paper had first been accidentally set in type with 'Shapley – of no consequence' at its head.

Among those mentioned as having been opposed to Van Maanen's findings was Heber D. Curtis of the Lick Observatory, who in 1914 had begun a study of the spirals, which he already suspected to be 'inconceivably distant galaxies of stars or separate stellar universes'. But how to measure their distances? One answer was found almost by chance as a consequence of a discovery made twice over. In 1917 Curtis discovered a nova in one of the spirals he was examining, but did not announce the fact. Then in the same year George W. Ritchey, working at Mount Wilson, was photographing spirals in the hope of finding rotations and proper motions, when in one nebula (NGC 6946) he too found evidence of a nova. This led him to examine old plates and so immediately to discover a handful of others that had previously gone unnoticed. Astronomers elsewhere did likewise, and quickly the number reached into the 'teens. Curtis was delighted with the implications of the apparent brightness of the novae, extremely small by comparison with novae in the Milky Way. He said that with two exceptions the average difference was ten stellar magnitudes, making them a hundred times more distant, and thus far outside the limits of the Galaxy, whether on Kapteyn's or any other model then thought acceptable. (The two exceptions were later recognized as supernovae.)

There were two markedly different opinions, therefore, at this time about the spirals. Shapley and Van Maanen placed them so near to the Sun that they could be treated as subsidiary to the Galaxy (Milky Way). Curtis placed them so far away that their physical sizes, as deduced from the apparent sizes, must be comparable with the Galaxy's. If one were to include the globular clusters in the overall system of the Galaxy, as Shapley did, the case was less clear, but even then Curtis' evidence, if accepted at its face value, would have been difficult to incorporate.

In 1920, it seemed that matters were coming to a head when Curtis and Shapley agreed to debate the question of the scale of the Universe at a meeting of the National Academy of Sciences in Washington. In the event, the meeting seems to have been in a low key, with the two men at odds over what they were meant to be discussing. Shapley took seven out of nineteen pages of his script before reaching the definition of a light year, a fact conceivably related to the presence at the lecture of those who might choose to appoint him director of the Harvard College Observatory, for which he was then applying. He concentrated on the size of the Galaxy, and said little about the spirals, while Curtis made them his theme. Their differences of opinion made more of an impression on the astronomical world when the two went into print with their views, shortly afterwards. In the light of later developments we may say that, despite many imperfections in the data available, each was broadly right on his main theme – Shapley on the Galaxy and Curtis on the spirals – but that Shapley was greatly disadvantaged by his acceptance of Van Maanen's work on the spirals.

OUR SPIRAL GALAXY. DARK MATTER

In 1871 the Swedish astronomer Hugo Gyldén, working in Stockholm, examined large numbers of large proper motions, that is, of stars that were likely to be relatively close to us. He discovered that they are not symmetrically arranged, but seemed to be concentrated in one half of the sky, and that there they drift in a roughly similar direction. At right angles to the greatest motions the stars had negligible motions. He interpreted this as a sign of rotation in the Galaxy. Quite independently of this discovery, when studying stellar motions spectroscopically – that is, using the Doppler effect – Benjamin Boss, Walter S. Adams and Arnold Kohlschutter found a similar asymmetry in the arrangement of high-velocity stars. In 1914 they announced that they had found that three-quarters of those recorded seemed to be approaching the Sun.

Adams continued this study, and it was also taken up by Jan Hendrik Oort (1900–92), who had studied with Kapteyn, and who there in 1926 completed his doctoral thesis on this very subject while working with Van Rhijn – after Kapteyn's death. In 1922, however, he published an interim study showing that the Sun was approaching high-velocity stars in one half of the sky (more strictly between galactic longitudes 310° and 162°) and receding from those in the other half. Oort found another peculiarity: below 62 km/s, the radial velocities were random, while above this value they showed the asymmetry. He attempted a gravitational analysis of the Galaxy, and assumed, mistakenly, that the high-velocity stars had entered the system from outside. He assumed that 62 km/s was the velocity above which a star would be able to escape from the system's gravity. From this he was able to deduce an average stellar mass of 0.65 solar masses. This was hard to reconcile with the dynamical equilibrium of stars within the system, however, which required masses nearly eight times as great.

The general picture Oort drew from the velocities was that the high-velocity stars are moving round the centre of the main system, and also with respect to the globular clusters. The Sun he took to be a member of a local system (a cloud of stars) moving in much the same way, but somewhat faster.

In all this, Oort was assuming a Kapteyn model of the Galaxy. When he examined the radial velocities of the globular clusters, he found that not only did they favour one part of the sky but that they too had their own asymmetry in velocity. A Galaxy arranged along the lines of Kapteyn's model would not be able to hold them gravitationally, and yet they certainly seemed to belong to the Galaxy, judging by their symmetrical arrangement about its principal plane. The explanation, Oort believed, was that the Galaxy is more massive – as much as 200 times more – than the visible matter in it (that is, mainly visible stars) would lead us to believe. This was the beginning of a new interest in a difficult but vital part of astronomy, the study of dark matter.

Oort's tentative explanation of the invisible material was that it was obscured by matter in the galactic plane. By 1932 he was insisting that we may deduce the existence of two or three times the visible mass, using dynamical arguments. By this time astronomers were well accustomed to the idea that ionized atoms were to be found in clouds in outer space. Johannes Franz Hartmann (1865–1936) had discovered a spectral line for ionized calcium in 1904 by studying the spectrum of delta Orionis, and by the 1920s lines of ionized sodium and titanium were also recognized. In anticipation of future developments, however, it should be added here that the 'missing mass' is not only of dust and small particles but contains a contribution by stars of very low mass and low luminosity belonging to the halo of the Galaxy. Such stars might be black holes or burned-out dwarf stars, of a type to be mentioned again in due course.

We have already encountered early arguments – for instance Michell's and Laplace's – suggesting that under certain conditions light will not be able to escape from massive bodies. In other words, these bodies will be invisible, and the only way of detecting them might be through their gravitational effect on neighbouring visible bodies. 'Dark matter', which may be invisible for many reasons, assumed increasing importance over the course of the twentieth century. E. E. Barnard's fine collection of photographs of the Milky Way star clouds, often showing dark areas, had provided evidence of a general sort.

The first really penetrating theoretical study of the problem of the interstellar medium was Eddington's, in response to important spectrographic studies of calcium lines by the Canadian astronomer John Stanley Plaskett (1865–1941), director of the Dominion Astrophysical Observatory in Victoria, British Columbia. Plaskett had concluded from line displacements that clouds of calcium were not moving with the stars whose light allowed us to detect it, and Eddington agreed, but went further in maintaining that there was a continuous cloud occupying the entire Galaxy, and more or less at rest in relation to it. He calculated the energy density of starlight in space, and so the temperature of the gas, which he found to be high enough for the gas to be doubly ionized.

This work of Eddington's, done around 1926, stimulated more intensive observational work, for example by Otto Struve at Yerkes, who soon came to Plaskett's defence on the question of whether the calcium was in discrete clouds, or was all-pervading. Over the next few years, as Struve pursued the problem further, he withdrew his arguments, and with Trumpler's work at Lick on the globular clusters, mentioned previously, the issue was settled in Eddington's favour. (Trumpler's results were published in 1930.)

As we have already seen, the spectrographic route was not the only approach to the problem. Before Oort, Kapteyn had compared the observed mass-to-light ratio in the Sun's neighbourhood within the Galaxy, and had compared it with what was calculated assuming a Galaxy in equilibrium. He had concluded that the mass of dark matter could not be excessive, but now Oort's work, also using dynamical arguments, had dispelled that idea.

A new model for the Galaxy was needed that would cover its form, its internal movements, and its relationship with the globular clusters and possibly the spiral nebulae. Many had speculated about a spiral structure for the Galaxy itself, but this was singularly difficult to prove. Bart J. Bok, another Groningen astronomer, who had spent a short time with Shapley at the Harvard College Observatory while working for his doctorate, had applied Kapteyn's numerical methods to the Galaxy, but after attempts spanning more than a decade he had failed to produce any conclusive proof of spirality.

After Oort's dissertation of 1926, it became clear that rotation must be taken into account, but Bertil Lindblad (1895–1965) of Sweden had been laying the foundations for this even earlier, in a mathematical treatment of stellar statistics begun in earnest around 1922. Lindblad's was a system of interpenetrating subsystems of stars, all moving in elliptical orbits around the galactic centre, as do the planets in our solar system, following Newton's laws. The galactic case is of course very much more complex. (We recall that Kapteyn's model was ellipsoidal.) Over the years, Lindblad became a leading authority on the dynamics of systems with spiral arms. His theories provided a stimulus to Swedish astronomy, and led many at the observatories of Lund, Uppsala and Stockholm to specialize in the same branch of study. More immediately it led Oort to work out a new model for differential galactic rotation, and to estimate the distance and direction of the galactic centre. In 1927 he made out the distance to be about 6.3 kpc, only a third of Shapley's value, but of the same order of magnitude.

Others joined in the mapping of velocities in the Galaxy. At the beginning of the century the two great American telescopes, at Lick and at Yerkes, were manned respectively by William Wallace Campbell (1862–1938) and Edwin Brant Frost (1866–1935), who were unequalled in their use of the spectrograph to measure stellar (radial) velocities. Their careers ran curiously parallel courses, even to the extent that both became blind – Campbell partially and Frost totally. Still more relevant to the Oort problem than the materials they had assembled were data obtained by Plaskett, who was highly experienced in measuring stellar velocities with the spectrograph. Stars of types O and B are intrinsically very bright, and so can be seen at great distances. When he heard of Oort's work, he had extensive data on stars in these classes, that he and J. A. Pearce had long been collecting (especially of types between O5 and B7), and so could immediately test Oort's theory, which had been based on relatively sparse data. Agreement with Oort's parameters was surprisingly close.

Oort's discoveries, of galactic rotation and of the enormous importance of invisible matter, led to considerable new activity in the development of galactic models. On the question of dark matter, the situation was coming to a head in the early 1920s. Kapteyn's suspicions have already been mentioned. James Jeans estimated that there must be three dark stars in the Universe to every bright star (1922). The first study that could be called definitive was yet again by Oort, in 1932, and this set a value for the density in space determined by dynamical arguments. This, known as the 'Oort limit', he set at roughly one solar mass per ten cubic parsecs. (By 1965 he had increased his estimate by fifty per cent, and he thought forty per cent of the total to be invisible stars and gas.)

An approach distinct from Oort's was developed by Fritz Zwicky (1898–1974), an astronomer born in Bulgaria but of Swiss parentage. In 1933 Zwicky found a very surprising result when he compared the ratio of mass to luminosity (taking the Sun as a standard of measurement) in ordinary single galaxies with the figure for galaxies in a cluster (the

Coma cluster, containing more than a thousand bright galaxies). The second figure, based on the spread of velocities of the component galaxies, was fifty times greater than the first. Although the analysis was crude, it was clear that non-luminous matter on a cosmic scale was of even greater significance than had previously been supposed. In 1936 Sinclair Smith repeated the procedure for members of the rich and nearby Virgo cluster of galaxies, and found there a mass per Galaxy a hundred times that implied by the luminosities. Zwicky believed that in addition to intergalactic material there must be intergalactic stars and even dark dwarf galaxies. The Universe had evidently been behaving as a dark horse throughout all of human history.

During the period following the Second World War, evidence was found to suggest that in some systems the mass-to-luminosity ratio might be as much as 1000, and that single galaxies might have still more dark matter than had been suspected. As we shall see later, the steady-state cosmology of this time called for a field of dark matter being constantly replenished as the Universe expanded, out of which stars formed; and so the whole issue became in some quarters highly topical. Even so, in singling out this theme for discussion, it has to be said that in the 1950s and 1960s there were many astronomers strongly opposed to the idea that dark matter could be of much overall importance in the Universe. In the 1960s those astronomers who considered the nature of clusters of galaxies, for example, were split into two groups, one arguing that galaxies in clusters were bound together by dark matter, the other that this was not so, and that the clusters were relatively short-lived, each expanding from some sort of explosion peculiar to its own system. Opinion is still divided, but there seems to be a consensus that mass-to-luminosity increases steadily with the size of the system considered, and that the hypothesis of dark matter is here to stay.

To return to the form of our Galaxy as it was understood in the two decades after 1930: when further inroads were made into the problem, it was as a consequence of closer studies of spiral galaxies beyond our own, galaxies such as M31 and others nearby. In this work, Walter Baade (1893–1960) produced the crucial evidence. Born in Westphalia in Germany, Baade had acquired experience at the Bergedorf Observatory of Hamburg University. A meeting with Shapley in 1920 had led him to a study of globular clusters, for which his telescope was barely adequate, and in 1926–7 he was able to enlarge his experience when, on a Rockefeller Fellowship, he visited the large Californian telescopes. After his return to Bergedorf he became a close friend of the telescope-builder Bernard Voldemar Schmidt (1879–1935), an eccentric Estonian genius whose best ideas are said to have come to him when in a state of complete intoxication. Their friendship grew on a sea voyage to the Philippines, to observe an eclipse, and resulted in Schmidt's designing a new optical system of the greatest importance for astronomy.

This uses a thin correcting plate across the upper end of the reflecting telescope, and makes possible high apertures (low f/numbers), and therefore shorter photographic exposures. Schmidt was fond of pointing out that he made his telescopes single-handed. In fact he had lost an arm as a boy. Schmidt telescopes would eventually be almost indispensable aids to the survey of the sky that led to the proof of spiral arms in our Galaxy.

In 1931 Baade accepted an invitation to join the staff at Mount Wilson. During the Second World War he used the 100-inch refractor to study M31 and its satellite galaxies M32 and NGC 205. The fact that the skies over Los Angeles were blacked out during wartime allowed him to photograph individual stars in the inner regions of M31, and brought about the discovery that the brightest stars there were red, and not blue, as in the spiral arms. The paper in which he announced this discovery gave the names type I and type II to stars of types that had in effect been hinted at long before, without any analysis of internal characteristics, in our own Galaxy in Kapteyn's two 'star streams', and that had been the subject of Oort's work. Baade's type I was of O- and B-type stars, highly luminous and blue, while type II were the brightest red stars, also found in globular clusters.

By June 1950, at the dedication of a Schmidt telescope in Michigan, Baade expressed his conviction that our Galaxy is a spiral of a type known as Sb (using Hubble's classification), simply because its nucleus resembled that of M31, which was of that type. Type I stars might serve as markers for the spirals, but we can only plot them in three dimensions if we know their distances, and this requires a knowledge of their absolute magnitudes. William W. Morgan of Yerkes and Jason J. Nassau of Warner and Swasey Observatory were at this time studying that very problem, and shortly afterwards announced that on the basis of forty-nine estimated distances (out of 900 stars of these types examined) it seemed that our Sun is on the outer edge of a spiral arm of the Galaxy. Another clue had become available, since O- and B-type stars were by this time known to be associated with large regions of ionized hydrogen (H II). By the end of 1951, they, and Stewart Sharpless and Donald Osterbrock, had used H II as a tracer outlining two spiral arms, one through the Sun and the other beyond the galactic centre. Tracing the second was a remarkable achievement. To all intents and purposes, the problem that had been attacked for so long using star-counting techniques had yielded at last to quite a different technique.

But only just in time. In the same year, 1951, radio emissions were detected from neutral hydrogen, and they very soon supplemented the evidence that had been based on the visible emissions from ionized hydrogen. The discovery was made almost simultaneously in the United States, the Netherlands and Australia. The whole of the Galaxy immediately became open to observation. Oort and Bok and their numerous colleagues and students quickly threw themselves into a study of the Galaxy's structure as marked out by these radio sources, and within a year the spiral structure of the Galaxy was established beyond all doubt. Oort had by this time been long working in Leiden, and of course supplementary information was needed for those parts of the Milky Way appearing only in the southern sky. Some of the earliest radio work was in fact done in the 1950s by astronomers in Sydney, Australia.

The *origin* of the spiral structure of the Galaxy was not satisfactorily explained until the 1960s, when a theory of density waves was developed by Lin and Shu. One of the great advantages of the radio measurements was that the *intensity* of the emissions from the neutral gas gives a measure of the concentrations of gas, and so of the gravitational field. Unfortunately, however, there is an important missing ingredient in all this, namely molecules, dust and gas that provide a site for star formation. This material has been detected in various ways since the early 1960s through radio emissions.

From the 1970s onwards, it became possible to study the rotation of external galaxies in more detail, using in particular the radio emission (the 21-cm line) from neutral hydrogen from relatively nearby galaxies. In 1970, using evidence of the rotation of the two satellite galaxies NGC 300 and M33, K. C. Freeman deduced from the measured rotations that there must be dark matter at least as massive as the Galaxy, and that it must be distributed in a way quite different from the visible matter in those galaxies. The measurements were done with a single dish antenna. When data from multiple, synthesizing, radio telescopes became available – first from Owens Valley and more particularly from Westerbork in the Netherlands – it became clear that the total masses of galaxies are roughly proportional to their radii – not a result to be expected from a simple spherical or elliptical model of the Galaxy, for example. Massive invisible haloes were deduced from the patterns of the observed rotations, and these haloes assume great importance, gravitationally speaking, in the *outer* parts of the galaxies. Improvements in knowledge of the patterns of rotation in galaxies provide perhaps the most reliable way of estimating dark matter in galaxies, and the great technical improvements made in the early 1980s on such telescopes as that at Westerbork much advanced this type of study.

THEORIES OF GALACTIC EVOLUTION

One of Van Maanen's lines of defence, when he produced what purported to be evidence that the nebulae were relatively near, was to show that this idea fitted well with a theory of how the nebulae actually form and evolve. In this he followed a lead given by the Cambridge mathematical physicist James Jeans (1877–1946) in an important study published in 1919: *Problems of Cosmogony and Stellar Dynamics*. Jeans had retired from teaching in 1912, on the grounds of ill health, but he continued to make fundamental contributions to theoretical mechanics and physics until, from the late 1920s until his death, he turned to writing for a wider public – which he did with great acclaim. In his 1919 monograph Jeans explained how a spherical mass of gas would contract under gravity and flatten with spin until it became unstable. It would then throw out filaments of matter from its edge, and they would eventually form themselves into spiral arms. One of the main issues was whether the chief motions observed in the spirals were rotational or (as Jeans thought) outwards along the arms, with condensations occurring in the arms and so giving birth to giant stars.

In 1921 Van Maanen supplied what he thought to be evidence for the second view. Jeans was so impressed by the evidence that he went so far as to toy with the idea of modifying Newton's law of gravitation, since the two did not seem to be reconcilable. His enchantment with Van Maanen's data did not last long, however, once Hubble began to produce new evidence as to the vast distances of the spirals, based on the discovery of cepheids in them (1924). Van Maanen's claimed rotations remained a problem, as explained earlier. The instincts of those who chose to solve the problem by simply ignoring it turned out to be sound.

The crucial positive evidence – like that long afterwards which discredited Van Maanen's measurements – was supplied by Hubble. Edwin Powell Hubble (1889–1953) was the son of a Missouri lawyer. Introduced to astronomy at the University of Chicago by G. E. Hale, after a degree in mathematics and astronomy he went on to Oxford for a time, and there took a degree in law. A successful athlete and boxer – surely the only astronomer to have entered the ring with the great French champion Georges Carpentier – he returned to Kentucky where he practised law until joining the staff of the Yerkes Observatory of his old university in 1914.

In 1917 Hubble completed a doctoral dissertation on the classification of nebulae, and was soon offered a post at Mount Wilson by Hale. This first attempt at what was essentially a revision of Max Wolf's was not particularly noteworthy, and there were several alternatives in play at the time. He returned to the problem later, and resisted the temptation to pay too close attention to the nebular evolution theory of Jeans. This is ironical in view of the fact that his later scheme was rejected by the International Astronomical Union in 1925 because it employed terms suggestive of a physical theory of evolution. That it did so is hardly surprising, for as Jeans later pointed out, it agreed almost exactly with his own ideas.

Lundmark thought that it plagiarized his own scheme, and a bitter dispute arose. The reason for the similarities was surely that Jeans' and even earlier writers' ideas were in the thoughts of both. At all events, Hubble's later version of his classification of nebulae (developed in 1923) provided a basis for that which is still in use.

He distinguished between regular (with a clear nucleus) and irregular (relatively scarce, lacking a nucleus), and divided the regular nebulae into normal spirals, barred spirals and ellipticals, each class being represented by subclasses that seemed to be in an evolutionary sequence. The diagram from Hubble's *Realm of the Nebulae* (1936) is reproduced here in figure 16.2.

FIG: 16.2 *The evolution of nebulae (Hubble)*

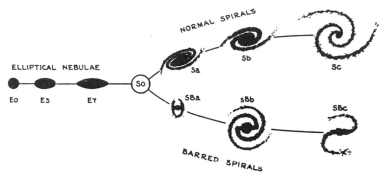

When the United States eventually entered the First World War, in 1917, Hubble enlisted in the infantry, and was not free to move to Mount Wilson until two years later. Working there with the 60-inch telescope, he was at once in a position to record numerous new objects. He noted that diffuse nebulae (as we now know, *within* our Galaxy) are made luminous by nearby blue stars of high surface temperature. (We recall William Herschel's example of the illumination of a nebula by a nearby star.) Hubble gave the phenomenon a theoretical basis, but his reputation was made in a very different way.

By 1923, now also using the 100-inch telescope, he was able to resolve the outer regions of the spirals M31 and M33 into stars, of which altogether thirty-four varied in brightness following the pattern of cepheid variables. At last temperature. (We recall William Herschel's example of the illumination of a nebula by a nearby star.) Hubble gave the phenomenon a theoretical basis, but his reputation was made in a very different way.

By 1923, now also using the 100-inch telescope, he was able to resolve the outer regions of the spirals M31 and M33 into stars, of which altogether thirty-four varied in brightness following the pattern of cepheid variables. At last it was clear that M31 lies well outside the limits of the Galaxy. By the end of 1924 he put its distance at 285 kpc, of the order of ten times the greatest diameter of the Galaxy as then widely accepted.

During the late 1920s Hubble made estimates of the distances of further nebulae, and tried to extend the ladder of distances, that is, by finding other criteria than cepheids, which can only be seen up to say 4000 kpc. He very tentatively proposed using a 'brightest star' criterion in late-type spirals, objects that seem to be about 50 000 times as luminous as the Sun. It hardly mattered what the objects were. (In fact in 1958 Allan Sandage showed that they are clouds of ionized hydrogen.) By averaging the properties of such bright stars visible in galaxies of the great Virgo cluster, which must be of comparable distance, he found data that he could apply to still more remote galaxies. In this way Hubble built up an incomparable body of data as to galactic distances. By 1929 he had distances for eighteen galaxies and four members of the Virgo cluster.

In this way Hubble provided what proved to be an essential key to a subject that for more than a decade had been theoretically prepared for it. The subject in question was a type of cosmology that had grown, quite unforeseen, out of Einstein's general theory of relativity.

The Renewal of Cosmology

THE ORIGINS OF RELATIVISTIC COSMOLOGY

The ancestry of almost all modern theories of the overall structure of the Universe can be traced back in part to the ideas of Albert Einstein, developed in the decade ending in 1915. A paper he published in that year contained his general theory of relativity in a developed form. Einstein's two theories of relativity, the special theory and the general, are too complex to be as much as sketched in our all-too-short account, but central to both is the idea that the physical laws governing a system of bodies should be essentially independent of the way an observer studying those bodies is moving. In the special theory (1905), he considered only frames of reference moving at constant relative speed. Into this theory he introduced a very important principle, that the measured velocity of light in a vacuum is constant, and does not depend on the relative motion of the observer and the source. He drew several important conclusions. One was that if different observers are in relative motion, they will form different conclusions about the relative timing and separation of the things they observe; and that instead of distinguishing sharply between space and time co-ordinates, we should consider all together, as space-time co-ordinates. Another conclusion was that the mass of a body increases with its velocity, and that the speed of light is a mechanical upper limit that cannot be crossed. A third important conclusion was that mass and energy are equivalent and interchangeable.

All of these principles were anticipated to a greater or lesser degree by earlier physicists, but it was Einstein alone who bound them into a single and elegant physical system. The conversion of nuclear mass into nuclear energy is of course a fact of modern life, but understanding the conversion of mass to energy has also been of the greatest importance to an understanding of the production of energy in stars – a subject already touched upon.

Einstein's general theory of relativity was very much more his own creation than the special theory had been. In it he considered how laws of physics are changed when referred to frames of reference *accelerating* relatively to one another. In his special theory, space-time was somewhat like the simple space of Euclidean geometry. It was 'flat'. You can work out separations in space-time by simply applying an extension of Pythagoras' theorem. From the 1830s onwards – and there are even earlier traces of similar ideas – mathematicians had been developing theories of non-Euclidean geometries where space was said to be 'curved', by analogy with the geometry on the surface of a sphere – where Pythagoras' theorem only works for infinitesimally small figures.

Einstein required the space-time in his general theory of relativity to be curved. He laid down principles by which the curvature (which might vary from place to place, as it does, say, on the surface of an artichoke) is produced by matter. One of his most brilliant and important ideas was that concerning the behaviour of particles moving freely through this curved space-time. They move, he said, along *geodesics*. A geodesic is in space-time analogous to the shortest line between two points in Euclidean space: on a two-dimensional flat surface this is a straight line; if we confine ourselves to the surface of a sphere, the shortest distance (or the longest) will be along a 'great circle' between the points. In Einstein's theory, the space-time geodesics are such that a particle that is moving *freely under gravity* follows them. There is no need for an extra law of force, to account for gravity. Gravity is built into the geometry. There are also in general relativity certain special kinds of geodesics, of null length, and these are the tracks of light. Again, the geometry takes care of the physics, as it were, and this in a very elegant way.

We shall later have occasion to mention various models of the Universe in which assumptions made about the distribution of matter have consequences for 'the geometry of space-time'. The Universe, the sum total of all matter, may of course change – time and change are aspects of space-time. One of the most conspicuous properties of Einstein's theory of gravitation is that small-scale gravitational problems cannot be solved in principle until the geometry of the space-time is known, and this requires a knowledge of *the whole material system*. The new theory of gravitation therefore became implicated in developing a cosmological view quite inevitably. The first relativistic model for the Universe was in fact announced by Einstein on 8 February 1917. It was called a 'cylindrical' model, but the word here has an extended meaning, based on a mathematical analogy. (It treated space as a three-dimensional surface of a cylinder in four dimensions.)

One often speaks of 'the geometry of space-time', and it is occasionally useful to think of this as an oblique reference to the set of rules for calculating intervals between points (points with space and time co-ordinates). Using familiar geometry, as mentioned earlier, to find the distance between points with known co-ordinates, we may make use of the theorem of Pythagoras. The non-Euclidean geometries (whether of space alone, or of space-time) use more complicated principles in place of this, more complicated rules of calculation; and of necessity the masses and energies of the system enter into the rules.

Einstein was by no means the first to use non-Euclidean geometries in physics. Four earlier instances deserve brief mention. Nicolai Lobachevskii (1792–1856), one of the three mathematicians chiefly responsible for the final development of non-Euclidean geometry (the others were János Bolyai and Carl Friedrich Gauss), had proposed an astronomical test for the curvature of space – unrealistic since so few parallax measurements were available to him. The German mathematician Lejeune Dirichlet studied the law of gravitation in non-Euclidean space towards the end of 1850. The astronomer Karl Schwarzschild, towards the end of the century, was able to use similar arguments to Lobachevskii's, with later parallax measurements, to set lower limits to a parameter known as the space curvature, for two different sorts of geometry. And A. Calinon had as early as 1889 gone so far as to suggest that the discrepancy between our space and Euclidean space might vary with time. Einstein, however, was the first to make use of the idea that gravitation is explicitly related to the geometrical structure of what is called 'Riemannian' space-time. (The name comes from the fact that he followed methods established by the mathematician Georg Friedrich Bernhard Riemann (1826–66) for an analytical treatment of non-Euclidean geometry.)

When Einstein's general theory was first published, the western world was divided by war, as it was to be again within scarcely more than two decades. The inter-war period was a golden age for the development of an extraordinary number of new and exciting cosmological ideas, but even before, as we have seen, a few of the best astronomers were coming round to the idea that the nebulae were island universes. In 1914, for example, Eddington had been limited to confessing that direct evidence on the nature of the spirals was entirely lacking. He could not say whether they were within or without our own stellar system, but he was of the opinion that the island Universe theory was 'a good working hypothesis'. He still spoke of the structure of the *sidereal* Universe: the significant unit member of the Universe was for him the *star*, and not until he was within twenty pages of the end of his *Stellar Movements and the Structure of the Universe* did he touch briefly on the nature of the spirals. We explained in the last chapter how this intuition was confirmed by careful observational work in the 1920s. The new relativistic cosmology was waiting in the wings, fully prepared for the new empirical discoveries.

OLD ASSUMPTIONS QUESTIONED

The general theory assumes importance chiefly on the large and very large scale. Of the numerous astronomical predictions made by Einstein and others, differences between them and the Newtonian theory – on which it was to some extent modelled – are slight, but when very large stellar and galactic masses are concerned, and distances comparable with those between galaxies, then predictions are usually very different. One important stimulus to Einstein's researches was an awareness of differences over the scale of the solar system. One puzzling fact was that, while Newton's theory – when perturbations of planetary motions by other planets were taken into account – could account for most planetary observations to a high degree of accuracy over centuries and even millennia, the perihelion of Mercury was found to be advancing at a rate that could not be explained. The smallness of the advance gives a measure of how extraordinarily refined astronomical techniques had become. Newcomb had derived a figure of forty-three seconds of arc per century for the unexplained part of the advance of the perihelion – a quantity so small that it would take over eight millennia to move through a degree. Einstein's theory yielded a figure only a second of arc smaller than this.

There were other problems, some of them touched on in earlier chapters. When applied to cosmological problems assuming an infinite Universe, ordinary Newtonian theory, based on the familiar (Euclidean) geometry, seemed to lead to inconsistencies. In fact Carl Neumann and Hugo von Seeliger – to name only two – tried to modify the Newtonian law of gravity to remove these difficulties. In doing so, strangely enough, they introduced what was effectively a cosmical repulsion (one that they supposed worked against the much more powerful gravitational attraction), which has its counterpart in later relativistic cosmology. These theoretical difficulties might seem remote from astronomy, but they were at least as important to future cosmology as the realization that the spirals were comparable in status to the galaxy.

Accounting for the behaviour of matter of small and finite average density uniformly spread over an infinite Universe led all too easily to paradox. There were paradoxes of infinite gravitational force (or potential), but there were others, not at first thought to connect with the gravitational properties of the world, such as the old paradox of the dark night sky. Some tried to avoid these problems by modifying physical laws slightly, while others – such as the Swedish astronomer Carl Charlier in 1908 and 1922 – wished to preserve the laws, and accordingly modified the assumptions generally made as to the distribution of matter in the cosmos. Charlier replaced the notion that matter is on average homogeneous throughout the Universe with an assumption – reminiscent of Lambert's – that the Universe is arranged hierarchically, as a series of systems within systems. Independent empirical evidence for such an idea was virtually non-existent, although it remained an open question as long as the status of globular clusters and spirals was uncertain. Yet a third way of modifying the traditional treatment of large-scale gravitational problems was to dispense with the idea that the space we inhabit obeys the rules of ordinary Euclidean geometry. As we know, Einstein followed this course.

In all these cases, the approach was truly cosmological, in the sense that it concerned the totality of matter in the cosmos. Moreover, what was gradually dawning on those concerned with these questions was something that had already come under discussion in the post-Copernican period, namely that there is a strong element of convention in the path one chooses to follow in drafting theories. If one happens to feel strongly about Euclidean geometry, and if the empirical evidence seems to be placing it under threat, then – as Henri Poincaré realized – you may be able to retain the geometry if you are prepared to modify, for example, the laws of optics. Those who feel strongly about the need to preserve the simplicity of the Newtonian law of gravitation may likewise be able to do so by modifying the geometry. Cosmology, true to its ancestry, showed the way to a more liberal treatment of the whole notion of scientific truth, and the message has not been lost on the other physical sciences in the intervening years.

Newtonian and Euclidean principles were for the great majority of practising scientists unquestionable truths, and resistance to changing either was very strong indeed. To take another example of an almost universally held belief from the first three decades of our century, and of course before: almost everyone took the Universe to be on average *static*. The generally unchanging pattern of stars and galaxies seemed to be guaranteed by centuries of observations – it seemed only marginally less obvious than that classic of the obvious, the darkness of the night sky. In the absence of large proper motions, who would have expected large radial velocities? The post-war period was a time when increasingly numbers of large radial velocities were being found among the nebulae, *as judged by Doppler shifts*, but so strong was the conviction that we inhabit a static Universe that long after this discovery there was a veritable industry in finding alternatives to the usual interpretation of the spectral shifts, that is, as indicating a real velocity. As we shall see, even Hubble, a central figure in this episode, had his doubts.

Explanations of the anomalous advance of Mercury's perihelion were, roughly speaking, of three sorts. Some thought to postulate invisible or barely visible matter, such as asteroids round the Sun (Le Verrier, 1859) or zodiacal light. Others tried modifying Newton's law of gravitation. Asaph Hall was probably the first to do so in this context, in 1894. Ironically, in view of his cosmological objections to Newton's law, it was von Seeliger who first proposed the zodiacal light hypothesis, in 1906. A third group attempted to introduce other physical forces than gravitation – electrical forces, for example. These various hypotheses were subject to intense discussion, especially between the years 1906 and 1920, and many new ideas were tried out. Not only was gravitational absorption, for instance, discussed by theoretical astronomers, but they attempted to detect it in elaborate experiments. It even provided a cosmological apprenticeship for Willem de Sitter (1872–1934), who in 1909 and 1913 made a critical study of the principle.

During these early years of the century there was far more professional contact between the applied mathematicians and the astronomers than is often supposed. Einstein was much helped on astronomical questions by Erwin Freundlich. De Sitter and Arthur Eddington in particular both helped to develop the general theory of relativity and to integrate it into astronomy. Both were eminently qualified for the task. De Sitter was yet another of Kapteyn's many influential pupils, perhaps the greatest of all. He worked under David Gill at the Royal Observatory in Cape Town (1897–9) and then returned as assistant to Kapteyn. He moved to Leiden in 1908. Eddington, as we have seen, had served at the Royal Observatory at Greenwich, where he remained from 1906 to 1913. He had led an eclipse expedition to Brazil in 1912, and it is not surprising that he was the man who led one of the two British expeditions in 1919 that offered the first empirical support for one of Einstein's predictions – that light grazing the Sun's surface would be deflected by a specified amount. Both De Sitter and Eddington were experienced in the statistical analysis of proper motions and star counts in the Kapteyn tradition, and in galactic modelling on that basis, and both were fully aware of the latest astronomy of large scale. It was, however, within the context of Einstein's general theory that each made his first important contribution to cosmology proper.

From about 1911, De Sitter occupied himself with the potential repercussions of relativity theory on practical astronomy. He interested himself in a variety of fundamental problems, such as the interrelationship of Mach's principle and general covariance, the Ritz theory of light emission, and the astronomical relevance of the relativity of time. He was regularly in correspondence with Eddington in these years, and met and discussed problems of common interest with Einstein and the physicists Paul Ehrenfest and Hendrik Lorentz in Leiden. Einstein valued this contact, because it allowed him – through two papers by De Sitter in 1916 – to make his ideas known in Britain. Eddington pacified his colleagues with the message from De Sitter that Einstein was anti-Prussian.

MODELS OF THE UNIVERSE

Eddington appreciated the revolutionary nature of Einstein's new work as soon as he learned of it from De Sitter. He threw himself wholeheartedly into a study of the mathematical underpinning of the general theory, and wrote a masterly *Report* on it in 1918, the proof sheets of which were read by De Sitter. We may see this as a test run for Eddington's *Mathematical Theory of Relativity* of 1923, a work described by Einstein in 1954 as the finest presentation of the theory in any language.

Towards the end of his report, in the context of De Sitter's model – which we shall mention again shortly – he was able to refer to 'the very large observed velocities of spiral nebulae, which are believed to be sidereal systems', and to add that it 'is not possible to say as yet whether the spiral nebulae show a systematic recession, but so far as determined up to the present receding nebulae seem to preponderate'. This reference to the fact that many nebulae had been found to have velocities away from the Sun, and that the recession might be 'systematic', was a portent of things to come.

As briefly explained, one of Einstein's new principles in the general theory was that gravitational masses affect the entire system. Conversely, he believed that gravitational behaviour required matter – for how can the geometry be modified without it? – and that there should be no solution of the field equations (the equations that forge a link between matter and geometry) that describe a Universe empty of matter.

In 1916 Paul Ehrenfest had suggested to De Sitter that some of the difficult problems associated with infinity in understanding our Universe might be avoided if one took instead a *closed* model. This refers to a type of non-Euclidean space-time that can be thought of as analogous to a sphere in ordinary space, which has a closed and finite surface, but which allows one to move endlessly around a great circle on its surface, and is in that sense infinite. In 1917 Einstein tried to find a static model with the same sort of spatially finite character, but could not do so without introducing a notorious 'cosmical term'. This was considered to be a universal constant of unknown but necessarily very small value. He gave various interpretations of this constant, and for a decade and more there were many surprisingly strong views aired as to its virtues and vices. De Sitter retained it, but referred to it as 'a term which detracts from the symmetry and elegance of Einstein's original theory, one of whose chief attractions was that it explained so much without introducing any new hypothesis or empirical constant'. What they did not realize was that the cosmical constant was a veritable Trojan Horse, carrying within it a solution to a cosmological phenomenon as yet undiscovered.

Einstein was very soon shown to be mistaken in supposing that his general theory (or rather field equations) would not allow a solution for which the model Universe was empty of matter. De Sitter found a trio of solutions. (He stipulated that his model should be *isotropic* – meaning that appearances be the same in all directions – and *static*. He also demanded that the spatial part of space-time satisfy certain conditions of constant curvature.)

One of these models was Einstein's own. It had matter in it of finite density (related to the cosmical term) but zero pressure. Another had zero density, zero pressure and zero cosmical term. The third was a solution we now know by De Sitter's name. Density and pressure were both zero, and the model had a strikingly interesting property: *it suggested that an observer should see a reddening of distant sources of light within it* – overlooking the fact that the model did not actually allow for any mass-like objects to send out light. This seemed highly relevant to the real world, despite the fact that the model seemed inadequate to represent mass in the Universe, for red shifts in the spectra of the spirals were then being found in increasing numbers.

Despite its potential relevance to the observed world, De Sitter's model alarmed some of his readers, including Einstein, for another reason. It revealed a 'horizon' for every observer in it, a certain distance at which any finite space-time interval between two events would correspond to an infinite value of their time-co-ordinate. As then interpreted, Nature would appear to be there at rest. In Eddington's words, 'the region beyond . . . is altogether shut off from us by this barrier of time'. Horizons of various similar kinds have played an important part in cosmological discussions ever since De Sitter's example. Several initial confusions as to the nature of clock-time and co-ordinate time needed to be sorted out, but here again De Sitter did a useful service, as did Eddington.

The reddening of distant sources, known as the 'De Sitter effect', was not strictly interpretable as a Doppler effect. Had it been so, it would have seemed immediately to indicate a general recession of the spirals. However, both De Sitter and Eddington did argue, along a different route, for recessional movements. They calculated that a number of particles initially at rest in the De Sitter world would tend to scatter, up to a certain limit, at which their velocities would be comparable with that of light. The conclusion was much criticized in the 1920s, and there was in the same decade much more polemic over the interpretation of other points in the same theory. Interest in De Sitter's solution soon waned, however, simply because it was supplemented by others that were no less exciting and seemed much more plausible.

THE EXPANDING UNIVERSE

Around the time that Einstein was putting together his general theory, there was a consensus among the best astronomers that Kapteyn's picture of the stellar Universe, or Karl Schwarzschild's rewriting of it, was more or less acceptable. Through their statistical methods, especially as extended in scope by Oort, astronomers were becoming 'universe-minded' just as general relativity entered its cosmological phase. The evidence being gathered with the help of the new American telescopes suddenly turned the highly mathematical investigations into more than academic exercises. We have seen how in 1918 Shapley argued for a vast increase in the diameter to be assigned to the Galaxy and in the distance of the Sun from its centre. Distances for the spirals were beginning to come in, but what of their motions?

H. C. Vogel had observed Doppler shifts in starlight as early as 1888, but it was Vesto Melvin Slipher (1875–1969) of the Lowell Observatory (Flagstaff, Arizona) who, towards the end of 1912, had been the first to measure the radial velocity of a spiral nebula. A graduate of Indiana university, he had been at Flagstaff since 1903. Slipher became highly practised in spectrographic work, spending several years measuring planetary rotation periods by this means. (He was the first to detect bands in the spectra of Jupiter's satellites.) He now found that M31, the Andromeda nebula, was approaching the Sun at about 300 kilometres per second – the highest radial velocity then recorded. By 1914, Slipher had thirteen velocities – or spectral shifts, if we wish to reserve judgement, and consider the possibility that they were the De Sitter effect – and by the end of 1925 he had forty-five. Apart from a few spectral shifts to the violet, indicating approach, all were to the red, and some implied velocities so great that it seemed evident to those who pondered the mechanical side of the case that the nebulae concerned must be well outside the gravitational influence of our own Milky Way system.

When Eddington had asked whether there was 'systematic' recession of the nebulae he seems to have had in mind a double effect: the De Sitter effect and the scattering of nebulae of which we have spoken in the same connection. There were astronomers who simply hoped to derive the *Sun's* velocity from Slipher's results: Wirtz and Lundmark were among these, but they soon realized that they were dealing with a more mysterious phenomenon. In 1925 Lundmark actually used data from forty-four nebulae to write down a law connecting velocity with distance. Hubble, on the other hand, was soon in possession of the best data on distances, and by 1929 he had information enough to allow him to announce a very simple law. It seemed that *the Universe of galaxies is expanding, and that the radial velocities of the galaxies are simply proportional to their distances from the sun.*

A word is in order here to emphasize that the apparent tendency of the spirals to recede from the Sun does not mean that the Sun has been reinstated as the centre of the Universe. A lump of dough filled with currants, for example, expanding as it is baked in the oven, will be such that *every* currant will perceive the recession of the others.

The law of universal expansion, with velocity proportional to distance, has become associated with Hubble's name, and his exceptionally thorough observational work was certainly important in making it possible, but he was in reality playing only one part in a complex intellectual operation. The very interpretation of a spectral shift depended heavily on an underlying theory – whether general relativity or some other. Hubble gave relatively little attention to the latest theoretical developments, although at first he thought that with his systematic recessional velocities he had found evidence for the De Sitter Universe. As he wrote later, when commenting on the vital need to obtain reliable distances: 'This fact, together with perhaps a natural inertia in the face of revolutionary ideas couched in the unfamiliar language of general relativity, discouraged immediate investigation'. In fact in a paper published by Howard Robertson in 1928, we find a claim that there is a linear relationship between assigned velocities and distances of the extragalactic nebulae. Even earlier, Georges Lemaître had put forward a similar idea. We must return, then, to the new theories.

After De Sitter's trio of solutions to Einstein's equations, he, Eddington, Ludwik Silberstein, Hermann Weyl, Richard Tolman and others began to investigate the physical aspects of the De Sitter model. At this time there was a young Russian applied mathematician making considerable progress with Einstein's equations. Aleksandr Aleksandrovich Friedmann (1888–1925) was one of the founders of modern theoretical meteorology and aeronautics, and was not without practical experience, having served as an aviator-meteorologist at the northern front in the war. Returning to Petrograd as head of the mathematics department in the Academy of Sciences there in 1920, a post he vacated for that of director of the observatory shortly before his early death five years later, he soon turned to the cosmological problem in general relativity. In 1922 he published an outstanding paper, in which he drew attention to the possibility of a cosmological model that was non-static. In this model, the curvature of space – the non-Euclidean analogue in three dimensions of the curvature of a two-dimensional spherical surface – changed with time.

In 1924, the year before his premature death, Friedmann investigated yet further possibilities, cases of stationary and non-stationary worlds having the geometrical property known as negative curvature. From these beginnings he derived a complete set of new models. He showed how it was possible to introduce matter into these models, and so cast off the embarrassment of De Sitter's empty world.

Friedmann's work received surprisingly little attention from the scientific community. Einstein criticized it in a short note, and then retracted his criticism, which had been based on an arithmetical error of his own. Tragically, Friedmann's fame was posthumous, and came in the wake of a renewal of interest in these matters aroused by the work of Georges Lemaître (1894–1966) and H. P. Robertson.

Lemaître was a Belgian Jesuit, educated in engineering, mathematics and physics at the university of Louvain, before and after the war. He served in the Belgian army, and in fact won the Croix de Guerre. In 1923–4 he studied with Eddington in Cambridge, and when he left there it was to stay for nine months at the Harvard College Observatory. From America he wrote his first paper on cosmology, raising objections to De Sitter's 1917 model. The model lacked matter, of course, but Lemaître disliked it for a different reason: since space had no curvature in the model, its extent was infinite. The education of cosmologists had come a long way when one of them could be so readily dismissive of that idea, which is a part of our everyday Euclidean geometry.

Lemaître objected further against De Sitter's model that it was presented in a way to suggest that the Universe has a centre. He went on to develop a model of his own, reminiscent of Friedmann's, but Lemaître – just as Silberstein had done the year before, in another connection – derived a formula for the red shift of spectra proportional to distance. A 'Hubble's law' of sorts is seemingly everywhere we look. If it did not attract much attention, perhaps this has something to do with the fact that those who were working in general relativity were constantly exploring various mathematical devices for transforming one model into another, with apparently different properties from the first. Thus even Hubble and Milton Lassell Humason, of the Carnegie Institution of Washington, writing in 1931, emphasized that the strange behaviour of the distant nebulae may be 'only apparent', and that it may in some way be an illusion inherent in observation over vast distances. Humason was obtaining many new values for the red shifts, and – assuming that they meant real velocities – he obtained values up to a seventh of the velocity of light, an astonishing statistic in its own right.

If the Universe really is expanding, does this not mean that there was a time in the past when it was a small compact mass? Friedmann's and Lemaître's models seemed to allow for this possibility. But was an expansion from such an 'initial singularity' not simply an illusion created by the mathematics? Lemaître had been ordained abbé in 1923. His science had strong theological relevance for him. An initial singularity was not something to be avoided, but a positive merit, a token of God's creation of the world. In 1927 we find him investigating more complex models, taking radiation pressure into account, for example. This is characteristic of his style. Where Friedmann and most of those who worked in general relativity during its first two decades were at heart mathematicians, Lemaître was a physicist, and he showed this clearly in his subsequent writings.

It is curious to learn that in 1927 Lemaître met Einstein, who told him that while his paper was mathematically sound, he did not believe in the expanding Universe that it entailed. Lemaître said later that he had the feeling that Einstein was not aware of the latest astronomical facts, and indeed Einstein does not seem to have accepted a non-static Universe until a visit to California in 1930, when he discussed the matter with Hubble.

From Einstein on this occasion Lemaître learned for the first time of Friedmann's earlier work, and thereafter he was always very reticent about his own achievements. When he read a statement by Eddington, however, advocating more attention to non-static relativistic models, he was prompted to write to remind his former mentor of the 1927 paper. Eddington was at the time working with his research student G. C. McVittie on the problem of the instability of the Einstein spherical world, and he saw at once the implications of Lemaître's work, which he had apparently quite forgotten: the Einstein model is intrinsically unstable, and had our Universe ever resembled it, we ought indeed to be expecting the very expansion suggested by the red shifts of the distant galaxies.

Eddington convinced De Sitter of this, and through these two widely respected figures the astronomical world learned of the important theoretical developments that had lain dormant for three years and more. Einstein gave the expanding Universe his blessing, and many popular works brought it to the notice of the press, for whom 'relativity' was becoming something of a cult.

The age of the Universe was and remained a problem. 'Hubble's law', when naively stated, makes the velocities of the galaxies proportional to their distances. The multiplying factor of the distances has the dimensions of (1/time). We call this the Hubble factor, H. On a simple interpretation, $(1/H)$ was taken to be a measure of the age of the Universe, the time that has elapsed since the galaxies were all together.

Eddington developed Lemaître's model further, and a new 'Lemaître–Eddington model' became a standard interpretation of the latest data from Mount Wilson. This pictured a Universe evolving from a stagnating Einstein world of indeterminate age. Accepting Hubble's data for the expansion, that is, the ratio of velocity to distance, Eddington supposed that it began about two billion years ago. The same rough answer was available to various different relat-

ivistic models. When, at a later stage, there were claims
that – based on radioactivity in rocks – the Earth was *four*
billion years old, say twice as old as the 'Hubble time', the
Lemaître–Eddington model seemed to have a distinct
advantage over other models. It was very useful to conceive
of the initial 'Einstein state' as having existed for an arbit-
rarily long duration (see figure 17.1).

FIG: 17.1   *Early models of the expanding Universe*

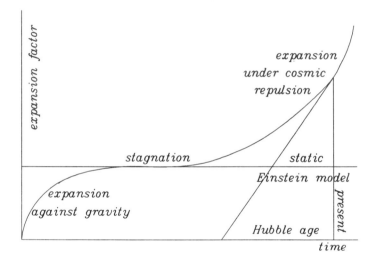

The situation changed around 1952 when Walter Baade,
using the 200-inch Hale telescope at Palomar, and A. D.
Thackeray, using the 74-inch Radcliffe telescope at Pretoria
(then the largest in the southern hemisphere), discovered
that Hubble had made a mistake in regard to galactic distances. The 'Hubble time' expanded overnight by a factor
of five or ten, and the Lemaître–Eddington world no longer
had the edge over its rivals. (It experienced a revival in the
1960s, when it was used to explain the concentration of
the red shifts of quasi-stellar objects in the neighbourhood

of 2. The *changes* in the wavelengths, that is to say, were about double the original wavelengths, a colossal shift.)

THE PHYSICS OF THE UNIVERSE

Lemaître soon lost faith in his theoretical model, and began to work on other things, but his paper turned the interests of several other theoreticians to examine new physical aspects of the expansion. Richard Chase Tolman (1881– 1948) is a good example of a new type of cosmologist. A graduate of the Massachusetts Institute of Technology who had also studied in Germany, he taught mathematical physics and physical chemistry at the California Institute of Technology, Pasadena. The author of the first American textbook on (special) relativity, Tolman took a strong interest in the work of Hubble – who in cosmological terms was one of his nearest neighbours – and he wrote a brilliant and influential study of ways in which thermodynamics could be introduced into relativistic cosmology. Others with similarly broad physical interests were Eddington, McVittie and William McCrea. Most mathematical cosmologists tended to concentrate on the geometrical aspects of their subject, taking Einstein's theory as a datum. But then there were the observing astronomers. Hubble was not exactly typical. While the relativists had been waiting for reliable distances, intense discussions of the significance of the red shifts had been going on, and it seems to have been Tolman who made Hubble nervous of treating the red shifts as indicating velocities. Most working astronomers had no such scruples, and as a result, the 'expanding universe' became for many people a by-word for the most exciting astronomical discovery of all time.

At the other extreme of theoretical sophistication comes Eddington, who had a remarkable way of uniting seemingly all branches of mathematical physics – with the result that many of his colleagues were deeply suspicious of him. His ambition was to unite general relativity with quantum theory. In 1931, Lemaître, on reading an address by Eddington on this theme, replied to it. Eddington – an active member of the Society of Friends (Quakers) – had written that 'the notion of a beginning of Nature' was repugnant to him. Lemaître explained why he believed that quantum theory suggests a beginning of the world very different from the present order of Nature. Thermodynamic principles, he said, require that (1) energy of constant total amount is distributed in distinct units (quanta); and (2) the number of distinct quanta is ever increasing. If we go back in the course of time we must find fewer and fewer quanta, until we find all the energy of the Universe packed in a few *or even in a unique quantum.*

So was born the idea of the primeval atom, a unique atom that might be seen as having an atomic weight equal to the entire mass of the Universe. Highly unstable, it 'would divide in smaller and smaller atoms by a kind of super-radioactive process', and 'some remnant of this process might, according to Sir James Jeans' idea, foster the heat of the stars until our low atomic number atoms allow life to be possible'. In November of the same year, 1931, he returned to the theme, explaining in broad outline how cosmic radiation, with its enormous energy content, might be understood as 'glimpses of the primeval fireworks of the formation of a star from an atom of atomic weight somewhat greater than that of the star itself'.

To support his ideas, Lemaître needed a theory of nuclear structure applicable to atoms of extreme weights. He also needed better information about cosmic radiation, about which very little was then known. He thought his hypothesis of super-radioactive cosmic radiation was supported by A. H. Compton's claim that the radiation consists of charged particles, and his faith in it was increased when it was gradually revealed that the energies involved were much greater than had at first been suspected. There is something almost prophetic about all this. He died in 1966, a year after the discovery of the cosmic microwave background radiation by Arno Penzias and Robert Wilson. (His successor at Louvain, O. Godart, was able to keep him informed about it.) Cosmology had changed considerably in the meantime, but Lemaître clearly saw that there was much more to cosmology than gravitational geometry. Star formation, Galaxy formation and the relative abundance of the various chemical elements, all figured in his scheme, but again in a sense he was too far ahead of his time.

No real progress on this front was possible before the late 1930s. Only when in the late 1940s George Gamow (1904–68), with a young associate Ralph Asher Alpher, began to study nucleosynthesis in the primeval explosion of the Universe, were plausible estimates of the abundances of elements possible. Gamow, born in Odessa (Russia), had studied relativistic cosmology briefly under Aleksandr Friedmann. His talents in nuclear physics brought him to the attention of Niels Bohr in Copenhagen, and he spent time there, in Cambridge (England) and in Paris, before settling in the United States in 1934. In 1938 Gamow interpreted the Hertzsprung–Russell diagram of stellar evolution, and the mass-luminosity relation in stellar theory, in terms of what was then known of nuclear reactions. He arranged a conference on this theme, and partly as a result of it, Hans Bethe went on to discover – later in the same year – the carbon cycle, the key to understanding the generation of energy in stars. (Carl Friedrich von Weizsäcker developed similar ideas independently at about the same time.)

Gamow and his colleagues were at first able to explain abundances of nuclei of only the lightest elements convincingly. Alpher, assisted by Robert Hermann, did nevertheless throw up a prediction that radiation surviving from the early history of the Universe should still exist, with a present temperature of about 5 K. Such radiation was later detected, as we shall see, by A. A. Penzias and R. W. Wilson in 1965. Data on the abundances of the elements are among the most reliable we have about the history of the universe.

On a question of terminology: Lemaître's notion of a 'primeval atom' was not identical with that of a primordial state of matter ('ylem') envisaged by Gamow—neutrons, protons and electrons in a sea of radiation – but both were more precise than the phrase 'big bang' that is applied fairly indiscriminately to any theory that the Universe began suddenly from some primordial state. The expression was not used before 1950, when Fred Hoyle coined it for use in a talk on the BBC in which he was advocating a theory of a very different kind. Like such pejorative terms as 'Quaker' and 'Methodist', the expression is now embraced by believers – and in this case by believers in a great range of creeds. And one final point that cannot be emphasized too often: in picturing relativistic theories of a 'Big Bang', one should not think of a primeval atom, fireball, or whatever, exploding outwards into pre-existing empty space. According to them, *space itself is expanding, along with the matter in it* – the dough and the currants together, to use our previous analogy.

ALTERNATIVE COSMOLOGIES

The 1930s saw a movement in cosmology of great value to scientific practice generally. It was not for nothing that Eddington's general writings aroused great interest among the philosophers, especially about the nature of theoretical entities. Of course some of the problems raised came directly from fundamental physics and the theories of relativity, but the question of whether or not the spectral red shifts were true Doppler shifts indicative of velocities was seen to depend on what was meant by distance, and as soon as this question was pondered, the entire network of interrelations between observational data and the concepts of cosmological theory was seen to be highly problematic. In no other branch of science was so much care given to the analysis of the concepts employed in it, and here the names of Eddington, E. T. Whittaker, R. C. Tolman, E. A.

Milne and G. C. McVittie are among those deserving mention. Milne is an especially interesting case, for with the help of W. H. McCrea, and later G. J. Whitrow, he showed how Einstein's ideas could be side-stepped, and how a form of 'Newtonian cosmology' might be revived.

E. A. Milne first systematically investigated his 'kinematic relativity' in 1932, and with its help constructed a world model in the familiar space of ordinary (Euclidean) geometry. It contains a system of fundamental particles moving in uniform motion relative to each other, all of them having been together at the zero of time. From the point of view of one of these particles – or rather of an observer associated with the particle – the others seem to recede from it. It is easy to see how, at any given time, the whole system can be contained within a sphere, and that this sphere, whose size is determined by the fastest particles, will expand in time. Milne presented his model in two different ways. He found that he could change the scale of time in a particularly interesting way (making one sort of time vary logarithmically with the other) that turned it into a stationary system, each particle then being associated with a fixed point of space. On doing so, however, space ceases to be Euclidean – it is rather what is called 'hyperbolic' – and the timescale is not what would be kept by an atomic clock.

Milne developed these ideas further, first with W. H. McCrea and then with G. J. Whitrow, with consequences that closely resembled those of general relativity. This may be broadly explained by reference to an important step in Einsteinian relativistic cosmology taken by H. P. Robertson in 1929 (and independently by H. G. Walker some years later). The expression 'space-time metric' refers to a formula by which intervals between two events (points in space-time) may be evaluated in one of the various Riemannian (non-Euclidean) geometries. It is an extension of the theorem of Pythagoras in ordinary geometry, as we have already briefly explained. Robertson considered what would be the consequence if he placed two restrictions on the new geometry. He required that space-time be (1) *homogeneous*, with all places alike, and (2) *isotropic*, that is, the world should look the same in all directions from all points at a given (cosmic) time. The resulting metric, which we shall not give in detail here, was interesting because it included a certain *time factor*, a term by which the spatial interval between events had to be multiplied. In other words, by taking Einstein's theory, and adding the two conditions of cosmic symmetry, an 'expanding universe' of sorts follows quite naturally. As we recall, it had indeed been found by several theoreticians before the vital observations were produced.

As a shorthand for the expansion factor referred to here, we shall use the usual expression for it, $R(t)$. This term, which depends on the time, is sometimes spoken of as an expanding radius of space; but here we shall not consider its various interpretations. We must be content to make the general point that, in Milne's theory, *exactly the same* metric could be obtained as that found earlier (by Robertson and Arthur Walker) for Einstein's brand of relativity.

In both types of theory, the metric is really only the beginning of a cosmology, however. How the Universe expands, that is, how the expansion factor $R(t)$ evolves in time, can only be judged by relating the matter in the observed world, with all its physical properties, to the model. In 1934 Milne and McCrea found that, with a simple mathematical trick for treating an unbounded system, Newtonian *mechanics and gravitation* led to precisely the 'Friedmann–Lemaître' equations for a model without pressure (with or without cosmical repulsion of the sort Einstein had introduced with his cosmic term). On a local scale, the model should behave mechanically more or less exactly as in general relativity. On a larger scale, the fitting together of the local pieces of space was done differently in the two theories. One important lesson, though, was that there seems to be something very natural about the expanding Universe, even when understood in a 'classical' way.

As we saw when considering the early history of relativistic cosmology, the form of the expansion factor varies from model to model. It varies, for instance, as between the Einstein model and De Sitter's, as between Lemaître's and Eddington's variant of it, and so forth. An interesting theoretical model was introduced in 1932 by Einstein and De Sitter jointly. In this, the expansion factor was proportional to the two thirds power of the time. This has the effect of making the present age of the Universe only one-third of the Hubble time. Another model introduced by P. A. M. Dirac in 1938 was interesting because it did not require space to be curved and it made the expansion factor proportional to the one-third power of the time, so that the age of the Universe is just a third of the Hubble time. Without attempting to judge on the merits of the various alternatives, we can only repeat that it is obviously wrong to treat the Hubble factor as indicating the age of the Universe in the way that was usually done in the years following the discovery of the general recession of the galaxies. (As for opinion in the 1930s on the question of the actual age of the Universe, this is a separate issue, but we recall the difficulties into which Hubble's observations had led astronomers.)

Other alternatives to Einstein's general theory of relativity were the theories of gravitation developed by G. D. Birkhoff, A. N. Whitehead and J. L. Synge. All of them had cosmological implications. They were symptomatic of a period of great intellectual vitality, motivated perhaps by a desire to create something comparable with what Einstein had produced. Some ideas of a very different kind were then being put forth by Hermann Weyl (1930), Eddington (1930) and Dirac (1937–8), who seemed to many to be suggesting that cosmological observation was superfluous, and that all could be deduced from the constants of physics. Eddington, for instance, thought that all the *dimensionless* constants (pure numbers) obtained by suitably multiplying and dividing powers of the constants of physics (the mass of a proton, the charge on an electron, and so forth) turn out to be close to unity or of the order of 10 raised to the power 79. This vast number he thought might characterize the number of particles in the Universe. Eddington's monumental study *Fundamental Theory* (published from his notes in 1946, two years after his death) has left many with a sneaking suspicion that there may be a numerological link between the physics of the very small scale and that of the cosmos.

Eddington was actually able to derive a value for 'Hubble's constant' close to Hubble's. When the observed value was revised, seemingly refuting his methods, it was found that he had made a mistake, and that he should have obtained a constant more or less equal to the revised figure. With later revisions, however, no such rescue has proved possible.

In the short term, what mattered more to observational cosmology than Eddington's brilliant number magic was the ongoing drive, by Tolman, McCrea, McVittie, Otto Heckmann and the rest, to derive theoretical relations between observable quantities. By means of these – for example between apparent magnitude and red shift, between apparent magnitude and counts of galaxies, and so forth – the various cosmological models could in principle be put to observational test. In practice, doing so was much delayed by the coming of the Second World War.

THE CREATION OF MATTER

In 1918 and 1925, to remove the paradox of the dark night sky, W. D. MacMillan had put forward the idea that radiation, when travelling through empty space, may suddenly disappear, and reappear in the form of hydrogen atoms. R. A. Millikan, a pioneer of cosmic radiation and one of MacMillan's colleagues at Chicago, thought that this might be the explanation of the origin of cosmic rays, but that idea at least was soon disproved by A. H. Compton. There was nothing particularly mysterious about the 'creation' of atoms on this theory, for there was a named source of the energy responsible for them. Tolman and others likewise considered the conversion of radiation into matter, which did not at first attract much attention.

There were others, however, whose suggestions as to the creation of matter were far more radical. In 1928 James Jeans, searching for an explanation of the spiral character of the nebulae, suggested that the centres of the spirals might be places at which matter is 'poured into our Universe from some other, and entirely extraneous, spatial dimension'. He added that to us, therefore, they appeared to be 'points at which matter is being continually created'. Milne later noted that his kinematic relativity conformed with the speculations of Jeans.

In 1937 the Cambridge physicist P. A. M. Dirac, picking up some of Eddington's ideas, argued that large numbers of the sort that occurred in Eddington's theory, if they were proportional to the epoch, should increase in time. (The 'age of the universe' is to that extent a very unusual sort of physicial 'constant'.) He thought that the number of protons and neutrons in the Universe might therefore grow – implying continual creation – and that the gravitational constant might likewise change. In Japan, at much the same time, mathematical physicists of the Hiroshima school, investigating the De Sitter model, claimed that the number of particles in it increases with time. In Germany, Pascual Jordan took over some of Dirac's work and claimed that stars, perhaps in the form of supernovae, and even entire galaxies, might be the created matter. He thought that Dirac might have been afraid of violating the principle of energy conservation, and he was careful to provide a potential source for the new matter. It was, he said, provided by a loss of gravitational energy from the universe.

Despite the fact that their authors were of the highest credentials in fundamental physics, these speculations were little heeded at the time, perhaps because they were mostly put forward in a half-hearted way. Dirac soon abandoned the whole idea, although Jordan stuck to it resolutely into the 1950s. Continual creation sprang to the attention of a wide public after the war, when Hermann Bondi, Thomas Gold and Fred Hoyle made it a part of a new class of cosmological theories. By this time, attitudes to the Universe had changed very substantially from those of the 1920s and 1930s. The old assumption that it was an overall static entity had gradually been whittled away, and there was wide agreement that the red shifts do indicate a real expansion of the system of nebulae. And gradually more and more physics was woven into the story.

As we have seen, Lemaître, Gamow and others had shown that cosmology might hold the key to the origin of the elements. When Eddington had been writing his work on the constitution of the stars, the importance of hydrogen and helium was scarcely perceived, but after Eddington, McCrea had developed the first acceptable model of a large hydrogen star. Hoyle introduced Bondi and Gold – who had both come to Britain from Nazi-occupied Austria – to astrophysical questions, first when all three were working at the Admiralty during the war, and later when they were all in Cambridge together, especially from 1948. They were dissatisfied by Lemaître's model, which seemed to overcome the embarrassingly short figure for the 'age of the universe' then obtained from Hubble's work, but only by sleight of hand – that is, by the indeterminate 'stagnation period' (see figure 17.1 above). Gold had the idea that if there were a process of continual creation of matter it should be possible for the Universe to continue in a steady state, despite its general expansion, so that the 'age' problem would disappear.

From these beginnings, Bondi and Gold moved in one direction and Hoyle in another, although they shared many ideas. Bondi and Gold presented their work as a consequence of a very general law that they called the 'perfect cosmological principle', according to which the Universe 'presents on the large scale an unchanging aspect'. Milne – whose work was of much influence here – had earlier worked with a 'cosmological principle' that made the large-scale aspect of the Universe independent of the place of observation; but now it was also the *time* of observation that was said to be immaterial. This meant that the average densities of matter and radiation remain the same, and that the age distribution of nebulae remains constant. In steady-state cosmology the average age of objects is only a third of the 'Hubble age', so even in this theory it seemed that our Galaxy was much older than average. Since the data then available for distances led many to think that our Galaxy was unusually large by comparison with other galaxies, however, it was thought not improper that its age might be greater than the average.

Hoyle's methods seemed less radical – the earliest publications by the three all came in 1948. He took Einstein's general theory of relativity as his starting point, and was influenced by a principle first enunciated by Hermann Weyl in the 1920s. (According to this, there is a preferred velocity at each point in space-time, and only a set of fundamental observers sees the Universe in the same way.) Hoyle now simply added a term to the field equations that was interpreted as indicating the creation of matter. This seemed to violate the conservation of energy. In 1951, McCrea showed how Einstein's equations could be salvaged by reinterpreting Hoyle's new 'creation field', and in the 1960s Hoyle revised his earlier treatment, in collaboration with J. V. Narlikar. The heyday of steady-state cosmology, however, was the late 1940s and 1950s. For a time it became a national talking point, with many of the English clergy writing letters to the press and preaching sermons on the audacity of those who seemed to wish to deny the Christian account of a single creative act. Matters were not helped by Hoyle's use of the pejorative phrase 'big bang' for rival models that began from a compact origin at a particular point in the past.

The hopes of the steady-state cosmologists were raised for a time as the result of work by Hoyle, Geoffrey and E. Margaret Burbidge and William A. Fowler (of the California Institute of Technology) on nucleosynthesis in stars – that is, the transformation of chemical elements into others by means of nuclear reactions. Their work was largely prompted by steady-state cosmology, with which it fitted well, but in time it became clear that the new theories of nucleosynthesis could be adapted to other cosmological models.

The first observational evidence that worked against the steady-state theory came from counts of celestial radio sources, done primarily by Martin Ryle in Cambridge, with several colleagues in Britain, also by B. Y. Mills and J. G. Bolton in Australia, and by M. Ceccarelli in Italy. It seemed that the number of faint radio sources was far too large, by comparison with the number of bright sources, for the steady-state theory's predictions to be acceptable. The explanation that came most easily to hand was one in terms of a 'big bang' model: this was that the (intrinsically) bright sources are relatively near, since radiation from distant bright sources has not had time to reach us. In other words, on this evolutionary model, the further we go out into space the less evolved, and thus intrinsically fainter, will be the sources we observe.

Ryle's first evidence, with what to some was its clear message, was presented in 1955 at the same time as his latest survey of 1936 radio sources. He had then been working seriously at the cosmological problem for about two years. The human effect of his 1955 claim, presented by Ryle and his student Peter Scheuer, was electric. Ryle had kept his cards close to his chest with good reason, but even he was unprepared for the onslaught of the steady-state theorists, and the popular press in Britain made much of the controversy. Pawsey and Mills in Sydney soon found radio results of their own that were hard to square with Ryle's, and for seven or eight years afterwards there were independent judges of the situation who considered that the observations might be subject to errors. In fact what revisions were made tended at first to make the case less decisive.

The situation in cosmology around the year 1960 was therefore by no means clear to those who had no commitment to a particular cause. There were several models over and above those mentioned here – for example the oscillatory models that expanded (for billions of years) before contracting, and then perhaps entered again into an expanding phase. Such models had bizarre thermodynamical properties that made them generally unpopular with astronomers. There were attempts to develop cosmologies based on the different theories of gravitation founded, as already mentioned, in the 1920s by the mathematicians A. N. Whitehead (extended by J. L. Synge) and G. D. Birkhoff, but they were isolated, and attracted ever-diminishing attention. Einstein's general relativity was still far from being widely accepted, and many who worked entirely within it admitted to having doubts. When the tide turned, it turned rapidly, and this largely as a result of a series of new discoveries made with the radio telescopes that were transforming the entire face of astronomy.

Radio Astronomy

THE FIRST ATTEMPTS AT RADIO ASTRONOMY

Just as William Herschel, with the help of a thermometer, discovered infrared radiations from the Sun that were invisible to the human eye, and just as photography allowed an extension of the spectrum beyond the violet, so radio receivers have made possible the detection of electromagnetic radiation between wavelengths of roughly 1 mm and 30 m. (The wavelength of yellow light, by comparison, is a little under 0.0006 mm.) The bridge between the theory of light and classical theories of electricity and magnetism was one effected only gradually in the course of the nineteenth century, but the most significant single contribution was that made by James Clerk Maxwell (1831–79). Maxwell united electricity and magnetism into a single theory, and his equations made clear the possibility of propagating electric waves in air as an extension of the case of light radiation. In 1882 the Dublin physicist George Francis Fitzgerald (1851–1901), basing his argument on Maxwell's theory, claimed that the energy of varying currents might be radiated into space, and a year later he described an apparatus (a magnetic oscillator) by which such an effect might be produced. This led to further theoretical work, but the practical repercussions of what was done were slight. There were attempts by O. J. Lodge to detect electromagnetic waves in wires, while Joseph Henry (1797–1878), Thomas Alva Edison (1847–1931) and others had all shown that it was possible to send electromagnetic action over appreciable distances, but these were usually taken to be no different from local electromagnetic effects, 'inductions'. David Edward Hughes (1830–1900) showed that signals from a spark transmitter could be detected at distances of several hundred metres by a microphone contact to which was connected a telephone – by which the signals were heard. In 1879 and 1880 he demonstrated these experiments to representatives of the Royal Society and the Post Office, who thought that they were merely induction effects, so that he refrained from publishing them until long afterwards.

The world of physics was finally convinced in the late 1880s, with some outstanding work by Heinrich Rudolf Hertz (1857–94). Hertz had worked with Helmholtz but was now at Karlsruhe. He had spent much time in attempting to justify Maxwell's equations on theoretical grounds, and in carrying out a number of related experiments. It is not clear whether he knew of the work of his predecessors when in 1888 he produced electric waves. He did so with an open circuit connected to an induction coil, and detected the waves with a simple loop of wire with a gap in it. This, however, in a practical sense marks the birth of the new radio technology.

For many years advances were slow. Detection was difficult, and a great breakthrough here came in 1904 with John Ambrose Fleming's invention of the 'thermionic valve'. Fleming's credentials were excellent: he had worked with Maxwell in Cambridge, he was for a time consultant to the London branch of Edison's company and was inspired by some of Edison's work, he had experimented with radio transmission from the 1880s, and he had helped design the transmitter used by Guglielmo Marconi to span the Atlantic in 1901.

Although radio, therefore, is usually regarded as a product of the twentieth century – since most of us are vaguely conscious of the beginning of regular sound broadcasting in the 1920s – it was moderately well established before the turn of the century. Since the connection between optical and electromagnetic radiations was by this time an accepted part of physics, it is not surprising that several astronomers considered the *sun* as a potential source of radio waves. As early as 1890, Edison, with a colleague A. E. Kennelly, discussed the design of a suitable antenna (they settled for a cable on poles around a massive iron core). Oliver Lodge set up an ambitious receiver in Liverpool in the period 1897–1900, but any signals there might have been were swamped by electrical interference from the city. Other unsuccessful attempts were made by J. Wilsing and J. Scheiner in Potsdam in 1896, and C. Nordman in 1901 – unfortunately during a minimum of solar activity. Nordman set up good apparatus high in the French Alps, and used a long antenna. Had he persevered for longer than a day he would very probably have been successful; but his results, like the rest, came to nothing, and the idea that the Sun is a transmitter went into almost total eclipse.

The first unequivocal detection of radio waves from the Sun was to be by James Stanley Hey on 27 and 28 February 1942, investigating what was thought to be the jamming of British radar by the Germans. That the Sun was the source was confirmed by stations at various towns in Britain. Later it was found that the observatory at Meudon in France had (visually) recorded strong solar flares at the same time. It was also appreciated that although the emission was, as Hey said, a hundred thousand times more than had been expected, it would not have been registered had the Sun been as near as our nearest stellar neighbour. Only exceptional stars can reach out to us across galactic distances at radio wavelengths.

Hey's discovery is often called fortuitous, but one should hesitate before calling it that, in view of Hey's record. He it was who first discovered radar reflections from meteor trails – investigated for the same reason. He it was who first discovered a radio Galaxy, Cygnus A, after the war. Only gifted observers make chance discoveries.

GALACTIC RADIO WAVES

Radio astronomy was well and truly incorporated into astronomical practice only in the late 1950s, but its successful practice goes back as far as 1932. In that year, Karl G. Jansky (1905–50), an American radio engineer at the Bell Telephone Laboratory, was studying interference on the newly inaugurated trans-Atlantic radio-telephone service when he discovered that much of the interference came from extra-terrestrial sources. His receiver – set up at Holmdel, New Jersey – was working at a frequency of 20.5 MHz (that is, in the 15 m band), and he had a steerable antenna with moderately strong directional properties. His records show that he distinguished between three main sources of noise: local thunderstorms, distant thunderstorms, and a background hiss that varied with the time of day. At first he thought he had detected a radio emission from the Sun, but as the Sun changed its celestial position he realized that he was concerned with quite another source. The year was a minimum of solar activity, which made it easier for him to study the hiss of background 'static'. He did so for a year, and established that the signals were associated with the Milky Way, and were strongest in the direction of the centre of the Galaxy. (He later made another conjecture, that the signals were from the solar apex, the point towards which the Sun is heading in space, but in this he was mistaken.)

Jansky published his findings over the next three years, and then moved on to other research. His work attracted little attention. Grote Reber (1911–), another radio engineer, from Wheaton, Illinois, built himself a paraboloidal antenna of a sort now familiar, and in his spare time repeated much of Jansky's work. Working at higher frequencies and with a more strongly directional antenna, in 1939 he was able to detect strong radiation at 160 MHz. Between 1940 and 1948 Reber produced contour maps of intensities from the sky, paralleled only by work of which he was then unaware, by J. S. Hey and others (at 64 MHz) working at the Army Operational Research Group in England, using the newly developed radar equipment mentioned in our previous section. Reber's antenna has been preserved and now stands at the entrance to the US National Radio Observatory at Green Bank, West Virginia.

Theories explaining the 'galactic static' were numerous. Several astronomers proposed new classes of stars, rich in radiations at radio frequencies. Nuclear physics was giving increasing attention to the synchrotron as a means of accelerating charged particles, so it is not surprising to find a theory — first put forward in 1950 by K. O. Kiepenheuer (1910–75) — that radio waves are transmitted by very-high-velocity electrons (close to the speed of light) spiralling through the magnetic field of the Galaxy, the latter resembling a gigantic synchrotron. (The motion of an electron in a magnetic field follows a helical path — not a flat spiral but like the sort of spring one might make by coiling a wire round a pencil.) A similar sort of theory was proposed in the same year by H. O. G. Alfvén and N. Herlofson, who thought that the scale of celestial synchrotrons might be much smaller, comparable in size with the solar system, perhaps, so that radio waves are received from innumerable discrete stellar sources rather than only from galactic centres.

Vitaly L. Ginzburg and Josef S. Shklovsky in the Soviet Union immediately began to develop the synchrotron theory, as applied to radio galaxies, and their detailed predictions have proved highly successful, in particular those relating to the polarization of radiation. This entire subject was nowhere more dramatically applied than in the case of the Crab Nebula, which has provided astronomy with a testing ground for numerous theories of stellar evolution.

In 1921, and again in 1939, photographic plates of the Crab Nebula (taken in 1909, 1921 and 1938) were carefully compared by John C. Duncan of the Mount Wilson Observatory, and he eventually decided that it was expanding at about a fifth of a second of arc per year. Its size implied that it was about 800 years old. In 1937 Nicholas Mayall at the Lick Observatory measured the velocities of expansion of parts of the cloud, using spectroscopy and the Doppler effect. These velocities were greater than any previously recorded, well over 1000 km/s. In 1942, Oort and Mayall published a joint paper (in the United States) arguing for the identity of the Crab Nebula with the supernova of AD 1054, a date that matched the data rather well. (Their paper was accompanied by a survey of relevant information from chronicles by the Dutch orientalist J. J. L. Duyvendak.) From the two sets of measurements it was possible to say that the nebula was about 5000 light years away, and this figure allowed a first estimate of the energy presently being radiated.

In 1949, as we shall see, John G. Bolton and Gordon Stanley in Australia found that it is a powerful radio source. In 1954 Viktor A. Dombrovsky and M. A. Vashakidze in the Soviet Union studied the polarization of light from the object – I. M. Gordon and V. L. Ginzburg had independently predicted that this should be observable. With supplementary information from Palomar provided by Walter Baade, and studies done in Leiden by Oort and Theodor Walraven, it seemed clear that both the light from the Crab Nebula and its radio waves were being generated by electrons or other charged particles moving at almost the speed of light and emitting synchrotron radiation.

NEW RADIO TECHNIQUES

Radio astronomy received a great impetus with the ready availablity after the Second World War of large quantities of surplus equipment and antennas, not to mention high expertise in radio communications and radar. An early result of great significance was obtained by J. L. Pawsey (1908–62), who in 1946 found that the brightness temperature of the Sun at metre wavelengths is of the order of a million kelvins (K). This fitted well with optical observations of the corona, and the corona was thus seen to be all-important in this radio work – and to be opaque to radiations from below it at these wavelengths.

One of the most vital early problems in radio astronomy was to improve resolution, that is, to narrow down the angle of view. One early technique was to use an eclipse of the Sun, and make use of the moment when the Moon blocked out a varying fraction of the Sun. This was done, for instance, by the Americans R. H. Dicke and R. Beringer in 1945, by A. E. Covington in Canada in 1946, by the Soviet astronomers S. E. Khaikin and B. M. Chikhachev, working from a naval vessel off the shore of Brazil in 1947, and by W. N. Christiansen, D. E. Yabsley and B. Y. Mills in Australia in 1948. But something better was clearly needed, a method that could be put to daily use.

Two broadly similar solutions were soon found, both depending on the phenomenon known in optics as 'interference'. (This has nothing to do with radio interference commonly known as 'static', of course.) When two waves interact they may reinforce one another, or they may cancel, that is, 'interfere'. This may be seen on the surface of water, for example, where under certain conditions the places of reinforcement and cancellation of ripples (and stages between these extremes) may even be made to remain stationary, producing the familiar phenomenon of 'standing waves'. In the case of light there are many ways of obtaining analogous results, the classic experiment being that of Thomas Young (1801). In this, two slits in a plate serve as the two light sources, these being supplied with light by a single slit between them and a lamp. 'Interference fringes', alternating lines of light and dark, are in this way produced on a screen.

A variant of the arrangement, requiring only a single slit, was devised by H. Lloyd, who viewed that slit very obliquely in a mirror, so producing in effect a second slit.

This 'Lloyd's mirror' technique was translated into radio terms, and used for several years especially in Australia (by L. L. McCready, J. L. Pawsey and R. Payne-Scott), but it eventually gave way to another type of interferometer, analogous to that designed for light waves long before by A. A. Michelson. Here the light wave is split (part is reflected by, and part passed through, a half-silvered mirror), sent along two different paths, and finally recombined to produce an interference pattern. Ryle and D. D. Vonberg in Cambridge developed the radio analogue of this technique in the 1940s, using antennas separated by distances of up to 140 wavelengths. By the following decade it was being adopted by groups in many different countries.

Of course there were many variants on the theme, but radio interferometers of various sorts, often with large arrays of component antennas, became commonplace. They were relatively inexpensive, although usually too large to be steerable. One simply waited for the sky to move into position. A method known as 'aperture synthesis' was developed at Cambridge, and this allowed the positions of radio sources to be found with an accuracy approaching that of optical positions.

The resolution of a telescope increases with its aperture, but it also depends on wavelength, and a radio telescope has to be much larger, to match optical results. Two radio telescopes at very large distances can be linked by cable, however, in ways that could compensate for other shortcomings of radio telescopes. (There are also techniques for recording observations and combining them later, so avoiding the connecting cable.) 'Very-long-baseline interferometry' (VLBI) achieved spectacular results on a relatively small budget.

Another new technique developed around 1950, by J. P. Wild and L. L. McCready, was important for other reasons: they developed radio spectrographs, scanning a wide range of frequencies (around 70 to 130 MHz) over a time lasting less than a second. This allowed them to analyse bursts of emission from the Sun in ways that greatly improved the understanding of its physical constitution. The ability to improve on the response time was to prove of the greatest importance to astronomy generally, for as we shall see, it made possible the detection of pulsars.

Hey's discovery of a discrete radio source in Cygnus, while he was still working at his military establishment after the war, has already been mentioned. Very soon afterwards, Ryle, in Cambridge, using his different type of interferometer, found another intense localized source in the constellation of Cassiopoeia. In Sydney, in 1949, John G. Bolton and Gordon Stanley discovered a source in Taurus, and their suggestion that it was related to the Crab Nebula – believed, as we saw, to be the remnants of a supernova of AD 1054 – was generally accepted. In view of these and other exciting discoveries that were then being made, Bernard Lovell – who was at the time working on cosmic radiation, radar reflections from meteors, aurorae, the Moon, even the planets, as well as radio waves from space and the Sun – felt acutely that *steerable* antennas were called for. In 1948 he launched plans for the first giant paraboloid, of a kind with which most people are now familiar.

His tenacity eventually yielded results. Finally completed in 1957, at Jodrell Bank near Manchester, the dish was seventy-six metres across, and weighed 1500 tonnes. It was carried on turrets that had carried fifteen-inch guns on the battleships *Royal Sovereign* and *Revenge*. This extraordinary engineering venture – which once nearly came to grief in a hurricane – seemed at times to be doomed to failure through lack of funds when it suddenly entered the public consciousness. In short, the telescope proved to be useful for locating by radar the carrier rocket of the first Soviet Sputniks (the first and second were launched in 1957). Three years later, the American Pioneer V was released from its carrier rocket by a signal from Jodrell Bank. This use, for which of course the telescope was never intended, prompted the gift of money that was the telescope's final salvation. An earlier campaign in which the press had tried to shame the British government ('Schoolboys send pocket money to save our face', *Sunday Dispatch*) had not been successful.

The 'Mark I' telescope was later modified, but the pattern was set. Another vast project, the Parkes telescope in Australia with a sixty-four metre dish, followed in 1961. Dutch radio astronomy had for several years been working along lines set largely by Jan Oort's pioneering work on the structure of the Galaxy. During the war his Leiden student H. C. van de Hulst had calculated that neutral hydrogen should send out radiations at a wavelength of 21 cm. This was first detected at Harvard in 1951 by Harold I. Ewen and his professor, E. M. Purcell, using a crude but sensitive receiver. Almost simultaneously a radio telescope was completed at Leiden, where Van de Hulst confirmed his own prediction, and at Sydney, Australia. More comprehensive surveys of the sky at the 21-cm wavelength followed, the first using a twenty-five metre steerable dish at Dwingeloo (1956). One of the most important projects of the period was the synthesis radio telescope at Westerbork (1970), making use of no fewer than twelve parabolic steerable dishes of this size, movable over a baseline 1620 metres long. One last example of a fully steerable radio telescope, completed now in 1972 at Effelsberg in Germany: this has a 100-metre dish and a surface accurate enough for observations at wavelengths almost as short as 1 cm.

To appreciate the great change that was here taking place in astronomy, the world over, it will be as well that we put the optical tradition in perspective, at the risk of giving a bare catalogue of events. Optical astronomy did not suddenly fade away.

The flagship of this subject in the post-war period was of course the great 200-inch Hale reflector at Palomar, in the United States, brought into service in 1948. Extremely important work was done with it from the very beginning – Baade's work on star populations has already been mentioned. Allan R. Sandage (1926–) was an assistant to Hubble as a graduate student from 1950, and joined the staff of the Mount Wilson and Las Campanas Observatories in 1952. At Hubble's death he inherited the mission to map the distances and expansion rate of the galaxies. He produced data, including a figure for the deceleration of the expansion, that greatly influenced cosmological thinking during the 1950s and 1960s. (The 'deceleration parameter', a measure of curvature in velocity–distance graphs for the galaxies, was found by more refined correlations of galactic magnitudes with red shifts. Rival values moved to and fro for over twenty years and more, until eventually B. M. Tinsley persuaded the profession that the *evolution of galactic sources* made a direct measurement of the deceleration virtually impossible for the time being.) Later from Palomar came Jesse Greenstein's studies of white dwarf stars, work by Eric E. Becklin and G. Neugebauer in the infrared, and observations of remnants of supernovae by Baade, Zwicky and Minkowski.

The Soviet Union put into commission an alt-azimuth instrument of larger aperture (six metres, or 236 inches) in 1975, but this was used under less favourable conditions, and is only numerically superior. (It is located in the Caucasus mountains in southern Russia. The mirror alone weighs 70 tonnes: it was ground and polished in Leningrad, beginning in 1968.) The importance of matching site to wavelength was not always well appreciated, but in the second half of the twentieth century it has led to the establishment of mountain observatories in Arizona (Kitt Peak), Hawaii (with the Canada–France–Hawaii Telescope and the NASA and United Kingdom infrared telescopes), Chile (one of three major institutions there is the European Southern Observatory at La Silla), Australia (including the Anglo-Australian Telescope and the United Kingdom Schmidt Telescope), and Spain (including a German–Spanish Center at Calar Alto, and the Herschel and Newton reflectors at La Palma).

So much for our brief survey of the optical scene. By the early 1990s there were still only about a dozen optical telescopes with mirrors more than three metres in diameter. At radio wavelengths, on the other hand, where much valuable work was possible even in climates where water vapour blocked certain wavebands, large radio telescopes quickly multiplied. We shall confine attention to a few of the more ambitious schemes.

The National Radio Astronomy Observatory (NRAO) was funded by nine American universities and the National Science Foundation. Its first telescope went into use in 1959, a 25.9-metre dish telescope that could be linked with two similar movable telescopes forming an interferometric array of baseline 2700 metres. Other fully steerable dishes followed there. When NRAO built a dish accurate enough to be used at millimetre wavelengths, it was sited at Kitt Peak, to avoid atmospheric absorption of the signals. By the mid 1980s the annual budget of NRAO was around $15 million, which covered the running of the highly sophisticated VLA ('Very Large Array') on the plains of San Augustin near Socorro, New Mexico (operational in 1975, completed 1982). This $80 million array includes twenty-seven parabolic dishes of 25 m diameter in a Y-shaped pattern, the arms of the Y covering 61 km. It can be compared with a telescope of 27 km diameter for angular resolution, which for radio wavelengths is roughly equivalent to the optical resolution of the best visual telescopes, and so allows optical, infrared and radio images to be compared. A large Multi-Element Radio-Linked Interferometer Network (MERLIN) was later built in Britain, and others have followed elsewhere.

The VLI was actually preceded by a technique alluded to earlier, known as Very-Long-Baseline Interferometry (VLBI), which uses baselines of *thousands* of kilometres, and so allows resolutions of a thousandth of a second of arc.

These are unmatched by any other instrument. With them came the discovery of 'superluminal' radio sources, that appear to be expanding at several times the speed of light. The first observations of this phenomenon were made in 1967 by David S. Robertson in Australia and A. T. Moffat at Owens Valley in California. Three years later Irwin Shapiro and colleagues, using facilities in southern California and in Massachusetts, discovered a more spectacular example, 3C 279 (source number 279 in the Third Cambridge Catalogue), which was apparently expanding at ten times the speed of light.

The largest radio telescope in terms of its reflecting area is at Arecibo, Puerto Rico, and has a diameter of 305 metres. Its fixed bowl is built into the ground of a mainly natural valley, and a cage with antenna and other materials is slung from three colossal pylons. The angle of view can be adjusted somewhat by shifting the position of the antenna in relation to the bowl, so that planets and asteroids can all be observed. The Arecibo telescope owes its existence (1960–3) to a desire by the US Department of Defense to track Soviet (and other) satellites. A contract was given to Cornell University, but it was soon realized that the theory of the ionosphere on which it was founded was mistaken. It is now an arm of the National Astronomy and Ionosphere Center.

Since its inception by Hey and others during the war, studies of the solar system with large radio telescopes have depended heavily on radar techniques. Soviet astronomers soon made this their speciality – dare one say that some good came out of the Cold War, on both sides? Using an array of eight 16 m dishes in the Crimea they were first to receive reflected signals from Mercury, Venus and Mars, in the early 1960s. Later the Soviet Academy of Sciences built a vast telescope (RATAN), with nearly nine hundred panels arranged in a circle about 576 metres across, and so comparable with the VLA in collecting power.

THE DISCOVERY OF QUASARS

The number of discoveries made with radio telescopes is legion, but one of the most exciting in its day was that of 'quasi-stellar radio sources', which soon became known as quasars. The first radio galaxies had been discovered in the early 1950s, but it was more than ten years before positions had been found with an accuracy that allowed them to be matched with visual objects. The astronomers at Jodrell Bank were nevertheless showing a strong interest in a number of radio sources with very small angular diameters. In 1960, Allan R. Sandage of the Mount Wilson and Palomar Observatories took photographs of three regions containing such sources, and Thomas A. Matthews and J. G. Bolton at the Owens Valley Radio Observatory found in each case that the only visual object within the error rectangle was a star. At the end of that year, Sandage announced in an unscheduled paper at a meeting of the American Astronomical Society that the photographic plates showed what seemed to be a bright star at the precise position assigned to a strong radio source, 3C 48. It was accompanied by a faint luminous wisp. If this was indeed a star, then it would be the first distant radio star to be discovered. When its spectrum was analysed, however, it turned out to have numerous emission lines and to be quite unlike that of any star then known. Sandage's talk was given a summary paragraph in the popular magazine *Sky and Telescope* in March 1961, but its significance was not properly appreciated, even by those responsible for the work.

Early in 1963 an even brighter star was identified with another radio source (3C 273) that had been very accurately located by Cyril Hazard, then at the University of Sydney, and his colleagues Mackey and Shimmins. Working at the Parkes telescope, they used an occultation by the Moon to pinpoint the star's position and form, to a precision of about a second of arc – in fact the object turned out to be double. (A group of observers headed by H. P. Palmer at Jodrell Bank had determined as early as 1960 that its angular size was less than four seconds of arc.) Maarten Schmidt (1929–) of Mount Wilson and Palomar then obtained a spectrum for it, and found hydrogen lines in it that were shifted towards the red by an amount that was so large (sixteen per cent) that it seemed more appropriate to a distant Galaxy. This allowed him to spot oxygen and magnesium lines similarly displaced. His colleagues Jesse L. Greenstein and Thomas A. Matthews, armed with this information, then re-examined the spectra for 3C 48, and found that it had a red shift of thirty-seven per cent, suggesting a remarkable velocity of recession.

In 1964 Margaret Burbidge and T. D. Kinman began recording spectra with the 120-inch reflector at the Lick Observatory, and C. R. Lynds and his colleagues did so using the 84-inch reflector at the Kitt Peak National Observatory. By the end of 1965 ten such objects were known, and thereafter the number increased rapidly. As it did so, even more extraordinary red shifts were measured: already by the end of 1966, three were known of greater than 200 per cent.

These discoveries set astronomers the difficult problem of deciding whether the red shifts were 'cosmological' – that is, indicative of objects moving with the general expansion of the world of galaxies – or due to some new intrinsic properties, or perhaps simply due to colossal local velocities such as might follow from some sort of galactic explosion. A few influential physicists and astronomers, including James Terrell, Geoffrey and Margaret Burbidge, and Fred Hoyle, argued that they were within, or relatively near to, our Galaxy. Martin Rees and Dennis Sciama argued against this conclusion. If the red shifts were a local velocity effect, then should we not see approaching objects, with spectra shifted towards the blue? They made an analysis of the numbers, and argued that there were too many faint sources with large red shifts for them to be compatible with the steady-state cosmology that Hoyle was still defending in a rear-guard action. A considerable body of evidence eventually accumulated that the quasar red shifts were indeed cosmological: for example, the nearby quasar 3C 206 is seen in a *cluster* of galaxies, which all have identical red shifts to that of the emission lines of the quasar.

On a question of terminology: 'quasar' was very quickly seen to be an ambiguous term, when quasi-stellar galaxies and 'interlopers' were identified, and when optical studies gradually revealed the fact that only about one quasar in ten emits massive energy at radio wavelengths. (Sandage began finding 'radio-quiet quasars' in 1965.) 'Quasi-stellar object' (QSO) is a safer but uninformative name. The term 'quasar' is now usually reserved for a massively energetic star-like source with large red shift that is thought to be cosmological.

Now the great majority of galactic red shifts measured with the help of the Schmidt telescope at Palomar were of less than twenty per cent. If the red shifts of the quasars were cosmological, then the objects were obviously very distant, and correspondingly bright – each pouring out around a hundred times more energy than an entire typical Galaxy. (It was found, however, that the most powerful radio galaxies, such as 3C 295, may have luminosities comparable with theirs.) A problem henceforth confronting astronomers was to explain the source of energy in these, the most luminous known objects in the universe.

Radio studies in the 1970s and 1980s showed that many of them have (at radio wavelengths) a double structure typical of many radio galaxies, and the optical 'star' then usually coincides with a powerful and compact radio component. Radiation from quasars was shown to be often partly polarized and to include X-ray emissions. It was found that many are of variable output (in radio and optical terms) and that the timescale of the variation is of the order of a year. Measurements of their angular sizes with the help of interferometers with an intercontinental baseline – giving results of around a thousandth of a second of arc – showed that, despite their luminosity, the diameter of a typical Galaxy is around ten thousand times as great as that of a quasar. It was found that many quasars are embedded in galaxies, and twenty years after the first was discovered, it was shown that 3C 273 is embedded in a Galaxy with a nebulosity similar to that of many giant elliptical galaxies.

The theory was thus developed that quasars are galactic nuclei. In due course this line of research linked up with another, stemming from a discovery made by Carl K. Seyfert of the Mount Wilson Observatory in 1943. Seyfert had found galaxies with bright compact nuclei and unusual spectra. (We mentioned these in passing, in chapter 16.) He noted that the spectra of the nuclei of his galaxies seemed to suggest emission lines of hot ionized gas streaming out at velocities of thousands of kilometres per second. Twenty years later, optical detectors were developed that made closer study of Seyfert galaxies possible. They were found to vary in brightness with time. Many of their properties were later found in quasars. There are galaxies with similar properties to Seyfert's but with nuclei less active than theirs that it is possible to resolve. (Contrast quasars, where the nucleus is more luminous than the galaxy.) These N galaxies, so-called, include the BL Lacertae objects, a subgroup lacking strong emission lines in their spectra. The object BL Lacertae (Lacerta is a constellation) was once considered a variable star, but was later found to be a radio source with an elliptical nebulosity around it. There is a whole class of objects with these properties, assumed to be peculiar types of active galactic nucleus. The study of galaxies with active nuclei became an important new development in the astronomy of the 1970s and after.

It was soon realized that the enormous luminosity of quasars gave them a very special place in cosmology. When we observe a distant Galaxy we are observing it as it was in the past. There are few observable galaxies known at distances greater than 1200 megaparsecs, but those are being observed as they were when their light left them around 3.6 billion years ago. There are quasars, however, that *seem to be at more than ten times this distance*. Here one cannot simply divide the distance by the velocity of light and produce a time of light travel, for the cosmological model accepted is of crucial importance. In broad terms, however, we can say that as we look further and further into space, red shifts do not simply go on increasing. In fact few are known above 350 per cent, which sets an upper limit to the age of the objects, say about eighteen billion years. This is a substantial fraction of the age assigned to the Universe based on most expanding models, which in the 1980s was usually set at about twenty billion years. But again this figure depends, as already explained, on the model accepted. Which model to accept was a critical and hotly debated question in the early 1960s. By 1965 there was much evidence inclining the astronomical community to the view that the Universe was evolving, but there was no direct evidence that it had passed through the hot dense phase of its early existence that had been under discussion for so long. This evidence was provided from an unexpected quarter in 1965.

THE COSMIC MICROWAVE BACKGROUND

The story begins around 1930, with the work of Tolman on thermodynamics and radiation in an expanding Universe. In 1938, von Weizsäcker tried to explain the production of the heavy elements from hydrogen, in an early 'superstar' stage of the Universe – that is, before its expansion. In 1948, Gamow pointed out that according to general relativity the Universe could never have existed in a static high-temperature state. He proposed instead that the elements were formed, and radiation emitted, during the early and very rapid expansion. A theory of the formation of the galaxies followed. He and his collaborators calculated that the density of radiation in the early Universe was much greater than that of matter, but he did not consider the possibility of remnants of this phase, in the form of persisting radiation, surviving to the present day.

As mentioned previously, in 1949, Alpher and Herman – following the likely change in temperature over the history of the Universe up to the present – predicted a universal background temperature of 5K. They noted that there were no observational data on the present density of radiation overall. Four years later in a classic study with J. W. Follin they extended their work on physical conditions in the initial stages of the expansion, but they did not revise the earlier calculation, although the Soviet astronomers A. G. Doroshkevitch and I. V. Novikov later did so, and decided that the present temperature of background radiation throughout the Universe is close to zero.

Gamow's ideas had a small following, but it would be a mistake to pretend that he had presented the astronomical community with a clear goal. In 1950 C. Hayashi criticized his ideas and calculated that in the first two seconds of the expansion of the Universe the temperature would have been above the threshold for the creation of electron–positron pairs. Other calculations showed that, although helium would have been produced in the first phase, it was impossible to account for the heavier elements in the way Gamow had suggested. Added to this, new theories of element creation within stars were soon being very successfully developed, and that very fact led to a general neglect of his theories. The work was waiting to be rediscovered as surely as the radiation that occurs as an essential characteristic of it.

In the late 1950s plans were being made at the Bell Laboratories in the United States, especially at Holmdel, New Jersey, to work on communications satellites. Initial tests were to be made with the inevitably weak radio echoes from a balloon. This required a receiving system with very low noise. Use was made of a so-called 'travelling-wave maser' working at very low temperatures (those of liquid helium) and a 6-m horn reflector – vaguely resembling a squarish ship's horn. Although not large by the standards of radio astronomy at that time, its properties were accurately measurable, and when no longer needed for echo work it was transferred to radio astronomical projects, in charge of which were Arno Penzias (1933–) and Robert W. Wilson

(1936–). They hoped to be able to calibrate radio sources more accurately than had previously been done. Even when used in 1961 with the Echo satellite by a colleague, Ed Ohm, the temperatures obtained for the system – partly due to internal 'noise' and partly from outside – had been consistently 3.3°K more than expected, and now Penzias and Wilson too found an excess over expectation. They thought it was an antenna problem, but almost a year passed by, and even clearing out a pair of roosting pigeons made no difference. No matter which part of the sky the antenna pointed to, the radiation was there.

Whereas most experimenters would have assigned the unexplained readings as due to unknown instrumental errors, they persevered tenaciously. They discussed the matter with R. H. Dicke, who was then considering an oscillatory model of the Universe with hot phases, and who was expecting something of the sort they had found. Neither group knew then of Gamow's theory. In fact Dicke, at Princeton, had published a prediction of a background radiation at 10K at a wavelength of 3 cm. His group – P. J. E. Peebles, R G. Roll, D. T. Wilkinson – had the misfortune to be embarking on a programme to search for the background radiation with some highly sophisticated radio apparatus they had developed when they heard of the work done at the Bell Laboratories. With an accompanying letter from them, Penzias and Wilson published their own findings at last in 1965. In 1978 the two shared a Nobel prize for physics for their work.

What Penzias and Wilson had found, working at a wavelength of 7.3 cm – two hundred times shorter than that used in the pioneering work of Karl Jansky at the very same laboratories – was that *even a seemingly empty part of the sky gives off radio waves*. The fact could be explained in terms of a 'big bang', say ten or twenty billion years ago. The idea is that the energy of the primeval explosion has become diluted as a result of the general expansion of the Universe, so that it now corresponds to radiation from what physics terms a 'black body' (a perfect radiator) at a temperature of about 3 K. (They quoted 3.5 ± 1.0 K.)

Penzias and Wilson were fortunate in their wavelengths. There is a 'window' of wavelengths, roughly between 1 and 20 cm, through which radiation from the 'primeval fireball' (to take one of several names for it) can be observed at the Earth's surface. At longer wavelengths extra-galactic signals are submerged by those from our own Galaxy, while at shorter wavelengths the Earth's atmosphere radiates too strongly. And theirs was not the only possible route to the discovery. In retrospect it is possible to say that others – for example Haruo Tanaka in Japan (1951) and Arthur E. Covington and W. J. Medd in Canada (1952) – had anticipated them, but the uncertainty of their data was high, and its accuracy, judged by later work, poor. In retrospect it was possible to make sense of an explanation offered in 1940 by Andrew McKellar of the Dominion Astrophysical Observatory in Victoria, Canada, for some puzzling absorption lines found at Mount Wilson: he had thought that they might be due to absorption by cyanogen molecules in space at a temperature of 2.7K. He even predicted the existence of another absorption line, and found it, but his ideas were little known, explicitly rejected by some, and like the others just mentioned had no wider theoretical repercussions.

In 1965, however, cosmologists reacted with great excitement, for now the case was different in one vital respect: thanks to Dicke, the evidence was seen to have a bearing on a crucial question, one that was being asked by more and more astronomers. Here at last was evidence that could greatly narrow the choice of cosmological theories. The steady-state theories could still perhaps be defended. Some of their supporters considered the possibility that *new radiation* comes into existence with newly created *matter* throughout space, but this was not how the discovery of the 3 K background radiation was generally perceived. From this time onwards, most cosmologists turned their attention to the investigation of an evolving Universe with a hot big bang initial phase whose evolution is determined by the laws of the physics of fundamental particles.

This new style in cosmology, what one might call the Lemaître–Tolman–Gamow style – was vigorously pursued in the 1960s and 1970s by such theoreticians as Fowler, Wagoner, Thorne, Sachs, Wolfe, Sacharov, Weinberg, Schramm and Steigmann. Many new kinds of data were appearing with an important bearing on this theoretical stream, not only from the radio astronomers, but for example from X-ray and gamma-ray astronomy. But parallel to this new theoretical field of interest another was developing fast, with a more mathematical focus. Studies were being made of so-called 'horizon' effects in cosmological models, physical effects with strange topological properties that led some to countenance the idea that the Universe might not be homogeneous on a large scale. We shall return to these questions when we have examined the new evidence acquired at these other wavelengths.

As already intimated, not all steady-state cosmologists gave up their ideas readily. In 1975 we find Fred Hoyle and Jayant Narlikar defending an expanding steady-state model, with a 'scalar–tensor' version of the general theory of relativity resembling one developed by Jordan in 1939. Particles are no longer said to be created, but existing particles are said to change in mass. The 3K cosmic radiation was presented as starlight from an earlier phase turned into heat, having been scattered from atoms of very great size at the epoch when most particles are near to zero mass. These ideas have so far not acquired a large following.

THE DISCOVERY OF PULSARS

Two other important astrophysical discoveries were made around the time of the verification of the microwave background radiation, both having interesting aspects in common with it. In the first case, that of cosmic masers, we see the principle that discovery requires recognition and recognition requires some sort of theoretical understanding. In the second case, the discovery of pulsars, we see yet again the principle that one does not brush aside the unexplained, however insignificant it may seem.

The *maser*, a device for amplifying microwaves of very narrowly defined frequency, was invented in 1954 by Charles Townes and his associates at Columbia University, New York. (Its name is an acronym for 'microwave amplification through stimulated emission of radiation'. The same principle was later used at optical wavelengths for the better known *laser*, for 'light amplification through stimulated emission of radiation'.) In 1964 a group at Berkeley, California, led by Harold Weaver, studying the Galaxy at microwave frequencies, found an extremely puzzling set of spectral lines, which at first they called 'mysterium'. It was soon identified with radiation resulting from a change occurring in the hydroxyl radical, a pairing of an oxygen atom with a hydrogen atom, and from similarities with what can be produced in the laboratory it was accepted as an example of a cosmic equivalent of the maser. Many others have since been recognized, some found at MIT and Jodrell Bank even earlier, before they were understood, some involving other groups of atoms. They have been found in cool clouds in the vicinity of hot ionized gas, and also associated with certain types of stars that are strong in infrared radiation.

Pulsars have a more specific point of discovery, beginning with an observation by a graduate student at Cambridge University, S. Jocelyn Bell. She noticed 'a bit of scruff' on a 120-metre paper roll recording one complete sweep of the sky over the antenna. She was attempting to detect the way in which gas emitted by the Sun affected signals from radio sources, and here man-made radio noise was a problem. The recording had been made on 6 August 1967, and it was in October that she noticed its unusual character, and that the signal, looking like a series of rapid pulses, remained at the same point of the sky. With Anthony Hewish, her thesis adviser, and three other colleagues, she carefully observed the signal further, and the pulses were found to hold their spacing very accurately indeed (about 1.3s). Before Christmas, Jocelyn Bell had found a second pulsating source, with a period only marginally shorter than the first. At the Cambridge seminar at which the first pulsars were announced it was mooted that they might be due to distant civilizations, and their provisional names, LGM 1–4, alluded to 'Little Green Men'. One of the most surprising things about them was the sheer precision of their timing. The periodicity of the first is now confidently quoted to eight places of decimals.

It was Thomas Gold, then at Cornell University, who seems first to have recognized that pulsars were the long sought-for neutron stars, discussed on theoretical grounds since the 1930s. Other extensive theoretical studies of neutron stars have been made since the discovery of the first pulsar (named CP 1919, that is, Cambridge Pulsar at right ascension 19h 19m). Theory predicts that the magnetic fields of neutron stars are a thousand billion (10^{12}) times that of the Earth, and that these play an essential part in the production of radiation from the stars. In radio pulsars it is supposed that there is a rapidly rotating neutron star with a synchrotron mechanism – analogous to that used for accelerating particles spiralling through a controlled magnetic field in the laboratory. This produces radio waves from relatively near to its *surface*. As we shall see, X-ray-emitting pulsars were to be found in 1971 by satellite observation, and in these cases theory suggests that the radiation comes from the *poles* of the star. There are pulsars known at other wavelengths. In all cases the pulsation we observe is simply a lighthouse effect, a pulse occurring every time the peak of the emission flashes across us.

The incidental value of historical records in astronomy has already been illustrated in connection with the Crab Nebula, the remnant of a supernova recorded in China on

4 July 1054. This was eventually found to contain a pulsar at its centre, rotating about thirty-three times a second, one of the most rapid known. Its rapidity is a consequence of its relative youth – theory predicts a slowing down with time, as energy is dissipated from the star, and the historical record is an important datum for understanding the mechanism. The rates of change are extremely small but measurable – witness the accuracy with which that of CP 1919 is quoted. Not all changes are smooth, however. Discontinuities in timing are often observed, and have been found various explanations, for example in terms of fractures of the star's crust. While there is of course no unanimity over the theories of neutron star mechanisms, the nucleus of generally accepted theory has explained a striking number of phenomena. The state of matter at the centre of neutron stars presents, however, some of the most difficult questions in fundamental physics, as may well be imagined when bearing in mind that the mass in a cubic centimetre there is of the order of twelve million tonnes.

Observatories in Space

AIRBORNE OBSERVATORIES

Balloons, filled by hot air, hydrogen and later helium gas, have been a scientific object in their own right since the time of the experiments by the brothers Montgolfier, beginning around 1782. The two brothers, Michel Joseph and Étienne Jacques, began their experiments with hydrogen but achieved real success only with heated air. The first human flight was made on 20 November 1783. We have encountered one astronomer's travels in a manned balloon in the case of Janssen, who used it to make a successful escape from Paris in 1871, but he had no intention of observing his eclipse from the balloon itself. The first serious demands on balloons for what turned out to be astrophysical purposes were made by those studying the physics of fundamental particles. In 1911–12, the Austrian physicist Victor Franz Hess (1883–1964) made a series of ten balloon ascents. On 7 August 1912, with the balloon commander and a meteorologist, he made his most valuable flight. Travelling from Aussig on the Elbe to Pieskow, their flight lasted about six hours and took them to an altitude of more than five kilometres. Hess took readings from three electroscopes during the journey to measure the intensity of the radiation that caused ionization of the atmosphere. (There was a widespread belief that its source was rocks in the Earth's crust.) He found that while it fell for the first 150 metres it rose steadily thereafter as the balloon gained in altitude. He had earlier found that at a given altitude it remained the same, night and day, and so was not due to the direct rays of the Sun. Hess published his findings later in the year, and concluded that there was a highly penetrating radiation entering the atmosphere *from above*.

Hess was effectively the founder of cosmic ray astronomy. (The phrase 'cosmic ray' was coined by R. A. Millikan in 1925.) In 1936 he shared the Nobel prize for physics. His results had been verified by W. Kọhlhörster long before then – in fact in 1913 Kohlhörster had ascended by balloon to continue the measurements to an altitude of over nine kilometres – but not until the mid 1920s had they gained wide acceptance. After the *Anschluss* (the annexation of Austria by Nazi Germany in 1938), Hess's strict Catholicism led to his being dismissed his professorship at Graz, and he moved to the United States, where he continued his work – for example from the tower of the Empire State Building.

A notable use of unmanned balloons to carry instruments to investigate radiation from space was made by the German physicist Erich R. A. Regener (1881–1955), an authority on the physics of the atmosphere and stratosphere. (In 1909 Regener had obtained a rather accurate value for the charge on the electron, and it is for this that he is now usually remembered.) To measure cosmic radiation in the early 1930s he sent up balloons of rubber, and later of cellophane, some of them reaching as high as thirty kilometres. In 1933 he made a personal ascent, and on this occasion he found a connection between an eruption of the Sun and an unusually high degree of ionization of the atmosphere. This was an important discovery, for it showed that the *stars* are one source of cosmic radiation. Especially after the war, unmanned balloons developed for meteorological purposes were used as a standard means of transporting cosmic ray instrumentation.

Conventional aircraft have been used with much success to avoid the blocking effect of the Earth's atmosphere at short wavelengths. Serious work began in 1966, when Frank Low and Carl Gillispie made measurements of the brightness temperature of the Sun (for instance at 1 mm wavelength) from a Douglas A3-B bomber, flown for fourteen missions. A year later the National Aeronautics and Space Administration (NASA) funded studies of the planets from a Convair 990 aircraft, named 'Galileo'. This crashed with great loss of life in 1973. Its successor, Galileo II, likewise met with misfortune, for it was lost in a fire on the runway in 1985, but this time without human injury. NASA funded the Lear Jet Observatory from 1968, which took a 30-cm reflector to over fifteen kilometres altitude. This was the prototype for perhaps the most successful facility of this type, namely NASA's Kuiper Airborne Observatory (KAO), carried in a modified C-141 four-engined military jet transport.

Operating from 1974 onwards, the KAO typically carried a crew of three, up to seven experimenters, the telescope operator, the tracking telescope operator and a computer operator. A flight might be six or seven hours in duration. It was usually from Moffett Field in California, but in some cases from Hawaii, Australia and Japan. Great technical problems were overcome in this project, especially in achieving stability – through gyroscopes and the adoption of special flying techniques – and in avoiding air turbulence across the field of view of the instruments.

Airborne observatories were flown by several nations in the 1970s, including the United Kingdom, West Germany, India and Japan. Cosmic radiation studies were much pursued, and planetary studies too. From the KAO came a notable series of spectral studies of planetary atmospheres. The rings of Uranus were discovered from it in 1977 – simultaneously with a ground-based discovery at the time of a stellar occultation – and like other airborne observatories it yielded up much information about the heat emitted from planets. It was found, for instance, that Jupiter, Saturn and Neptune all emit more heat than can be accounted for in terms of the reflection of solar heat, showing that they possess active *internal* sources of heat. Uranus was found to have none of any significance. This type of high-altitude astronomy was soon to be put into eclipse, however, by rocket-launched probes.

ROCKET-BORNE OBSERVATORIES

One of the secondary effects of the many successes in astrophysics and cosmology in the period before the Second World War was that academic work in celestial mechanics declined. When a resurgence of interest in the subject came in the post-war period, it was for reasons having at first little to do with astronomy, and it took place far from centre stage. This was the 'Sputnik era', when for the first time observatories were carried by rockets beyond the limits of the Earth's atmosphere. To give the era such a title is of course to oversimplify events, but for the world at large the 'space race' had more to do with the transporting of warheads than of telescopes, and Sputnik was the first rocket to be capable – more or less – of remaining in orbit round the Earth.

Military rockets have a long history, with Chinese origins. The first rockets to be used with success in modern warfare were those designed by the English engineer William Congreve (1772–1828). They were used in many campaigns in the Napoleonic wars and were copied by most European armies soon after, largely as a consequence of the writings of Jacques-Philippe Mérignon de Montgéry, who in 1825 produced a well documented history and theory of the rocket as a weapon of war. Rocketry remained an appendage of gunnery, albeit with a theory of its own that might have been developed more enthusiastically by artillery experts had propellants been more reliable. Konstantin Eduardovich Tsiolkovsky (1857–1935), a Russian schoolteacher, made several important contributions to the theory; Robert Hutchings Goddard (1882–1945) of Worcester, Massachusetts, did likewise, and launched his first successful rocket powered by liquid fuel in 1926. Hermann Oberth (*b*. 1894) in Germany was an admirer of Tsiolkovsky's work, and organized enthusiasts into a society for space flight to which the young Wernher von Braun belonged. Limited by the Treaty of Versailles at the end of the First World War to artillery of small calibre, the Germans put much research effort into rocketry, and called on the services of several members of the society. It was von Braun who led the remarkable German military programme that culminated in the attacks on southern England by supersonic V-2 rockets in 1944 and 1945. A captured stock of these was shipped to the United States, and twenty-five of them were earmarked for scientific purposes.

At first the rockets were used for research into the upper atmosphere: these objects, fourteen metres long and fourteen tonnes in weight, could reach to an altitude of 120 kilometres. Radio telemetry techniques were developed to transmit data to Earth during flight, one of the early difficulties – hardly a serious problem for the German rocketeers – being that of reducing the speed on impact, and so salvaging the instruments. As early as October 1946, the first spectra of the Sun were obtained from one of these modified V-2s, spectra that were potentially valuable since they could be obtained from above the ozone layer that cuts out much of the ultraviolet radiation. The group responsible, at the US Naval Research Laboratory, was headed by Richard Tousey. The rocket ascended to eighty kilometres, and it became immediately clear why so many attempts to obtain ultraviolet spectra of the Sun from balloons in the 1920s and 1930s had failed: the height of the ozone layer had been greatly underestimated.

It was to be several years before controls were developed that would allow pointing of the instruments accurately enough for most serious astronomical purposes, but when this was done, it became possible to obtain a solar spectrum up to X-ray wavelengths. Where the first half of the twentieth century had seen phenomenal astronomical gains won by building ever-larger reflectors, one can say that in the second half of the century, in view of developments then taking place in radio astronomy, still larger gains were made by a vast extension of the wavelengths of radiations received. And in due course, knowledge of the solar system snowballed by the simple device of visiting the objects of interest.

On 4 October 1957 the Soviet Union launched Sputnik, the first artificial satellite to orbit the Earth. The United States had long been preparing Vanguard for an entry into space, and while conscious of Soviet ambitions, was taken unawares by the successful launch. Only a month later Sputnik 2 carried the dog Laika into orbit.

Both countries had large numbers of space technologists, but suddenly the whole enterprise became a focus of national pride. New educational schemes were set up by the US National Science Foundation and the National Aeronautics and Space Administration (NASA), and in the name of 'national defense and space exploration' many hundreds of government employees, industrial research workers, college teachers and students were introduced to the skills needed for orbit and trajectory calculation – celestial mechanics in a new guise.

By this time, astronomers had been profiting for more than a decade from budgets that had originally been justified in terms of national security. While they were trying to determine the nature of the cosmos, and in some cases whether there is life elsewhere in it, they were often obliged to work with others who seemed to be seriously threatening life here. Even astronomical observation was under threat. There were plans, for example, to fill the upper atmosphere with vast quantities of copper needles for purposes of radar screening, and these would have amounted to a permanent 'fog' above all radio telescopes. Astronomers on both sides of the Iron Curtain were often used cynically to provide an intellectual cover for the development of new weapons. It was fortunate that in so many cases the astronomical mask became the reality, astronomical aims were acknowledged as worthy of vast financial support, and in the space of twenty or thirty years the face of astronomy became transformed out of all recognition.

One of the most important of early discoveries was that there is a dense distribution of very energetic charged particles surrounding the Earth. The so-called 'Van Allen belts' are named after their discoverer, James Alfred van Allen, one of a team working at the Johns Hopkins Applied Physics Laboratory, Baltimore. Something had jammed the radiation counters on Explorer 1, the first US artificial Earth satellite. Van Allen redesigned the equipment and was later able to chart the doughnut-shaped distribution of charged particles.

Their discovery confirmed to some extent ideas about the movement of charged particles in the Earth's magnetic field developed in the course of studies of the polar aurorae, the spectacular lights in the night sky seen in latitudes between about 60° and 75°. In 1896 the Norwegian physicist Olaf K. Birkeland suggested that the aurorae might be caused by electrically charged corpuscular rays shot out by the Sun and drawn in by the Earth's magnetic field near the poles. In the 1930s F. C. M. Störmer (1874–1957), a Norwegian mathematician, tried to calculate the paths the particles should follow, but his theory was only superficially successful, and for many decades the theory of what has become known as the 'solar wind', responsible for the outer portions of the Van Allen belts, failed to explain the aurorae well. In the 1930s, however, the British geophysicist Sidney Chapman made much progress when he showed that the magnetic storms that disturb radio and telephone communications on the Earth may be due to clouds of ions ejected from the Sun. Others later found that a similar explanation might explain the observed rise and fall in cosmic ray intensity.

In 1957 Chapman was working at the High Altitude Observatory in Boulder, Colorado, when he developed the idea that the Earth's orbit around the Sun lies within the Sun's corona – and that the latter indeed fills the entire solar system. With E. N. Parker he developed a mathematical explanation of the phenomenon, an explanation that covered many other phenomena that had long been puzzling, such as why the tail of a comet points away from the Sun: it is blown back by the high-speed streaming of hydrogen in space. Because of the Sun's rotation, its lines of magnetic force are spirals, and this is one of several complications that have to be taken into account in the theory. But how to confirm it? Only by means of the satellite was it possible to chart the material. Many early vehicles were equipped to record charged particles in space, and they – Lunik 1, Lunik 2, Mariner 2 and the satellite Explorer 10, for instance – quickly showed the existence of the solar wind. They and others showed that the magnetic field had the expected spiral pattern, apart from irregularities. They showed that the energy involved in expanding the solar wind is not great – and that an insignificant million tonnes of hydrogen per second are thrown out, this act requiring only a millionth of the Sun's regular energy output.

The Sun had been one of the chief objects of rocket-aided study in the pre-Sputnik era. Between 1949 and 1957 solar spectra of high resolution were obtained at all optical wavelengths using rocket-borne instruments. One surprising discovery was that the far ultraviolet and X-ray radiation from the Sun was extremely variable. In 1956 work was done by the staff of the Naval Research Laboratory, Washington, on early type stars in the Galaxy, and there was a suspicion that X-rays were being detected from *outside* the solar system. (We shall have more to say about this in the following section.) In 1962 there came the launching of the first *Orbital* Solar Observatory (OSO-1), one of a series of eight, covering seventeen years in all – three-quarters of a full double cycle of solar activity, observed at various wavelengths simultaneously almost without break. Using a coronagraph of the type developed by Bernard Lyot in 1930, an image of the Sun's corona, free from scattering by the Earth's atmosphere, was seen continuously for several months out to a distance of ten solar radii from the limb — far better than during the most favourable eclipse as seen from the Earth.

The largest of the solar observatories was a manned example, and was given the name Skylab. It carried eight large telescopes, one with a coronagraph, and its crew brought back many thousands of photographs of the Sun's atmosphere from their three missions (May 1973 to February 1974). A later satellite, SMM (Solar Maximum Mission), was sent up in 1980 to examine the Sun at the maximum of its cycle of activity. It was necessary to carry out a repair, and this was done by astronauts James Nelson and James van Hoften using a remotely controlled manipulation system from the Space Shuttle Challenger on 11 April 1984. (The problems that Dondi had experienced in repairing his astrarium, or Herschel his mirror, were by comparison mundane.) In 1985, astronomers were reminded of their debts to their military patrons. An American satellite named Solwind had been launched in 1981 to supplement the findings of SMM, and to monitor the Sun through a complete sunspot cycle. Among other things it revealed the presence of five sun-grazing comets, all previously unobserved. In September 1985, however, its life was brought to an abrupt end when it was used as a target for an American anti-satellite weapon (ASAT). It was important to remember who was the paymaster.

Great advances have been made in knowledge of the Moon since the end of the 1950s, thanks largely to the competition between the superpowers, and the desire to prove military superiority, albeit in a peaceful guise. In September 1959 a Soviet probe, Luna 2, crashed on the far side of the Moon, and in the following month Luna 3 sent back pictures of that hidden face. One of the most surprising discoveries here was that the far side of the Moon is lacking in large maria ('seas'), the vast plains of basalt that create the 'man in the moon' image visible to us on the near side. In 1964 and 1965 American Ranger probes sent back pictures taken on the approaches to their crash landings. The first soft landing was by Luna 9 in 1966. Five Lunar Orbiter missions prepared the way for the most spectacular landing of all, that of the American astronauts, and obtained extremely thorough and scientifically valuable photographic information of almost all of the lunar surface.

Undoubtedly the best known of the lunar expeditions were the Apollo missions that first placed men on the Moon's surface. The first human being to orbit the Earth had been the Soviet cosmonaut Yuri Gagarin, whose spacecraft was launched on 12 April 1961, and who parachuted safely to Earth after a single orbit. (He died in a training flight on a humble jet aircraft in 1968.) There was a strong popular feeling for human involvement in space travel, although with the passing of time the American episode has perhaps been reduced in most human memory to the walk on the lunar surface on 21 July 1969 by Neil A. Armstrong (1930–), and to his words: 'That's one small step for a man, one giant leap for mankind'. (The first reported version, where 'a man' became 'man', made less good sense.) Armstrong was accompanied on the Moon by Edwin E. Aldrin, Jr. The spacecraft was Apollo 11, and this remained in orbit round the Moon while they were landed by the lunar module Eagle.

For all this, we should not forget the sheer scale of the manned operations of the period from 1968 to 1972. No fewer than nine manned American spacecraft orbited the Moon in that period, all named Apollo (and numbered 8 and 10–17). Six of them landed astronauts (not 8, 10 or 13) and between them they sent back over 380 kg of lunar samples. A scientifically invaluable visit was by Apollo 12's lunar module, which on 20 November 1969 landed astronauts, allowing them to examine Surveyor 3, a craft that had soft-landed in April of the same year. An analysis of flakes of paint from the probe later yielded useful evidence of the so-called 'solar wind', a most unexpected bonus.

Three unmanned Soviet vehicles (Lunakhods) were landed on the Moon, but they sent back only 0.3 kg of samples. In the Luna missions, a series of detectors was set in orbit round the Moon to pick up signals from a sensor on the surface. The same kind of geological sensing, with an automated station on the planet's surface, has since been used for the geological analysis of Mars, as part of the Viking project.

The samples of lunar rocks have been analysed by numerous laboratories in many countries, and their comparison with meteorites has provided far more information about the origins of the Earth–Moon system than had ever been available before. Three theories in particular were previously current: that the Moon and Earth were a naturally formed planetary pair, that the Moon was captured before it crystallized, or that the Moon was a detached part of the Earth's mantle. In the event, the evidence allowed none of these theories to survive unscathed.

Soviet scientists had much success with the inhospitable atmosphere of Venus, which is hot (470°C) and mostly of carbon dioxide. They have parachuted instruments to its surface from various probes (the types called Venera and Vega), some in the course of a mission to Halley's comet, and these have transmitted valuable signals carrying information on conditions prevailing on Venus, before falling silent, usually within an hour. A rendezvous with Halley's comet around the time of its nearest approach to the Earth in 1985–6 was the aim of no fewer than five satellite launches, from Europe, the Soviet Union and Japan. Giotto was damaged in its encounter, but passed on to meet up with a second comet. (Interest in comets around this time was enhanced by a debate on the possibility that, being rich in carbon compounds, they were responsible for bringing these to the Earth and so initiating living forms.) The Soviet Venera 4 and the American Mariner 5 arrived at Venus in 1967 within thirty-six hours of one another. Venera 9 and

10 in late 1975 sent back the first pictures of the rock-strewn landscape of Venus before the television camera perished.

The planetary surfaces, features of which even the best optical telescopes could not always resolve well, have been closely studied with the help of numerous missions. Mariner 2, for example, carried 18 kg of scientific instruments within 3000 kilometres of Venus in 1962, after a journey of four months. The first close-up pictures of Mercury's surface were obtained by Mariner 10 in March 1974, and showed heavily cratered regions. The Mariner series produced the first scientifically valuable missions to Mars. The Mariner and Viking probes marked the end of an era of phantasizing over the possibility that a civilization had produced the 'canals' on Mars. With one exception, Lowell's lines were shown to be the products of optical illusions. The exception was a giant canyon that he had called Agathodaemon, but that his compatriots preferred to rename Valles Marineris after its second discoverers.

The 1970s did not put an end to the desire for intelligent company in the Universe – on the contrary. The Pioneer spacecraft 10 and 11 carried gold-anodized aluminium plaques with a message that it was hoped might one day reach the gaze of intelligent beings beyond the solar system. These plaques, six by nine inches in size, were designed to be the longest-lived works of humankind – say some hundreds of millions of years – and to locate our solar system in space with respect to fourteen pulsars. They carry images of a Tarzan- and Jane-like couple with what were supposed to be 'panracial characteristics', and much implicit information to intrigue the finder – such as that earth-folk have not yet invented clothing, children or the metric system. Another human project in much the same spirit was realized in 1974, when a complex message in binary code was transmitted towards the Great Cluster in Hercules from the vast antenna at Arecibo. Among other information, this carried the chemical formulae for components of the DNA molecule. Not a century had passed since Camille Flammarion had published his monograph on Mars (1892) in which he waxed eloquent about the discovery of new worlds, their inhabitants surrounded by the 'work and noise of peace', and with whom we might some day unite. He had spoken of the Earth as a mere province of the Universe, and of 'unknown brothers' living in its infinite depths. He at least would have approved of the fact that by the end of the twentieth century there were to be groups of astronomers devoting a substantial part of their energies and considerable sums of money to the search for extraterrestrial intelligence.

The first planetary survey with great visual appeal was the near approach to Jupiter by the message-bearing Pioneer 10 in 1974. This crossed the asteroid belt and transmitted data successfully (data that were reconstituted into images) even after experiencing high-velocity impacts with small particles in Jupiter's neighbourhood. A year later, Pioneer 11 sent back fine images of Jupiter before moving on to Saturn (August 1979). Passing through the rings safely, it sent back images of Saturn. The Voyager probes, launched in August and September 1977, both passed near Jupiter and Saturn and returned extremely fine and beautiful images of Saturn's satellites and rings, of their structure, of Cassini's division and of marks crossing the rings like bent spokes. The trajectories of the Voyager probes were so calculated that the near-encounter with the first planet pulled the probes round in the direction of the next. In the case of Voyager 2, this sling-shot technique was used beyond Saturn: after launch (5 September 1977) and visiting Jupiter (July 1979), Saturn (August 1981), Uranus (January 1986) and Neptune (August 1989), the probe left the solar system for good.

As a measure of what had become possible by the 1980s using interplanetary probes, the case of Voyager 2 is particularly instructive. When it flew past Uranus on 24 January 1986, for example, in one brief encounter lasting only a matter of hours, the planet that for two centuries had been no more than a tiny disc of light was revealed as being at the centre of a complex system of rings (two new ones were added to nine found in 1977) and satellites (ten previously unseen were added to the five already known). Its rotational period was accurately found for the first time, likewise its axis of spin, tilted at a surprising 98°, and its strange magnetic field, tilted at a considerable angle (60°) to the rotational axis. This field was thought to be generated by a super-pressurized ocean of water and ammonia between the molten core and the atmosphere of the planet. The atmosphere, mainly of hydrogen and helium, at a low temperature (−219°C), experiences winds at well over 500 kilometres per hour. Its Moon Miranda is no less startling to the imagination: 500 kilometres across, it has canyons twenty kilometres deep, terraced layers and cliffs that are sheer ice as high as sixteen kilometres. By comparison, the worlds depicted in *Star Trek* are positively suburban.

The number of probes and satellites involved in these various enterprises is now to be numbered in hundreds, and what began as a nine-day wonder has become an event that fails to make even the inside pages of most of the world's press, except in the case of disaster. Disaster can come in many forms. NASA's Galileo planetary explorer, for example, launched in October 1989, had by the time of its first circuit of the Earth (December 1992) cost something in the region of $1.5 billion, and yet its main radio reflector was jammed by a piece of epoxy resin a millimetre or two in thickness. Speaking generally, the space enterprise has served astronomy well. Giotto, Ulysses, Galileo, Phobos, Vesta and the rest have created a situation that might well make us forget that the telescope was ever needed for planetary work. In fact before the advent of rocket-borne probes, most professional astronomers were content to leave the physical appearances of the planets to their amateur colleagues. I remember being told in 1957 by W. H. Allen − author of a standard work of astronomical reference, *Astrophysical Quantities* — that in his view the sum total of all planetary studies to date was 'not worth a row of beans'. Whether his judgement was true or false, in the era that was then beginning it would soon be possible to produce satellite photographs of planetary surfaces of such fine detail that they might easily be confused with photographs of the Earth's surface. More important, it has become possible to frame plausible theories of the mantles and cores of planets, their atmospheric, magnetic, geological, seismic and other properties, and their likely evolution. Rockets have in fact provided planetary astronomy with very many of its missing links.

Images that do not require reconstitution from data obtained by scanning, and so are instantaneously available, have a much stronger popular appeal than composite images, and the numerous Soviet Vega probes, the first and second having been launched in December 1984, have made use of this facility. As explained, both sent landing stations into the atmosphere of Venus, and then passed on to within 10 000 kilometres of the core of Halley's comet (March 1986).

The sky as a whole has been surveyed at infrared wavelengths by various means. In the mid 1970s, for instance, several American rockets were launched for this purpose, making flights of only short duration, and not as a means of placing instruments in orbit. The first almost complete survey of the sky was achieved by the Infrared Astronomical Satellite (IRAS), launched in 1983 as a joint project of the United States, the United Kingdom and The Netherlands. The satellite carried a 57-cm telescope cooled by liquid helium to less than 3K, so that the telescope's own heat emission could be cut down to negligible proportions. IRAS made a number of important discoveries, including clouds in the Galaxy that emit infrared radiation, a cloud around the star beta Pictoris that has the appearance of a planetary system in formation, and a series of six hitherto unknown comets. (That the birth of a planetary system had indeed been observed around beta Pictoris was challenged two years after it was announced, and the discussion ended on an inconclusive note.)

Observation beyond the other end of the visible spectrum, that is, at ultraviolet wavelengths, is needed for very hot sources – say hotter than the Sun's photosphere. This includes the Sun's chromosphere, and interstellar gas heated by nearby hot stars. It also includes very massive stars – some of even more than a hundred solar masses. Only in 1981 did this class of object enter astronomy. Satellites dedicated to observing the sky in the ultraviolet had by then included the second Orbiting Astronomical Observatory (OAO-2) launched in 1968, with no fewer than eleven telescopes – seven built at the University of Wisconsin and four by the Smithsonian Astrophysical Observatory. Its successor was OAO-3, launched in 1972, and named after Copernicus (*d.* 1473).

As we have already seen, by the late 1970s astronomy was becoming increasingly international in scope, and one outstanding achievement of the new phase was the International Ultraviolet Explorer (IUE) launched in 1978. This highly successful satellite carried a 45-cm reflector and could record stars down to magnitude 16. Always visible from the Goddard Space Center near Washington or the European Space Agency's station near Madrid, it was controlled alternately from one or the other. Earlier ultraviolet telescopes had revealed an extraordinarily high rate of loss of mass by very massive stars – say a loss of one solar mass in a million years, for a star of thirty solar masses. IUE made possible a closer study of this problem of mass loss, which radically changed the theory of stellar evolution then generally accepted, as regards stars in the upper part of the Hertzsprung–Russell diagram. Stars with phenomenally large luminosities, even a million times that of the Sun, could of course not previously be accounted for in terms of a model that entailed instability in stars above 60 solar masses. Mass loss was in the 1980s seen as the answer to instability. It has been seen, too, as possibly offering assistance in understanding the Wolf–Rayet stars that have been so difficult to fit into the H–R diagram. (As mentioned previously, these stars were described for the first time in 1867. They have peculiar spectra, are very hot and luminous, and eject shells of gas at high velocities.)

SATELLITES AND X-RAY ASTRONOMY

In the decades beginning in 1970, no branch of astronomy benefited more from rocket-borne satellites than high-energy astrophysics. This study makes use of radiations of very short wavelength (below half a nanometre, but not as short as gamma-rays, with wavelengths a thousandth of a nanometre; a nanometre is one billionth part of a metre). The name 'high-energy astrophysics' comes from the fact that photons, packets of radiation, have at these wavelengths considerably more energy than photons of visible light.

The Sun was known to emit X-rays, and in 1962 the most powerful source of X-rays in our sky, Scorpio X-1, was found in the constellation of Scorpio. This discovery is another in a series on the fringe of professional astronomy. Bruno Rossi, a professor of physics at MIT, was also chairman of a company (the American Science and Engineering Corporation) founded by a former student, Martin Annis. The company had taken on an Italian cosmic ray physicist to start a programme in space science, and together with another MIT physicist, George Clark, they designed instruments for X-ray observations of the Sun, Moon and certain stars (such as supernovae). NASA turned one proposal down, but the Air Force supported an attempt to study X-ray fluorescence radiated by the Moon, and in the course of this it was discovered that the background X-radiation from the sky completely obscured any lunar fluorescence there might have been. X-ray sources were found, but it was their intensity rather than their existence that was surprising. In fact the team knew of an earlier unpublished report of a suspicion that an X-ray source had been found. This, by Herbert Friedman and James Kupperian Jr in 1957, was not published since it could not be confirmed by a later flight.

It had never been expected that individual stars in the galaxy, which is what these X-ray sources are, could have such energy as they proved to have. Some of them pour out say as much as 100 000 times the energy of our Sun. On 12 December 1970 another American satellite, now specifically intended for the study of X-ray stellar sources, was launched from Kenya to commemorate the country's independence. (It was called Uhuru, the Swahili word for 'freedom'.) The event opened an era in which the positions of the most powerful X-ray sources were charted. Among 339 sources identified with the help of the Uhuru satellite, Cygnus X-1 was discovered, one of the brightest in the galaxy. This source is associated with an optically visible blue giant star of twenty solar masses and an invisible companion that was later estimated at 8.5 solar masses – which some have argued must be a black hole, since its mass exceeds the theoretical limit for a neutron star.

Another satellite with the same aim, HEAO-2 (successor to the similar HEAO-1), was launched in 1978 and eventually renamed Einstein, to celebrate the centenary of his birth (1979). (In a previous historical age one had to discover something new in the sky to dispense a favoured name; now one had only to put something there.) The American and European satellites of the preceding period could not produce direct images, and were not accurate to more than about a degree. HEAO-2 carried a wide range of instruments, was capable of direct imaging and could place X-ray sources to an accuracy of about two seconds of arc, usually allowing them to be identified with optical sources.

Yet another satellite observatory for high-energy astronomy launched at much the same time as HEAO-2 was the Japanese Hakucho ('Cygnus') X-ray satellite, launched in 1979 and joined four years later by satellite Tenma

('Pegasus'). The age of the satellite continued, checked but not stemmed by an economic recession. And we have reached the stage when merely to catalogue the host satellites would be as idle as the cataloguing of telescopes in, say, Herschel's day. Each played its part in creating a substantially new vision of stellar structures and their role in the cosmic pattern. It was soon widely agreed that the X-ray sources correspond with systems including highly compact stars of one sort or another – white dwarfs, neutron stars, or black holes – all with very high gravitational energy that can be converted to high-energy radiation (largely X-rays and gamma-rays). Much of the debate over the existence of black holes has centred on the powerful source, Cygnus X-1. In this and many other cases, a binary system is found to be involved, the compact star component seemingly drawing in material from its partner, a normal star. This takes place in various ways, according to the size and proximity of the components.

X-ray sources were found to vary in output, often periodically, but occasionally with violent bursts lasting only hours or days. In some cases periodic bursts last only seconds. Even in some of these cases (known as 'bursters') it was found that a binary system was involved, but the systematic eclipsing that had formerly been offered to explain the fluctuations in light intensity was no longer enough to explain the more complex patterns of X-ray pulsation. Relatively simple cases were found, with explanations that included a single star's rapid rotation, perhaps with an unsymmetrical disc around it, or a stream of ejected matter; but in other cases it was found that pulsation is rapid. A high concentration of X-ray bursters (stars showing themselves in sudden outbursts lasting perhaps a few days, but not periodic) has been found in the galactic plane. At an early stage of satellite exploration, globular clusters were also found to be powerful sources of X-rays.

Gamma-ray bursters have been found, but not distributed in the same way. Those recorded seemed at first to be relatively near and faint, perhaps isolated neutron stars. The first were found by American military (Vela) satellites from the late 1960s. They had been deployed to detect Soviet nuclear explosions, which would have been puny by comparison. Some gamma-ray bursters were found to be associated with optically observable objects. A remarkable event observed from no fewer than nine satellites on 5 March 1979 involved a very short pulse followed by a series of about two dozen pulses at eight-second intervals. There has been much discussion of a possible mechanism, perhaps involving a neutron star, but what made the incident especially interesting was that the source seemed to coincide with a supernova remnant in the Large Magellanic Cloud beyond our galaxy.

Built by five European research institutes (the European Space Agency, France, The Netherlands, Italy and Germany), Cos-B was a satellite with a gamma-ray detector that operated for what was then an unusually long period, from 1975 to 1982. With it, a map of sources was produced, many of these very powerful, and of the two dozen most powerful almost all were in or very near to the plane of the Galaxy. This coloured the views of astronomers for several years, until in 1989 NASA launched an enormous orbiting probe (seventeen tonnes, sixty times as massive as Cos-B). Known as the Gamma Ray Observatory (GRO), the ambition was to bring gamma-ray astronomy to a state that X-ray astronomy had attained two decades earlier. Supernovae in galaxies, pulsars and active galactic nuclei were on the agenda, but also gamma-ray bursters. It was expected that these would be found concentrated in the plane of the Galaxy, where neutron stars are to be found; but not so. They appeared over the whole sky. Gamma-ray sources had seriously unsettled astronomers once again.

'Pulsar' is a term that has been used for pulsating stars of various sorts, as we have already seen. Neutron stars in rapid rotation seemed to offer the best explanation of what was observed. How complex the situation might be, even when a neutron star is postulated to account for pulsation, became evident when a pulsar of period less than a tenth of a second was found in the Large Magellanic Cloud. This must count as the most extraordinary of all X-ray sources to have been discovered by satellite recording, for it apparently radiates more X-rays than all the sources in our Galaxy combined. While many a modest lump of rock in the solar system bears a human name, the object AO 538–66 seems to have confounded the collective imagination of astronomers. It appears to include an ordinary star of about a dozen solar masses, perhaps combined with a black hole (needed to explain its energy) or neutron star (needed to explain the pulsation). The decision between the two alternatives, black hole or neutron star, is one based ultimately on the mass deduced for it. If greater than roughly three solar masses, theory favours a black hole. In the first two decades of X-ray astronomy only a handful of candidates for this distinction were found, and each case was the subject of much disagreement. The theory of black holes, however, may be found an application on a much larger scale, that is, it may be applied to the *nuclei of galaxies*, where the mass is of the order of a billion stars, and where the black hole consumes matter (gas) equivalent to several solar masses annually.

NOVAE AND SUPERNOVAE

X-ray astronomy soon proved to offer a way to a more complete understanding of the evolution of stars, and this through the study of supernovae.

We have already seen that the word 'nova' was at first reserved for any star showing a sudden increase in brightness, and that the 'new star' of 1054, recorded in China, and those of 1572 and 1604, made famous by Tycho Brahe and Johannes Kepler, would now be classified as *supernovae*. For them a completely different mechanism is responsible. In the case of a nova, only the outer layers of the star seem to be involved in the sudden flaring up, a relatively small fraction of the mass of the star is lost, and indeed some of the mass involved comes in any case from an adjacent star. Accepting an idea mooted by Robert Kraft in 1964, novae, without exception, are now thought to be members of close binary systems, for example a white dwarf with a cool companion. The change in absolute magnitude is of the order of ten magnitudes or less.

By contrast, a supernova is an explosion on a much greater scale, involving most of the material of the star. It was impossible to appreciate the difference between the two classes of phenomena without first knowing the change in intrinsic brightness, and this required a knowledge of distances. The turning point in an understanding of the situation came with observations of novae (some later reclassified) in the great nebula in Andromeda, M31, and in particular a star later designated S Andromedae. First seen by C. E. A. Hartwig of the Dorpat Observatory on 20 August 1885 – unless we count L. Gully of Rouen, who saw it three days earlier but thought it was due to a blemished telescope – it brightened from the ninth to the seventh (apparent) magnitude, before fading rapidly. It disappeared from view after 7 February, but not before its spectrum had been studied by at least five astronomers. Among these, Huggins recorded bright emission lines and Copeland bright bands. These were the first steps towards understanding the extraordinary events that had taken place in 1054, 1572 and 1604.

In 1895 a similar 'nova' was found by Williamina P. Fleming at the Harvard College Observatory, in an unresolved nebula (NGC 5253) in Centaurus. It was later named Z Centauri. Williamina Fleming found it from its peculiar spectrum, and this was classified by Annie Cannon as of spectral class R. She thought it resembled that of S Andromedae – a decision that was revised long afterwards by Cecilia Payne-Gaposchkin, who in 1936 realized that the spectral lines were unusually bright and wide.

For the time being, novae were classified loosely in a scheme devised for variable stars generally by Pickering (1880, revised 1911). On this scheme there were simply 'normal novae' and 'novae in nebulae'. In addition, Pickering's classification of variables included stars of the 'U Geminorum type', which are now generally called 'dwarf novae'. Their outbursts are typically at intervals of a few months. The prototype star was discovered by J. R. Hind in 1855–6. Not for forty years was another found – this time by Miss L. D. Wells of the Harvard College Observatory (SS Cygni). By 1922 the similarities of the spectra of this still small group to the spectra of novae was recognized by Adams and A. H. Joy.

Other novae in spiral nebulae were found in 1909 (Max Wolf) and 1917 (G. W. Ritchey), the latter prompting astronomers to study much more carefully photographs that had been taken with the Mount Wilson telescopes. As a result, many more were found in this same category.

The last step to appreciating their great brightness came with the realization that the spirals were indeed very distant 'island universes'. Only then, in the mid 1920s, was the highly energetic character of the novae that could be seen in them perceived. Data remained in short supply. (By 1937, only five had been studied spectroscopically.) The typical star in this class, when it is identifiable on earlier plates, is found to increase in brightness by at least fifteen magnitudes. In the sudden explosion, more energy is released than our Sun has radiated in its entire lifetime of four or five billion years.

In 1925 Lundmark distinguished between 'upper-class' and 'lower-class' novae, and Baade and Zwicky in 1934 substituted the name 'supernova' for the former, the extremely bright novae in distant galaxies. It was left to Baade, in 1938, to insist that here we are not only dealing with widely differing luminosities but with completely different classes of object.

The study of novae and supernovae could hardly have been advanced at all without photography, since the stars concerned are almost always inconspicuous before the outburst, but with the help of old plates the pattern of their changing brightness can often be pieced together reasonably completely. The typical light curve shows a rapid rise followed by a slow decline, and there is an element of luck in catching the star during its rise. The first photograph of a nova's spectrum during the rise to maximum was that for the nova in Perseus in 1901. That it was largely an absorption spectrum almost persuaded the Harvard astronomers that they had photographed the wrong star. Not until 1918 was a *spectrum* found for a nova *before* its outburst (nova Aquilae). The plates were from 1899.

Numerous attempts had been made to explain novae. Newton, for instance, had a collision theory; Laplace believed in some sort of surface conflagration; W. Klinkerfues argued for tidal eruptions through the near-approach of another star; Seeliger thought a collision between a dust cloud and a star was involved; and Lockyer held that collisions between meteorites in two intersecting streams were responsible. At least this last theory could be ruled out once photographs had shown stars in place before the outbursts. What Nova Aquilae 1918 provided was a great deal of new information. From the spectral information W. S. Adams and J. Evershed drew the conclusion that a shell of gas was being ejected from the star at high velocity, and that some of the spectral peculiarities were due to the fact that we see superimposed spectra from the near and far parts of the shell, spectra that undergo Doppler shifts in opposite senses.

Broadly speaking, this fits into the general picture pieced together later in the century, of a complex interaction between a relatively cool star and an adjacent dwarf star. A disc of accreted material forms round the cool star. In the case of an ordinary nova this is drawn to the dwarf and triggers certain highly energetic nuclear reactions there that explode the outer layers of the star. In the case of the dwarf nova it seems that matter is drawn at supersonic speeds from the dwarf star, not completely into the other, but into the surrounding accretion disc. The disc is now heated at the place where the incoming material hits it, and can do so repetitively, with the release of much less energy than in the 'ordinary' case. In both cases the accretion disc plays a crucial role. Observation of this, using the ultraviolet and X-ray regions of the spectrum, has been made possible only with the advent of satellite observatories.

After 1937 it was realized that there are two different classes of supernovae. A supernova in that year, analysed by Zwicky and Baade, had an intrinsic brightness about a hundred times greater than that of the entire galaxy in which it was found (IC 4152). R. Minkowski studied its spectrum, and so was led in 1941 to announce that fourteen supernovae he had studied were distinguished clearly by their spectra into two classes, distinguished primarily by the presence or absence of certain hydrogen lines. One class ('type II') is reminiscent of ordinary novae, but of course much brighter, and one ('type I') is considerably brighter than type II, and with very unusual emission bands. Work on the Crab Nebula provided much valuable data here, and we have already seen how radio studies added to this. In addition, in 1964 the Crab Nebula was found to be a powerful emitter of X-rays: in a pioneering rocket observation conducted by Herbert Friedman and associates from the US Naval Research Laboratory, the *gradual* occulting effect of the Moon (gradual because the Nebula is not a point source) was used to prove that this highly photogenic Nebula really was the X-ray source. Only later was the additional complication of the pulsar at its centre appreciated.

In the 1950s, William A. Fowler and Fred Hoyle suggested a mechanism to explain the source of energy. Their complex picture of the star is not one for a brief explanation, but it involves a star built up of a series of shells, like the layers of an onion, each shell the product of nuclear reactions at a particular stage in the long history of the star.

As heavier elements are created, temperatures increase and alternate with gravitational contractions until a stage is reached when the star is in equilibrium, with a mixture of iron and nickel and other moderately heavy elements at its centre. Gamma-ray photons then step in, and enter into certain nuclear reactions with the iron and nickel, in processes that require heat. The star cannot remain in equilibrium, and so collapses. The temperature rises, and after further stages have been gone through, nothing but protons, neutrons and electrons remain. The protons absorb the electrons, and the core contracts further very rapidly until a point is reached where a certain nuclear force that causes neutrons to repel one another brings the contraction to a halt. The star is then very compact, with the neutrons of the order of a ten-millionth of a millionth of a centimetre apart (10^{-13} cm). This is the theoretical state of the neutron stars that we have referred to so often. The core collapse is said to take only a matter of minutes. As it takes place, the outer layers fall inwards, are accordingly compressed, and heated. Reaction rates are accelerated, and the layers explode. And this is the supernova explosion, according to Fowler and Hoyle.

Other theories were offered later, although in broad outline this pioneering study managed to hold the allegiance of a majority of astronomers very successfully for a long period. The variant theories have much in common. They explain the formation of very heavy elements such as uranium by the nuclei in the outer shells capturing neutrons during the final neutron phase. In this phase too, in the very high temperatures involved – say ten billion degrees – extremely large numbers of neutrinos are generated. (Neutrinos are fundamental particles with spin, but with little or no mass, and no charge.) Now on 23 February 1978 a supernova was observed in the Large Magellanic Cloud, by an astronomer working at Las Campañas Observatory in Chile. This, the brightest supernova since 1595, was the only one that had ever been identified with a star known before its outburst (Sanduleak-69 202). It gave rise to the first burst of neutrinos ever to be detected from outside the solar system. On the basis of the neutrinos received, it has been said that, for a second or so, the luminosity of the star in neutrinos alone equalled the luminosity of the rest of the Universe combined.

Supernovae explosions spread heat energy and the products of nucleosynthesis around the Universe, and so influence the evolution of the galaxies in which they are situated. Especially important here are the heavy elements, and in the 1980s much attention was given to their theoretical role in the formation of new stars. A particularly important supernova was found in 1986 in the radio galaxy Centaurus A, important since it is the nearest radio galaxy to Earth. For its X-ray, gamma-ray and radio properties to be better understood, its distance was needed, and an amateur astronomer in New South Wales made possible a much improved estimate of this. The Rev. Robert Evans, who had already discovered a dozen or so supernovae in galaxies with his 40-cm reflector, noticed a bright star in Centaurus A. After half an hour it had not moved, so was not an asteroid in the line of sight. Evans rang the Siding Springs Observatory, and within three hours it was confirmed as a supernova. (It later proved to be of type I.) Its especial interest lies in the fact that it was discovered before maximum brilliance was reached, so allowing it to be monitored through the critical stages. The Galaxy was in this way shown to be at a distance that places it among the Local Group of galaxies.

THE HUBBLE SPACE TELESCOPE

The numerous early satellite observatories were short-lived – their lives were in months more often than years. The first plans for a versatile optical observatory that could be kept working in space for many years were being laid at the time of building the OAO spacecraft in the early 1970s.

The key figure in its conception was Riccardo Giacconi, director of the research team working towards it, at first within the private company American Science and Engineering (1958–). Giacconi was made director of the independent Space Telescope Science Institute, which had limited proprietary rights to the data obtained and the co-ordination of the various tasks. Their realization was spread over many other organizations, and the boundaries of responsibility shifted as the project continued. The Institute was operated by the Association of Universities for Research in Astronomy, Inc., on behalf of NASA. This venture is somewhat reminiscent of Tycho Brahe's, advancing learning through the astute combination of technology and capitalism, with the support of an enlightened state. Occasional government irritation is a part of the formula.

The Space Telescope, eventually named after Hubble, was to have been put into orbit in 1985 by NASA's Space Transportation System – the 'Space Shuttle'. The first serious delay in the launch followed the temporary suspension of the Space Shuttle program in the wake of the tragic explosion of the shuttle *Challenger* on 28 January 1986. The telescope project was plagued with misfortune, however, even before the instrument left the ground. One of its most important parts was a device on which light falls, converting it to electrical signals that are then reconstituted as a picture. Such CCDs ('charge-coupled devices') have effectively replaced photographic plates in this kind of work, and by chance it was discovered that the CCD chip's sensitivity to blue light depended on the area on which the light fell and the extent to which the chip had been exposed previously. At least this problem was discovered in time, but unfortunately it was one of several.

The advantage of the shuttle system is that a payload can be put into orbit and can be repaired and updated to a certain extent for many years – visits to the Hubble telescope at three-year intervals over fifteen years were planned at one stage. In this case the instruments are carried within a tube roughly 4 m diameter and 13 m long, and the basic telescope is a Cassegrain of modified design, with 2.4-m aperture. This would have been a sizeable instrument even for a ground-based observatory, but above the atmosphere it is in principle far superior even to a ground-based 4-m telescope.

When calls for proposed scientific programmes to be carried out on the satellite were made, the project was oversubscribed six times. A selection was made, and the Hubble Space Telescope was provided with its secondary instrumentation. Many of the component parts were so arranged that they could be moved into the focal plane of the main instrument when needed, there detecting wavelengths from the far ultraviolet to the infrared.

The 2.4-m primary mirror was planned to be the most precise large optical surface ever made, so it was all the more tragic that after launch the mirror was found to suffer from certain distortions that might have been discovered earlier had the testing apparatus itself not been faulty. Other faults developed, in the gyroscopes and in the solar panels, which were not rigid enough. Despite these painful problems the telescope has returned much useful information not available from any other source.

Communication to and from the telescope was planned to be through a network of communications satellites, stationary above the Earth. Data transmitted through them passed to the Goddard Space Flight Center, and so to the Space Telescope Science Institute, from which commands could be returned via the same links. From this Institute the data were analysed by resident astronomers and distributed if necessary to others elsewhere. A European Coordinating Facility assisted in various ways, providing instruments and astronomers.

Progress was much slower than anticipated, and when the telescope launch finally took place in April 1990, the object launched – even disregarding the cost of launching it – qualified as the most expensive scientific instrument in human history. Many lessons had been learned, not all of them about astronomy. The slow progress and cost overruns on the project had led to delays in other space-science programmes, such as the Advanced X-ray Astronomy Facility (AXAF); but if harsh words were spoken, they were certainly nothing new to astronomy.

By the time of the Hubble launch, the HIPPARCOS satellite of the European Space Agency was beginning a successful astrometric programme, the first satellite to be aimed at measuring star positions. Here too there were problems: two of its gyroscopes failed and the apogee booster motor failed to put it into the desired 'geostationary' orbit, so that it crossed the Van Allen belts four times a day, making it unusable for a third of the time. While its name pays tribute to an activity pioneered by Hipparchus more than two millennia before, the acronym has another interpretation: high precision parallax collecting satellite. Whereas air turbulence means that observations with the very best ground-based telescopes can be accurate to no more than a tenth of a second of arc, the precision of HIPPARCOS is one or two milliseconds of arc—say the angle subtended by a centimetre at a distance of 2000 kilometres. Parallaxes and proper motions, and all that depend on them—such as the scale of the Universe, no less—have been obtained with unprecedented accuracy as a result. In addition, the satellite measured star brightnesses to high accuracy, allowing for improvements in the light curves of many known varieties of variable star, not to mention the discovery of new varieties.

Such achievements as these come at considerable cost. The cost of the Hubble Space Telescope, all told, approached one per cent of the United States' annual budget for national defence. In cosmic terms, of course, that is a mere nothing.

Macrocosm and Microcosm

ASTRONOMY AND NEIGHBOURING THEORIES

Throughout the long history of theorizing about the Universe, there have always been important principles at stake far removed from the province of the observatory. There have always been considerations of simplicity, harmony and aesthetics, often masquerading under the name of philosophy, and often dictated by strongly held religious belief. The steady-state theories, for example, were often charged with having taken away all sense of purpose from the Universe, and having put it into a monotonous, endless and *pointless* state. If such notions are not exactly visible in the mathematical equations of rival theories, when those who put forward a theory enter into a debate of this kind, they often make it clear that some such ideas were present at the theory's conception. It has been said that a Universe beginning in a primordial fireball and dwindling into nothingness is intrinsically even less *attractive* than the steady-state idea. If nothing else, this shows that we cannot discount the place of the human psyche in modern cosmology.

It is easy to forget how deep are the potential connections between cosmology and the other sciences. To begin with theories of stellar evolution: the *parts* of the Universe are certainly in a process of change. Fred Hoyle, in a retrospective view offered in 1988, went so far as to say that to deny evolution in the sense given to the term by the 'big bang' cosmologists 'is no loss at all, for the issue of whether galaxies have grown a little brighter or a little fainter, or a little larger or a little smaller, over the past [ten billion] years is singularly uninteresting'. 'The only evolutionary processes of real subtlety in astronomy', he went on, 'are those which relate to stars, and strikingly enough stellar evolution proceeds essentially without reference to cosmology.' To think about evolution cosmologically, he thought, one should look at problems of a 'superastronomical order of complexity', such as the problem of the origin of the biological order; and to have any hope of solving such a problem as this, a particular type of cosmology (the steady-state theory, as he believed) may be required. In short, no cosmologist can afford to be that and nothing else.

The fact that all cosmology, all astronomy, should ideally be made to connect with other acceptable sciences – and that their histories must therefore also be linked – is not a new idea, but in an increasingly complex intellectual world the principle of the division of labour tends always to force them apart. Those in the past who have tried to draw them together have often been undisciplined phantasists, but a few have had the highest formal credentials – Kepler, Newton, Einstein and Eddington, for example. Their attempts were often greeted by their colleagues in embarrassed silence, until of course they seemed to have succeeded. Generality has always been acknowledged as one of the highest of scientific virtues. One of the most remarkable aspects of astronomy in the twentieth century was the high degree to which it was shown that it could ally itself with an ever-increasing number of branches of science. Biology apart, we have seen how theories of gravitation, optics, electromagnetism and spectroscopy became allied, and that the last-named introduced considerations of physics at the sub-atomic scale. From the 1930s thermodynamics played an increasingly important role, and theories of stellar structure and the transformation of chemical elements made use of quantum physics. We have mentioned too some attempts to link the physics of the very large and the very small through the fundamental physical constants. However, from the 1960s and later came a number of exciting ideas that bound together quantum physics and relativistic cosmology, not to mention thermodynamics, in entirely new ways. At the heart of this new movement is the notion of a black hole.

SPHERICAL MASSES

We have already tentatively introduced the notion, but without explaining how it insinuated itself into the affections of astronomers. (As for the phrase 'black hole', John Wheeler coined it only in 1968.) It is necessary at the outset to make the point that while one may use ordinary words to describe the strange properties of black holes, the essential geometrical ideas can only easily be conveyed by using analogies that are potentially misleading. We shall need to have a rough idea of what is meant by a *singularity*, in the sense in which the word is used in geometry. Suppose one quantity depends mathematically on another – for instance y may be equal to x^2 or z may be equal to $1/x$. Producing a simple graph of the relationship in the first case will give rise to no great problems, but in the second we shall find that when x is zero the function z behaves badly. In this case we may loosely say that it 'becomes infinite' when x is zero. In other cases of 'bad behaviour' we may find that a relationship we wish to represent in a graph is discontinuous, or that we cannot specify its direction unambiguously. The values of x where the trouble occurs are the 'singularities' for the function in question.

We came across one sort of singularity in certain relativistic models of the expanding Universe that – when naively interpreted – seemed to point back to a Universe of zero dimensions. Another sort was in the case of Schwarzschild's treatment of the gravitational field surrounding a spherical mass. We discussed this briefly in connection with the Sun, in chapter 16. (The Schwarzschild singularity is at a distance from the centre equal to twice the solar mass, if we choose suitable relativistic units.)

It was in the context of Schwarzschild's account of this gravitational field that Eddington, in 1924, found a way of avoiding the singularity problem – although he did not place any emphasis on what he had done. The device is simply to choose a different co-ordinate system, to redefine x, in the simple example given above. For example, if x is redefined as $x+1$, the singularity will be moved elsewhere, where it might be physically less troublesome; but if x is redefined as $1/x$, the singularity will disappear. The second case may seem like cheating, and it certainly looks trivial, but in a physical theory co-ordinates are not always directly interpreted. For instance, we might wish to interpret x as a length, but it might make sense in a particular theory to treat it as something that indicates a length very indirectly. Co-ordinates are not sacrosanct. This was well appreciated by those who manipulated the co-ordinate systems used in De Sitter's cosmological model, and Lemaître in 1933 first stressed this point in regard to the Schwarzschild singularity. He and Eddington had found co-ordinate systems that took them through the troublesome region. Later J. L. Synge (1950) provided an improved co-ordinate system, and at least four other mathematicians did the same independently over the next ten years or so.

As a result of their work some very strange results were obtained concerning 'Schwarzschild geometry'. One of the most surprising concerned the case where there was no mass in the central 'star'. Instead of the geometry being that of a point mass at zero distance from the centre, it turned out to be more like a wormhole joining two universes, a wormhole capable of expanding and recontracting. It is not possible here to go into more detail, except to make the point that work done on the Schwarzschild geometry was making it abundantly clear that space-time geometry has many strange and unexpected properties when gravity becomes very strong, and that it must be taken into account in the physics of stars that collapse under gravity, and in the physics of black holes.

Suppose we turn to astronomy now, to remind ourselves of the objects that were considered to be candidates for a relativistic treatment of the sort here sketched. Had we asked an astronomer around the year 1970 for a historical summary of the situation, we should have had information given to us about at least five categories of object, four of which had not been observed in the first instance in any direct sense at all, but had been put forward on theoretical grounds. Here is what we should have been told:

(1) *White dwarf* stars are of radius about 5000 kilometres, of about one solar mass – we recall that Chandrasekhar before 1930 had shown that their masses must be below a certain limiting value not much greater than the Sun's mass – and of density about one tonne per cubic centimetre. They are collapsing under gravity but are supported by an outward pressure of degenerate electrons, following Pauli's exclusion principle. (R. H. Fowler was perhaps the first to appreciate this possibility, in 1926.) They have stopped burning their nuclear fuel, and are gradually cooling as they radiate heat. In 1949 S. A. Kaplan showed that relativity predicts that they are unstable if they are under a certain radius – he said 1100 kilometres.

(2) *Neutron stars* are of much the same mass but with much smaller radius, say about 10 kilometres, making the density enormous – around 100 million tonnes per cubic centimetre, comparable with the density of an atomic nucleus. They are supported against gravitational collapse by two forces, the pressure of neutrons and certain nuclear repulsive forces ('strong-interaction forces'). The energy they radiate is partly heat and partly at the cost of their energy of rotation. They were implicit in the theory of L. D. Landau (1930), as mentioned earlier, and were called upon by Baade and Zwicky (1933–4) to explain supernovae. The outburst of a supernova was said to be the consequence of a normal star collapsing to a neutron star. In 1939 J. R. Oppenheimer and G. Volkoff used relativity to

work out the process in detail, and in doing so set the stage for later relativistic theories of star structure. (Oppenheimer himself became engaged on the atomic bomb project during the Second World War.) A certain star was singled out by Baade and Minkowski in 1942 as the remnant of the Crab supernova. After Gold's suggestion (1968) that pulsars are rotating neutron stars, it was shown in the following year by W. J. Cocke, H. J. Disney and D. J. Taylor that the same star is in fact a pulsar. This seems to tie up the presumed connection conclusively.

(3) Although the idea of a *black hole*, a region in which there is so much mass that light cannot escape, is one with a long history, within general relativity the foundations of the theory of black holes were laid only in 1939. It was then that J. R. Oppenheimer and Hartland Snyder showed how, when all thermonuclear sources of energy are exhausted, a sufficiently massive star will collapse. The collapsing sphere will cut itself off from the rest of the Universe. There is a 'horizon', a 'surface of the black hole', that will not reflect light from outside it and through which light from the 'collapsed star' cannot pass to the outside. No interaction with the outside world is possible. (The account, which considered the viewpoints of both internal and external observers, differs radically from any account that would be given in Newtonian theory.)

(4) *Supermassive stars* were conceived by Fred Hoyle and William Fowler in 1963, and S. Chandrasekhar and R. P. Feynman very soon afterwards developed a theory of their pulsations and instability. The original idea was that they might be associated with the nuclei of galaxies, and be the energy sources for quasi-stellar object, quasars, newly discovered. They would be of hot plasma and be less dense than ordinary stars, supported by the pressure of light (photons) mostly trapped within. They would have masses between a thousand and a billion solar masses. ('Plasma' is a word used for a fluid containing a large number of free

negative and positive electrical charges, such as our ionosphere or the gases where a fusion reaction takes place.)

(5) *Relativistic star clusters*, clusters so densely packed that
Newtonian physics cannot be used to explain their behaviour, were analysed in 1965 by Ya. B. Zel'dovich and M. A.
Podurets. In 1968 J. R. Ipser showed how, when the cluster
becomes great enough, it can begin to collapse to form a
black hole. It should be noted that the greater the mass
packed into a black hole, the less tightly it needs to be
packed to create a black hole, and the less the density of
the end-result.

So much for the situation around 1970. Now although
our astronomer of that time knew that any star containing
more than about three solar masses should theoretically
collapse to form a black hole, the number of references to
the idea in astronomical literature was relatively small. The
reason was that there are many stars whose masses were
known to be in excess of three solar masses. Gravity is there
balanced by the outward pressure from nuclear reactions,
and it was widely taken for granted that when the nuclear
process stopped there would be an explosion that would
reduce the central mass to less than the critical mass – perhaps producing a white dwarf. In short, it was supposed by
many that black holes need not be taken too seriously.

Not all were of the same mind. In 1964 Zel'dovich and
O. Kh. Guseynov began a search for black holes, looking
through catalogues of stars known from spectroscopic evidence to be binaries, but where only one component can be
seen. The masses could be estimated, and the two sought
examples where the mass of the invisible companion is
more than three times the mass of the Sun. For several
years the evidence remained extremely tentative, and there
was much controversy over its interpretation, but slowly
attitudes were changing, and the very dispute helped to
create an atmosphere favourable to the idea that black holes

might exist. Other reasons were about to intrude, and while they came from a highly theoretical direction, they had entirely unexpected consequences as to the ways in which black holes might be *detected*. The two central figures in this story are Roger Penrose and Stephen Hawking.

ZEL'DOVICH, PENROSE AND HAWKING

Black holes were a subject of intensive mathematical study before astrophysicists fully appreciated how important they might be. The number of people involved was not small, but in our all-too-short account we must confine attention to three outstanding figures: Yakov Boris Zel'dovich (19??–), Roger Penrose (1931–) and Stephen William Hawking (1942–). Zel'dovich was the head of an Institute for Physical Research at the Soviet Academy of Sciences in Moscow, a man with enormous energy, who led a team of very talented younger physicists working on black holes, and especially their interaction with light. In their work of the early 1960s they viewed black holes in a way hinted at by the name they then gave them, 'frozen stars', stars that had contracted until brought to a halt at the Schwarzschild radius.

Penrose was an applied mathematician, and was at the time at Birkbeck College, London University. (He had studied at University College there, and had later worked at Cambridge and in the United States.) Around 1965 he showed how it was possible to introduce new kinds of co-ordinates ('Eddington–Finkelstein co-ordinates') in which the star's collapse does not slow down but continues all the way to a singularity – a region of zero volume with matter of infinite density – leaving behind it a 'horizon' at the Schwarzschild radius. In fact it was in response to this new kind of language that John Wheeler, in 1968, chose the words 'black hole' to describe the curved, empty, space-time left behind with the horizon.

At much the same time, Denis Sciama (1926–) was leading a group doing research on relativity and cosmology in the Department of Applied Mathematics and Theoretical Physics in Cambridge. Among a succession of doctoral students there were George Ellis (from South Africa), Hawking, Brandon Carter and Martin Rees. In Cambridge at this time were Hoyle and Narlikar, and there were active contacts between Sciama's group and Bondi, Penrose and others in London. It was when returning from a meeting in London that Hawking conceived an idea that gave his doctoral thesis its importance. In it he applied Penrose's so-called 'singularity theory' not to a star but to the Universe as a whole. The central idea was that of reversing the time and treating Penrose's point singularity as the *beginning of the universe* rather than as the end of a star's collapse.

In 1970 Hawking and Penrose wrote a joint paper claiming that a Universe containing the expansion and matter that ours is observed to have must have begun in a singularity. This was not readily accepted at first, but most of the community of cosmologists was gradually converted. Actually Hawking later changed his position on the question of the initial singularity, when he found ways of taking quantum mechanics, the theory of the very small, into account.

Hawking had studied physics at Oxford before his move to Cambridge in 1962. His doctoral study had hardly begun when it became clear that a motor-neuron disease he had developed (ALS, 'amyotrophic lateral sclerosis') was serious, and was leading to a situation where he would soon be largely incapable of speech and movement. His subsequent battle with fate would have been noteworthy enough on its own account, but he was at the same time making some of the most remarkable scientific advances of the century. These were to bring him wide acclaim. He was to succeed to the Lucasian chair in mathematics at Cambridge in 1979, a latter-day successor to Newton, and to become one of the best-known scientific figures of his time.

In the 1960s black holes had a theory of dynamics developed for them, an important pioneer in this respect being the mathematical physicist Werner Israel. Israel – born in Berlin, brought up in South Africa, a graduate student in Ireland, but by this time working in Canada – had unravelled some of their physical properties in 1967, but was dealing then only with *static* black holes. He believed that they have to be spherical, and that only spherical objects can collapse to form them. Penrose and Wheeler argued that the demand for perfect sphericity was not particularly stringent, since in contracting, the star would give off gravitational waves and become more perfectly spherical, until in the end it would be a truly perfect sphere.

Israel had shown that the collapsing stars may have external fields (gravitational and electrical) that are determined by their masses and charges. In Cambridge, Carter (1970) and Hawking (1971–2) modified and extended the principle, adding angular momentum as a third property, and showing that the shapes need not be spherical. Between them, the three had shown that black holes could have no other distinguishing characteristics than those mentioned. In a phrase coined at the time, they 'had no hair' to distinguish them. The chemical nature of the matter going into them, for example, ceases to be relevant to what we can learn about them from the outside.

The chances of detecting black holes seemed slim, and yet it was realized that they might conceivably make up a significant fraction of the Universe. In 1966 Zel'dovich and Novikov wrote of what would now be termed black holes that might have formed as perturbations in the matter of the Universe as it began to expand. A black hole can have any mass: the smaller the mass, the greater the pressure needed on it, but great pressures were no problem in the early Universe, on this theory. Some of the mini-black-holes might by now have sucked in matter and radiation to such an extent that they have become as massive as a million galaxies, they thought. Hawking later argued (1971) that others might have remained unchanged, and still be only a few millionths of a gram in mass.

In 1969 Penrose showed that a black hole could lose energy and slow down, and so might energize electromagnetic radiation (light, radio waves, and so on) nearby. But what of its size? It was in the course of pondering this question at the end of 1970 that Hawking made one of his most fruitful discoveries. If nothing can get out of a black hole, then the surface area of its 'horizon' could not decrease, and if anything – matter or radiation – were to fall into it, or it were to combine with another black hole, then the area would actually increase. This in itself may seem an innocuous enough point, but its implications, in Hawking's thoughts, were dramatic.

The non-decreasing behaviour of the black hole's area was reminiscent of the behaviour of the physical quantity known as entropy. This is a concept used together with others such as temperature, pressure, heat energy, and so on, in specifying the thermal state of a system. It may be measured in terms of the heat that we need to add in order to transform a system from a given state to the state considered. Conversely, it can be seen as a measure of the 'quality' of heat energy, that is, as the amount of energy in a system that is available for doing useful work. Another way of considering entropy is as a measure of disorder, for instance among the atoms making up a system. The 'second law of thermodynamics' states that the entropy of a closed system, one that does not interact with its surroundings, never decreases, or (on some accounts) that its decrease is extremely improbable. It is a generalization of the principle that heat cannot, of itself, pass from a colder to a warmer body. Without outside interference, a glass of water does not suddenly start to boil in one region while forming ice at another, to provide the heat for the boiling. Some might wish to say that there is a remote possibility that this might happen, but all will agree that the probability is extremely small. Open a bottle of perfume in one corner of the room, and an hour later molecules of scent will be detectable in all parts of the room. The probability that they will all later simultaneously find their way back into their original highly ordered state, in the bottle, is so small that it can be ignored. The entropy of the system has increased.

Do black holes violate the law? What if matter with a great deal of entropy were to fall into a black hole? The entropy outside the hole would decrease. And inside? We cannot look inside to see, but might it not be that there is some indirect way of judging? A Princeton research student, Jacob Bekenstein, had suggested that the area of the event horizon of a black hole could be a measure of its entropy. Since this would increase as matter fell into the hole, we are encouraged to think that the second law of thermodynamics is preserved for the total system.

Hawking here raised the objection that if a black hole has entropy it should have temperature, and in that case it must emit radiation. But black holes, by definition, emit nothing. In 1972, with Carter and an American colleague Jim Bardeen, Hawking dismissed Bekenstein's idea, but later saw how he could make use of it. It was not until Hawking was visiting Moscow in 1973 that he was made aware of a proof by Zel'dovich and Aleksandr A. Starobinsky (published in 1971) that a *rotating* black hole might create and emit particles. When Hawking later tried to improve the mathematics of the proof he found to his 'surprise and annoyance' that even *non-rotating* black holes should create and emit particles at a constant rate. He thought there was an error in his working until he realized that the emission was of precisely the amount required by thermodynamics, that is, such as would prevent violations of the second law. The black hole behaves as though it has a temperature: the higher the mass, the lower the temperature.

Hawking's explanation of this escape from the supposedly absolute security of the black hole was that the particles are coming from just outside its event horizon. There are electrical and gravitational fields there, which we should be inclined to set at zero were it not that quantum mechanics forbids one to do so. Heisenberg's 'uncertainty principle'

(1927) in quantum mechanics denies to us the possibility of measuring the position and momentum of a particle simultaneously, with complete precision. The more certainty we have in the one, the less we shall have in the other. Niels Bohr in the same year (in his 'complementarity principle') extended this to experimental knowledge of other aspects of physical situations. In the case of black holes there will be a certain minimum of uncertainty in the value of the field. Hawking proposed that we think of the quantum fluctuations in the value of the field as pairs of particles of light (or gravity) that appear together at some time, move apart, and then – in reuniting – annihilate one another. In some cases, however, one of the two 'virtual particles', a particle or its 'antiparticle', might be captured by the black hole, and if the other is not captured too, and has positive energy, it might escape. In this case it will have seemed to come from the black hole, and since the particle of negative energy that enters the black hole will reduce the mass of the latter, this completes the illusion – if that is the best way of describing an aspect of a theoretical discussion – that the black hole is emitting particles.

As a black hole loses mass, its temperature rises and its emissions increase. This process is thought likely to accelerate until the thing finally explodes with extraordinary violence. It is possible to calculate what might be called the 'evaporation time' of such objects, and in the case of a black hole of mass equal to that of the Sun, this will be far in excess of the supposed age of the Universe. The primordial black holes proposed by Zel'dovich and Novikov, on the other hand, could have evaporated by now had their masses been less than a billion tonnes. Those of somewhat greater mass might still be radiating X-rays and gamma-rays profusely – black holes, but white hot, as Hawking has remarked. His calculations based on the observed background radiation of gamma-rays set an upper limit to this phenomenon, if it is a real one. It emerged that primordial black holes cannot make up more than a millionth of the matter in the Universe, although they would be expected to congregate near large masses, such as the centres of galaxies.

In this context, the extent and nature of the background gamma-ray radiation became a question of great concern. If the early Universe had been significantly irregular, it would have produced many more primordial black holes than seem to be set by the limit stated here. A Universe that was smooth, uniform and at high pressure would be expected to produce fewer than a Universe irregular or at low pressure.

These ideas were received with general incredulity when Hawking first tried them out on his colleagues, but after initial scepticism they have been widely accepted. Their implications are of enormous importance, for previously it had been thought that a black hole was a one-way sink to another world, as it were, whereas after Hawking it became clear that there is a process of cosmic recycling going on through their agency.

From a wider theoretical viewpoint we can say that with Hawking's work something still more fundamental was happening: general relativity itself was being modified, or extended, in the sense that quantum physics was now shown capable of eliminating some of the singularities that relativity predicted. But would singularities reappear in a combined quantum theory of gravitation?

From 1975 Hawking turned to problems of quantum gravity, using the so-called 'sum-over-histories' approach devised by the American mathematical physicist Richard Phillips Feynman (1918–88). Feynman said that instead of thinking of a particle as we ordinarily do, as something having a single history, taking it along a particular path in time, we should consider it to have every possible history, taking it over every possible path in space-time. Not all paths will be equally probable, however, and ways of computing the probabilities have been offered, using the rules of quantum mechanics. With each 'history' there are two numbers associated, one representing the intensity of a wave, the other the phase of the wave, and at each point in space-time the probability of finding the particle there is found by summing all the waves for all possible histories. In most places the probabilities more or less cancel out, but there will be places where they build up significantly. As an example, there are very high probabilities that an electron going round the nucleus of an atom will be in one of a limited number of orbits.

In Einstein's theory of gravitation a free particle follows a 'geodesic' line in curved space-time, the equivalent of a straight line in Euclidean space, the shortest distance between two points. Applying Feynman's sum-over-histories approach here, what corresponds to a clear-cut history of a single particle involves the whole of space-time, the whole of the history of the Universe. In the classical theory of general relativity, various models had been developed that described the history of the Universe following on from its initial state. In Hawking's new theory of quantum gravity we cannot be specific about the way the Universe began, although the probabilities of some of the results ensuing will be greater than that of others.

If the Universe resembles a black hole, then everything may be pictured as expanding outwards from the initial singularity (commonly referred to as 'the beginning of time'), and after a certain stage may collapse back on itself, in what is now graphically referred to as the 'big crunch' (ending at 'the end of time'), mirroring the 'big bang'. There were no boundaries as far as the spatial arrangement of the accepted models was concerned: these were finite in space, without edges, in the way the two-dimensional space that surfaces a ball is finite and unbounded. Hawking disliked the fact that there seemed to be time boundaries, however, for what seem at first to have been aesthetic reasons. Whatever his reasons, testable predictions ensued.

How the boundaries in time were dissolved is something that requires a careful mathematical description, but Hawking has provided a graphic analogy with circles of latitude on the Earth's surface. These are to be envisaged as representing the spatial size of the Universe, and distance from the north pole as representing time. The Universe begins as a point (the north pole) and grows as a circle of latitude, until it reaches the equator, after which it shrinks until reaching the south pole, where it becomes a point again. The Universe has zero size at the poles, but these are not singularities any more than the Earth's poles are singular. This is to picture the Universe as completely self-contained, and having no edges or boundaries, as it were. There is no place in the model for negative time, no time before the big bang, none after the big crunch. But is the model an acceptable one?

Under his 'no boundaries' assumption, Hawking found – approximately in 1981, and more precisely with the help of Jim Hartle of the University of California in 1982 – that, out of the histories that the Universe may have followed, one particular group is very much more probable than the others. It was calculated that there is a very high probability that the present rate of expansion of the Universe is almost the same in all directions – a result that seemed to be borne out by observations of the microwave background radiation. This led on to an investigation of the sizes and consequences of departures from uniform density in the early stages of the Universe. The uncertainty principle suggested the absence of complete uniformity at the beginning, ten or twenty billion years ago, and that during expansion the initial irregularities would have been amplified, that is, the initial fluctuations would have led to the existence of clouds, stars and galaxies, and as a side-show, human beings. How could astronomers detect that primordial granulation of matter?

The key was likely to be in the microwave background radiation, which it was realized might somehow show ripples as residuals of the big bang. Considerable effort was spent on the search for such irregularities, from the 1970s, and numerous claims were made that were short-lived, judging by the astronomical consensus. That there is massive inhomogeneity in the distribution of *galaxies* cannot be doubted, and an aside on this narrower problem is in order here, for clearly the two must be somehow connected.

In our neighbourhood, galaxies occur at intervals of a few million light years. In 1975, a study by G. Chincarini and H. J. Rood claimed to have detected a lumpiness in matter over a unit of about twenty million light years. At an IAU symposium held in 1977, several astronomers reported spaces more or less empty of galaxies, voids that were several hundred million light years across. A year later, S. A. Gregory and L. A. Thompson gave evidence for a large assemblage of galaxies known as the Coma supercluster, surrounded by relatively empty space. Gradually a picture was built up of a world of galaxies that is very lumpy indeed, and by the late 1980s it began to seem that the distribution followed something like that of a foam of soap bubbles, some large, some small, each with galaxies around its surface and a space relatively empty of galaxies within. The details of this three-dimensional picture were largely due to Margaret Geller, John Huchra and Valerie de Lapparent of the Smithsonian Astrophysical Center. One curious structure they later found was a 'wall' of galaxies stretching over at least half a billion light years (see figure 20.1).

But what of the smoothness of the cosmic background microwave radiation? In 1992, some exciting new evidence on inhomogeneity in that was produced by the NASA Cosmic Background Explorer satellite (COBE). John Mather had been a young graduate student at Columbia University in 1974 when he sent a modest design for such a probe to NASA, who decided to back the idea. The plan was to search for possible variations with direction in the strength of signals of the sort first found by Penzias and Wilson in 1964. One of the biggest problems facing Mather and his colleagues – George Smoot was leader of the team – at the Goddard Space Center was to prevent detectors in

FIG: 20.1   A 'wall' of galaxies

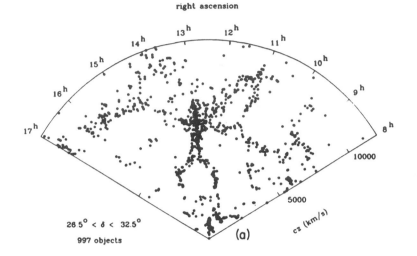

their satellite from picking up microwave radiation from within our own Galaxy. Three different detectors were used to monitor radiation levels at slightly different wavelengths, and the results were then compared by computer.

In December 1991 they had data showing, as it seemed, clear signs of fluctuations, of patchiness in their radiation map, as they looked out into the past, so to speak. Within four months they had run a number of convincing checks and counter-checks. The difficulties were immense, for the ripples detected were only denser than the background radiation by one part in a hundred thousand. Even greater, therefore, would have been the difficulties of detecting them from below the Earth's atmosphere, as various groups of astronomers, notably the Jodrell Bank group working in Tenerife, planned to do.

The COBE results were surprising in one important respect: they indicated that ripples were formed within 300 000 years of the big bang, which implied that the Universe was much denser than was generally believed. (The greater the mass, the more quickly would fluctuations have formed. Abundances of extra-terrestrial deuterium detected in 1973 in Jupiter, the Orion nebula, interstellar clouds and elsewhere had been thought to indicate a low-density universe.) This was precisely the sort of evidence needed by those who had argued for a 'big crunch' version of the universe's future, as opposed to the 'heat death' models that go on expanding and thinning out for ever, producing a world filled with an ever-diminishing haze formed from the ashes of burnt-out stars.

A more massive Universe than had generally been supposed fitted well with a set of cosmological models that gained favour in the 1980s and that stemmed from the so-called 'inflationary model' due to Alan Guth of the MIT in 1980. Guth had suggested a very rapid initial expansion taking place then at an increasing rate (rather than the decreasing rate that is accepted as that of the present epoch). A Universe that was 'inflationary' in this sense would at first be governed by a unification of the strong and weak nuclear forces and the electromagnetic force; but with expansion would come a 'phase transition', a time when the strong force would be differentiated from the others.

The inflationary theory was developed in the 1980s by Hawking and others, especially by the young Soviet cosmologist Andrei Linde, and independently by Paul Steinhardt and Andreas Albrecht of the University of Pennsylvania. Their new inflationary model involved a *slow* dissolution of the symmetry condition, as opposed to Guth's original idea that this happened very rapidly.

One of the stranger consequences of the theories of the inflationary Universe was that many versions of it envisaged our visible 'Universe' as a single island among many in a steady-state background. The death of the steady-state model has perhaps been greatly exaggerated.

By those who think otherwise, the situation immediately after the origin of the big bang has been pictured as comparable to that with supercooled water, still liquid below its freezing point, and in a false state of equilibrium, full of energy waiting to be triggered off. The idea is that in passing to the lower energy state, the released energy powers the extraordinarily rapid inflationary expansion of the Universe. Linde's idea was that the breakthrough might have come in different places at different times – it is usual to draw an analogy with champagne bubbles. Inflationary models generally were successful at explaining how irregularities in the early Universe could have been very effectively smoothed out by the later expansion. Other variants were later added – for example by Linde in 1983 ('the chaotic inflationary model'). They differ notably in the temperature variations they require the microwave background radiation to have. The chaotic versions picture the Universe as having some regions in expansion, others in contraction, some at high and some at low temperature. It was suggested that we might inhabit, as it were, a bubble that has not merged with another bubble, as it might in principle have done; or that there might be laws forbidding the merging of such disparate regions.

From such ideas came a stream of new cosmological models. Hawking radiation, which had originally been related to the horizon around a black hole, was now found a new role. With Gary Gibbons, Hawking showed that this sort of radiation might originate with other kinds of horizon – for instance the sort that prevents regions of the Universe communicating with light signals when they are moving apart (with the expansion of space) faster than the speed of light. They showed that this might have been significant in the early history of the Universe. Again, these ideas were developed, and led to a multiplicity of new alternatives. The observational evidence was hardly equal to the task of deciding between them, but the COBE results at least marked a new phase in this process.

After the COBE findings, perfect uniformity was no longer the desideratum it had once been, although the observed uniformity was still very considerable. The COBE data, announced at a meeting of the American Physical Society in April 1992, were taken to corroborate the class of models beginning with a big bang, but could not put an end to the steady refinement of more conventional observational testing, which had been proceeding modestly for more than half a century, with the correlation of number counts, galactic magnitudes, red shifts and so forth.

Just as with the last few pieces of a jigsaw puzzle, so new data relevant to any crucial problem always arouse more excitement than the mass of data to which everyone has grown accustomed. The difference is that in the cosmological case there is no such thing as completion. One of the outstanding problems remaining in the last decade of the century is the missing mass. The candidates for this are by their nature elusive: burnt-out stars, black holes and dark matter of various sorts, from the conventional to the very exotic. And here we can do no more than mention in passing that the exotic candidates include wimps (weakly interacting massive particles), axions and photinos, which have been proposed by different physicists as particles moving among us but not detectable by standard methods.

MAN AND UNIVERSE

Just as in the late 1940s the steady-state theories aroused much hostility from theological quarters, so in the 1980s. A number of theological asides by cosmologists, who had been excited by some of the new ways of describing the Universe, led to another series of exchanges redolent of those of the 1950s. New theories were occasionally described as indicating 'God's way of thinking', or 'the details of God's act of creation'. Despite many parallels with earlier history, there have been a few detectable changes in emphasis. Humankind having throughout history been steadily marginalized in spatial terms – as for instance by Aristarchus and Copernicus – is still generally kept at centre stage by an insistence on a division of responsibility: the scientist explains *how* things happen, *how* the Universe evolves, while theologians now generally insist that they only ask *why*. Theologians have long known that it is safest to avoid being drawn into discussions of the literalness of scriptural authorities. In a Judeo-Christian context, for instance, it is now often said that the book of Genesis says no more than that the world exists because God wills it to be, for human reasons, and that God is not to be regarded as a cosmic impresario, or as the one who pressed the button starting off the cosmic fireworks. Wise or unwise, timid or brave, this is not in the tradition that kept astronomy and theology in close contact for more than three thousand years, during which time theologians were generally proud to insist that they were concerned with a *complete* picture, and one in which God is a creator of the physical world. Times have changed in this respect.

Oddly enough, a number of scientists during the last years of the twentieth century offered something akin to traditional theology, and often did so without any theological motive. In a rather peculiar sense, they were trying to explain the properties of the Universe in human terms. On the face of it, this attempt must seem doomed to failure, but after a little thought, the surprise element in the proof disappears, almost to the point of seeming a sham. If, in the middle of the Thai jungle I meet by chance a Cambodian cigarette manufacturer whom I discover to have studied history with a close friend of mine in England, I am taken by surprise. If I meet him at my friend's house in England, in an alternative world, I am not. That there are likely to be certain intimate relationships between ourselves and the kind of Universe we observe need not necessarily surprise us. We often come across arguments in history that have gone something like this:

A: 'By God's good grace, animals act considerately towards their offspring.'
B: 'No. Only those species of animals with this property survive to produce later generations.'

There now exists a very extensive literature in which questions are discussed in very similar terms:

A: 'Only by God's good grace do we inhabit a Universe perfectly suited to our needs, that is, satisfying the conditions necessary for our existence.'
B: 'God may well be responsible, but at all events we should not be surprised that we encounter conditions suited to our existence. If they did not exist, we should not exist.'

Whether or not such a counter-argument has any point of contact with theology, it certainly touches upon modern cosmology.

Throughout history there have been those who have argued that the form of the world as described by us is dependent on principles that we ourselves create. The mind is in some respects the maker of Nature – as the philosopher Kant, among many others, insisted. What we see is likewise governed to some extent by what we are, as biological entities. Astronomers have long been conscious of the limitations of the human senses, and of the fact that there is a world of radiations – Herschel's infrared, Ritter's ultraviolet, and so forth – of which we can only be aware with the help of instruments. Our relationships with the Universe depend on the character of life-forms – of language, of our physiological nature and so forth – but there are others that have to do with our very material existence. The elements needed for the existence of the living organisms known to us include nitrogen, oxygen, phosphorus and above all carbon. At the primordial explosion of the Universe, hydrogen and helium were synthesized, and only after a very long process were they converted into the heavier elements, chiefly in the interiors of stars. The death of stars is followed by the dispersion of those elements, and by their conversion into planets and life-forms. All told, a time of the order of at least ten billion years is needed before this can happen, and it can therefore be seen as not in the least surprising that we – a carbon-based form of life – find ourselves in a Universe more than ten billion years old. To this extent there is nothing inherently occult or mystical in saying that there is a connection between humankind and Universe.

Throughout history there have been attempts to assess the probability that life exists elsewhere in the Universe, and these attempts have usually involved some sort of linkage between our biological properties and conditions elsewhere, astronomically judged. During the nineteenth and twentieth centuries there was a three-cornered interplay between theories of animal evolution, geological transformation of the Earth and the age of the Sun – all of which could be related to time. Thomas Chamberlain, for instance, believed in some sort of atomic power supply for the Sun simply because the known alternatives would make the Sun's life too short for the other processes. Some attempts have been made in the present century to relate intelligence to biology, thus carrying the argument one stage further. We have mentioned briefly some attempts made by Eddington, Dirac and others in the 1930s to find relationships between the fundamental constants of physics. Some of Dirac's ideas suggested that there might be a slow change with time in some of the 'constants' of Nature, and Robert Dicke (1916–) spent several years examining astronomical and geological evidence for such change.

In 1957 Dicke pointed out that the number of particles in the observable part of the Universe, and Dirac's numerical coincidences, were not random but were conditioned by biological factors. He argued that if the Universe were appreciably older than it is, 'all stars would be cold', and under those conditions humankind would simply not be around to survey the scene. In 1961 and later, he presented his ideas with more detailed calculations, proposing what became generally known as the *(weak) anthropic principle*: the conditions necessary to produce life involve certain relationships between the fundamental constants of physics. For example, a slight change in the charge on the electron, with other physical constants held unchanged, would have made the mechanism of nuclear fusion in stars unworkable, and so would have made our form of life impossible. Martin Rees, a Cambridge astronomer of wide experience, from the 1970s spent much time investigating the detailed and quite dramatic consequences for the evolution of the Universe if even only the gravitational constant were to be slightly changed. Life might have appeared, but not at all as we know it.

Another kind of argument entirely is that the physical nature of the Universe is such that it must have within it, at some stage, living beings, and humankind in particular. Brandon Carter gave the name *strong anthropic principle* to the principle that the Universe must be knowable and must at some stage 'admit the creation of observers within it'. Those who have favoured this strong principle have often proceeded by trying to describe in very broad but acceptable physical terms all possible worlds, some inevitably without, and some capable of supporting, life. The idea is then to try to pinpoint structural qualities necessary if observers are to be generated. We must of necessity inhabit one of the narrow group of possible worlds so described. However remote the possibilities of proceeding in this way might seem, much of the work done in cosmology and astrophysics in the second half of the twentieth century, while done without any thought for this general problem, fell naturally into the scheme of the argument. Fred Hoyle, for example, was led to consider the anthropic principles quite naturally as a consequence of his discovery of a series of coincidences in the relations between certain properties ('nuclear resonance levels') of biologically important elements.

As a footnote to the long-standing debate over the probabilities of life on planets elsewhere in the Universe, it might be added that the importance of planetary satellites to the calculation was curiously ignored until recently. If our own Moon did not exist, the orientation of the axis of the Earth would not be stable, and would have been subject to chaotic variations over the course of time. These would have brought about climatic changes, which might not have prevented, but would certainly have affected strongly, the development of biological organisms.

For most of human history, or at least as long as astronomy has been conceived to be based on laws, the world has been seen as one of *change*. In its essential *patterns of change*, however, it has been seen as a world of order and *permanence*. Leibniz, for example, touched a raw nerve when he accused Newton of implying that God was a clockmaker who needed to wind up the clock of the Universe from time to time. On the face of things the 'clock' was the Universe itself, running down, according to Newton's laws, because of friction. The real imperfection, however, in the eyes of the two parties to the debate, was in a law, in this case the law of the perfection of the Universe. To Christian believers, the issue was whether Newton's laws were responsible for a maligning of the Creator. The easiest way out of the difficulty is to redefine perfection, to reinterpret the Universe and its laws – theologically, for example.

Towards the end of the nineteenth century, the idea of the 'heat death' of the Universe, to which thermodynamics seemed to point, stimulated a vigorous debate about God's plan for the world, a debate that still continues. This is one of many theological discussions based on scientific discovery where theologians and others thought that they could easily interpret the newly described Universe. Again the crux is one of interpretation. How, it was often asked in this case, could a beneficent God have created a decaying Universe? The answer was simple: find a new theological interpretation. History shows that the exercise has never been very difficult, and interpreters have almost invariably found what they set out to find. In this particular case, for example, we have the American psychologist William James – brother of the novelist Henry James – saying that while the ultimate state might be the extinction of everything, 'the *penultimate* state might be the millennium'. He thought that the 'last expiring pulsation of the Universe' might be at hand, and added 'I am so happy and perfect that I can stand it no longer'.

The great majority of astronomers are of a more sober cast of mind, content to see themselves as offering a scientific account of the world we inhabit, an account essentially no different from those offered by the other physical sciences. Astronomy has been many things in its time, not least a prototype for the more fundamental sciences. It provided them with their methods, with laws of motion and with empirical data, only later to borrow from them with interest. Astronomy remains what it has been throughout its long history, a playground for unrestrained metaphysical and theological speculation, but this can hardly now be said to lie at its core. When modern astronomers in one breath profess themselves to be agnostic and in the next describe the ripples revealed by COBE as 'traces of the mind of God', this tells us only that theological sophistry is not what it was. The overall response is hardly surprising, bearing in mind how entangled are astronomy's roots with those of religion. It is one of the ironies of history that the study of such a vast and impersonal subject matter should from beginning to end have been so intimately bound up with principles of human nature.

BIBLIOGRAPHICAL ESSAY

INTRODUCTION

This bibliography is highly selective – it would not be difficult to amplify it a hundredfold – and confines itself mainly to books rather than articles. To avoid undue bias, several works without equivalents in English have been included. Although it makes occasional references to scientific texts – especially to some that have become classics, and to others towards the end where histories are as yet few – it is aimed at history, and not at astronomy as such. This book was meant to be more or less self-sufficient, but for relative strangers to astronomy two suggestions are offered for further reading. We might divide astronomy texts very roughly into two classes: mathematical 'spherical astronomy', where at an introductory level a standard text of a century ago differs little from its recent equivalents; and physical astronomy, where current wisdom changes almost daily. In the former category, W. M. Smart's (1931) *Text Book of Spherical Astronomy* (Cambridge: Cambridge University Press, 6th edn, 1977) is still widely used and recommended, but it must be remembered that it is of necessity a text requiring some mathematical knowledge. As a single source of information on main-line physical astronomy, beautifully illustrated, and readable with a minimum of prior scientific knowledge, one could seek out the latest edition of *The Cambridge Atlas of Astronomy* (Cambridge: Cambridge University Press). This originated in *Le Grand Atlas de l'Astronomie*, ed. by Jean Claude Falque under the general editorship of Jean Audouze and Guy Israel (Paris: Encyclopaedia Universalis, 1983), and the second English-language edition appeared in 1988. If a star atlas is needed (in the more conventional sense of the word), the choice is great, but the 18th edition of *Norton's* might be found useful, for general astronomical instruction as well, and its maps are now brought up to the millennium: Ian Redpath (ed.), *Norton's 2000.0 Star Atlas and Reference Handbook* (London: Longman, and New York: Wiley, 1989).

Since astronomy makes extensive use of data from the past, simple forms of historical study are found even in the ancient world. One of the first histories that dedicated historical scholars will wish to consult is P. J. B. Riccioli's truly monumental *Almagestum novum* (Bologna, 1653) but this is in Latin, and many of even the best libraries will not have it. Other classical works of history are those by J. B. J. Delambre, beginning with his *Histoire de l'astronomie ancienne* in two volumes (Paris, 1817) and including further volumes on medieval (1 vol., 1819), 'modern' (2 vols, 1821) and eighteenth-century astronomy (1 vol., 1821). These are a fundamental source, and have been reprinted (in French, New York and London: Johnson Reprint Corporation, 1965–9). Delambre's work was motivated by the need to use history as an astronomical tool, but ended by setting high standards in intellectual history as such. Not in the same class, but certainly encyclopaedic, is Pierre Duhem's weighty *Le Système du monde. Histoire des doctrines cosmologiques de Platon à Copernic* (Paris, 1913–59, 10 vols). Excerpts from this work have been issued in an English translation by Roger Ariew under the title *Medieval Cosmology* (Chicago: University of Chicago Press, 1985).

Not new, but valuable still, is Harlow Shapley and Helen E. Howarth, *A Source Book in Astronomy, 1900–1950* (Cambridge, Mass.: Harvard University Press, 1960). It should be supplemented by Kenneth R. Lang and Owen Gingerich, *A Source Book in Astronomy and Astrophysics, 1900–1975* (Cambridge, Mass.: Harvard University Press, 1979). The selection of extracts reflects not only on the astronomical taste of the editors but on the availability of sources.

As a modern example of the successful use of history for astronomical purposes, see David H. Clark and F. Richard Stephenson, *The Historical Supernovae* (Oxford: Pergamon,

1977). Robert R. Newton has published much along similar lines, as for example *Ancient Planetary Observations and the Validity of Ephemeris Time* (Baltimore and London: Johns Hopkins University Press, 1976). He has a somewhat astringent way with past astronomers, and his colleagues may well shudder at the thought that they will be similarly handled in two thousand years' time.

Articles on the history of astronomy, as a part of the history of science, will be found in the many journals – several hundreds – devoted to that wider subject. Most of the journals supported by national organizations are now effectively international in character, for example *Isis*, the journal of the American History of Science Society, and the *British Journal for the History of Science*, both written almost entirely in English. Articles in five or six world languages will be found in the journal of the International Academy of the History of Science, *Archives internationales d'histoire des sciences*, currently published by the Institute of the Italian Encyclopaedia, Rome. A specialist publication of great value is the *Journal for the History of Astronomy* (Chalfont St Giles, Bucks: Science History Publications), which since 1979 has published *Archaeoastronomy* as a supplement.

Among international bodies responsible for organizing conferences of relevance to the subject, the IAU (International Astronomical Union, with closed membership) publishes extensively (largely with Kluwer, Dordrecht). The IAU has a Historical Commission. The IUHPS (International Union for the History and Philosophy of Science, with countries as members) has large meetings every four years, and since it often publishes papers given at them, can provide an idea of work in progress. One of the most useful of all bibliographical sources, however, is the ongoing *Isis Critical Bibliography*. First published as occasional supplements to the journal *Isis*, a series of large bound volumes edited by Magda Whitrow was begun by the History of Science Society, combining in the first instance bibliographies for the years 1913–65 (London:

Mansell, 1971–6, 3 vols). These and later volumes list a very large proportion of books and articles in the history of science, from many countries.

There are various surveys that simplify the task of navigating through the wider subject, for example: David Knight, *Sources for the History of Science* (London: The Sources of History, 1975); S. A. Jayawardene, *Reference Books for the Historian of Science* (London: Science Museum, 1982). Biographical information can often be found in detail in various national biographies (such as the British *Dictionary of National Biography*). Biographies of scientists, together with partial bibliographies, have long been available in German in an important compendium usually known simply as 'Poggendorff'. Johann Christian Poggendorff's first two volumes of the *Biographisch-Literarisches Handwörterbuch* appeared in 1863, and in various forms this work has been published ever since. Another German venture worth remembering is the *Encyklopädie der mathematischen Wissenschaften* (Leipzig: Teubner, 1898–1935), of which section VI includes two volumes of astronomical material, edited by Karl Schwarzschild, Samuel Oppenheim and Walther von Dyck. George Sarton, in his *Introduction to the History of Science* (Washington: Carnegie Institution, 1927–1948, 3 vols in five parts, often reprinted) made an ambitious attempt to give a bio-bibliographical outline of the subject for the whole of history, but only reached to the second half of the fourteenth century. These weighty volumes are now dated, in fact and in tone, but are still useful.

By far the most important single biographical source of scientists, entirely in English, is now C. C. Gillispie (ed.), *Dictionary of Scientific Biography* (New York: Charles Scribner's Sons, 1970–8, 15 vols). An abbreviated version appeared in one volume in 1981, but errors crept in during the abbreviation, and it is best avoided. The same publisher published a series of illustrative volumes to supplement the *Dictionary*, under the general title *Album of Science*, and all are of some relevance to our theme: John E. Murdoch (ed.),

Antiquity and the Middle Ages; I. Bernard Cohen (ed.), *From Leonardo to Lavoisier, 1450–1800*; L. Pearce Williams (ed.), *The Nineteenth Century*; Owen Gingerich (ed.), *The Physical Sciences in the Twentieth Century*. For bare outlines of the biographies of some 30 000 scientists, half of them living (in 1968), see Allen G. Debus (ed.), *World Who's Who in Science* (Chicago: Marquis Who's Who, 1968).

There are several modern histories of astronomy covering long periods of history. For the ancient world, the most complete survey of astronomy with a mathematical content is Otto Neugebauer, *A History of Ancient Mathematical Astronomy* (New York: Springer-Verlag, 1975, 3 vols), referred to below as *HAMA*. Willy Hartner, *Oriens–Occidens*, in two volumes (Hildesheim: Olms, 1968, 1984) covers such a great range of history that it may be placed in this category. Others are mentioned below, including volumes of an important series as yet incomplete, *The General History of Astronomy*. Published under the auspices of the IAU and the IUHPS, this will eventually comprise four volumes, each in more than one part: (1) O. Pederson and J. D. North (eds), *Antiquity to the Renaissance*; (2) R. Taton and C. Wilson (eds), *Planetary Astronomy from the Renaissance to the Mid-Nineteenth Century*; (3) M. A. Hoskin (ed.), *Stellar Astronomy, Instrumentation and Institutions from the Renaissance to the Mid-Nineteenth Century*; (4) O. Gingerich (ed.), *Astrophysics and Twentieth-Century Astronomy to 1950*. Another ambitious project, to publish a series of authoritative works on the history of science generally, is being planned by the International Academy of the History of Science and the Istituto della Enciclopedia Italiana jointly. This should begin to appear around the turn of the century.

Finally, since it is of relevance to all early chapters, a warning should be given about the tangled history of star names. R. H. Allen's *Star Names, their Lore and Meaning* (1899, repr. New York, Dover, 1963) has done valiant service, but is not altogether reliable and should at the very least be supplemented by the excellent but very brief work by P. Kunitsch and T. Smart, *Short Guide to Modern Star Names and their Derivations* (Wiesbaden: Harrassowitz, 1986), where further bibliography will be found.

CHAPTER 1

A historical awareness of the astronomical concerns of prehistoric peoples is itself ancient, but little work of lasting value was done before the end of the nineteenth century. Norman Lockyer's works are still of some value, and their shortcomings may often be put down to excessive enthusiasm. See *The Dawn of Astronomy* (London: Macmillan, 1894, repr. Cambridge, Mass.: MIT Press, 1964) and *Stonehenge and Other British Stone Monuments Astronomically Considered* (London: Macmillan, 1909), both since reprinted (Cambridge, Mass.: MIT Press). Works on prehistoric astronomy should be read together with works giving the archaeological context – such as R. J. C. Atkinson, *Stonehenge* (London: Penguin Books and Hamish Hamilton, 1956, 1979), for that monument.

Archaeoastronomy first reached a high standard with the work of Alexander Thom, whose books include *Megalithic Sites in Britain* (Oxford: Oxford University Press, 1967) and *Megalithic Lunar Observatories* (Oxford: Oxford University Press, 1971). Several articles by him and his son A. S. Thom appear in the *Journal for the History of Astronomy* and its supplement (see above). The astronomy of Stonehenge is the subject of a forthcoming book of my own. An important symposium held in 1972 led to the publication of F. R. Hodson (ed.), *The Place of Astronomy in the Ancient World* (Oxford: Oxford University Press, for the British Academy, 1974). E. C. Krupp (ed.), *In Search of Ancient Astronomies* (London: Chatto and Windus, 1979) gives a good overall survey at an introductory level of European, American, Egyptian and other early astronomy by largely non-literate peoples. It includes a useful bibliography, although many items in it are from the lunatic fringe.

CHAPTER 2

Pyramidology has long had its astronomer-adepts, most of them best forgotten. Of interest for its own sake, or rather because it was written by no less a man than the Astronomer Royal for Scotland, is Piazzi Smyth, *The Great Pyramid. Its Secrets and Mysteries Revealed* (London and New York: Bell, 4th edn, 1880, repr. New York: Outlet Book Co., 1990). Note also Lockyer's *Dawn of Astronomy* (bibliography to chapter 1) as an influential source from the same period. A more balanced view of Egyptian technical competence, which has long been exaggerated, may be had from O. Neugebauer's *HAMA* (see introductory bibliography), his *The Exact Sciences in Antiquity* (2nd edn, Providence, RI, 1957; reissued New York: Dover, 1969), and his *Astronomy and History. Selected Essays* (New York: Springer-Verlag, 1983). With R. A. Parker he produced a magnificently printed study that will be available in larger libraries: *Egyptian Astronomical Texts* (Providence, RI, 1960–9, 3 vols). More readily available is Parker's topical essay in volume 15 of the *Dictionary of Scientific Biography* (see above), which includes further bibliography.

CHAPTER 3

Works on Babylonian astronomy include B. L. van der Waerden, with contributions by Peter Huber, *Science Awakening*, vol. 2, *The Birth of Astronomy* (Groningen: Wolters Noordhoff, 1950, and Leyden and New York: Oxford University Press, 1974), chapters 2–8; A. Pannekoek, *A History of Astronomy* (London and New York: Dover, 1961, repr. 1989, translated from the Dutch edition of 1951), chapters 3–6; O. Neugebauer, *The Exact Sciences in Antiquity* (reissued New York, 1969), and his collected essays (see bibliography under chapter 2); H. Hunger and D. Pingree, *MUL.APIN: An Astronomical Compendium in Cuneiform* (Horn, Austria, 1989). These works, as well as O. Neugebauer, *A History of Ancient Mathematical Astronomy* (see above), contain numerous references to fundamental work by such scholars as T. G. Pinches, J. N. Strassmaier, J. Epping, F. X. Kugler, A. J. Sachs and A. Aaboe. Of the many excellent introductions to the history of the region, Joan Oates, *Babylon* (London: Thames and Hudson, 1979, 1986) is readily available.

CHAPTER 4

To get an idea of how essentially primitive was Greek cosmology before Eudoxus, even in the brilliant *Timaeus* of Plato, see the translation of that, with running commentary by F. M. Cornford, in his *Plato's Cosmology* (London: Routledge, 1937, often repr.). D. R. Dicks, *Early Greek Astronomy to Aristotle* (London: Thames and Hudson, 1970) re-examines some of the traditional views of its subject. Pioneering work by Delambre has been mentioned earlier, and Neugebauer's *HAMA*. Among the various classic works by Thomas L. Heath, an important instance, far more general than its title suggests, is *Aristarchus of Samos. A History of Greek Astronomy to Aristarchus, together with his Treatise on the Sizes and Distances of the Sun and Moon* (Oxford: Clarendon Press, 1959). J. L. E. Dreyer, *A History of Astronomy from Thales to Kepler* (London, 1912, repr. New York: Dover, 1953) is still valuable over a wide historical range. (We have mentioned Dreyer's astronomical work at several points in our text.)

The technicalities of Aristotle's homocentric astronomy are well treated in the works here mentioned. His complete writings, as befits their enormous influence over two millennia, are readily available in numerous editions and translations. For a general commentary on his system of natural philosophy see G. E. R. Lloyd, *Aristotle: The Growth and Structure of his Thought* (Cambridge: Cambridge University Press, 1978).

To appreciate properly the great stature of Ptolemy it is almost essential to see his own writings, in particular his

Almagest, which is now available in a fine English translation by G. J. Toomer (London: Duckworth, 1984, and later New York: Springer-Verlag). This could be supplemented for Ptolemy's physical views by Bernard R. Goldstein (ed.), *The Arabic Version of Ptolemy's Planetary Hypotheses* (Philadelphia: American Philosophical Society, 1967). See also Goldstein's collected papers, *Theory and Observation in Ancient and Medieval Astronomy* (London: Variorum Reprints, 1985). An excellent guide to the intricacies of Ptolemy's *Almagest* is Olaf Pedersen, *A Survey of the Almagest* (Odense, Denmark: Odense University Press, 1974). At a simpler level, the work by Pedersen and M. Pihl, *Early Physics and Astronomy* (London: Macdonald, and New York: American Elsevier, 1974; republished Cambridge: Cambridge University Press, 1993), takes the story to the middle ages in an eminently readable way.

A. Bouché-Leclercq, *L'Astrologie grecque* (Paris, 1899, repr. Brussels: Culture et Civilisation, 1963) has been superseded in a number of minor respects, but it remains the best single source on Greek astrology, with no equivalent in English. In German, Wilhelm Gundel and Hans Georg Gundel, *Astrologumena: die Astrologische Literatur in der Antike und Ihre Geschichte* (repr. Wiesbaden: Steiner, 1966) and F. Boll, C. Bezold and W. Gundel, *Sternglaube und Sterndeutung: Die Geschichte und das Wesen der Astrologie* (repr. Darmstadt: Wiss. Buchgesellschaft, 1977) continue the history into later periods. In English, Jim Tester, *A History of Western Astrology* (Bury St Edmunds, Suffolk: Boydell Press, 1987) is useful and eminently readable. Of classic astrological texts, the most important is Ptolemy's *Tetrabiblos*, and this is available in parallel translation in a standard Loeb edition (Heinemann and Harvard University Press).

Astronomy continued to be cultivated in Byzantium, and a series entitled *Corpus des Astronomes Byzantins* is dedicated to publishing relevant texts, for example (no. 3 in the series) Alexander Jones (ed.), *An Eleventh-Century*

Manual of Arabo-Byzantine Astronomy (Amsterdam: J. C. Gieben, 1987).

CHAPTER 5

The history of Chinese science generally was placed on a completely new footing with the life work of Joseph Needham, whose many-volume *Science and Civilization in China* (Cambridge: Cambridge University Press, 1954–) will be found in many large libraries. Colin A. Ronan, *The Shorter Science and Civilization in China* (Cambridge: Cambridge University Press, 1978–86) is an abridgement of Needham's original volumes, now in three volumes. Works peripheral to Needham's are numerous, but note Ho Peng Yoke, *Li, Qi And Shu: An Introduction to Science And Civilization in China* (Hong Kong: Hong Kong University Press, 1985) and Wang Ling, Joseph Needham and Derek J. De Solla Price, *Heavenly Clockwork* (Cambridge: Cambridge University Press, 1960, 2nd edn with supplement by J. H. Combridge, 1986). Also valuable is N. Sivin, *Cosmos and Computation in Early Chinese Mathematical Astronomy* (Leyden: Brill, 1969). See in addition articles in Hartner's *Oriens–Occidens* (cited in our introduction), and others by Yasukatsu Maeyama (in English) in the *Archives internationales d'histoire des sciences* for 1975 and 1976 and in *Prismata, Festschrift für Willy Hartner*, edited by W. Saltzer and Y. Maeyama (Wiesbaden: Steiner, 1977).

Although earlier Japanese work is inaccessible to most ordinary mortals for reasons of language, it is worth mentioning that Japan now publishes an excellent journal in the history of science in English, *Historia Scientiarum*, alias *Japanese Studies in the History of Science*. Articles in this by K. Yabuuchi, especially on Chinese and Japanese calendars, are of high importance. An excellent survey of Japanese astronomy, paying much attention to the historical background, is Shigeru Nakayama, *A History of Japanese Astron-*

omy. *Chinese Background and Western Impact* (Cambridge, Mass.: Harvard University Press, 1969), and this includes a useful bibliography to that time. Most general accounts of the Jesuit period are written without much astronomical awareness. An exception, which if it were not so well illustrated might encourage the reader to learn Dutch, is Noël Golvers and Ulrich Libbrecht, *Astronoom Van de Keizer. Ferdinand Verbiest en zijn Europese Sterrenkunde* (Leuven: Davidsfonds, 1990). Even without the language one cannot fail to enjoy the magnificently illustrated catalogue prepared for a Brussels exhibition held under the title *China, Hemel en Aarde. 5000 Jaar Uitvindingen en Ontdekkingen* [China, Heaven and Earth: 5000 years of Inventions and Discoveries] (Brussels: Vlaamse Gemeenschap, Trierstraat 1000, 1988). In one of the volumes in the series *Studia Copernicana* (vol. 6; see bibliography to chapter 11) Nathan Sivin shows that the failure of the Jesuits to promote Copernicus was a consequence of their having – under the shadow of the Church – garbled his work so. See *Copernicus in China* (Warsaw: Polish Academy of Sciences, 1973).

CHAPTER 6

Standard works on Maya writing are J. E. S. Thompson, *Maya Hieroglyphic Writing* (Norman: University of Oklahoma Press, 1960) and his *A Commentary on the Dresden Codex* (Philadelphia: American Philosophical Society, 1972). Anthony Aveni has written extensively on astronomy, and books he has edited include: *Archaeoastronomy in Pre-Columbian America* (Austin: University of Texas Press, 1975); *Native American Astronomy* (Austin: University of Texas Press, 1977); *Archaeoastronomy in the New World* (Cambridge: Cambridge University Press, 1982). For a history of the first Spanish contacts with the New World see T. Todorov, *The Conquest of the Americas* (New York: Harper and Row, 1982). For the Maya civilization see M. Coe, *The Maya* (New York: Praeger, 1973).

General studies of the Aztecs are A. Demarest, *Religion and Empire* (Cambridge: Cambridge University Press, 1984) and M. Léon-Portilla, *Aztec Thought and Culture* (Norman: University of Oklahoma Press, 1963). N. Davies, *The Ancient Kingdoms of Mexico* (Harmondsworth: Penguin Books, 1982) is readable and accessible.

For the fall of the Incas, see J. Hemming, *The Conquest of the Incas* (New York: Harcourt Brace, 1970). A general study of Inca culture is A. Kendall, *Everyday Life of the Incas* (London: Batsford, 1973). Concepts of cosmic religion are touched upon in a chapter by L. Sullivan in R. Lovin and F. Reynolds, *Cosmogony and Ethical Order* (Chicago: University of Chicago Press, 1985).

Accounts at a general and introductory level are included in E. C. Krupp (ed.), *In Search of Ancient Astronomies* (London: Chatto and Windus, 1979) and – for the cultural background to all these groups – C. A. Burland, *Peoples of the Sun. The Civilizations of Pre-Columbian America* (London: Weidenfeld and Nicolson, 1976).

CHAPTER 7

The gargantuan task of listing the scattered historical sources has been undertaken in different ways. David Pingree is compiling a *Census of the Exact Sciences in Sanskrit* (Philadelphia: American Philosophical Society, beginning 1970), and see S. N. Sen, *Bibliography of Sanskrit Works on Astronomy and Mathematics* (New Delhi, beginning 1966). There are a number of points on which historians of Indian astronomy are divided. Three important histories, of which the first is the most readily accessible, are: D. Pingree, 'History of Mathematical Astronomy in India', in *Dictionary of Scientific Biography*, vol. 15 (New York: Scribner's, 1978), pp. 533–633; R. Billard, *L'Astronomie indienne* (Paris: Ecole française d'extrème Orient, 1971); and S. N. Sen, 'Astronomy', in *A Concise History of Science in India* (New Delhi, 1971), pp. 58–135.

The classic source of Indian cultural (and astronomical) history, as seen by an *outsider*, is Albiruni's. For an English translation of the Arabic text see Edward C. Sachau, *Alberuni's India* (Delhi: S. Chand, 1964). A similarly rich early source is *Māshā'allāh*, the Astrological History, translated and edited by E. S. Kennedy and D. Pingree (Cambridge, Mass.: Harvard University Press, 1971). Anyone with an interest in the growth of a modern western awareness of Indian and Chinese astronomy will consult J. B. Biot, *Etudes sur l'astronomie indienne et sur l'astronomie chinoise* (Paris, 1862; repr. Paris: Blanchard, 1969).

On astronomical activities by Europeans in India, a subject as yet little investigated, see S. M. Razaullah Ansari, *Introduction of Modern Western Astronomy in India during the 18th–19th Centuries* (New Delhi: Institute of History of Medicine and Medical Research, 1985). On Persian astronomy see articles in W. Hartner's *Oriens–Occidens* (see our introductory bibliography) and his chapter 'Old Iranian Calendars', in *The Cambridge History of Iran*, vol. 2, *The Median and Achaemenian Periods*, ed. Ilya Gershevitch (Cambridge: Cambridge University Press, 1985), pp. 714–92. For further studies of relevance to the present chapter, see our next bibliography (King/Saliba and Kennedy).

CHAPTER 8

B. Lewis (ed.), *The World of Islam* (London: Thames and Hudson, 1976) is a beautifully illustrated collection of authoritative chapters on the main aspects of Islamic faith, people and culture, eastern and western. Medieval Islamic astronomy has been studied extensively, and two important collections of papers, the first more easily found than the second, are: David A. King and George Saliba, (eds), *From Deferent to Equant: A Volume of Studies in the History of Science in the Ancient and Medieval Near East in Honor of E. S. Kennedy* (New York: New York Academy of Sciences, 1987); E. S. Kennedy, with colleagues and former students, *Studies in the*

Islamic Exact Sciences (Beirut: American University of Beirut, 1983). Willy Hartner, *Oriens–Occidens*, 2 vols (Hildesheim: Olms, 1968, 1984) contains important studies spanning three continents. On observatories see Aydin Sayili, *The Observatory in Islam* (Ankara, 1960, repr. 1980), and on mathematical and instrumental techniques see David A. King, *Islamic Mathematical Astronomy* and *Islamic Astronomical Instruments* (London: Variorum, 1986 and 1987 respectively).

Various journals have specialized in Arabic science. Two substantial journals that seem to be well established are *Arabic Sciences and Philosophy* (Cambridge: Cambridge University Press) and *Zeitschrift für Geschichte der Arabisch-Islamischen Wissenschaften*, ed. F. Sezgin (Frankfurt, annually from 1984). Most of the articles in the first, and many in the second, are in English, and astronomy is a recurrent theme in both. Sezgin has also published an important series of bio-bibliographical aids, *Geschichte des arabischen Schrifttums*, of which volume 6 deals with astronomy (Leiden: Brill, 1978). The same publisher issues the magnificent *Encyclopaedia of Islam*, an international work of scholarship that has long been available in French and English editions. A second edition of this is more than half complete, and should be finished around the turn of the century.

CHAPTER 9

Western Islam has been fortunate in its historians, largely centred in Barcelona. The best literature is largely in Spanish. Classics in the field include two works by J. M. Millás Vallicrosa: *Estudios sobre Historia de la Ciencia Española* (Barcelona: CSIC, 1949); and *Nuevos Estudios sobre Historia de la Ciencia Española* (Barcelona: CSIC, 1960); both reissued together, with introduction by Juan Vernet (Barcelona, 1987). Of the many valuable works by Vernet, a general history of the European debt to Spanish Arab culture has been translated into French: *Ce que la culture doit aux Arabes d'Espagne* (Paris: Sindbad, 1978). The most recent general work of the Barcelona school with a strong bearing on astronomy is Julio Samsó, *Las Ciencias de los Antiguos en al-Andalus* (Madrid: Mapfre, 1992).

On specific technical matters, much can be learned from O. Neugebauer, *The Astronomical Tables of Al-Khwārizmī, trans. with Commentaries on the Latin Version edited by H. Suter*, (Copenhagen: Hist. Filos. Dan. Vid. Selskab, 1962). This could be supplemented by Bernard R. Goldstein, *Ibn al-Muthanna's Commentary on the Astronomical Tables of al-Khwārizmī. Two Hebrew Versions*, ed. and trans., with Commentary (New Haven and London: Yale University Press, 1967), and G. J. Toomer, 'A Survey of the Toledan Tables', *Osiris*, **15** (1968): 1–174. A useful edition close to the early printings of the Alfonsine tables (John of Saxony's version) is E. Poulle (ed), *Les Tables Alfonsines, avec les canons de Jean de Saxe* (Paris: Editions du CNRS, 1984). A survey of the history of the tables generally will be found in J. D. North, 'The Alfonsine Tables in England', repr. as ch. 21 in his *Stars, Minds, and Fate* (London and Ronceverte: Hambledon, 1989). Articles in various languages on the Alfonsine theme are in M. Comes, R. Puig and J. Samsó (eds), *De Astronomia Alphonsi Regis. Proceedings of the Symposium on Alphonsine Astronomy held at Berkeley, August 1985* (Barcelona: Inst. 'Millás Vallicrosa', 1987). A finely illustrated work on science in Muslim Spain (Al-Andalus), with authoritative but readable topical essays, was produced under the direction of J. Vernet and J. Samsó in connection with an exhibition held in Madrid in 1992: *El Legado Científico Andalusí* (Madrid: Ministerio de Cultura, 1992). There are plans to publish this in English ('The Andalusian Scientific Legacy').

CHAPTER 10

For a scholarly overview of medieval science, including much astrology and some astronomy, from the perspective of surviving manuscript evidence, see Lynn Thorndike, *A History of Magic and Experimental Science* (New York and London: Columbia University Press, 1923–58, 8 vols). Extracts from numerous texts will be found in Edward Grant (ed.), *A Source Book in Medieval Science* (Cambridge, Mass.: Harvard University Press, 1974). The medieval handling of the physics of the Aristotelian cosmos is a recurrent theme in Grant's collected essays: *Studies in Medieval Science and Natural Philosophy* (London: Variorum Reprints, 1981); see also his *Much Ado About Nothing: Theories of the Infinite Void* (Cambridge: Cambridge University Press, 1981). The collection in J. D. North, *Stars, Minds, and Fate* (London and Ronceverte: Hambledon, 1989) covers many medieval and Renaissance astronomical themes, and includes an elementary introduction to the astrolabe. Illustrations of astrolabes may be found in many museum catalogues. The best collections are not usually confined to the medieval west. Important collections are held by the Museum of the History of Science (Oxford), National Maritime Museum (Greenwich), British Museum (London), Smithsonian Museum (Washington), Adler Planetarium (Chicago) and the Museo di Storia della Scienza (Florence). These all have a wide variety of medieval astronomical instruments. An outstanding and copiously illustrated catalogue of an exhibition that drew on over seventy collections (notably that of the German National Museum in Nuremberg) is Gerhard Bott (ed.), *Focus Behaim Globus*, 2 vols (Nürnberg: Germanisches Nationalmuseum: 1992). The first volume includes topical essays (all in German), some on astronomical themes, some on the terrestrial globe of Martin Behaim, the focus of the exhibition. An important work on astronomical instruments, covering the eleventh to the eighteenth centuries, is Ernst Zinner, *Deutsche und niederländische Astronomische Instrumente des 11.–18. Jahrhunderts* (Munich, 1956, rev. edn 1967). Among pictorial surveys of a general sort are Henri Michel, *Scientific Instruments in Art and History* (London, 1966), Harriet Wynter and Anthony Turner, *Scientific Instru-*

ments (London: Studio Vista, 1975), Anthony Turner, *Early Scientific Instruments. Europe 1400–1800* (London, 1987), and Gerard l'E. Turner, *Antique Scientific Instruments* (Poole, Dorset: Blandford, 1980).

The processes of higher education are the subject of Hilde de Ridder-Symoens, *A History of the University in Europe*, vol. 1, *Universities in the Middle Ages* (Cambridge: Cambridge University Press, 1991). For more detail on Oxford see J. I. Catto and T. I. R. Evans (eds), *The History of the University of Oxford*, vol. 2 (Oxford: Oxford University Press, 1992). On the general position of astrology in medieval and Renaissance culture, Theodore Otto Wedel, *The Mediaeval Attitude Toward Astrology, Particularly in England* (New York: Archon Books, 1968) and Don Cameron Allen, *The Star-Crossed Renaissance: The Quarrel about Astrology and its Influence in England* (N. Carolina: Duke University Press, 1941), although now dated, make easy reading and are still important. For a different perspective, see Eugenio Garin, *Astrology in the Renaissance: the Zodiac of Life*, trans. from the Italian by C. Jackson and J. Allen (London: Routledge and Kegan Paul, 1983). 'Astrology at the English Court and University in the Later Middle Ages' is the subject of Hilary M. Carey, *Courting Disaster* (London: Macmillan, 1992). Several medieval themes are discussed in Patrick Curry (ed.), *Astrology, Science and Society: Historical Essays* (Woodbridge, Suffolk: Boydell Press, 1987).

For an introduction to the main doctrines of medieval astrology, as they were inherited from Islamic writers, and for Chaucer's literary use of them, see J. D. North, *Chaucer's Universe* (Oxford: Clarendon Press, 2nd edn, 1990). This work also explains the principles of the astrolabe and Chaucer's equatorium. For an extensive history of the medieval equatorium see E. Poulle, *Equatoires et Horlogerie planétaire du XIIIe au XVIe siècle* (Geneva and Paris: Droz, 1980, 2 vols). For a shorter account in English, see the description of the albion and its historical context, in J. D. North, *Richard of*

Wallingford (Oxford: Oxford University Press, 1976, 3 vols). These volumes contain the earliest description of a mechanical (and astronomical) clock, and also (in appendix 31) a fuller outline of medieval planetary theory than could be given in the present work.

Far more extensive in scope than its title suggests, and containing the answers to most questions the reader is likely to have on the relations of astronomy and the ecclesiastical calendar, is G. V. Coyne, M. A. Hoskin and O. Pedersen, eds., *Gregorian Reform of the Calendar . . . 1582–1982* (Vatican City: Pontifical Academy of Sciences and Specola Vaticana, 1983).

CHAPTER 11

Copernicus' own writings are available in many editions, but the most complete is that by the Polish Academy of Sciences, of which the first volume was a reproduction of Copernicus' manuscript of the *De revolutionibus* (Warsaw and Cracow, 1973). The series includes Latin texts and English translations. Another translation is A. M. Duncan, *On the Revolutions of the Heavenly Spheres* (Newton Abbot and London: David and Charles, 1976). The *Commentariolus* is well translated by Noel Swerdlow (*Proceedings of the American Philosophical Society*, **117** (1973): 423–512), and is available with translations of two other minor Copernican works (the *Letter against Werner* and the *Narratio Prima* of Rheticus) in Edward Rosen's *Three Copernican Treatises* (New York: Dover, 3rd edn 1971). The latter contains a Copernicus bibliography with over a thousand items and short critical – often hypercritical – descriptions of each. A Polish bibliography produced in 1958 by H. Baranowski contained nearly four thousand items, and that number was greatly swelled by the 500th anniversary celebrations in 1973. The fortunes of the various early editions and copies of the *De revolutionibus* are the subject of the article that provides a title for Owen Gingerich's *The Great Copernicus Chase and other Adventures in Astronomical History* (Cambridge, Mass.: Sky Publishing, 1992). The most complete single history analysing the mathematical aspects of Copernicus' work is Noel M. Swerdlow and Otto Neugebauer, *Mathematical Astronomy in Copernicus' De Revolutionibus* (New York: Springer, 1984, 2 vols). A good general biography in English is M. Biskup and J. Dobrzycki, *Copernicus, Scholar and Citizen* (Warsaw, 1973). A widely known but somewhat fanciful account is in Arthur Koestler, *The Sleepwalkers. A History of Man's Changing Vision of the Universe* (New York and London: Grosset and Dunlap, 1959). This book does more justice to Kepler than to Copernicus and Galileo. Among the better collections of studies dating from the 1973 celebrations is Jerzy Dobrzycki (ed.), *The Reception of Copernicus' Heliocentric theory. Proceedings of a Symposium held by the IUHPS in Torun, 1973* (Dordrecht and Boston: Reidel, 1973).

A valuable series of monographs going far beyond Copernican studies proper is produced by the Polish Academy of Sciences under the title *Studia Copernicana*. A later parallel series is published in Leiden with the qualification 'Brill Series', and this includes Janice Adrienne Henderson's, *On the Distances between Sun, Moon and Earth According to Ptolemy, Copernicus and Reinhold* (Leiden: Brill, 1991). For an example of a writer representing Copernicus' work as a relatively superficial geometrical transformation of Ptolemy, see Derek J. de S. Price, 'Contra Copernicus', in M. Clagett (ed.), *Critical Problems in the History of Science* (Madison: Wisconsin University Press, 1959), pp. 197–218.

CHAPTER 12

The works of Tycho Brahe were edited in fifteen volumes by J. L. E. Dreyer (Copenhagen, 1913–29). Dreyer wrote a standard biography which is still valuable: *Tycho Brahe. A Picture of Life and Work in the Sixteenth Century* (1890; repr.

New York: Dover, 1963). Dreyer's general history (see bibliography to chapter 4) is by no means entirely superseded, and is strong on this period. He issued a facsimile of Tycho's work on instruments, which was later translated into English and edited by H. Raeder, E. Strömgren and B. Strömgren as *Tycho Brahe's Description of his Instruments and Scientific Work as given in Astronomiae Instauratae Mechanicae* (Copenhagen, 1946). The best biography of Tycho is now Victor E. Thoren, *The Lord of Uraniborg: A Biography of Tycho Brahe* (Cambridge: Cambridge University Press, 1990).

Much of the important work of Alexandre Koyré centred on this period, and the following remain influential: *Etudes Galiléennes* (Paris: Hermann, 1966); *Galileo Studies*, trans. by John Mepham (Hassocks, Sussex: Harvester Press, 1978; not identical to the French work); *From the Closed World to the Infinite Universe* (Baltimore and London: Johns Hopkins University Press, 1970); *The Astronomical Revolution: Copernicus–Kepler–Borelli*, trans. by R. E. W. Maddison (London: Methuen, 1973).

Fundamental texts on the discovery of the telescope were long available only in Dutch editions, but see Albert van Helden, *The Invention of the Telescope* (Philadelphia: American Philosophical Society, 1977). For a general history see Henry C. King, *The History of the Telescope* (New York: Dover, 1979).

On Galileo, see the forthcoming study by A. C. Crombie and A. Carrugo, which will cover both life and work authoritatively. The works of Galileo were published in an Italian edition by A. Favaro (Florence, 1890–1909) and have been reprinted with additions (Florence: Barbèra, 1929–39, 1965). English translations are numerous, beginning with Thomas Salusbury's of 1661. Relevant to astronomy are, for example: Stillman Drake and C. D. O'Malley (trans.), *The Controversy of the Comets of 1618* [texts by Galileo Galilei; Horatio Grassi; Mario Guiducci; Johann Kepler] (Philadelphia: University of Pennsylvania, 1960); S. Drake, *Dialogue Concerning the Two Chief World Systems* (Berkeley,

Cal., 1953, rev. 1967); S. Drake, *Discoveries and Opinions of Galileo* (New York, 1957). The last includes a work on sunspots and parts of the *Sidereal Messenger*, among other things. A complete translation of the latter with comment is by Albert van Helden: *Sidereus Nuncius, or the Sidereal Messenger* (Chicago: University of Chicago Press, 1989).

On Harriot, see John W. Shirley (ed.), *Thomas Harriot, Renaissance Scientist* (Oxford: Oxford University Press, 1974) and also articles in J. D. North and J. J. Roche (eds), *The Light of Nature. Essays . . . presented to A. C. Crombie* (Dordrecht: Nijhoff, 1985). Shirley is author of the standard work *Thomas Harriot: A Biography* (Oxford: Oxford University Press, 1983). Since Harriot was interested in astronomical techniques of navigation, this is a suitable point at which to mention David Waters, *The Art of Navigation in England in Elizabethan And Early Stuart Times*, 3 vols (Greenwich: National Maritime Museum, 1978). A natural supplement to this is E. G. R. Taylor, *Mathematical Practitioners of Tudor and Stuart England* (Cambridge: Cambridge University Press, 1967).

A new edition of the complete works of Kepler has long been in preparation (Munich, 1938–). English translations are patchy. Kepler writes at a fairly difficult mathematical level, but less difficult works are: Johannes Kepler, *Kepler's Somnium: the Dream, or Posthumous Work on Lunar Astronomy*, trans., with a commentary by Edward Rosen (Madison: University of Wisconsin Press, 1967); and *Mysterium Cosmographicum: The Secret of the Universe*, trans. by A. M. Duncan, with introduction and Commentary by E. J. Aiton (New York: Abaris Books, 1981). Lengthy translations from the *Astronomia nova* are available in Koyré's *Astronomical Revolution* (see above) and a full version appears in William H. Donahue, *Johannes Kepler, New Astronomy* (Cambridge: Cambridge University Press, 1992). A standard life of Kepler is Max Caspar, trans. C. D. Hellman, *Kepler* (London and New York, 1959; originally 1938). Caspar prepared a basic bibliography (1936), which was revised by Martha List

(Munich, 1968); but the most easily accessible recent bibliographies in English are O. Gingerich's in the *Dictionary of Scientific Biography*, vol. 7 (1973), pp. 308–12, and that in J. V. Field (below). A number of good articles appeared in Arthur and Peter Beer (eds), *Kepler: Four Hundred Years. Proceedings of Conferences Held in Honour of Johannes Kepler* (Oxford: Pergamon, 1975). Judith V. Field, *Kepler's Geometrical Cosmology* (London: Athlone Press, 1988) adopts a well-balanced view of the mathematical mysticism in Kepler. Related subjects are dealt with in: Owen Gingerich and Robert S. Westman, *The Wittich Connection: Conflict and Priority in Late Sixteenth-century Cosmology* (Philadelphia: American Philosophical Society, 1988); Curtis Wilson, *Astronomy from Kepler to Newton: Historical Studies* (London: Variorum Reprints, 1989); Albert van Helden, *Measuring the Universe: Cosmic Dimensions From Aristarchus to Halley* (Chicago: University of Chicago Press, 1985).

On Hevelius, see the *Dictionary of Scientific Biography* (vol. 6, 1972, pp. 360–4). On seventeenth-century solar theories, see Yasukatsu Maeyama, 'The historical development of solar theories in the late sixteenth and seventeenth centuries', *Vistas in Astronomy*, **16** (1974): 35–60. A comprehensive survey of star maps from the advent of printing to 1800 is found with a number of reproductions of maps, rare and not so rare, in Deborah J. Warner, *The Sky Explored. Celestial Cartography, 1500–1800* (Amsterdam: Theatrum Orbis Terrarum, 1979).

Many of the literary consequences of advances in cosmology are more widely covered than his title suggests by Francis Johnson, *Astronomical Thought in Renaissance London* (Baltimore: Johns Hopkins University Press, 1937). As a measure of the progress of astrology at this time and later, see Bernard Capp, *Astrology and the Popular Press: English Almanacs 1500–1800* (London: Faber, 1979). That astrology was still very much alive, and even developing new mathematical techniques, is evident from J. D. North, *Horoscopes and History* (London: The Warburg Institute, 1986). The perennial question of truth and hypothesis, as seen through astronomers' eyes, is the subject of N. Jardine, *The Birth of History And Philosophy of Science: Kepler's* A Defence of Tycho Against Ursus *and Essays On Its Provenance and Significance* (Cambridge: Cambridge University Press, 1984).

CHAPTER 13

The cosmological views of Descartes are well surveyed in Eric J. Aiton, *The Vortex Theory of Planetary Motions* (London and New York: Macdonald and American Elsevier, 1972). Those who wish to study Descartes closely should make use of Gregor Sebba's excellent annotated bibliography covering the period 1800 to 1960: *Bibliographia Cartesiana* (The Hague: Nijhoff, 1964), supplemented by the *Isis Critical Bibliographies* (see introductory bibliography). Newtonian bibliography to a somewhat later period is covered by Peter and Ruth Wallis, *Newton and Newtoniana, 1672–1975* (London: Dawson, 1977), and this may be supplemented by the *Isis* lists and the excellent biography by R. S. Westfall, *Never at Rest. A Biography of Isaac Newton* (Cambridge: Cambridge University Press, 1980). Newton's *Principia* is readily available in English translation in numerous editions and printings, many based ultimately on Andrew Motte's of 1729 (for example Florian Cajori's 1934 revision frequently reprinted by the University of California Press at Berkeley). Newton's mathematical papers, where much theoretical astronomy resides, have been edited, translated and commented upon in a masterly way in D. T. Whiteside, *The Mathematical Papers of Isaac Newton* (Cambridge: Cambridge University Press, 1967–80, 8 vols).

As an indication of contemporaneous practical astronomy, a useful counterweight to Newtonian abtractions, see John Flamsteed, *The Gresham Lectures of John Flamsteed*, edited and introduced by Eric Forbes (London: Mansell, 1975). Flamsteed's instrumentation is dealt with in Allan Chapman, *Dividing the Circle. The Development of Critical Angular Measurement in Astronomy, 1500–1850* (New York and

London: Horwood, 1990). A working tool for future historians of practical astronomy, which takes its title from the place where it was prepared, is Derek Howse, *The Greenwich List of Observatories. A World List of Astronomical Observatories, Instruments and Clocks, 1670–1850* (Chalfont St Giles, Bucks: Science History Publications; being *Journal of the History of Astronomy*, **17** (4) (1986): 100 pp.).

CHAPTER 14

The papers in Michael A. Hoskin, *Stellar Astronomy. Historical Studies* (Chalfont St Giles, Bucks: Science History Publications, 1982) shed light on many aspects of the astronomy of this period. On instrumentation, the literature is very extensive, but see again the last items under chapter 13, and H. C. King, *The History of the Telescope* (New York: Dover, 1979). Maurice Daumas, trans. M. Holbrook, *Scientific Instruments in the Seventeenth and Eighteenth Centuries* (London: Batsford, 1972) covers a wider field than astronomy, but reflects well on French practice. The notes in the English version are chaotic, and the original French edition may be preferred. For the work and influence of Short, see David J. Bryden, *James Short and his Telescopes* (Edinburgh, 1968). For a conspectus of the all-important English trade in instrument-making at this time, see E. G. R. Taylor, *Mathematical Practitioners of Hanoverian England* (Cambridge: Cambridge University Press, 1966).

On Halley, see C. A. Ronan, *Edmond Halley–Genius in Eclipse* (New York, 1969; London, 1970), and for his comet see the bibliography in Bruce Morton, *Halley's Comet, 1755–1984: A Bibliography* (Westport, Conn.: Greenwood Press, 1985). A historical introduction to astronomy in the southern hemisphere, although not strong on Australasia, is David S. Evans, *Under Capricorn: A History of Southern Hemisphere Astronomy* (Bristol: Hilger, 1988).

A key text in the new cosmology is Thomas Wright, ed. and introd. by Michael Hoskin, *An Original Theory or New Conception of the Universe* (London: Macdonald, 1971).

This edition includes *A Theory of the Universe* (1734). Immanuel Kant's cosmological views are usually submerged beneath a sea of irrelevant philosophy. For an English text, see his *Universal Natural History and Theory of the Heavens*, trans. with introduction and notes by Stanley L. Jaki (Edinburgh: Scottish Academic Press, 1981). Lambert's contribution to the unfolding picture of island universes can be seen at first hand in Johann Heinrich Lambert, *Cosmological Letters on the Arrangement of the World-Edifice*, trans. with introd. and notes by Stanley L. Jaki (New York: Science History Publications, 1976). William Herschel's collected papers were edited by J. L. E. Dreyer and printed with other biographical material in *The Scientific Papers of Sir William Herschel* (London: Royal Astronomical Society, 1912, 2 vols). Long excerpts from the papers are included, with historical commentary by M. A. Hoskin and astrophysical notes by K. Dewhirst, in *William Herschel and the Construction of the Heavens* (London: Oldbourne, 1963). See also relevant studies in Hoskin's *Stellar Astronomy* (listed above). Full-length biographies are available by J. B. Sidgwick (1953), Angus Armitage (1962) and Günther Buttmann (Stuttgart, 1961, in German). Buttmann's biography of John Herschel is available in English as *The Shadow of the Telescope* (New York: Scribner's, 1970; London: Lutterworth, 1974).

On the possibility of a plurality of worlds, see Steven J. Dick, *Plurality of Worlds: The Origins of the Extraterrestrial Life Debate from Democritus to Kant* (Cambridge: Cambridge University Press, 1984) and Michael J. Crowe, *The Extraterrestrial Life Debate 1750–1900: The Idea of a Plurality of Worlds from Kant to Lowell* (Cambridge: Cambridge University Press, 1986); and, for reprints of two important texts in the debate, John Ray, *Wisdom of God Manifested in the Works of Creation* (New York: Garland, 1979) and Bernard Le Bovier De Fontenelle, *Conversations On the Plurality of Worlds*, trans. from the French by H. A. Hargreaves (Berkeley: University of California Press, 1990).

The extraordinary additions made, especially by French astronomer-mathematicians, to Newtonian planetary theory in the eighteenth and early nineteenth centuries is central to the classic study by Robert Grant, *History of Physical Astronomy From the Earliest Ages to the Middle of the Nineteenth Century* (London: Robert Baldwin, 1852). It is the theme of the last of Delambre's histories (see introductory bibliography). Perhaps the most important single work on Laplace is the entry in the first Supplement to the *Dictionary of Scientific Biography* (vol. 15, 1978), pp. 273–403. (That the entry is seven and a half times as long as that on Einstein, and eighty-seven times as long as that on Richard of Wallingford, illustrates the law that a creature's size varies as the cube of the gestation period.) Harry Woolf, *The Transits of Venus. A Study of Eighteenth-Century Science* (Princeton: Princeton University Press, 1959) provides some of the background to the scaling of the solar system. For a good general survey of Newtonian mechanics, and theories that preceded and followed it, René Dugas, trans. J. R. Maddox, *A History of Mechanics* (London: Routledge, 1955) is still worth reading. It was written at a moderately technical level. A companion volume by the same author covers mechanics in the seventeenth century. A journal worth combing systematically for articles on this general theme is *Archive for History of Exact Sciences*. The subtitle 'A new perspective on eighteenth-century advances in the lunar theory' well describes Eric G. Forbes, *The Euler–Mayer Correspondence (1751–1755)* (London: Macmillan, 1971), which might lead interested readers into the deeper and richer waters of Euler's collected works.

CHAPTER 15

A bibliography of over 1400 entries is to be found in David H. DeVorkin, *The History of Modern Astronomy and Astrophysics* (New York: Garland, 1982). A convenient and brief source of biographical information on the important astronomer

Bessel is Jürgen Hamel, *Friedrich Wilhelm Bessel* (Leipzig: Teubner, 1984). A classic historical work, in whose title the word 'popular' does not mean 'commonplace', is Agnes M. Clerke's *A Popular History of Astronomy During the Nineteenth Century* (London: Black, 1893; 4th edn, 1902). Her historical interests give added value to her general astronomical contributions, such as *The System of the Stars* (1890) and *Problems in Astrophysics* (1903). At a more elementary level, Camille Flammarion's *Astronomie populaire* (Paris, 1880) is instructive. Since so many instrumental advances were made in Germany at this time it is useful to consult in the same vein such a work as Rudolf Wolf, *Geschichte der Astronomie* (Munich, 1877), or for a modern bird's-eye view, Dieter B. Herrmann, *Geschichte der Astronomie von Herschel bis Hertzsprung* (Berlin: VEB, 1975), transl. and revised by K. Krisciunas as *The History of Astronomy from Herschel to Hertzsprung* (Cambridge: Cambridge University Press, 1984). This last popular work has a useful bibliographical guide to further sources. (The numerous references to Friedrich Engels need not be taken too seriously.) It is impossible in a short space to pay much attention to autobiography, but I make an exception for *The Reminiscences of an Astronomer* (Boston, 1903), in which Simon Newcomb brings to life a vanished era.

On instrumentation, as well as the birth of astrophysics, see Owen Gingerich (ed.), *Astrophysics and Twentieth-century Astronomy to 1950: Part A.* [vol. 4A of the *General History of Astronomy*, ed. M. Hoskin] (Cambridge: Cambridge University Press, 1984). This includes a checklist of refractors and reflectors, 1850–1950. H. C. King's history of the telescope (see the bibliography to chapter 11 above) may be supplemented by D. Howse's 'Greenwich list' of observatories and instruments (see bibliography to chapter 13). Histories of observatories are legion, but can mostly be tracked down using the former. Pulkovo was important enough to be singled out here: see A. N. Dadaev, *Pulkovo Observatory. An Essay on its History and Scientific Activity,* trans. by Kevin Krisciunas (Springfield, Virginia: NASA, 1978). Krisciunas later wrote a survey of observatories, especially useful for the later period (it has a good chapter on Pulkovo, and others on Harvard, Lick, Yerkes, Mount Wilson and Palomar, for instance): *Astronomical Centers of the World* (Cambridge: Cambridge University Press, 1988). While not having the space to list many individual histories of observatories, I mention for its easy narrative style the history of Lick (founded 1888) by Donald E. Osterbrock and others, *Eye on the Sky: Lick Observatory's First Century* (Berkeley and London: University of California Press, 1988).

A comprehensive study of spectroscopy that well summarizes the essentials of the original scientific sources is J. B. Hearnshaw, *The Analysis of Starlight. One Hundred and Fifty Years of Astronomical Spectroscopy* (Cambridge: Cambridge University Press, 1986). See also R. L. Waterfield, *A Hundred Years of Astronomy* (London and New York: Macmillan, 1938). For solar astrophysics, see A. J. Meadows, *Early Solar Physics* (Oxford: Pergamon, 1970). For photography, see Gérard de Vaucouleurs, *Astronomical Photography, from the Daguerrotype to the Electron Camera* (London, 1961) and Dorrit Hoffleit, *Some Firsts in Astronomical Photography* (Cambridge, 1950). A fine collection of photographs taken by the Schmidt telescopes at Tautenberg and the European Southern Observatory, with a biography of Bernhard Schmidt, is S. Marx and W. Pfau, trans. from the German by P. Lamle, *Astrophotography with the Schmidt Telescope* (Cambridge: Cambridge University Press, 1992; originally Leipzig: Urania Verlag).

Hale's remarkable achievements are reported with a wealth of illustration in Helen Wright, Joan N. Warnow and Charles Weiner, *The Legacy of George Ellery Hale. Evolution of Astronomy and Scientific Institutions in Pictures and Documents* (Cambridge, Mass.: MIT Press, 1972).

For a history of our understanding of meteorites, which made rapid progess in the nineteenth century: John G.

Burke, *Cosmic Debris: Meteorites in History* (Berkeley and London: University of California Press, 1987).

CHAPTER 16

In addition to the bibliography of the previous chapter (Clerke, Gingerich (ed.), Hearnshaw, Meadows), for stellar physics the following original texts are a useful historical source: J. Scheiner, trans. by E. B. Frost from the German edition of 1890, *A Treatise on Astronomical Spectroscopy* (Boston, 1894); Arthur Stanley Eddington, *The Internal Constitution of the Stars* (Cambridge: Cambridge University Press, 1930); Ejnar Hertzsprung, ed. D. B. Herrmann, *Zur Strahlung der Sterne* (Leipzig: Ostwalds Klassiker, 1976); Martin Schwarzschild, *Structure and Evolution of the Stars* (Princeton, 1958, repr. New York: Dover, 1965); Subrahmanyan Chandrasekhar, *An Introduction to the Study of Stellar Structure* (New York: Dover, 1967).

The source books mentioned in the introductory bibliography are useful. On solar theories: Karl Hufbauer, *Exploring the Sun: Solar Science Since Galileo* (Baltimore: Johns Hopkins University Press, 1991). On the famous chart showing stellar evolution, see B. W. Sitterly, 'Changing interpretations of the Hertzsprung–Russell diagram, 1910–1940', in *Vistas in Astronomy*, **12** (1970): 357–66. For background to the question of geochronology see Francis Haber, *The Age of the World, Moses to Darwin* (Baltimore: Johns Hopkins University Press).

Not easily found, but useful, is W. Strohmeier 'Variable Stars, their Discoverers and First Compilers, 1006 to 1975', *Veröffentlichungen der Remeis-Sternwarte Bamberg* (no. 129, 1977). The role of the Harvard College Observatory in this story is so important that Solon I. Bailey, *The History and Work of Harvard Observatory, 1839–1927* (New York and London, 1931) is worth seeking out.

For the discovery of spirals, see Charles Parsons (ed.), *The Scientific Papers of William Parsons, Third Earl of Rosse, 1800–*

1867 (London, 1926). The background to the discovery is well depicted in Patrick Moore, *The Astronomy of Birr Castle* (London: Mitchell Beazley, 1971). An introductory history in which the illustrations more than compensate for the occasional slips (such as the labelling of a photograph of John Herschel with William's name) is Richard Berendzen, Richard Hart and Daniel Seeley, *Man Discovers the Galaxies* (New York: Science History Publications, 1976). This covers the period to the 1930s.

CHAPTER 17

For a comprehensive history of cosmology in the first half of the twentieth century, and the mathematical background to it, see J. D. North, *The Measure of the Universe. A History of Modern Cosmology* (Oxford: Oxford University Press, 1965, 1967; New York: Dover, 1990). A readable survey of changing conceptions of space will be found in Edmund Whittaker, *From Euclid to Eddington* (Cambridge: Cambridge University Press, 1949). Of the many biographies of Einstein, Abraham Pais, *Subtle is the Lord. The Science and Life of Albert Einstein* (Oxford: Oxford University Press, 1982) is one of the most comprehensive. Jeremy Bernstein, *Einstein* (London: Viking Penguin, 1976; 2nd edn, Fontana) concentrates on Einstein's physics for the general reader. For a selection of classical papers by H. A. Lorentz, A. Einstein, H. Minkowski and H. Weyl, see *The Principle of Relativity. A Collection of Original Papers On the Special And General Theory of Relativity*, trans. from the German (New York: Dover, no date). Texts of a cosmological character that should help to give an idea of the events of the first half of the century without being excessively difficult are these: Arthur Stanley Eddington, *Stellar Movements and the Structure of the Universe* (London: Macmillan, 1914), *Space, Time and Gravitation* (Cambridge: Cambridge University Press, 1920), *The Internal Constitution of the Stars* (Cambridge: Cambridge University Press, 1926), *The Expanding Universe* (Cambridge:

Cambridge University Press, 1933) and *Stars and Atoms* (Oxford: Oxford University Press, 1927), the last being his only popular account of his astrophysical work; James Jeans, *The Mysterious Universe* (Cambridge: Cambridge University Press, 1930); Edwin Hubble, *The Realm of the Nebulae* (Oxford: Oxford University Press, 1936); *The Observational Approach to Cosmology* (Oxford: Oxford University Press, 1937); Georges Lemaître, *The Primeval Atom: A Hypothesis of the Origin of the Universe*, trans. from the French by B. H. Korff and S. A. Korff (Toronto and New York: Van Nostrand, 1950). More difficult, but classics of the subject, are R. C. Tolman, *Relativity, Thermodynamics and Cosmology* (Oxford: Oxford University Press, 1934), Otto Heckmann, *Theorien der Kosmologie* (Berlin: Springer, 1968), and Herman Bondi, *Cosmology* (Cambridge: Cambridge University Press, 1960).

For a biography of Friedmann, see E. A. Tropp, V. Ya. Frenkel and A. D. Chernin, *Alexander A. Friedmann. The Man Who Made the Universe Expand*, trans. from the Russian edition (Nauka, Moscow) by A. Dron and M. Burov (Cambridge: Cambridge University Press, 1993). The best separate history of the paradox of the dark night sky is Edward Harrison, *Darkness at Night: A Riddle of the Universe* (Cambridge, Mass.: Harvard University Press, 1987). The subtitle of Robert Smith, *The Expanding Universe: Astronomy's 'Great Debate' 1900–1931* (Cambridge: Cambridge University Press, 1982) is self-explanatory. Edward R. Harrison, *Cosmology: the Science of the Universe* (Cambridge: Cambridge University Press, 1981) is only semi-historical, but simple and instructive. Norriss S. Hetherington, who also discusses previous centuries in his *Science and Objectivity: Episodes in the History of Astronomy* (Ames: Iowa State University Press, 1988), is rather harsh in his onslaughts on astronomers' professionalism and integrity, but raises some interesting questions. Pierre Kerszberg, *The Invented Universe: The Einstein–De Sitter Controversy (1916–17) and the Rise of Relativ-*

istic Cosmology (Oxford: Oxford University Press, 1989) is a detailed and professional historical study.

For a survey of astronomy in the Soviet Union from the Revolution to Stalin's purges, see E. Nicolaïdis, *Le Développement de l'astronomie en l'URSS, 1917–1935* (Paris: Observatoire de Paris, 1984). H. van Woerden, R. J. Allen and W. B. Burton (eds), *The Milky Way Galaxy: Proceedings of the 106th Symposium of the IAU held in Groningen, 30 May–3 June, 1983* (Dordrecht and Boston: Reidel, 1985) has a useful historical component. B. Bertotti and others (eds), *Modern Cosmology in Retrospect* (Cambridge: Cambridge University Press, 1990) is an invaluable document, for it includes many chapters of a scientific but autobiographical nature by those who contributed to cosmology in the earlier part of the century (including R. A. Alpher, R. Herman, H. Bondi, W. McCrea, F. Hoyle, R. M. Wilson, and M. Schmidt). Wolfgang Yourgrau and Allen D. Breck, (eds), *Cosmology, History and Theology. Based on the Third International Colloquium Held at Denver, 1974* (New York: Plenum, 1977) has a similar value, including as it does papers by H. O. Alfvén, P. G. Bergmann, W. H. McCrea, C. W. Misner, A. Penzias, Kenji Tomita and others who have been responsible for notable advances.

CHAPTER 18

The most thorough history of the first decades of radio astronomy is W. T. Sullivan III, *The Early Years of Radio Astronomy* (Cambridge: Cambridge University Press, 1984), which could be supplemented by the source material in his *Classics in Radio Astronomy* (Dordrecht: Reidel, 1982). D. O. Edge and M. J. Mulkay, *Astronomy Transformed* (New York and London, 1976) and also G. I. Verschurr, *The Invisible Universe Revealed* (Berlin, London, New York: Springer-Verlag, 1987) both deal with the various drastic changes in astronomical technique. Readable autobiographical material by pioneers in radio astronomy are J. S. Hey, *The Evolution of Radio Astronomy* (London, 1973) and Bernard Lovell, *The Voice of the Universe. Building the Jodrell Bank Telescope*, rev. edn (London and New York: Praeger Press, 1987). An institutional view of the new science – and much else – from a European perspective is given in Adriaan Blaauw, *Early History: The European Southern Observatory, from Concept to Reality* (Munich: ESO, 1991). Peter Robertson, *Beyond Southern Skies. Radio Astronomy and the Parkes Telescope* (Cambridge: Cambridge University Press, 1992) tells of the building and operation of the important Parkes radio telescope in New South Wales (completed 1961).

CHAPTER 19

The time has not yet come when one can list many true histories of the astronomy of the second half of the twentieth century. K. Krisciunas, *Astronomical Centers of the World* (Cambridge University Press, 1988) gives a sizeable part of his book over to space astronomy. Robert W. Smith et al., *The Space Telescope. A Study of NASA, Science, Technology and Politics* (Cambridge: Cambridge University Press, 1989) was published before the launch of the Hubble telescope, but is incidentally an important study of the background to many enterprises in this new category. Richard Hirsch, *Glimpsing an Invisible Universe. The Emergence of X-Ray Astronomy* (Cambridge: Cambridge University Press, 1983) is valuable, although thin on work outside the United States, which was by no means insignificant. Allan Needell, (ed.), *The First 25 Years in Space* (Washington: Smithsonian Institution Press, 1989) deals with the impact of the political, military, commercial and scientific aspects of space.

CHAPTER 20

Again, few historical studies of the modern period are as yet available, and the line between biography and hagiography is as thin as ever. John D. Barrow and Frank J. Tipler *The Anthropic Cosmological Principle* (Oxford: Clarendon Press, 1986) is not history, but it has a valuable historical element. Stephen W. Hawking and W. Israel (eds), *Three Hundred Years of Gravitation* (Cambridge: Cambridge University Press, 1987) does not pretend to be historical, despite its title, although Werner Israel writes on the evolution of the idea of dark stars and C. M. Will surveys experimental gravitation from Newton to the twentieth century. The rest of the volume will in the fullness of time be of extreme historical value. It has chapters by R. Penrose on quantum theory and reality, A. H. Cook on gravitational experiment, R. D. Blandford on astrophysical black holes, K. S. Thorne on gravitational radiation, and M. J. Rees on Galaxy formation and dark matter. S. K. Blau, A. H. Guth and A. Linde write on inflationary cosmology, while S. W. Hawking writes on quantum cosmology. Stephen Hawking, *A Brief History of Time from the Big Bang to Black Holes* (London: Bantam Press, 1988) hardly needs an introduction, since it has broken numerous best-selling records. Steven Weinberg, *The First Three Minutes* (New York and London: Basic Books, 1976) had much popular success a decade earlier, with the physics of the big bang. His *Gravitational Cosmology* (New York: Wiley, 1972) was a successful textbook that a reader might use to gain an entry to the historical field, although D. W. Sciama, *Modern Cosmology* (Cambridge: Cambridge University Press, 1971) and Michael Rowan-Robinson, *Cosmology* (Oxford: Oxford University Press, 1977) both make fewer mathematical demands. An earlier classic work that long held the field as a textbook: G. C. McVittie, *General Relativity and Cosmology* (London: Chapman and Hall, 2nd edn, 1965). Ya. B. Zel'dovich and I. D. Novikov, *Relativistic Astrophysics, vol. 1, Stars and Relativity; vol. 2, The Universe and Relativity* (Chicago: University of Chicago Press, 1971–74) throws much light on important Soviet work. I make no attempt to identify the best source material for the last quarter of the century.

Twenty-seven interviews with cosmologists are to be found in Alan Lightman and Roberta Brawer, *Origins. The Lives and Worlds of Modern Cosmologists* (Cambridge, Mass.: Harvard University Press, 1990), which includes a good select bibliography of recent cosmology.

On theological attitudes with respect to cosmology one might consult essays by Charles Misner, Philip Hefner, David Peat, Arthur Peacocke, or Stanley Jaki in Yourgrau and Breck (see bibliography to chapter 18), or the earlier work by Jaki, *Science and Creation: From Eternal Cycles to an Oscillating Universe* (Edinburgh and London: Scottish Academic Press, 1974), which marries theology to the history of certain cosmological views. For eighteen historical views that between them cover a broader spectrum, that is, the natural sciences and Christian theology over two millennia, see David C. Lindberg and Ronald L. Numbers, *God and Nature* (Berkeley and London: University of California Press, 1986). It becomes clear, reading this, that cosmology has had more than its due share of influence on the shaping of theology, and that for some strange reason it continues to have it.

INDEX

Personal names are treated differently according to historical period. In the Middle Ages, surnames generally take second place, but conventional names are occasionally kept. Initials have been preferred to forenames from the 18th century onwards. Short forms of Arabic names are used. Early institutions are listed, but those of modern times only selectively.

674 · Index